The Urban Book Series

The Urban Book Series is a resource for urban studies and geography research worldwide. It provides a unique and innovative resource for the latest developments in the field, nurturing a comprehensive and encompassing publication venue for urban studies, urban geography, planning and regional development.

The series publishes peer-reviewed volumes related to urbanization, sustainability, urban environments, sustainable urbanism, governance, globalization, urban and sustainable development, spatial and area studies, urban management, transport systems, urban infrastructure, urban dynamics, green cities and urban landscapes. It also invites research which documents urbanization processes and urban dynamics on a national, regional and local level, welcoming case studies, as well as comparative and applied research.

The series will appeal to urbanists, geographers, planners, engineers, architects, policy makers, and to all of those interested in a wide-ranging overview of contemporary urban studies and innovations in the field. It accepts monographs, edited volumes and textbooks.

Indexed by Scopus.

Guido Cimadomo • Quazi Mahtab Zaman
Eduardo Jiménez-Morales
Editors

Urbanism at Borders

Navigating New Frontiers in a Globalized World

Springer

Editors
Guido Cimadomo
Department of Art and Architecture
Universidad de Málaga
Malaga, Spain

Eduardo Jiménez-Morales
Department of Art and Architecture
Universidad de Málaga
Malaga, Spain

Quazi Mahtab Zaman
Scott Sutherland School of Architecture and
Built Environment
Robert Gordon University
Aberdeen, UK

ISSN 2365-757X ISSN 2365-7588 (electronic)
The Urban Book Series
ISBN 978-3-032-09593-0 ISBN 978-3-032-09594-7 (eBook)
https://doi.org/10.1007/978-3-032-09594-7

This Springer imprint is published by the registered company Springer Nature Switzerland AG
The registered company address is: Gewerbestrasse 11, 6330 Cham, Switzerland

This book is dedicated to all those whose lives are shaped by the presence of borders—where movement is restricted, territory contested, and daily life shaped by forces beyond individual control. It honours the resilience of communities living under siege, separation, or displacement and reflects on the spaces where walls are built, with the hope they will disappear sometimes in the future.

Contents

Editors and Contributors

About the Editors

Guido Cimadomo is an Architect and Associate Professor in architectural history, theory, analysis, and criticism at the Department of Art and Architecture, Universidad de Málaga, where he has been teaching since 2010. His research focuses on urban transformations, with particular attention to participatory processes, cultural and industrial heritage, and the impact of large-scale events such as world expositions on the city.

He is an expert member of CIPA Heritage Documentation, the ICOMOS scientific committee dedicated to the documentation of architectural heritage, a member of ISOCARP, and a board member of the Institute for the Study of International Expositions (ISIE), where he investigates the legacy and urban influence of international expositions. His academic career includes visiting research positions at the University of Oregon (USA, 2022), the University of Queensland (Australia, 2017), and the University of Seville (Spain, 2016), reinforcing his international engagement with urban and heritage studies.

A member of the Research Group HUM-696 UTOPIA, his current research explores the intersection between community-driven urban regeneration, digital documentation techniques, and the cultural narratives embedded in city-making. Cimadomo is the author of a broad number of journal articles, and editor of *Design, Architecture and Society in the Early Modern Age* (Ediciones Asimétricas, 2019), *Arquitectura española y tecnología. Siete episodios clave del siglo XX* (Recolectores Urbanos, 2021), and *Arquitectura y comunes urbanos. Redefiniendo la práctica arquitectónica* (Tirant lo Blanch, 2022).

As a professional member of the Board of Architects in Seville, he has designed and built sports and cultural facilities in Andalusia since 2003, several of which have been presented at international expositions (AzW Architektur Zentrum Wien, COAC Barcelona, Selinunte) and published in international volumes like *Architetture per un territorio sostenibile* (Skira, 2010) and *Convention Centers* (Braun Publishing, 2012).

He holds a degree in Architecture from the Politecnico di Milano (Italy, 1998) and a Ph.D. with International Distinction from the University of Seville with the thesis, 'Borderlands Transformations. Influence of Borders on the Territory' (Spain, 2014).

Quazi Mahtab Zaman is an architect and urban designer with an extensive international background in teaching, research, and practice.

He teaches the theory and practice of architecture and urban design at the Scott Sutherland School of Architecture and Built Environment at Robert Gordon University, Aberdeen, Scotland. He has also researched and taught at Cornell University, BRAC University, Bangladesh, and the University of Hong Kong. In addition, his research includes a Fulbright Fellowship at Cornell University's College of Architecture, Art and Planning; postgraduate research at Lund University, Sweden; and a Master's in Urban Design and a Ph.D. in Urban Design and Mass Housing at the University of Hong Kong.

Zaman founded the Borders Urbanism research initiative. As the chair, editor, and convenor of the Borders Urbanism research initiative and founder of the Border Urbanism Research Centre, he focuses on borders as a multidisciplinary research topic. He has expanded his urban design research interests to children in urban areas. In this work, he engages youth with a new participatory pedagogy tool, Stitching Urban Vision (SUV), that he developed to leverage 'communities as an extended class-rooms'. His research also addresses design to increase health conditions and considers the implications of congestion in developing countries. He serves on the editorial board of *MDP—Sustainable Journal* and the *Global Built Environment Review*. He is currently the International Director in Scotland for e-DAU (Team Designers Architects Urbanists Network based in Brazil).

In addition to numerous research papers and journal articles, he is the co-author of *Berlin: A City Awaits—The Interplay Between Political Ideology, Architecture and Identity* (Springer, 2020) and *Potsdamer Platz—The Reshaping of Berlin* (Springer, 2014), and he is also the co-editor of *Border Urbanism, Transdisciplinary Perspectives* (Springer, 2022) and *Transdisciplinary Urbanism and Culture—From Pedagogy to Praxis* (Springer, 2018).

He is a Fellow of the Institute of Architects Bangladesh (IAB) and a Professional Member of the Hong Kong Institute of Urban Design.

Eduardo Jiménez-Morales, M.Arch. (Universidad de Granada) and Ph.D. (Universidad de Málaga), is an Associate Professor in Urban and Regional Planning at the Department of Art and Architecture, Universidad de Malaga (Spain).

He is a member of UTOPIA (Urbanism, Tourism, Landscape and Architectural Innovation Research Group), and his most recent research focuses on the socio-spatial dynamics resulting from the functional reformulation of mature tourist destinations and the multi-scale impact of new urban tourism in the contemporary city.

His work as a scholar and architect has been recognised by the Universidad de Málaga (2021), La Caixa Foundation (Spain, 2020), the Southeast University Nanjing (China, 2020), the Ministry of Education (Czech Republic, 2017), the Andalusian Regional Government (Spain, 2010–2015), the Festival dell'Architettura (Italy, 2010), the Institute for Advanced Architecture of Catalonia (IAAC, 2010), or the Pasajes-iGuzzini Awards (Spain, 2005), among others. Eduardo is the author of a large number of refereed scientific publications in high-impact books and journals (ORCID iD: 0000-0003-3417-7990; Scopus iD: 56177911800) and he has completed research stages in several international institutions such as the École Nationale Supérieure d'Architecture Montpellier (France).

Contributors

Mehwish Abid McGill University, Montreal, QC, Canada

Mari Paz Agúndez Technische Universität Berlin, Berlin, Germany

Aleksandra Jevtović Department of Business Studies Blace, Toplica Academy of Applied Studies, Blace, Serbia

Aleksandra Stamenković Faculty of Digital Arts, Metropolitan University, Belgrade, Serbia

Felicia Anderson Structures Research Group, University of Cambridge, Cambridge, UK

Rosa Arma Research Centre for Architecture, Urbanism and Design (CIAUD), Lisbon School of Architecture, Universidade de Lisboa, Lisbon, Portugal

Francisco José Chamizo-Nieto Department of Urban and Regional Planning, Universidad de Sevilla, Seville, Spain

Guido Cimadomo Department of Art and Architecture, Universidad de Málaga, Malaga, Spain

Pina Giusi Ciotoli Independent Researcher, Rome, Italy

Alessandra Como Department of Civil Engineering, University of Salerno, Fisciano, SA, Italy

Marta Cornax-Martín Institute of Habitat Territory and Digitalisation, Universidad de Málaga, Malaga, Spain

Laura Coucill University of Salford, School of Science, Environment and Engineering, Salford, Salford, Greater Manchester, UK

Nuria Nebot-Gómez de Salazar Department of Art and Architecture, Universidad de Málaga, Malaga, Spain

Kezia Eka Sari Dewi Urban Design, Urbanism, Landscape and Planning Unit, KU Leuven, Leuven, Belgium

Omar Ferwati Goldsmiths, University of London, London, UK

Angela Fiorelli Università di Roma Tor Vergata, Rome, Italy

Bogdán Funk József Cziráki Doctoral School of Wood Sciences and Technologies, University of Sopron, Sopron, Hungary

Luca Gaeta Department of Architecture and Urban Studies, Politecnico di Milano, Milan, Italy

Jacopo Galli Urbicide Task Force, Università Iuav di Venezia, Venice, Italy

Iraj Esmaeilpour Ghoochani InsideOut Advisory Service, STJG (Stuttgarter Jugendhaus gGmbH), Stuttgart, Germany

Tom Jefferies School of Natural and Built Environment, Queen's University Belfast, University Road, Belfast, UK

Vilmos Katona Research Institute of Art Theory and Methodology, Hungarian Academy of Arts, Budapest, Hungary

Héctor Navarro Martínez Department of Architectural Composition, Universidad Politécnica de Madrid, Madrid, Spain

Pia Marziano La Sapienza Università di Roma, Rome, Italy

Metagora Studio, Concón, Chile

Jorge Minguet-Medina Department of Art and Architecture, Universidad de Málaga, Malaga, Spain

Luis Alonso Pérez Monge Universidad Politécnica de Madrid, Madrid, Spain

Escuela de Arquitectura, Universidad de Costa Rica, San Pedro, Costa Rica

Eduardo Jiménez-Morales Department of Art and Architecture, Universidad de Malaga, Malaga, Spain

Zbigniew Władysław Paszkowski Andrzej Frycz Modrzewski Krakow University, Krakow, Poland

Alona Martinez Perez De Montfort University, Leicester School of Architecture, The Gateway, Leicester, UK

Luisa Smeragliuolo Perrotta Department of Civil Engineering, University of Salerno, Fisciano, SA, Italy

Carlos Rosa-Jiménez Institute for Habitat, Territory and Digitalization, Universidad de Málaga, Malaga, Spain

Finzi Saidi Department of Architecture, University of Johannesburg, Bunting Road Campus, Auckland Park, Johannesburg, South Africa

Carlos Santamarina-Macho Independent Researcher, Valladolid, Spain

Eman Sherif Arab Academy for Science, Technology & Maritime Transport, Cairo, Egypt

Moriah J. Snowden Impact Collective, Atlanta, GA, USA

Chiara Toscani AA School of Architecture, London, UK

Irene Tuzi Humboldt University, Berlin, Germany

Paulina Valdivia Universidad Politécnica de Cataluña, Barcelona, Spain

9CERO4 Arquitectos, Viña del Mar, Chile

Ingrid Carolina Vargas-Díaz Independent Researcher, Malaga, Spain

Carla Zagal Universidad Politécnica de Cataluña, Barcelona, Spain

Quazi Mahtab Zaman Scott Sutherland School of Architecture and Built Environment, Robert Gordon University, Aberdeen, UK

Chapter 1
Thresholds and Territories: Rethinking Borders Through Urbanism

Guido Cimadomo, Quazi Mahtab Zaman, and Eduardo Jiménez-Morales

Abstract This volume examines the contemporary transformations of borders at a moment marked by democratic erosion, geopolitical instability, and mounting migratory pressures worldwide. Against a backdrop of restrictive migration regimes, resurgent conflict, and hardened territorial controls, borders are now understood as socio-spatial processes embedded in everyday life. Building on the foundations laid by Border Urbanism: Transdisciplinary Perspectives, this book expands the field by foregrounding the relational and experiential dimensions of bordering practices across multiple geographies, with particular attention to often overlooked contexts in the Global South. Contributors interrogate borders through architecture, urbanism, planning, ethnography, archival research, environmental justice, and forensic methodologies, revealing how historical legacies, conflict, informal urbanisation, economic forces, and regulatory frameworks continuously reshape spatial hierarchies and urban forms. The concept of the threshold serves as a critical interpretive lens, emphasising fluid, negotiated, and transformative spaces where exclusion, encounter, and adaptation co-exist. Across diverse case studies, the chapters illuminate how borders—geopolitical, urban, cultural, or symbolic—structure mobility, access, and socio-economic relations, while also generating opportunities for resistance, reimagining, and more equitable futures. The volume ultimately advocates a plural, flexible and interdisciplinary border theory capable of addressing emerging global tensions and the complex realities of contemporary border territories.

G. Cimadomo (✉) · E. Jiménez-Morales
Department of Art and Architecture, Universidad de Malaga, Malaga, Spain
e-mail: cimadomo@uma.es; eduardo_jm@uma.es

Q. M. Zaman
Scott Sutherland School of Architecture and Built Environment, Robert Gordon University, Aberdeen, UK
e-mail: q.m.m.zaman@rgu.ac.uk

G. Cimadomo et al. (eds.), *Urbanism at Borders*, The Urban Book Series,
https://doi.org/10.1007/978-3-032-09594-7_1

Keywords Border urbanism · Borderscapes · Socio-spatial constructs · Migration and urban conflict

The first quarter of the twenty-first century reveals a bleak panorama marked by the erosion of democratic values, the resurgence of armed conflicts across many parts of the world and the tightening of anti-migration policies. These include mass deportations and repatriations to third-country migrant centres that often fail to uphold the rights of displaced persons and refugees. At the same time, global economic tensions have escalated through rising tariff barriers between competing trade blocs. The hopes that had emerged with the end of the Cold War and the fall of the Berlin Wall in the final decade of the twentieth century have given way to widespread despair—affecting a significant portion of the world's population (IOM 2024)—and to the hypocrisy of many host countries. These nations, while rhetorically committed to universal human rights, often implement policies that starkly contradict those ideals. Europe, a continent that, during the darkest moments of its long and complex history, has been the origin of vast migratory flows, now faces serious challenges in keeping its doors open to the most vulnerable—those fleeing war, famine, or sexual and political or religious persecution. The recent approval of an initial list of seven so-called 'safe countries of origin', from which asylum requests will be fast-tracked, implicitly suggests that applications from all other countries will face greater scrutiny, increasing the image of a fortress Europe more centred in its safety and development. This points to a broader hardening of entry requirements to European territory (Ismail 2025). This landscape is also fuelling a renewed global arms race, sometimes masked under the label of defence, with clear implications for the material and symbolic meaning of borders—an element this volume continues to interrogate. While Europe currently grapples with the challenges of managing new migration flows and hardening its external borders, similar pressures—though often less visible in international discourse—reshape urban territories across the Global South. In regions such as South Asia, Latin America, and North Africa, conflict, displacement and informal urbanisation create complex and highly localised bordering dynamics that demand equal analytical attention.

This volume follows the opening contribution to the series *Border Urbanism: Transdisciplinary Perspectives* (Zaman and Hall 2023), which first proposed a shift in perspective: to approach borders not only as political constructs but also through the lenses of urbanism, architecture and adjacent disciplines. The late architect Ricardo Scofidio once remarked, in response to a *New York Times* initiative to reimagine the United States–Mexico border during intense migratory pressure from the Global South, 'Leave it to the security experts and the engineers'. His comment reveals a reluctance to engage with the issue's complexity and an acknowledgement of the increasing militarisation of border infrastructures (Hamilton 2006). The first volume in the series focused on themes such as the relationship between borders and the built environment, spatial segregation and the polarisation of border cities—emphasising the deep entanglement between territorial boundaries and the everyday lives of those who inhabit them. These are all critical issues for architects and urban planners (Cimadomo 2015), whose professional activity is inherently interwoven with broader discourses, political above all.

In recent years, border theory has evolved decisively towards understanding borders as socio-spatial constructs shaped by power, identity and everyday practice. They organise access, control and mobility—often in ways that are informal or implicit—while simultaneously shaping subjectivities and spatial hierarchies. As such, borders are enforced through policies or infrastructure and emerge through lived experience, regulatory frameworks and spatial imaginaries. Historical, social and ethnic roots remain critical in characterising contemporary borders, especially those in conflict zones. These roots form dynamic and ever-shifting relational frameworks, challenging the idea of borders as static, enduring lines (Brambilla 2014; Rajaram and Grundy-Warr 2007). **The concept of *borderscapes*** emerges from this shift, describing borders not as lines but as **multilayered spaces**—zones of interaction, tension and negotiation. Borderscapes are lived, contested and shaped by overlapping forces such as migration, memory, security and development. Far from being merely territorial dividers, borders must be seen as active processes that shape spatial configurations and urban life itself (Konrad and Amilhat-Szary 2023).

Whether open or closed, the fundamental character of borders profoundly impacts their surroundings. Today, the study of borders is increasingly recognised as a field that enhances our understanding of how spatial justice, urban form and the lives of marginalised and everyday citizens are affected—and potentially transformed—by bordering processes. The approaches and thematic variations within border studies are as wide-ranging as the human interests at stake: those that emerge from how people inhabit, negotiate and reimagine their spatial environments. To this end, this volume comprehensively explores the evolving urban fabric, examining the intricate interplay of contemporary conflicts, the emergence of new borders for refugees and migrants, the enduring weight of colonial legacies and postcolonial policies and grassroots struggles for more equitable social frameworks. Just as the concept of borderscapes has helped redirect attention towards previously overlooked dimensions of border studies, deciphering the complexities embedded along these lines now requires a pivotal shift in scale.

In this context, the threshold concept offers a productive lens through which to analyse borders not as boundaries that divide but as relational spaces of negotiation and transformation. Thresholds are spatially and temporally fluid, shifting according to social conditions, environmental pressures and cultural meanings. Unlike rigid borders, thresholds are lived and reinterpreted continuously—sites where exclusions may be reinforced and encounters and transitions occur. Recognising their mutable character enables new forms of spatial intervention and interdisciplinary research, inviting a compelling reframing of border thinking. At the same time, the boundaries between social, cultural and ethnic groups are increasingly fluid and situational, producing temporary yet impactful borders within the urban fabric. These internal divisions often emerge in response to shifting political and economic forces, giving rise to what can be described as *urban eccentricities*—unusual, uneven, or peripheral conditions that disrupt the expected logic of city development.

Such eccentricities may manifest in spatial fragmentation, socio-economic disparities, or regulatory grey zones. Economic dynamics' strong influence on urban form and everyday life highlights the urgent need for political frameworks capable of recognising and responding to these irregularities. A comprehensive understanding of borders today must, therefore, embrace a diversity of analytical approaches sensitive to these non-linear, often marginal, urban phenomena. Borders influence behaviours and experiences across a wide range of spatial scales—from the national to the urban-rural, from the inter-urban to the domestic. Whether they are geopolitical frontiers or subtle divisions within neighbourhoods, these borders shape access, mobility and social inclusion patterns. These forces intersect in complex ways in borderlands—whether peripheral or embedded within cities—demanding a layered understanding of how spatial divisions operate across contexts.

Economics, migrations, climate change and politics influence borders differently. This volume goes beyond the conventional discourse on borders, exploring often overlooked or unrelated dimensions such as tourism and housing. It becomes evident that architecture and urban planning deal with physical and intangible borders. By broadening the narrative to encompass these critical aspects, the aim is to enhance understanding and propose solutions to contemporary challenges, recognising that borders and their effects exist at any scale. Avoiding a possible structure based on varying scales, which in itself would not aid the task of delving into the issues related to contemporary borders, through diverse perspectives, the volume seeks to unravel hidden meanings, contributing to a nuanced comprehension of borders in a global context.

In this volume, authors examine borders' perception, interpretation and nature, shedding light on their transformative influence on historical, cultural and socio-economic interests. The book focuses on how urban border conditions determine the mobility of goods, resources and populations in several case studies, defining interactions and relationships ranging from geopolitical relations to the intricacies of everyday life. Besides, temporal issues arising from diverse urban conditions are also addressed, with each section contributing to a critical discourse on border urbanism and the nuanced interplay between conflict and coexistence resulting from the delineation, separation and segregation imposed by borders. The Italian architect Aimaro Isola stated in a recent interview on the occasion of his 100th birthday, reaffirming his lifelong belief in architecture as an ethical and humanistic practice: 'Too often, architecture has toyed with theory, forgetting that design must always and only be for life—for the beauty of a more ethical and liveable world. Against wars and the tyranny of power' (Gibello 2025).

1.1 New Conflicts, New Borders for Refugees and Migrants

The first section of this volume addresses the spatial, social and political repercussions of conflict-induced migrations and the formation of new borders within and beyond cities. As the world faces unprecedented levels of displacement due to war, climate change and political instability, the urban realm becomes both a stage for trauma and a laboratory for adaptation. This section interrogates how formal and informal borders emerge in cities affected by war, forced migration and humanitarian crises. They scrutinise the visible and invisible lines that dictate access, control, identity and, with increasing relevance, integration, using diverse methodological approaches—from spatial analysis and ethnography to interdisciplinary design research. A shared concern for justice, memory and urban resilience emerges in contexts where boundaries are drawn and redrawn through violence, negotiation, or necessity. In 'Urban Borders, War Borders, Reconstruction Borders: Retracing the Evolution of Land Ownership Structures as an Engine for the Reconstruction of Syrian Settlements', the post-war urban landscapes of Syrian cities are examined, focusing on how conflict reshapes land ownership and territorial structures. Through critical spatial analysis and mapping, Jacopo Galli examines calls for context-sensitive interventions that recognise war-altered urban conditions as dynamic border-making processes. His work sets the stage for the section by offering a conceptual link between war and the production of new spatial regimes. The third chapter, by Mari Paz Agundez, explores the low-income neighbourhood of Benzú in the Spanish exclave of Ceuta, highlighting the fragmented and overlapping influences of multiple agents—military, municipal, European and Moroccan—on this liminal urban zone. 'How the Multiscalar Agents Negotiating the Urban Structure of a City Impact the Region and the Border District of Benzú (Ceuta, Spain)' demonstrates how border governance often marginalises local communities while prioritising geopolitical control. Omar Ferwati's chapter 'Fronts and Counterfronts in Aleppo: Inflicting and Resisting Violence Through the City' deepens the exploration of war-impacted urban environments by analysing Aleppo as a spatial battleground. He conceptualises architecture as a weapon and a shield—used both to inflict violence and to resist it. Drawing from architectural theory, war studies and interviews, he proposes 'counter-fronts' as civilian responses to militarised space, offering a powerful lens into how civilians negotiate violence through space. The last chapter of this section presents a design-driven, interdisciplinary case study focused on Neukölln, a Berlin district shaped by the arrival of refugees. Angela Fiorelli and Irene Tuzi propose models of future urban coexistence, employing co-design and participatory research with children in 'An Interdisciplinary Approach to Make the City of the Future: A project for the Neukölln District in Berlin (Germany)'. In this way, they demonstrate how bottom-up processes can counteract marginalisation and foster new shared urban narratives. These chapters offer a multidimensional portrait of how conflict, migration and governance coalesce to redefine urban borders. Despite their diverse geographic settings, they share a common thread: the urban border is never static. It is contested, negotiated and often weaponised, but it also holds the potential for resistance and reinvention. Through both critical distance and participatory proximity, the section reveals that the making of borders can also be enacted through lived practices, spatial tactics and collective imaginaries.

1.2 Subtle Borders of Colonial and Postcolonial Policies

This section delves into the lasting presence of colonial and postcolonial structures in shaping contemporary urban borders. The borders discussed here can be considered as 'subtle' ones, as they are frequently drawn within cities, neighbourhoods and subjectivities, yet no less impactful than more traditional physical walls. Rather than focusing on overt geopolitical divides, these chapters explore how power, identity, race and memory are encoded in urban form, land use and community structures. Carlos Santamarina Macho examines how domestic US land policies—from the Jeffersonian grid to suburban fencing—have created 'inner borders' that structure urban life along class, race and ownership lines. In 'Inner Borders of the United States' Role in the Making of the American Landscape', the author traces how the frontier myth and property regimes have spacialised power, effectively internalising colonial logic and continuing to reproduce exclusion long after the physical American frontier—understood as the historical period of territorial expansion—closed. The following chapter, 'Identity Crisis: Social Schizophrenia and Blackness in "Post-Racial" America' by Moriah J. Snowden, keeps the focus on the United States, confronting the paradox of racial identity and interrogating the tension between national ideals of unity and the lived experience of Black Americans. Drawing from critical race theory and socio-spatial analysis, she explores how redlining, racial zoning and economic disenfranchisement function as everyday borders that limit mobility, opportunity and belonging. Snowden defines these psychological impacts as 'social schizophrenia', showing how it is spatially embedded in the urban landscape. Kezia Eka Sari Dewi traces Chinese settlements' spatial and cultural evolution in 'Reproduction and Transformation of Chinatown and "Chinese Kampung" in Postcolonial Indonesia: A Case Study of Semarang'. Through urban history and field observation, Dewi shows how former colonial spatial borders persist in new forms, often maintaining ethnic segregation under different names. Economic policies and cultural resilience emerge as key factors in shaping these communities' spatial identities. Mehwish Abid explores the transformation of domestic architecture in Lahore after the 1947 Partition of British India in 'Towards a Decolonial Home: Post-Partition Lahore, Pakistan (1947–1970)'. Using oral histories and archival imagery, she analyses how displaced populations reshaped colonial 'Model Town' houses into spaces of cultural continuity and decolonial expression. This chapter highlights the intimate scale of spatial resistance, revealing how personal narratives and material adaptations intersect to reconfigure inherited colonial structures. The last chapter of this session, by Finzi Saidi and Alona Martínez Pérez, critiques the spatial legacy of extractive colonialism and apartheid planning in Johannesburg. 'Navigating Contemporary Boundaries of Johannesburg Through Toxic Landscapes' exposes how urban mining operations, landfills and environmental degradation produce literal toxic borders in the city—separating wealth from poverty and access from abandonment. The chapter employs environmental justice theory and spatial forensics to connect historical exploitation with contemporary urban inequalities. While the geographical scope of this section ranges from the United States to South Africa, Indonesia and Pakistan, the chapters share an effort to deconstruct inherited spatial hierarchies and propose emancipatory reinterpretations of urban space. Through historical investigation, spatial critique and socio-political analysis, the contributors expose the persistence of segregationist logic in postcolonial urban contexts and uncover the reconfiguration of power relations through architectural, legal and cultural means. The research presented here shows how decolonisation will only be possible if it considers and counteracts the often hidden effects of subtle, intangible borders.

1.3 Barriers to Trade

Following the analysis of more intangible borders, this section investigates how trade, commerce and economic globalisation generate, reproduce and challenge urban borders. In fact, unlike traditional physical borders, commercial boundaries often emerge invisibly—through zoning laws, market access, ownership structures, or legal status. The following chapters explore how the circulation of goods and capital interacts with spatial inequalities, reconfiguring urban morphologies and fragmenting social ecologies. In the first chapter, 'The Role of Trade Centers Bordering the Public City Life: New Belgrade Case Study', Aleksandra Jevtovic and Aleksandra Stamenkovic analyse four trade centres in New Belgrade as pivotal public infrastructures from the Socialist era. Originally designed as spaces for collective life and social interaction—with an interesting relationship with Esmaeilpour Ghoochani's chapter in Part VI—these centres are now contested zones shaped by privatisation and shifting consumption patterns. Their work highlights how economic liberalisation has redefined the social meaning and accessibility of these public-commercial hybrids. Giusi Ciotoli offers a layered reading of Hong Kong's vertical urbanism, focusing on how economic imperatives have pushed the boundaries of density and architectural typology. The chapter 'Urban Transformations in Hong Kong' shows how trade, housing policy and urban infrastructures intersect to produce new spatial hierarchies. Methodologically, the chapter combines historical mapping with architectural analysis to examine how verticality becomes a spatial expression of economic flows and constraints. The last chapter of this section, written by Laura Coucill and Tom Jefferies, tackles one of the most politically charged trade borders of recent times: the United Kingdom–Ireland divide in the wake of Brexit. 'Current and Future Urbanisms of the United Kingdom and Republic of Ireland Border' interrogates how Brexit unsettles the border zone's spatial, economic and symbolic continuity. Several possible urban futures for this contested region are envisioned, highlighting the paradoxes of seeking seamless trade in a growing political disjunction. Far from being a neutral economic activity, trade is shown here to be a force that both binds and divides, challenging the assumption that trade brings integration and openness: shaping public space, social identity and national sovereignty, it offers insights into how commerce becomes a tool of both cohesion and separation, integration and exclusion, in the urban landscape.

1.4 Nature as an Intangible Border

This section explores the often-overlooked dimension of nature as a generator, mediator and signifier of borders in urban and territorial contexts. Rather than framing nature solely as a backdrop or a resource, the chapters here investigate how ecological systems, environmental disruptions and landscape imaginaries produce and negotiate spatial divisions. As in the case of trade examined in the previous section, nature can connect and isolate, functioning as a soft border that defines urban limits, separates or joins populations, or triggers contestation over land use and identity. As cities expand into fragile ecotones, post-Industrial voids, or conflict-ridden territories, the interface between urban and natural systems becomes a dynamic zone of tension and creativity. The authors in this section reveal the multiple scales—geographic, symbolic, architectural—at which nature articulates borders. Pia Marziano, Paulina Valdivia and Carla Zagal introduce the concept of the urban ecotone in 'Inhabiting the Border: Between Urban and Natural', drawing on comparative studies of informal settlements in Valparaíso (Chile) and Iquitos (Peru). They illustrate how precarious urbanisation interacts with fragile environments such as steep slopes and floodplains, showing that nature actively shapes the patterns and limits of settlement. Alessandra Como and Luisa Smeragliuolo Perrotta focus on the neighbourhood of Monteruscello in the extended periphery of Naples in 'Inhabit the Borders: A Landscape Design Project for the Neighborhood of Monterusciello (Italy)', where nature becomes a tool for social and ecological regeneration. They show how urban agriculture and landscape design can turn disconnected spaces into shared environments of care and identity. The chapter 'The Binational Municipal: An Alternative to the Architecture of a Controlled Border Landscape' by Felicia Anderson proposes an alternative architectural vision for the United States–Mexico border. Using speculative design and ecological analysis, she envisions a binational city that uses natural systems—especially rivers and green corridors—as shared infrastructure rather than as separation lines. Chiara Toscani's chapter 'Cities After Nature: Venice as a Tool for Reimagining Cities Beyond Nature and Culture' offers a philosophical and architectural meditation on the blurring of boundaries between nature and culture. Focusing on Venice as a singular case study, she reflects on how environmental challenges, such as rising water levels, dissolve traditional dichotomies between the built and the natural, proposing a hybridised approach to urban design instead. The last chapter of this section, 'Borderlands as Urban Systems: A Case Study of the Border Between Nicaragua and Costa Rica', by Luis Alonso Pérez Monge, moves to a broader territorial scale. He explores the border as a site of systemic conflict and ecological interdependence, proposing the 'Integral Cross-Border City' as a strategy to align human connectivity with natural flows. This section powerfully repositions nature as a central actor in creating, negotiating, and transforming borders. Across diverse geographies and scales, these chapters show that nature is never neutral—it enforces limits, stores memory and mediates relations between people and place.

1.5 Inequities of Neoliberal Policies

This section critically examines how neoliberal economic policies—and their spatial and social ramifications—create new urban borders and reinforce existing ones. While often studied with trade policies, the extent of such policies is broader and worth considering separately. Neoliberalism promises openness, flexibility and growth, even if its material manifestations often result in exclusion, fragmentation and inequality. The chapters explore how privatisation, deregulation, speculative development and tourism-driven urbanism reshape cities by generating invisible but powerful borders that separate functions, populations and narratives. Luca Gaeta opens the section with 'Uncharted Boundaries: The Territoriality of Everyday Life', arguing for a shift in how we perceive borders—not as static, official lines, but as constantly shifting and experienced through daily practices, the previously discussed thresholds. Drawing from Bourdieu's theory of *habitus*, he proposes the concept of 'practice on the move' as a lens for understanding the fluidity and fragmentation of contemporary urban life. Zbigniew Paszkowski's chapter 'The Transformational Polarisation as a Factor Stimulating Urban Development in European Cities' investigates how neoliberal policies have fostered spatial polarisation in post-Socialist urban contexts. Rather than being accidental, such polarisation is an active tool for economic development and real estate valorisation. In 'Clashes on the Sunny Side: Social and Territorial Exclusion in the Costa del Sol Touristic Paradise', Jorge Minguet analyses how tourism-based urban development in the Costa del Sol produces invisible but rigid class borders. He explores how gated communities, speculative housing and infrastructural asymmetries entrench exclusion, recasting leisure landscapes as sites of inequality. Carlos Rosa Jiménez examines similar dynamics in 'The Social Frontiers of the Sun-and-Beach Tourist Space: Duality and Socio-Spatial Fragmentation in Paradise'. In the closing chapter of this section, he identifies how tourism economies create dual urbanisms—one aesthetic and visible, the other hidden and functional—through erasure, segregation and the production of exclusive enclaves. Not surprisingly, the contributions of this section reveal how economic policies are enacted through spatial manipulation, displacing communities, eroding public goods, redefining belonging and finally putting inequality into the very structure of cities.

1.6 Community Struggles for Local Inequities

This section focuses on the role of communities in resisting spatial marginalisation and reclaiming agency over their environments. These struggles often occur outside traditional urban planning and policy frameworks, emerging instead from lived experience, local knowledge and grassroots creativity. The chapters highlight how residents mobilise intangible heritage, artistic expression and participatory design to counteract exclusion and contest-imposed spatial orders. Vilmos Katona and Bogdán Funk open the section with 'The Regeneration of Place with

Community-Based Architecture', proposing design interventions in *terrain vagues*—ambiguous, neglected spaces that can become opportunities for cultural renewal. Drawing on theoretical reflection and built examples, they explore how these margins can be transformed through collaborative and intercultural architectural practices. In 'Urban Evolution of the Informal Settlements in Ibagué-Colombia: Strategies and Policies', Ingrid Carolina Vargas-Díaz and Eduardo Jiménez-Morales examine the gradual formalisation of informal neighbourhoods. Their analysis shows how urban upgrading strategies that involve residents in decision-making can lead to more equitable spatial outcomes and help reknit fragmented urban fabrics. Iraj Esmaeilpour Ghoochani, in 'Artivists of the Neighbourhood, Unite! The Story of Stuttgart 21', analyses the cultural resistance to a significant infrastructure project in Germany. Through the lens of 'artivism'—activism through art—he demonstrates how public art and cultural performance can reclaim space and ignite civic debate. The last chapter in this section, by the research group formed by Francisco José Chamizo Nieto, Nuria Nebot Gómez de Salazar, Marta Cornax Martín and Carlos Jesús Rosa Jiménez, is 'Exploring the Significance of Intangible Cultural Heritage Within the Confines of the Neighborhood: An Examination of Participatory Techniques Used in Fontanalla, Malaga'. The authors detail how intangible heritage—songs, rituals, oral history—can be mapped and activated through participatory processes to restore neighbourhood identity and cohesion. This section makes clear that the struggle against urban inequity is not just fought in policy arenas or academic discourse—it is enacted daily by communities who confront borders through creativity, solidarity and persistence. The chapters illustrate how, once seen only as a zone of deprivation, the urban periphery can become the epicentre of transformation, emerging here as a site of political articulation and cultural innovation.

1.7 Change of Scale: Focus on Dwellings

This closing section brings the scale of analysis to the most intimate dimension of urban life: the dwelling. The home, often seen as a private and apolitical space, is revealed here as a contested zone shaped by regulatory norms, cultural expectations and socio-economic constraints. The chapters explore how formal planning systems intersect with gender, tradition and informality to define spatial practices within the home. Eman Sherif explores housing transformation in Bedouin communities in 'Privacy and Gender: Segregation in Bedouin Settlements in Dahab, Egypt'. The chapter shows how standardised state housing models fail to accommodate traditional privacy needs and gendered spatial practices, revealing a disconnect between imposed design and local norms. Héctor Navarro, in 'Towards Models of Non-Fragmented Cities: Continuity and Symbiosis in Brazilian Architectural Practice', presents architectural interventions that challenge urban fragmentation by rethinking spatial transitions from the dwelling to the neighbourhood. His analysis, grounded in practice and theory, shows how domestic-scale design can ripple outwards to shape more cohesive urban environments. Rosa Arma's '(Inter)Acting

Knowledges: Methodological Research Approach to Bairro da Torre Self-Produced Neighbourhood in Greater Lisbon' introduces a participatory research model for engaging with self-produced urban settlements. Through co-design, dialogue, and embedded fieldwork in a marginal neighbourhood in Portugal, Arma's concluding chapter brilliantly highlights how knowledge production can be shared and spatially enacted, with the dwelling as both the subject and site of collaboration. These chapters highlight how residents respond creatively—through adaptation, negotiation and participation—to resist imposed boundaries and reclaim agency over domestic environments. They also demonstrate that borders can be most powerfully felt at the scale of the home—where intimacy, identity and institutional influence collide, leading the dwelling to emerge as a battleground for spatial justice.

1.8 Conclusions

The research presented in this volume approaches the analysis of diverse border contexts through various methodologies, demonstrating the richness and complexity inherent in these environments. Through spatial and architectural analysis, participatory and co-design methods, ethnographic fieldwork, historical and archival research, critical and theoretical reflection, as well as environmental justice and forensic approaches, the contributions reveal not only the threats but also the opportunities that border spaces present—across all scales and in varied regions of the world. The authors collectively demonstrate a consistent view of borders as socio-spatial constructs rather than fixed territorial lines. The thematic structure used to organise the volume should be understood as flexible and open to overlap since it becomes clear that many sections share conceptual and methodological ground with others. For instance, the sections focused on trade and nature identify how borders link and divide, acting as permeable boundaries that shape the edges of urban space, distinguish or unite communities, and spark disputes over land use and collective identity. Why not adopt a more global perspective on a given context to propose direct improvements for the populations who inhabit these environments?

Moreover, the categories proposed here open up new ways of understanding borders—through less conventional approaches often tied to intangible or symbolic contexts, which are no less incisive for being less visible. For all these reasons, it is more necessary than ever to consider border territories—whether geographical or mental, religious or cultural, and therefore less obvious to critics and researchers—as complex arenas where numerous pressures converge and which require an integrated response, informed by successful practices from other disciplines.

Theories of *borderscapes* must, therefore, be expanded to encompass all the categories examined in this volume and those that will inevitably emerge from the re-escalation of political, religious and commercial tensions in the future. This volume also aims to affirm the need for greater flexibility in how borders are analysed, urging a move beyond entrenched stereotypes, categories and disciplinary habits. Only by incorporating the broadest possible range of experiences into the discourse can border theories—in the plural—be enriched. Using the plural is intentional: it rejects any effort to unify or simplify the complex realities at stake into a single, reductive narrative.

This volume ultimately shows that borders, far from being static or peripheral, are dynamic forces embedded in the everyday production of space. They emerge through conflict and displacement, are sustained by lingering colonial frameworks, and are reconfigured by trade, environmental pressures and neoliberal planning. Whether formal or informal, visible or latent, borders shape how cities and territories work, who has access, and under what conditions. They are enforced through policy and contested through lived experience, collective memory, and creative resistance. The chapters collected here offer critical readings and imaginative strategies—calling for a renewed urbanism that confronts inequality, embraces complexity and reclaims the threshold as a space of negotiation, adaptation and transformation.

References

Brambilla C (2014) Shifting Italy/Libya borderscapes at the interface of EU/Africa borderland: a "genealogical" outlook from the colonial era to post-colonial scenarios. ACME J Crit Geogr 13(2):220–245

Cimadomo G (2015) Spatial transformations of European borderlands. Fabrications J Soc Architect Hist Austr N Zeal 25(3):377–397. https://doi.org/10.1080/10331867.2015.1075463

Gibello L (2025) Amaro Isola: La bellezza contro le guerre e il prepotere degli individui. Il Giornale Dell'Architettura. https://speciali.ilgiornaledellarchitettura.com/2025/03/31/aimaro-isola-la-bellezza-contro-le-guerre-e-il-prepotere-degli-individui/. Accessed 18 Jun 2025

Hamilton WL (2006) A fence with more beauty, fewer barbs. The New York Times. https://www.nytimes.com/2006/06/18/weekinreview/18hamilton.html. Accessed 18 May 2025

IOM—International Organization for Migration (2024) Migration and migrants: a global overview. International migrants: numbers and trends. https://worldmigrationreport.iom.int/what-we-do/world-migration-report-2024-chapter-2/international-migrants-numbers-and-trends?utm_source=chatgpt.com. Accessed 18 May 2025

Ismail A (2025) EU issues list of "safe countries" for migrant returns that includes Egypt, Tunisia. Reuters. https://www.reuters.com/world/eu-issues-list-safe-countries-migrant-returns-that-includes-egypt-tunisia-2025-04-16/. Accessed 18 May 2025

Konrad VA, Amilhat-Szary AL (2023) Border culture: theory, imagination, geopolitics. Routledge, London

Rajaram PK, Grundy-Warr C (eds) (2007) Borderscapes. Hidden geographies and politics at territory's edge. University of Minnesota Press, Minneapolis

Zaman QM, Hall GG (eds) (2023) Border urbanism: transdisciplinary perspectives. Springer, Cham. https://doi.org/10.1007/978-3-031-06604-7

Part I
New Conflicts, New Borders for Refugees and Migrants

Chapter 2
Urban Borders, War Borders, Reconstruction Borders: Retracing the Evolution of Land Ownership Structures as an Engine for the Reconstruction of Syrian Settlements

Jacopo Galli

Abstract The destruction of entire parts of major cities in the Middle East and North Africa (MENA) region poses enormous design questions in terms of future reconstruction strategies. The *tabula rasa* conditions offer many chances but the ownership structure represents a major obstacle towards a sustainable reconstruction. The paper analyses the evolution of urban borders in Syrian cities before, during and after the civil war that has hit the country since 2011, retracing the processes of urban metamorphosis dictated by economic, social and historical causes. The comprehension of the complex concurrent factors that shaped current conditions is a starting point for the development of future strategies capable of acting as an alternative to current reconstruction laws that tend to intensify rather than diminish conflict. The homogenisation to unsustainable global standards of entire parts of Syrian cities seems to be the only possible destiny for future reconstruction. To tame and limit such tendency it is necessary to develop reconstruction strategies capable of intervening on city-making processes in light of the evolution of urban borders dictated by social, economic and political factors.

Keywords Borderscapes · Urban conflict · Memory · Reconstruction · Ownership-rights · Urbicide

---

J. Galli (✉)
Urbicide Task Force, Università Iuav di Venezia, Venice, Italy
e-mail: jacopo.galli@iuav.it

G. Cimadomo et al. (eds.), *Urbanism at Borders*, The Urban Book Series,
https://doi.org/10.1007/978-3-032-09594-7_2

2.1 The Tabula Is Never Rasa

In 1946, in a European continent almost totally covered by war rubble, the urban planner Hans Bernoulli wrote with unlikely ardour: 'Who intends to build the city that he and his age dream of, needs a free surface, a *tabula rasa* (1946). It is a barely concealed enthusiasm, immediately denied: 'The noble art that should proceed safely stumbles instead miserably in the ownership borders. Frightened upset, often even tired of the old disputes, finally allows that these borders and their rights dictate the law. The lion sees himself imprisoned among the meshes of a net, his roar should recall the mice to free him from that hateful tangle' (1946).

Bernoulli identifies in the ownership structures the founding character of every urbanisation process and of the future metamorphosis of existing urban forms. He hypothesises how in post-war reconstruction processes the choices made on the subject of ownership rights are decisive in strategically addressing any subsequent decision.

In this sense the most catastrophic destructions are paradoxically positive, allowing a wide range of possibilities and the radical modification of urban distortions:

> Where the ugly buildings of the last hundred years have been destroyed by the appalling tragedy of war, an unexpected opportunity is born, welcomed with the enthusiasm of laughter and tears, to rebuild a new, perfect, beautiful city. However, the first blow of a pickaxe is stopped by the invisible obstacle built by ownership borders, which persist even if the buildings have collapsed and the foundations disappeared: they want and must be taken into consideration. (Bernoulli 1946: 47)

In Bernoulli's vision, the erasing of the ownership structures and the transformation of the urban tabula rasa into a cadastral *tabula rasa* are seen as a unique opportunity for a radical rethinking of the deep character of the urban space. It is a positivist vision that reflects the conditions of Europe after the Second World War: a continent completely destroyed but at the same time pacified and with the simultaneous presence of a clear administrative and legislative structure and of a huge source of funding, the Marshall Plan (Schain 2001; Milward 2006; Holm 2016). The historical examples from WWII can only partially describe today's conditions in conflict-affected territories: large areas of Middle East and North Africa (Syria, Iraq, Yemen, Libya) are going through devastating wars that are leaving cities and territories completely destroyed and huge questions about design strategies that can be implemented in the short, medium and long term.

In an overall picture rich in problems and hopes, the choices made by the newborn national governments on the subject of the ownership structures for reconstruction shaped cities and entire nations; the long-term consequences are still visible today.

In Italy, the legislative decree *n. 154 Norme per i piani di ricostruzione degli abitati danneggiati dalla guerra* of 1 March 1945 confirmed the pre-war ownership structures and identified in the expropriation for public use the only instrument for the substantial modification of the urban form determined by the cadastral parcels (Mazzoleni and Bonfantini 2001). The reconstruction plans, with the prediction of moving whole settlements from geographically disadvantaged positions (Quilici 2001), imposed a shift in the problem of land ownership from the consolidated historical core to the agricultural territories adjacent to inhabited areas, predicting the speculative processes that will be the real protagonists of the post-war urban explosion.

In France (see Fig. 2.1), the policy of *remambrement urbaine* (Chabrol 2010) constitutes an opposite approach: the overall shape of the city and the future urban evolutions were considered a central factor in the reconstruction process favouring the total dematerialisation of properties (Voldman 1997; Gauthiez 2003). The nominal value of the land was replaced by the *Zones d'Égale Valeur* system with the aim of quantifying war damages based on the location, nature and urban context of each plot. The *remambrement* was applied on the ground through a process of progressive straightening of the land parcels that allowed the rebuilders, freed from the previous property borders, to concentrate on the urban design and on the single construction, in accordance with the changed urban needs (Chabrol 2010).

In England, the historical differences in the land ownership systems with the continental countries were confirmed by the *Town and Country Planning Act* of 1944 that avoided any form of nationalisation of suburban land, identified as development areas and at the same time underlined a public urbanisation system in which the surplus value created by urban development belongs to the whole community (Cullingworth and Nadin 1964). The *New Towns Act* of 1946 will transform the ownership structure in a real urban form by directing funds and reconstruction efforts towards newly founded peripheral urban areas.

Fig. 2.1 Saint-Malo in Normandy after the destructions suffered in August of 1944. Even after the complete removal of the rubbles, the plot scheme and overall ownership structure remain clearly visible and influence reconstruction plans. (Source: Cathedrale Saint Malo, public domain)

Today wars seem to have become a fluid condition that periodically characterises certain territories; 90% of the conflicts that occurred in the first decade of the twenty-first century occurred in territories already affected by civil wars in the previous 30 years (World Bank 2019). War becomes an endemic condition in which it is difficult to identify a clear beginning and an end and in which groups held together by variable interests confront each other without necessarily arriving at a clear distinction between winners and losers (Smith 2005). In this state of uncertainty, the configuration of the ownership structures becomes a preventive and long-term weapon whose consequences extend well beyond the war period. Abdulaziz Hallaj, a careful observer of the dynamics of the Syrian conflict, stated in a conference in Venice:

> This is another major problem that we will have to fight. Housing land and property issues will greatly decide how the city will evolve. In areas that have strong cadastral documentation, people will most likely rebuild along the same lines as before. In areas without strong records, there will most likely be a tabula rasa. Syria already has areas that have been prepared to become a tabula rasa; the government has already prepared some itself, while other actors in other parts of the country are setting tabulae rasae in preparation for the future. This is going to be a major challenge because people who used to have a community somewhere are fighting hard to return to that community. (Hallaj 2018: 81)

Hallaj's call requires the construction of a conceptual horizon that is able to re-read the evolution of ownership structures before, after and during conflicts in order to identify reconstruction strategies capable of adapting and reacting to different administrative choices. The starting point must be a careful analysis of the theatres of war, of their historical evolutions and of the distortions that may have led to triggering or exacerbating violence. The analysis foresees the understanding of the urban borders generated by the morphological evolution of the settlements in a long or short period, of war borders understood as battle lines but also as processes of redefinition of the urban form with the aim of bending it to war purposes and finally of reconstruction borders that emerge from the complex and intricate tangle of economic, political and social interests that allow the achievement of a new urban balance.

2.2 Urban Borders

A careful analysis of the morphological and numerical parameters of the urban space is a starting point in the construction of a cognitive process that allows the drafting of any kind of legitimate design proposals. The research group Urbicide Task Force, established at Università Iuav di Venezia, intends to initiate and support the necessary process for the development of the conceptual and operative tools necessary for a sustainable reconstruction.

The cities of Damascus and Aleppo, deeply involved in the Syrian conflict, allow us to explore urban borders and their evolution during and at the end of the conflict. The two cities were analysed by Urbicide Task Force (Albrecht and Galli 2019) in order to define their morphological characteristics through two main parameters: on the one hand a qualitative assessment of the urban fabrics through a broad definition of the urban form taking into consideration housing types, historical period of construction and morphologies (see Figs. 2.2 and 2.3), and on the other hand a quantitative assessment based on two different parameters, density (walkable surface over total area) and permeability (occupied area over total area).

The overall mapping of the parameters for each of the two cities makes it possible to understand and identify similarities and differences and to begin to build a network of boundaries that divide the urban spaces, either clearly or subtly. Both cities have a historical core located at the centre of the urban space with considerable dimensions (about 10% of the total). The first phase of modern development of the historical cores began in the 1930s with the 'parallel' plans of René Danger in Damascus (Njoh 2015) and Michel Ecochard in Aleppo (Fries 1994; Neglia 2001) showing the difficult attempt of coexistence and harmonisation between the intricate system of the historic city and the hygienist will of colonial projects (Driver and Yeoh 2000; Arnold 2000). The attempts of close proximity and uniformity between the two different urban fabrics, were often implemented through the disembowelment of historical parts (Wright 1991), which only partially succeeded in redefining the border of the consolidated settlement and more often transformed the borders in buffer zones for the positioning of larger scale services that are difficult to locate in the historic city.

The colonial city is an intermediate space between the historical core and the enormous post-Industrial development that has defined all global cities in a long

Fig. 2.2 A typological analysis of Aleppo documents the urban fabric and demonstrates the relationship between formal urban patterns of different densities (in the eastern part) and informal settlements (in the western part). (Source: Urbicide Task Force)

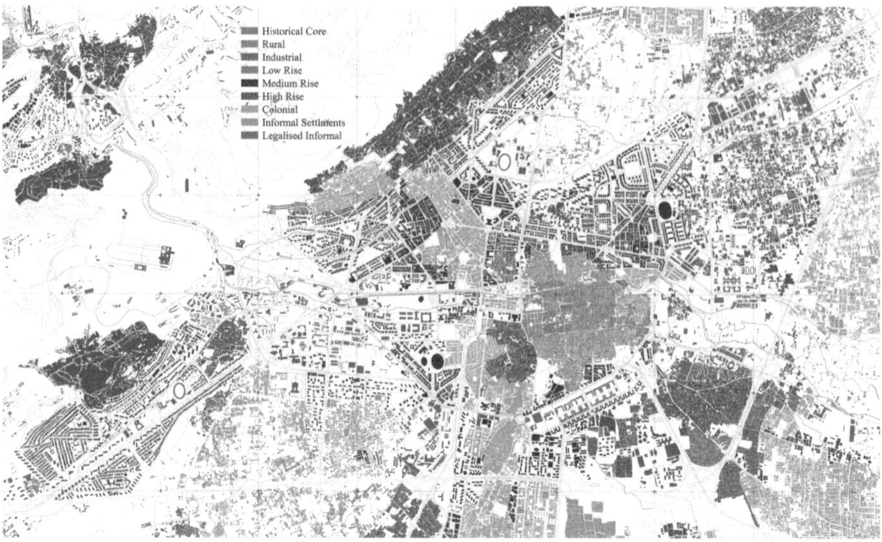

Fig. 2.3 A typological analysis of Damascus documents the urban fabric and shows the coupling of formal and informal settlements based on orography of the territory. (Source: Urbicide Task Force)

period that goes from the post-war years in Europe to the 1960–1970s in the Middle East, Asia and Africa. The great wave of development was characterised by the abandonment of the spatial parameters of the historical settlement in favour of a very broad and imprecise interpretation of the modernist creed. Construction on lots with variable heights transformed the cadastral subdivision into the main form factor of the urban settlement. Middle Eastern cities uncritically assumed the global model without taking into account the climatic and cultural specificities of the context (Ragette 2003) often leading to the break of social structures and to the crisis of ethnic and religious balances rooted in centuries (Burke and Yaghoubian 2006). The rapid construction of post-Industrial urban fabrics with different levels of density is coupled with the growth of informal settlements with very different characters: in Aleppo, neighbourhoods with high population density (GTZ 2009) absorbed urban growth that the building market failed to respond to; in Damascus, on the other hand, the orography of the city favoured a 'coupling' between planned areas and informal settlements that exploited the connections to infrastructural networks, in areas that, for safety reasons or environmental fragility, were commercially less interesting (Abdin 2017). Many peripheral areas of cities, often destined to agricultural production, were affected by migration from rural to urban areas, forming informal low-density settlements: this was the case of Qaboun or Douma in the peripheral areas northeast of Damascus (Ismail 2018). The informal settlement has played a central role in an urban development (Dupret et al. 2012; Schreiner 2016) unequal in terms of services and possibilities that is one of the long-term causes of the ongoing conflicts.

A careful analysis of the historical and political reasons that led to the formation of informal fabrics in Syria passes through a focus on the Ottoman and French colonial heritage, on the formation of urban elites, on the role of the popular masses in Hafez Assad's rise to power, on the growth of the state apparatus in the Baath period and on the rigidly bureaucratic approach to urban planning policies implemented by the party (Ghazzal 2012). Regardless of these complex political conditions, informal settlements are numerically important: in Damascus, about 1.3 million people (40%) out of a total of 3 million lived in 'zones of collective contravention' (*manāteq al moukhālafāt al-jāmia*) (Government of Syria 2010; Clerc 2014); in Aleppo, out of a total of 2.4 million, about half of the population lived in 22 recognised and mapped informal settlements (GTZ 2009). In both cities these areas mainly developed between the 1970s and 1980s due to the inability of Syrian urban policies to respond to the problem of growth, migration from rural areas and waves of refugees coming from regional conflicts (Palestinians, inhabitants of the Golan Heights, Iraqis etc.). The informal settlements should not be considered as an extreme attempt to build dwellings with emergency means but rather as singular or collective responses to legislative problems.

As an example of the complexity and sometimes senselessness of Syrian laws, in the novel *The Home That Was Our Country* (Malek 2017), the story develops around an apartment in the centre of Damascus where the protagonist's family, of good professional bourgeoisie, cannot evict a tenant who occupies it with impunity for more than 30 years. The analysis of informal fabrics must therefore proceed on two parallel tracks: on the one hand the morphological conditions that led to the pre-war urban configurations and on the other the political and cultural systems that generated and confirmed these structures. The sum of the two analyses allows the definition of visible and invisible urban borders that take the form of incurable cuts in the social body and in the urban form.

The generation of informal settlements starting in the 1970s proceeded through three parallel processes: the abusive occupation or squatting of non-urbanised areas, the purchase of agricultural parcels for the purpose of construction and the densification of existing settlements through the subdivision of properties and the addition of floors (Clerc 2014b). A comparative analysis of the only land register ever produced for the city of Aleppo in 1937 and of the 2018 map shows how the evolution of informal settlements has often taken place starting from a process of subdivision of agricultural land parcels generating a singular urban form that transformed the elongated shape of the plots for irrigated agriculture into row dwellings (see Fig. 2.4). In some cases, the intervention of a minimal planning system allows the construction of an infrastructure and basic services network and regularises by size and shape the different plots, while in others, characterised by larger agricultural plots, the metamorphosis from rural to urban maintains the external boundaries and subdivides the internal parcels through random processes. In the areas already inhabited in 1937, the memory of the cadastral parcelling remains almost totally intact, whether it is in the historical core or in small rural settlements incorporated in the post-Industrial fabric but still recognisable (see Fig. 2.5). The cadastral memory is instead totally lost in formally planned urban areas: the processes of

Fig. 2.4 A comparative analysis of the only land register ever produced for the city of Aleppo in 1937 and of the 2018 (in red destroyed buildings) map shows how the evolution of informal settlements has often taken place starting from a process of subdivision of agricultural land parcels generating a singular urban form that transformed the elongated shape of the plots for irrigated agriculture into row dwellings. (Source: Urbicide Task Force)

Fig. 2.5 A comparative analysis of the only land register ever produced for the city of Aleppo in 1937 and of the 2018 (in red destroyed buildings) map shows that in the areas already inhabited in 1937 the memory of the cadastral parcelling remains almost totally intact, whether it is in the historical core or in small rural settlements incorporated in the post-Industrial fabric. (Source: Urbicide Task Force)

expropriation of parcels or of transformation of land owners into shareholders for building development projects has completely disconnected urban form and original cadastral maps. The three contemporary processes of abusive occupation, subdivision of agricultural parcels and densification are very similar to the processes of formation of the historical city with the fundamental difference that while the refinement and improvement of the historical settlements takes place over a long period of time and through a complex trial and error mechanism (Mumford 1961), the process of formation of the informal city often occurs in a sudden way and in a short timeframe.

The morphological analysis of the urban fabric must necessarily take into account the political processes that led to the establishment and development of the settlements. The two peripheral districts of Damascus, Mezzeh 86 (Syrian Observer 2019) and Qaboun (Ismail 2018), are the perfect examples of the formation of urban boundaries due not only to physical and constructive reasons but to complex economic and political dynamics. Mezzeh 86 and Qaboun are both informal settlements positioned at the opposite sides of the city. Mezzeh 86 to the southeast was born from building permits issued directly by Rifat Assad, brother of former president Hafez, as a form of clientelism towards the Alawite families linked to the army. On the opposite northwest side, Qaboun hosts Sunni immigrants from the Idlib area and from the rural villages around Damascus who arrived in the capital for military service and then settled on a former agricultural land (Ismail 2018). The two districts are both characterised by an almost totally illegal but completely different economy: Mezzeh 86 lives off the economic remittances of the nearby military base while Qaboun, in addition to smuggling with neighbouring Lebanon, off small trade and transport services (mainly minibuses used as a collective taxis). Both districts are physically, socially and economically separate from the urban fabric of the Syrian capital, defining very clear social borders, albeit on opposing factions. As reported by Salwa Ismail:

> Qaboun and Mezzeh 86 embody the 'us' and 'them' division patterned around socio-economic lines, with overlapping affiliation to or membership in a sect, and also along rural-urban distinctions. Both quarters are on the margin of the city's economic and cultural life. The fact that Qaboun's population is predominantly Sunni has not translated into closer ties with the city's social elites, namely the Sunni Damascene merchants and entrepreneurs who are the regime's historical allies. The distinction between Sunnis and Alawites in reference to the migrant populations of Qaboun and Mezzeh 86, respectively, has to do with the functionalization of sect as an element in the institutions of coercion and with the articulation of this distinction with the social, spatial and economic realignments of the populist conjuncture. The sociospatial divisions, in which relations with the regime were a determining factor, overlapped with sectarian distinctions. However, the divisions are primarily, and more appropriately, political, being drawn on distinctions that set those who are co-opted by the regime against others who are marginalised or denied access to the apparatus of power (in particular the army and the security forces). Added to this is the enmity or antagonism that arose in processes of land expropriation that appeared to benefit one group at the expense of another. (Ismail 2018: 89)

The destiny of the two informal districts during the civil war followed a completely opposite path linked irremediably to the urban boundaries but above all to the settled social groups. The analysis of the history through the conflict allows to understand how new borders of war have overlapped or replaced existing urban borders.

2.3 War Borders

Mezzeh 86 has remained a loyalist area throughout the whole war, has not been a battlefield and has not suffered significant damage. However, it is interesting to note that the war has in any case influenced the neighbourhood through a process of densification and increase in volumes favoured by the lack of governmental control. Mezzeh 86 is certainly not an isolated case given that in 2011, the year of the beginning of the war, concrete consumption in Syria increased by 115% compared to the previous year (Syria Report 2011), an evidence of how, in a period of crisis for commercial activities, many invested in construction paradoxically seen as a safer investment. The increase in construction speed and the poor quality of the used materials have led to frequent collapses, fires problems linked to the abusive supply of electricity and structural problems as a consequence of the geological features of the soil (Syrian Observer 2019). The policy of the Syrian government towards informal settlements, before and during the war, reflects an opportunistic consideration of the urban value for the satisfaction of housing needs; in the early 2000s, the government through the *Ministry of Local Administration and the Environment* (MLAE) and in collaboration with the European Union carried out the *Informal Settlements Upgrading and Rehabilitation National Program* (Government of Syria 2010), which included improvements in services, intended as infrastructures as well as basic public services such as schools and religious buildings, and the simultaneous upgrading of the dwellings with the aim of improving their performances and safety. Syrian planning policies are closely linked to the pursuit of political objectives and in the same period Law 26/2000 (Ghadbian 2001) and Law 15/2008 (Cunial 2016) proposed a diametrically opposite approach defined as urban renewal. The urban renewal and the upgrading have coexisted within the Syrian cities with a logic linked to political patronage rather than spatial considerations exacerbating the differences in urban quality between the different districts.

During the civil war, Qaboun has taken a path diametrically opposed to Mezzeh 86: in July 2013, the neighbourhood was conquered during an offensive by the Free Syrian Army that, at the beginning of 2014, negotiated a truce with the regime guaranteeing access to water and electricity and safe passage to the nearby Tishreen hospital. The truce granted a relatively normal life to the inhabitants of the neighbourhood until 2016 with the end of the battle for Aleppo in the north of the country and the beginning of the loyalist offensive for the conquest of the Ghouta region northeast of the capital. The parallel stories of Mezzeh 86 and Qaboun show how it is impossible to construct an automatic correlation between areas of rebellion against the regime and informal areas (Clerc 2014). Even if illegal occupation is often associated with opposition to the government, a clear line cannot be drawn between informality and political opposition and the complexity of the social and urban fabric requires a careful and parallel study of the spatial features and political evolutions of each urban cell.

The period of government by the opposition forces in Qaboun and in other areas has contributed to the formation of a parallel geography: the ordinary administration (Baczko et al. 2013) was guaranteed in a short timeframe, building a parallel state with borders in continuous modification inside the urban fabric. The people involved in the provisional government and their origin immediately showed a number of contradictions: they were often young people (with a high percentage of women) with advanced education but coming from peripheral neighbourhoods and therefore cut off from the state patronage system (Leenders 2013). A clear break was also manifested between the provisional governments and activists outside the country, often members of an economically advantaged bourgeois elite with more cosmopolitan relations—the dialogue and construction of a common direction between internal and external front was particularly difficult (Fig. 2.6).

With the advance of the loyalist forces in the *Operation Damascus Steel* conducted by the Syrian loyalist army between February and April 2018, the Qaboun district became first a battlefield, was then surrounded and besieged with the complete blockage of food, water and medicine supplies and finally emptied through the establishment of humanitarian corridors to the Idlib area for civilians and fighters. The use of the strategic siege of neighbourhoods or entire cities (PAX 2018) by the regime shows the will to confirm and emphasise the homogeneous division of the population from an economic, religious and cultural point of view by accentuating the strong social and physical borders between neighbouring areas. The most substantial change in urban morphology brought by the war was the birth of a complex system of checkpoints that accentuated or partially deformed the pre-war urban borders.

Caerus Associates (2014) carried out a careful evaluation of the checkpoints present in the city of Aleppo since the beginning of the war: between 2012 and

Fig. 2.6 Destructions in the Darayya suburb in Damascus in autumn 2019. The collapse of the buildings allows to understand the property borders and the subdivision of the plots. (Source: Damascus Voice of the capital, public domain)

2014, 1562 checkpoints were mapped, 1054 controlled by the regime and 508 by rebel forces. The ratio of 1–3 does not reflect the territorial occupation of the city that at the time of the research was 70% occupied by the Free Syrian Army. The study clearly shows a very different will of the two factions: on the one hand the rebels who intended to give back immediately a resemblance of administrative normality and on the other the loyalist forces that, given the numerical inferiority in the first phase of the war, aimed at the control of the territory through the occupation of key strategic points. The checkpoints of the rebel forces were settled along the fighting fronts and were used for observation and war purposes while government checkpoints were also located within the occupied areas acting as demonstrations of the physical presence of the government. War strategies and urban strategies merged and overlapped, on the one hand confirming and underlining the cellular nature of some areas of the city (army barracks, university, planned residential neighbourhoods etc.) and on the other abandoning direct control of the territory in favour of an urban acupuncture with the aim of controlling only major infrastructures (see Fig. 2.7).

The new geography built by the checkpoints stabilised when the battles became forms of permanent siege to areas of the city (PAX 2018). The checkpoints were not intended any more as filters but were transformed into definitive blocks defining 'an area surrounded by armed actors with the effect of preventing the regular entry of humanitarian assistance and the exit of sick and injured' (UN OCHA 2016). According to the International Red Cross, urban conflicts with the siege tactic cause civilian victims eight times higher than those fought in the open field (Alzawqari

Fig. 2.7 Map of Aleppo based on UNITAR data; the red area is destroyed. Total destroyed area amounts to 24% of the city. Confronting the map with Fig. 2.2, it is easy to observe how the large majority of destruction occurred in informal settlements. (Source: Urbicide Task Force)

and Jaquemet 2018). The desire to increase the impact of the war on civilian population is testified by the naming of the campaigns *Starvation Until Submission* (Reuters 2013), which aimed at eliminating all food support for rebel areas. This war technique was preceded by a systematic selective destruction of bakeries in the rebel areas through the use of targeted air strikes (Caerus Associates 2014).

The use of war tactics based on the considerable increase of civilian victims and on terror as a form of government is part of an overall vision of the future Syria based on a social, ethnic and cultural homogenisation of the population (see Fig. 2.8). It is a process that from military warfare expands to demographic engineering and that imagines the country at the end of the conflict with a population decidedly lower than the pre-war 20 million, assuming as an advantage the 5.5 million emigrants and 6.6 internally displaced, and with a less complex ethnic composition clearly defined within urban and territorial borders (Gerlach 2019). The project of subdivision on an ethnic and religious base carried out through the progressive siege of cities and neighbourhoods was certified by Bashar Assad in 2017 in an official speech in which he hoped to build tomorrow's Syria, 'homogeneous and authentic in the true sense' (Gerlach 2019). The consequences on the urban fabric and on the borders of homogenisation strategies are rather unpredictable: on the one hand, the control of urban areas is simpler and allows the formation of direct clientele, pushing towards a process of concentration and increase in population density; on the other hand, the demographic pressure of the metropolises has been one of the triggers of initial protests that later turned into open conflict.

Fig. 2.8 Zoom of the northeast part of the city of Aleppo; the destruction (in red) is visibly more in the areas with informal urban pattern, mostly representing the parts opposed to the regime. (Source: Urbicide Task Force)

The first consequences on the urban fabric of the programmed homogenisation processes are already partly visible and show the overwhelming return of ownership structures as a battlefield at the end of the war: the European Institute for Peace witnessed 344 military clearance operations promoted by the regime only in December 2018 (Graham-Harrison 2019). These operations are selective demolitions that the army defines as clearance with the aim of returning to normal conditions through the displacement of rubble and the cleaning up of partially destroyed areas, but that in reality hide the will to define areas for future speculative financial exploitation. The areas affected by these operations are mainly places where ownership rights are not certain or are disputed and obviously are mainly located in areas like Qaboun, strongholds against the regime: the redesign of ownership borders as a weapon during and after the conflict.

2.4 Reconstruction Borders

In war areas, the ownership structure is used by authorities and fighting groups to pursue military ends and assert their authority and legitimacy during and especially after military operations. Expropriations for political purposes began well before 2011 in the Middle East and North Africa region addressing long-standing grievances: in the 1980s and 1990s, in the north of Iraq, the Iraqi government ousted tens of thousands of Kurds and Turks farmers and transferred their lands to Arab settlers; in 1970, the Libyan government legislated to redistribute tribal lands and in 1986 abolished private land ownership; none of these laws have yet been repealed. During and after armed conflicts in the region, expropriations were commonly used by non-state and state actors not only as a form of retaliation against perceived enemies, on an individual or collective basis, but also to assert their authority through enforcement of their legal order (World Bank 2019). Even the extremist group ISIS, in the period of territorial government in Iraq and Syria, articulated elaborate rules for land ownership justifying the expropriation of agricultural lands that belonged to apostates before the group captured them, and further regulations for the distribution of assets confiscated as charity for the poor (Revkin 2016).

It is significant that the first laws promulgated by the Syrian regime on the subject of reconstruction focus almost entirely on ownership structures. Law 10 was ratified on 2 April 2018 and amended on 11 November 2018, after protests by the international community and human rights associations (Albakkor 2018). The reconstruction law establishes a legal regime that allows the government to designate areas in any part of the country as development zones for reconstruction. In these areas the authorities have a week to request a list of property owners in the existing property records that must be provided within 45 days. The owners whose names do not appear in the land registry, mainly the people settled in informal areas or with ownership certificates for agricultural lands, will have a year to reclaim their property for the purposes of compensation, since all property owners will be displaced and their properties requisitioned following development plans (TIMEP

2018). Law 10 makes it almost impossible for owners living abroad or internally displaced to return to areas that were against the regime in order to reclaim their properties, for economic or personal security reasons. The regulation represents the extension of a series of laws aimed at redefining Syrian ownership structures: Decree n. 66 of 2012 (Rollins 2017) authorised the central government to intervene with specific regulations in two areas south of Damascus; Decree No. 63 of 2012 (Risk 2018) authorised the immediate expropriation of movable and immovable properties to those who were designated, with a very vague definition, as terrorists; Decree No. 11 of 2016, suspended and effectively cancelled properties in areas with emergency conditions; and Law No. 3 of 2018 (Haugbolle 2018) allowed the removing of war ruins, with another vague definition that favours targeted destructions.

The underlying intent of this powerful legislative apparatus is clear: to foster the ethnic and economic homogenisation of Syrian cities through either the complete expropriation of property or, in the best-case scenario, the transformation of the owners into shareholders of public–private partnerships. The bureaucratic apparatus and administrative complexity are seen as tools for the definition of homogeneous groups ('us' and 'them') in a conforming process that reaches the absurdity of a country with half of the population emigrated that imports inhabitants from the neighbouring Iran (Chulov 2017). The ultimate goal of demographic engineering is to change the religious balance between Sunnis, Shiites and minority groups within the country, a process already implemented in large and small urban centres and in different regions: Al-Qusair, Daraya, Wadi Barada, Aleppo, Homs etc. (Syrian Observer 2017). Obviously the first development zones identified by the government under Law 10 in November 2018 were Qaboun, Barzeh and Jobar (HRW 2018), three peripheral areas of Damascus that were the strongholds of the Free Syrian Army during the conflict. It is easy to match the ownership structures proposed in development zones to a speculative reconstruction that involves the economic exploitation of the most interesting areas and the use of urban planning as a tool for the definition and strengthening of insurmountable physical and social borders (Hallaj 2018). The country sketched by reconstruction laws is radically different from the pre-2011 one within a process already undergoing: Syria's pre-conflict urbanisation index was 53%, among the lowest in the region, while UN-Habitat estimates that the current rate is 73% (UN Data 2018), one of the highest. The war acted significantly on such a data at territorial scale: it favoured rural to urban migrations given that urban areas were less affected by quick changes in military control and that mass migrations to refugee camps were more common in villages that offered less housing options. At the urban level historical cores experienced a significant increase in population as both rich and poor strata of the population opted to move to the city centres that were perceived as less dangerous and less likely to be involved in open conflict.

The first project carried out following Decree No. 66 clearly shows the physical and urban consequences of reconstruction laws. Marota City, a completely new neighbourhood, developed at the outskirts of the Mezzeh district in eastern Damascus (Hanna and Harastani 2017), currently under construction in the area of Al Razi. The masterplan draws clear inspiration from the Dubai model (Elsheshtawy

2010), with a collection of architectural objects that maximise the developed volumes by raising up to 70 floors in an urban landscape where only a few years ago the minarets were the only tall elements above a dense urban fabric. Marota City abandons all reference to the urban quality of the historic Arab city, replaces the model of the covered road and the souq with the globalised shopping mall, abandons the typology of the court house or of the small multi-family building in favour of oversized single-floor apartments and has very bad sustainable performances and very high energy consumption due to the use of materials and construction techniques totally unsuitable to local climate. The project fits into the global trend of urban planning models that transform the citizen into a customer-consumer, a model whose impact is negative throughout the world but that is worse in reconstruction contexts in which resources are limited and the possibility of radical modification of the urban fabric larger (Fig. 2.9).

Observing the development of ownership structures makes it possible to understand and anticipate urban forms to come and might lead to a new design proposal alternative to the dubaisation model. Each innovative proposal cannot proceed without the careful analysis of borders before, during and at the end of the conflict to identify solutions specific to each examined area. The return to pre-war ownership structures is not a viable option because it will not be able to consider the complete history of urban morphologies linked to the historical events that shaped them. The option of a large-scale redrawing of property borders seems today the most probable

Fig. 2.9 Rendering of the volumes developed for the Marota City Masterplan used to publicise the intervention on the official Facebook page. (Source: Damascus Cham Holding Company Marota city, public domain)

considering how necessarily the reconstruction will take place with housing densities and building typologies radically different from pre-war ones. A possible alternative is a process of new foundation or re-foundation (Galli 2018) capable of identifying the most suitable measures of a totally renewed cadastral parcel. The size and shape of the lots can become important starting points for the reconstruction of the cities without a complete fixed design, which given the size and scope of destruction appears almost impossible, but through the establishment of open processes that give back to local communities and to single inhabitants a voice that today is completely denied. The new ownership structures can act as triggers of bottom-up processes, in which the community becomes the engine of the morphological change of the urban fabric and in which reconciliation is possible starting from the recognition of the historical and social phenomena that generated tensions and violence. The post-war *tabula rasa* can become a palette of future possibilities if it is understood as a complex overlapping of borders and narratives, invisible to the blind gaze of speculation, but identifiable through a patient work of cognitive analysis and strategic design.

References

Abdin Y (2017) The fragility of community security in Damascus and its environs. Int Rev Red Cross 99

Albakkor A (2018) New property law in Syria: for reconstruction or eviction? The Aleppo Project. https://www.thealeppoproject.com/new-property-law-in-syria-for-reconstruction or-eviction/. Accessed 26 Aug 2019

Albrecht B, Galli J (2019) Urbicide Mosul triggers for reconstruction. Anteferma Edizioni, Conegliano Veneto

Alzawqari S, Jaquemet I (2018) New research shows urban warfare eight times more deadly for civilians in Syria and Iraq. International Committee of the Red Cross. https://www.icrc.org/en/document/new-research-shows-urban-warfare-eight-times-more-deadly-civilians-syria-iraq. Accessed 26 Aug 2019

Arnold D (2000) Illusory riches: representations of the tropical world, 1840-1950. Singap J Trop Geogr 21(1):6

Baczko A, Dorronsoro G, Quesnay A (2013) The civilian administration of the insurgency in Aleppo, Syria. IPI Global Observatory

Bernoulli H (1946) Die Stadt und ihr Boden. Verlag für Architektur AG, Zurich

Burke E, Yaghoubian D (2006) Struggle and survival in the modern Middle East. University of California Press, Berkley CA

Caerus Associates (2014) Mapping the conflict in Aleppo Syria. American Security Project, Washington

Chabrol V (2010) Le remembrement comme vecteur d'une idée urbaine Reconstruire une ville après la Seconde Guerre mondiale. Histoire&Mesure XXV-1

Chulov M (2017) Iran repopulates Syria with Shia Muslims to help tighten regime's control. The Guardian. https://www.theguardian.com/world/2017/jan/13/irans-syria-project%20pushing-population-shifts-to-increase-influence. Accessed 26 Aug 2019

Clerc V (2014) Reconquérir ou reconfigurer les marges de la ville? Les contradictions des politiques de résorption des quartiers informels à Damas. In: Semmoud N (ed) Les marges et

la ville: entre exclusion et intégration. Cas Méditerranéens. Presses Universitaires François Rabelais, Tours
Clerc V (2014b) Informal settlements in the Syrian conflict: urban planning as a weapon. Built Environ 40(1):34
Cullingworth B, Nadin V (1964) Town and country planning in the UK. Routledge, London/New York
Cunial L (2016) Housing land and property (HLP) in the Syrian Arab Republic. Norwegian Refugee Council, Oslo
Driver F, Yeoh B (2000) Constructing the tropics: introduction. Singap J Trop Geogr 21(1):1
Dupret B, Denis E, Ababsa M (2012) Popular housing and urban land tenure in the Middle East: case studies from Egypt, Syria, Jordan, Lebanon and Turkey. American University in Cairo Press, Cairo
Elsheshtawy Y (2010) Dubai: behind an urban spectacle. Routledge, London
Fries F (1994) Les plans d'Alep et de Damas, un banc d'essai pour l'urbanisme des frères Danger (1931-1937). Rev Mondes Musulmans Mediterr:73–74
Galli J (2018) Siria 2018 il progetto di una rifondazione. In: Ferlenga A, Bassoli N (eds) Ricostruzioni. Architettura, città e paesaggio all'epoca delle distruzioni. Silvana Editoriale, Milano
Gauthiez B (2003) Espace urbain. Vocabulaire et morphologie. Éd. du Patrimoine, Paris
Gerlach D (2019) A turn towards Idlib in Syria—is the goal to 'achieve homogeneity' in the country? Washington Institute for Near East Policy. https://www.washingtoninstitute.org/fikraforum/view/a-turn-towards-idlib-in-syria-and-the%20homogenization-of-a-country. Accessed 26 Aug 2019
Ghadbian N (2001) The new Asad: dynamics of continuity and change. Syria Middle East J 55(4)
Ghazzal Z (2012) Shared social and juridical meanings as observed in an Aleppo 'marginal' neighborhood. In: Dupret B, Denis, Ababsa E (eds) Popular housing and urban land tenure in the Middle East: case studies from Egypt, Syria, Jordan, Lebanon and Turkey. American University in Cairo Press, Cairo
Government of Syria (2010) The Informal Settlements Upgrading Rehabilitation National Programme (ISURNP). Ministry of Local Administration, Damascus
Graham-Harrison E (2019) Assad demolishes refugee homes to tighten grip on rebel strongholds. The Observer. https://www.theguardian.com/world/2019/jun/08/syria-demolishes-homes-grabs-territory. Accessed 26 Aug 2019
GTZ Deutsche Gesellschaft für Technische Zusammenarbeit (2009) Informal settlements in Aleppo. Rapid profiles of all informal settlements in Aleppo. Aleppo urban Development Project, Aleppo
Hallaj A (2018) Syrian cities and the challenges of reconstruction. In: Galli J (ed) Syria the making of the future. From urbicide to the architecture of the city. Incipit Editore, Venice
Hanna E, Harastani N (2017) Is Marota city the type of reconstruction Syrians need? Shattuck Center at the School of Public Policy, Central European University, Budapest
Haugbolle S (2018) Law No. 10: property, lawfare, and new social order in Syria. Syria Untold. https://syriauntold.com/2018/07/26/law-no-10-property-lawfare-and-new-social%20order-in-syria/. Accessed 26 Aug 2019
Holm M (2016) The Marshall plan: a new deal for Europe. Routledge, New York/London
HRW Human Rights Watch (2018) Syria: residents blocked from returning government demolishes homes, denies property rights. Human Rights Watch. https://www.hrw.org/news/2018/10/16/syria-residents-blocked-returning. Accessed 26 Aug 2019
Ismail S (2018) The rule of violence: subjectivity, memory and government in Syria. Cambridge Middle East Studies, Cambridge
Leenders R (2013) Social movement theory and the onset of the popular uprising in Syria. Arab Stud Q 35(3) Special Issue: Perspectives on the Arab Uprisings
Malek A (2017) The home that was our country. Nation Books, New York
Mazzoleni C, Bonfantini B (2001) Cento anni di piani urbanistici. Edizioni della Triennale, Milan

Milward A (2006) The reconstruction of Western Europe 1945–51. University of California Press, Berkeley

Mumford L (1961) The city in history. Harcourt, Brace and World, New York

Neglia G (2001) Persistences and changes in the urban fabric of the old city of Aleppo. Environ Des: J Islamic Environ Des Res Centre 18(1/2)

Njoh A (2015) French urbanism in foreign lands. Springer, Berlin

PAX Siege Watch (2018) Siege Watch Tenth Quarterly Report Part 2—The culmination of "Surrender or Die". https://siegewatch.org/wp-content/uploads/2015/10/PAX-report-Siege-Watch-10b.pdf. Acccssed 26 Aug 2019

Quilici S (2001) Piani di Ricostruzione. In: Cento anni di piani urbanistici. Edizioni della Triennale, Milano

Ragette F (2003) Traditional domestic architecture of the Arab region. Edition Axel Menges, Felbach

Reuters (2013) Insight: starvation in Syria: a war tactic. Reuters World News. https://www.reuters.com/article/us-syria-crisis-hunger-insight-idUSBRE99T07I20131030. Accessed 26 Aug 2019

Revkin M (2016) The legal foundations of the Islamic state. The Brookings Project on U.S. Relations with the Islamic World Analysis Paper, No. 23

Risk R (2018) Syria's new housing law is a veiled attempt to displace tens of thousands of refugees-but even that won't help the regime win the war. The Independent. https://www.independent.co.uk/voices/bashar-alassad-syrian-civil-war-law-10-displacement-homes-papers-latest-a8377306.html. Accessed 26 Aug 2019

Rollins T (2017) Decree 66: the blueprint for al-Assad's reconstruction of Syria? New Humanit. https://www.thenewhumanitarian.org/investigations/2017/04/20/decree-66-blueprint-al-assad-s-reconstruction-syria. Accessed 26 Aug 2019

Schain M (ed) (2001) The Marshall plan: fifty years after. Palgrave, New York

Schreiner E (2016) Conflict urbanism: Aleppo seminar at Columbia University preparation materials

Smith R (2005) The utility of force. The art of war in the modern world. Allen Lane, London

Syria Report (2011) Construction activity leads to surge in demand for cement. Syria Report. https://www.syria-report.com/news/real-estate-construction/construction-activity-leads-surge-demand-cement. Accessed 26 Aug 2019

Syrian Observer (2017) Displacement and demographic change create new realities in Syria. The Syrian Observer. https://syrianobserver.com/EN/features/23831/displacement_demographic_change_create_new_realities_syria.html. Accessed 26 Aug 2019

Syrian Observer (2019) Mezzeh 86's contradictory relationship with authority… loyalty and conflict! The Syrian Observer. https://syrianobserver.com/EN/features/51470/mezzeh-86s-contradictory-relationship-with-authority-loyalty-and-conflict.html. Accessed 26 Aug 2019

TIMEP Tahir Institute for Middle East Policy (2018) TIMEP Brief: Law No. 10 of 2018: Housing, land, and property

UN Data (2018) Country profile, Syrian Arab Republic. UN DATA, New York

UN OCHA (2016) Humanitarian needs overview: Syrian Arab Republic. UN OCHA, New York

Voldman D (1997) La reconstruction des villes françaises de 1940 à 1954. Histoire d'une politique. L'Harmattan, Paris

World Bank (2019) Building for peace. Reconstruction for security, sustainable peace and equity in MENA. World Bank, Washington

Wright G (1991) The politics of design in French colonial urbanism. University of Chicago Press, Chicago

Chapter 3
How the Multiscalar Agents Negotiating the Urban Structure of a City Impact the Region and the Border District of Benzú (Ceuta, Spain)

Mari Paz Agúndez

Abstract Engaging in the urban negotiations of the Spanish exclave of Ceuta are various multiscalar agents, ranging from the widespread military presence, the imposing infrastructure of the *Fortress Europe*, Morocco's political interest over the border, along with the municipality's constant bargaining with the Spanish government. The significance of these intersecting power dynamics must be analysed based on the multiple levels of governance the actors represent. First, the lasting impact of the military and the exclave's colonialist past will be introduced as shaping the urban plan until present day. Furthermore, the European border infrastructure and its imposed control over the city will be expanded on. Finally, the focus will be on how interactions with actors at the local scale affect the urban fabric in a *trans-local* way. The low-income border district of Benzú, left isolated by the effects of conflicting governing interests, will be taken as a case study to illustrate how the border presence and its regulation impact the least powerful agents in the negotiation.

Keywords Ceuta · Spatial justice · Enclaves · Border infrastructure · Urban planning

3.1 Introduction

Together with Melilla, Ceuta represents the only territorial border shared by the European Union (EU) and the African continent. Located at the northernmost point of Africa, due to their border character both enclaves endure similar impositions derived by Fortress Europe, such as sharing a militarised land border with Morocco, each holding detention and infantile centres, as well as being subject to the same

M. P. Agúndez (✉)
Technische Universität Berlin, Berlin, Germany
e-mail: m.agundez@campus.tu-berlin.de

G. Cimadomo et al. (eds.), *Urbanism at Borders*, The Urban Book Series,
https://doi.org/10.1007/978-3-032-09594-7_3

migration phenomenon. To exemplify the impact of these transversal impositions on a local level configuring the city's daily life, this article focuses on the case of Ceuta.

The peninsula of Ceuta lies on the other side of the Strait of Gibraltar, which connects the Mediterranean Sea with the Atlantic Ocean—the 14 km stretch is the narrowest pass separating Africa and Europe. Due to its specific geopolitical situation, the exclave has become the first point of entry for irregular migration of sub-Saharan Africans in their pursuit to reach the EU, hindering the migration movements entering Europe.

Scholars have used metaphors to define this border complex, describing it as the *Fortress Europe* (Walters 2004) and referring to Ceuta as an *icon* of said Fortress (Castan Pinos 2009, 2014; Ferrer-Gallardo 2008, 2011). Given the fact that the EU-State border is crossed by hundreds of thousands of third-country nationals yearly (Geddes 2000; Bigo 1998), some find this term inappropriate; however, if the focus is on the security measures implemented at the border, aiming to block entrance from outsiders, the analogy to an enclosed mediaeval fortress is fitting (Castan Pinos 2009).

After Spain joined the European Union in 1986 and the Schengen Agreement came into force in 1995, the borders between Ceuta and Morocco were no longer simple borders between two regions or countries but a border enclosing the Schengen area from another continent (Ferrer-Gallardo and Albet-Mas 2013; Casas-Cortes et al. 2013). Consequently, Ceuta became a platform where a range of different actors, such as the European Union, the Spanish or the Moroccan government, the military and the municipality of Ceuta, as well as the local actors engage in the daily urban formal and informal practices—continuous *infrastructuring practices*, which classify Ceuta as an arrival infrastructure. According to Meeus et al. (2019), an arrival infrastructure is defined by the 'multiscalar territories through which migration is governed'. European migration policies interact with national and local policies and, in doing so, influence the shape and components of urban arrival infrastructures (Meeus et al. 2019). Ceuta poses an extreme case of the layering of government competencies, where the local scale, comprised of the municipality and the mixed population groups, including migrants settled in Ceuta, migrants awaiting to continue on to Europe, detained migrants, cross-border workers and locals, confront the macrocosmic scale, established by Fortress Europe along the Moroccan government, the Spanish government and its Ministry of Defence. As an agent manoeuvring the space, the municipality tries to improve the urban plan regarding the population groups and considering the social segregation and poor infrastructure. These attempts have been represented in the form of master plans throughout the twentieth century (nowadays P.G.O.U.); however, their implementation is predominantly thwarted by the military's claim to land and the security requirements. At the same time, the municipality is forced to accept the reality that Fortress Europe imposes on the city, present in the form of a highly securitised border, a temporary stay centre, the infrastructure hosting minors, the Schengen border at the harbour and the different bordering practices spread throughout the city (Sassen 2013; Brambilla 2014; Darling 2017). This multilevel governance ultimately shows the *trans-local* constellation the city is exposed to, where decisions made elsewhere on a larger scale of governance impact a city like Ceuta.

The intersections between these multiscalar agents occur on an urban level, limiting the socio-spatial mobility of the inhabitants, denying them quality housing and isolating them from the rest of the city both socially and economically. To illustrate this phenomenon, this article focuses on the case of Benzú, a district located along the northern part of the border with Morocco. Benzú used to host the checkpoint Benzú–Belyounech, one of two checkpoints along the border where there was a considerable exchange of goods with the neighbouring Moroccan border town Belyounech. This formed a mutually dependent reliance that reached beyond trade and into culture and infrastructure. Although the checkpoint was the primary source of commerce for Benzú, in 2003, the delegate of the Spanish Government moved to close this border checkpoint, effectively isolating Belyounech and stopping the development of the district. The reason behind the closing was the uncontrolled migration into the Schengen area; however, this top-down decision failed to consider the utter economic and socio-spatial isolation the neighbours of the border districts would be plunged into. The analogous border town relationship to that of Benzú–Belyounech is that of El Príncipe–Fnideq, located at the southern border-crossing of Tarajal. Despite similar socio-economic exclusion from the rest of Ceuta, El Príncipe has benefited from the closing of Benzú, for all border trade has shifted to the southern crossing, bolstering the growth of El Príncipe–Fnideq—a stark contrast to Benzú–Belyounech's stagnation. This demonstrates that borders embody the layering of government competencies (Brown 2010; Sassen 2013; Meeus et al. 2019), and should, therefore, be studied under a multiscalar approach.

Furthermore, borders are understood as a site of exclusion but also of crossing and encounter, both dividing and connecting (Mezzadra and Neilson 2013). Despite the wall's construction and the distance separating Ceuta's border districts from their Moroccan neighbour towns, on a local scale, there is a territorial continuity showcased by their mutual reliance and economic and cultural exchange. This border complex has been widely analysed by scholars such as Ferrer-Gallardo's study of the influence of *Schengenisation* on the securitisation of the European border (2008, 2011), Castan Pinos's investigation on the permeability of Fortress Europe (2009, 2014), or Ferrer-Gallardo and Albert-Mas' definition of Ceuta as a *limboscape* (2013); however, there is little research under a spatial perspective on how the local-scale urban life is impaired by trans-local regulations. Architect Buoli (2014) has studied the impact of cooperative practices on the landscape and Cimadomo (2023), extending the research's scale, has investigated the bottom-up initiatives generated at the USA–Mexico or the Palestine–Israel border. Expanding on this emerging spatial analysis and under a multiscalar approach, my research focused on Ceuta is key to shedding light on how interactions across different actors derived from borders configure the urban space.

By focusing on the case of Benzú, this paper describes the intersection of power dynamics generated by the multiscalar forms of governance taking place in this neighbourhood. This will be done in a twofold way: firstly, by showing how the military presence, although created in a colonial past, is still dominating the city and impairing its socio-spatial mobility; secondly, by investigating the infrastructure of Fortress Europe and how the conjunction between *hard* (physical border, military facilities, detention centres) and *soft* (by-products of border presence, interactions of population groups) infrastructure interact with each other, economically and socially isolating the micro-scale agents. Additionally, I will use the analogous border town relationship of El Príncipe–Fnideq and would introduce the extensive effort taken by Morocco to develop the northern Atlantic coast to contrast the economic and socio-spatial exclusion experienced by Benzú. Under an interdisciplinary approach, based on archival research, mapping, analysis of historic urban plans and participant observation, this paper aims to render different forms of control detaching Benzú from the city centre as well as its Moroccan hinterland.

3.2 Colonial Territories and Military Presence

Throughout the twentieth century, Spain underwent the military regimes of Primo de Rivera (1923–1930) and Franco (1939–1975). Both dictators used built infrastructure and urban development to enact power over Northern Morocco and establish Spanish sovereignty. This would prove the use of architecture and urbanisation as a tool to exercise control (Weizman 2007; Misselwitz and Rieniets 2009; Graham 2010).

Ceuta's natural border was adopted as the official border in 1860 after the Treaty of Wad-Ras (Remacha 1994). At the time of this treaty, there was an obvious strategic advantage to controlling the entrance to the Mediterranean Sea as well as to Africa, which boosted the Spanish colonial desire to spread their control over the North of Morocco. This form of influence was formally defined in 1912 after a treaty with France, converting this area of dominance into a protectorate, which would last until Morocco's independence in 1956.

In 1923, Miguel Primo de Rivera assumed control of Spain. Among his priorities was the colonisation and consolidation of the protectorate as Spanish land. Following the agenda of the regime and taking advantage of the socio-political conditions, in 1928, urban planner Gonzalez del Castillo proposed a linear city along the axis Ceuta–Tetouan, in which urbanisation was used as a means of colonisation, combining a military and a colonising linear city (Darias Príncipe 2008). The project aimed to structure the territory for future immigrants, classifying them according to their nationality and implementing territorial occupational patterns (ibid; Buoli 2016). The further development plans under Primo de Rivera's dictatorship—seeing that they had to solve the shortage of social housing for the working classes (Solá-Morales Rubio 1976)—would continue with the idea of class division, welfare, cleanliness and proximity to urban centres. Using zoning to define a social hierarchy, the 1927 proposal of Blanco Soler and Bergamín designated the expensive housing as facing the sea on the isthmus of the city, whereas the future districts, such as Benzú and El Príncipe, were located on Campo Exterior, hosting low-income houses at the continental side of the peninsula (García González and

Guerrero 2016). Further initiatives were proposed, such as the functional city by architects García Mercadal and Foertsch in 1929, following the CIAM (Modern Architecture International Congress) principles, or the Blein proposal in 1930, which relocated and grouped the barracks and the military headquarters to Campo Exterior. These initiatives had to consider the extensive land used by the military as a fundamental part of the city. However, these were ultimately dismissed, due to the impossibility of expropriating and relocating land from the military (García Mercadal 1931; Blein 1932; Bravo Nieto 2004).

After the civil war (1936–1939) and under Franco's regime, Pedro Muguruza was appointed Director General of Architecture to reorganise the public works of the Spanish protectorate and develop a new master plan for the city of Ceuta. Since none of the aforementioned plans had been implemented, the city continued to suffer from its previous problems, the shortage of housing and the dominance of the military zone. Muguruza's new plan again placed the future residential neighbourhoods in Campo Exterior (Muguruza 1946). Under fascism and influenced by the segregating ideas of his predecessors, he continued with the inherently racist practice of zoning. Furthermore, he markedly did not reference the lack of social housing nor the poor conditions of the housing stock, a problem that prevailed for over 40 years. The impact of Muguruza's plan irrefutably altered the enclave's urbanism and defined its current socio-spatial distribution, particularly, since its implementation and until 1992, with no other master plan having been enacted since (see Fig. 3.1).

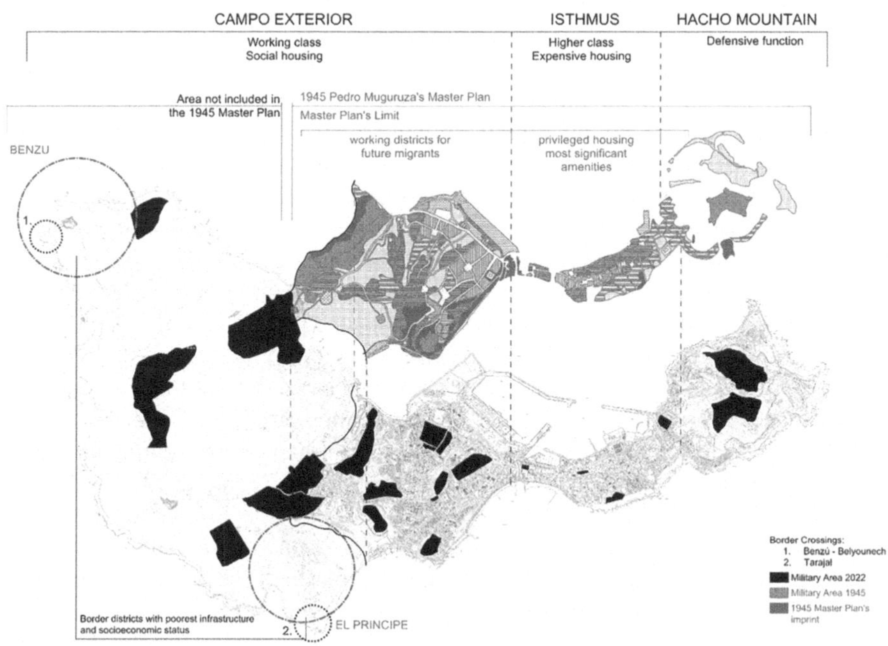

Fig. 3.1 The current socio-spatial division of Ceuta coincides with Muguruza's plan, superimposed in blue, with its limited scope and spatial hierarchisation. In black, territory controlled by the military. (Source: The author)

The 1992 master plan stresses the city's lack of housing and overcrowding as well as the obsolete conditions of residential areas such as the border districts of El Príncipe and Benzú, at the time two informal settlements comprising small shanty towns and other dwellings. These districts, largely populated by Muslims (up to 45%), concentrated an utter lack of facilities, rampant unemployment, high violence levels and marginalisation. Therefore, they had become isolated nuclei of dwellings from the city centre (Ciudad de Ceuta 1992). Despite the emphasis the municipality laid on the eradication of substandard housing, underscoring the urgency at the border districts, the situation did not improve, for the border districts had continued to grow informally and further segregated from the centre.

Thirty years later, the implementation of a new master plan is still pending, yet it has been partially approved (2022). The development of this plan began in 2008; however, the city of Ceuta does not possess the legal power to enforce the master plan by itself, for it has to comply with the regulations set by the supra-municipal agents, such as the Ministry of Defence, Development, Hydrographic Commission, Coastal Regulations, or the Harbour Council. One of its goals is to eliminate substandard housing and pays special attention to the socio-demographic trends between the city and the border districts, seeking an integration of the different social groups (Ciudad de Ceuta, 2021). The stress given in the 2021 proposal to improve the informal peripheral districts is not as determined as in 1992, which, together with the municipality's lack of power, raises questions about its eventual successful implementation.

The master plans and proposals described here are of considerable importance when examining the urban plan of Ceuta, for they designated the Campo Exterior as working-class districts and its urban border landscape as the epicentre of social segregation and economic injustice problems (see Fig. 3.1). Additionally, the strengthening of a constant, widespread military presence poses a persistent obstacle in structuring the urban landscape. The municipality's desire to tackle issues of housing and social integration while maintaining such a high volume of military on the small peninsula continues to represent a conflict of interest in the city's master plan. Throughout the history of Ceuta, particularly focusing on the twentieth century, the military regimes have habitually conditioned the city's development. Since the establishment of democracy in 1978 and until the present day, the Ministry of Defence and the military took control over the areas left by the regimes.

Currently, Ceuta occupies a surface of 18,626 ha and a third of this area belongs to the Defence Ministry. These areas are spread throughout the city and, as a result, the urban fabric has grown around these islands of military soil, creating an archipelago of extremely surveyed areas. Due to the strategic geopolitical location of Ceuta, and in particular the districts of Benzú and El Príncipe, located at the border, the urbanisation of the city is directly affected by the legal requirements of National Defence, compiled according to Spanish law of Zones and Facilities of Interest to National Defence, where security areas are defined as areas where no work or activity may be carried out without the express authorisation of the minister in charge (BOE 1978). The military facilities and their surrounding area, required to have a width of 300 m (Articles 3, 7 and 9 of the law), are regarded as security areas.

Within the urban plan, the entire border structure falls into this category. Consequently, the area included between the border and the 300-m offset line is subject to this law as well as all the military facilities spread along the city. Thus, demonstrating how, although the border is part of Fortress Europe's infrastructure, it also intersects with the military infrastructure as a form of control and governance. Moreover, the constant security derived from the National Defence requirements is imposed on the locals' daily life, consequently, according to Graham (2010), 'domesticating the security infrastructure'.

3.3 The Infrastructure of the Fortress Europe

Throughout the twentieth century, Ceuta was subject to the restrictions stemming from the historic presence of the army, which embodies the macro-scale agent enacting control. In 1986, after the end of the military regime and under a democratic government, Spain entered the European Union, which would represent the new macro-scale agent governing on an international level and setting the confines with which Ceuta would have to readapt its border infrastructure. Therefore, this confirms that the border is not merely a part of a municipal infrastructure but falls under European policies. Expanding on the commonly understood term of infrastructure of basic physical and organisational structures and facilities, such as buildings or power supplies, AbdouMaliq Simone goes further and extends the notion of infrastructure to activity of people in the city—the intersection and conjunctions of people become infrastructure; a platform reproducing life (Simone 2004). In defining the border infrastructure of Ceuta, both conceptions of infrastructure will be considered, combining hard infrastructure (see Fig. 3.1), such as the physical border, roads, military facilities and temporary stay centres, with soft infrastructure (see Fig. 3.2) and its resulting by-products, such as the presence of migrants in the city engaging in formal and informal practices, securitised areas and trade patterns across the border.

3.3.1 Hard Infrastructure

3.3.1.1 The Border

The *Schengenisation* of European territory provoked the strengthening of the European outer perimeter and, subsequently, the construction of a wall along the perimeter of Ceuta and Melilla. They were originally built in 1995 as single 2.5-m high fences. Since then, these militarised *borders* (Ferrer-Gallardo 2011) have increased their security measures, growing in height and width, doubling and even tripling the number of parallel fences (Mutlu and Leite 2012; Ferrer-Gallardo and Albet-Mas 2013). Currently, 12-m high wired walls separate European territory

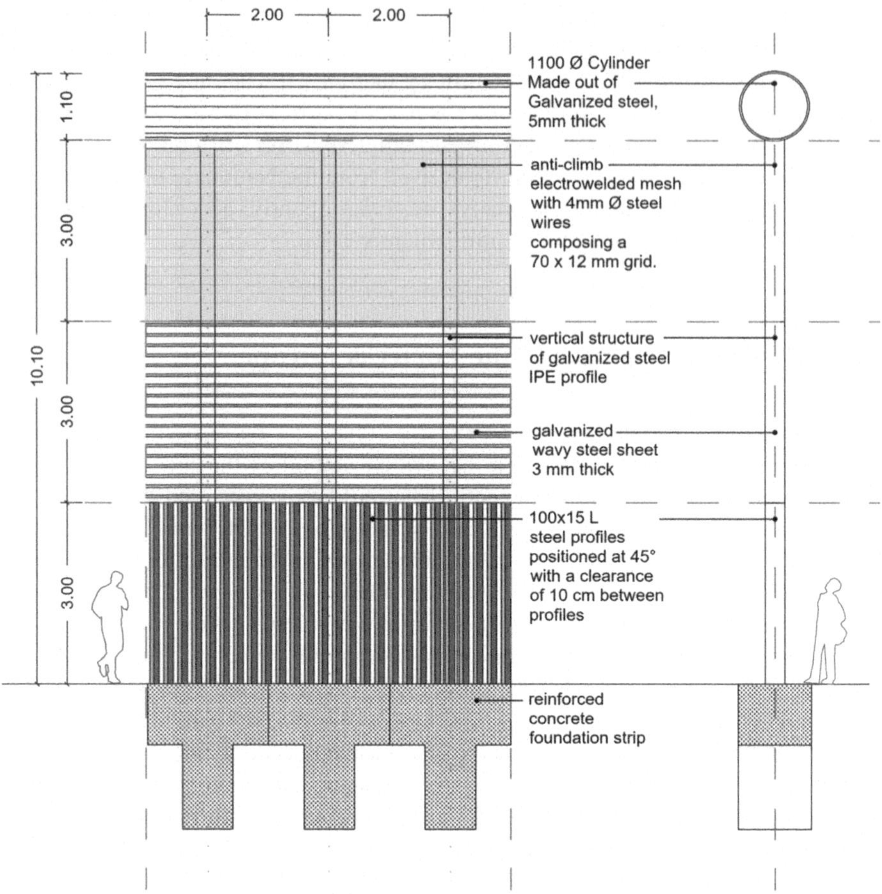

Fig. 3.2 Diagram of the current state of the border, explaining the different constructive methods to deter migration. (Source: The author)

from the African continent, spanning the entire geographical borders of both enclaves. A combination of three fences equipped with barbed and razor wires, wire nets, anti-intruder sensors and alarm systems constitutes the barrier. In Fig. 3.2, the construction of this weaponised barrier is detailed.

The border is constantly patrolled, not only by Moroccan and Spanish forces but also by the security towers spread along the fence. Migrants wait on the other side of the border until they gather a substantial number of people to massively jump the fence (Soddu 2006). Some hide in the Moroccan forests, where they live in makeshift settlements, fearing the raids of the Moroccan Army forces, who periodically crack down on camps (Amnesty International 2018; Bonamini 2018). These raids underline the role of Morocco as a third country involved in the forms of governance, participating in the control dynamics of the border infrastructure and manipulating the movements of migrants as political subjects.

3.3.1.2 The Temporary Stay Centre (CETI) and the Juvenile Centre

Another key element in the infrastructure of Fortress Europe is the temporary stay centre (*Centro de Estancia Temporal de Inmigrantes*, CETI). Located on the outskirts of the district of Calamocarro-Benzú, it was originally designed to be a temporary solution, covering the stay time from the asylum application until the completion of its processing. The ultimate conclusion of a migrant's stay in such a centre should, thus, enable their continuation further into the Spanish Peninsula and Europe (UNHCR 2018). However, the reality is often much different. As recent reports from the Spanish Ombudsman (2016) and Amnesty International (2018) have demonstrated, their stay in the centre is never clearly defined, often resulting in periods exceeding 18 months or even exceeding 3 years.

The architecture of the building reinforces the idea of control in several ways. First, to avoid interaction between the migrants with locals, the centre is placed on the outskirts, isolated from the city. Second, the building is designed in terraces. Resembling Foucault's *Panopticon*, the administration and management are elevated, ensuring constant surveillance over the lower terraces, where the migrants reside. Third, the building is constructed in modules, resembling a summer camp rather than an urban infrastructure, emphasising the temporary character of the migrant's stay and further isolating them from urban daily life.

The juvenile centre, ironically named *La Esperanza* (the hope), is also relatively displaced from the city. Unaccompanied minors arrived in Ceuta in the late 1990s and, despite their rising presence on the streets, in 2012 the sub-delegation opened a relocation centre with a hotel structure designed to lodge homeless people. Having no other infrastructure, the building promptly turned into a shelter for minors, illustrating that there had never been a constructive effort from the government to host minors (Fig. 3.3).

3.3.2 Soft Infrastructure

Officially, migrants and minors living in the detention centres are not detainees, therefore, their housing regimen is open. This means that during the day they can move relatively freely within the limits of the city and engage with its urban dynamics. Although the temporary stay centre and juvenile centre represent a hard infrastructure, the presence of migrants and minors and their interaction can be categorised as soft infrastructure. This population group, whether temporary or semi-temporary, represents a constant influence on the city and its urban negotiations. It is a result of the border and a defining feature of the border landscape. This group faces the infrastructure most as an obstacle and is the agent with the weakest power.

Since its origin, Ceuta has based its economy on trade; however, currently, the scope of this market has drastically reduced and its economy mainly depends on the exchange with its Moroccan hinterland (Ferrer-Gallardo 2011; Castan Pinos 2014).

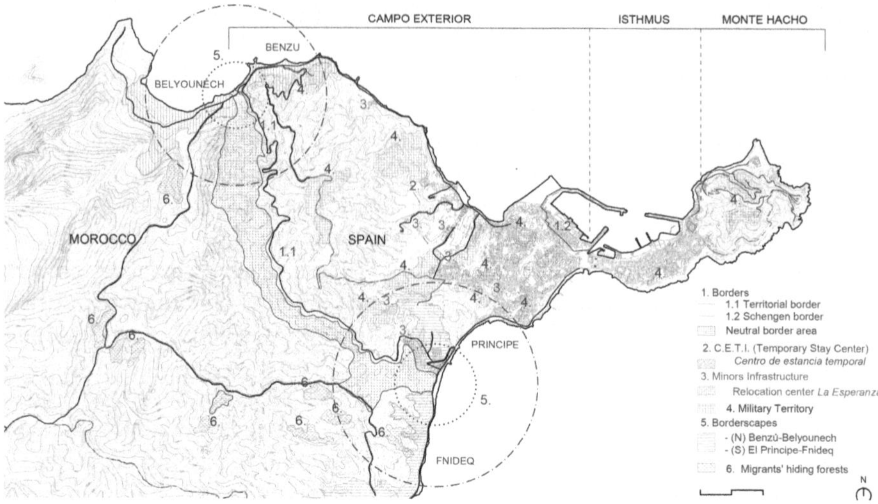

Fig. 3.3 Plan of Ceuta, depicting imposed hard infrastructure in relation to the districts. (Source: The author)

Due to the Protocol of Accession of Spain linked to the Schengen Agreement, Moroccan citizens living in the neighbouring province of Tetouan are exempt from the visa requirements (Ferrer-Gallardo 2011; Buoli 2014). This treaty fosters trade and, due to the lack of official customs, eases the exchange of goods without taxation. Benefitting from this legal gap, *porteadoras* (literally translated as *carriers*) cross the border daily, transporting goods as personal belongings (Buoli 2014). In many cases, the high volume of *porteadoras* attempting to cross the border has led to fatal progressive crowd collapses. According to Spanish authorities, these border rushes were sometimes a way of confronting the border agents of the Spanish military police who threatened to bar the entrance of the *porteadoras*. In Benzú, in the summer of 2002, these 'confrontations' escalated and, in an attempt to avoid more conflicts between the carriers and the Spanish police, Luis Vicente Moro, the government delegate in Ceuta, moved to temporally close the border-crossing of Benzú–Belyounech in October 2002 (Abad 2002, 2003). The presence of *porteadoras* not only had an impact on the border region's economy, but also contributed to maintaining a vibrant life in the border district.

3.4 The Closure of Benzú–Belyounech

Upon Spain's entrance into the European Union, the border between Ceuta and Morocco was subsequently militarised and experienced a heightened level of surveillance. Coinciding with the two border checkpoints are the two border districts of Ceuta: Benzú located at the northern crossing and El Príncipe, located at the southern crossing of Tarajal. Both neighbourhoods have been self-constructed by their dwellers and have grown informally, isolated from the city. As aforementioned, their economy relies primarily on the commercial trade with Morocco, which has structured two significant symbiotic relationships between neighbouring districts and their correlating border-crossings: Benzú and Belyounech along with El Príncipe and Fnideq.

Although the border setting demonstrates a disconnect between the two sides, there are, simultaneously, territorial processes that suggest their proximity and mutual inclusion (Ferrer-Gallardo 2011). The peripheral Spanish districts of Benzú and El Príncipe act as neighbourhoods of arrival (Meeus et al. 2019), inhabited largely by Muslims of Moroccan descent (the majority of which have been naturalised), who structure close-knit family orientated networks with their relatives in Morocco, thus, presenting a continuity on the territorial landscape. Ultimately these peripheral Moroccan-Spanish districts experience a history of mutually dependent relationships that stretch beyond their shared economic engagement into cultural and traditional aspects as well as joint infrastructure.

Besides the soaring tension with the carriers, the Spanish authorities considered the checkpoint of Benzú–Belyounech the main entrance point of illegal immigration and children's passage. As a means to control the influx of unaccompanied minors crossing the border, in 2003, the government permanently closed the border (Campos 2020). This top-down decision had two direct effects on the two areas surrounding the border-crossings: the initial isolation and decline of the northern commercial area of Benzú–Belyounech and the subsequent development and growth of the southern border region of El Príncipe–Fnideq (Soria Pedraza 2017).

3.4.1 The Stagnation of Benzú

Benzú had previously been viewed as the alternative border-crossing to release pressure of Tarajal's crossing. When the government moved to close it, the amicable commercial relations, traditionally maintained with the neighbouring Belyounech, ceased, isolating the town of Belyounech and limiting its growth along with the commercial and economic opportunities of Benzú (Campos 2015).

Although both urban centres have grown very close to one another, since the border closure, both Benzú and Belyounech have become divorced from each other and turned into communities at the end of a road defined by a wall (Soria Pedraza 2017). In the case of Belyounech, its isolation is also highlighted by geography. The mountains of Jebel Musa separate it from its own country, as the closest village to Belyounech is 20 km away. On the other side of the border, Benzú is the furthest district from Ceuta's city centre and, considering the reduced communication infrastructure, the most isolated (see Fig. 3.4).

Despite the border's closure, some processes suggest integration and rapprochement, highlighting the territorial continuity between the two districts (Ferrer-Gallardo 2011). Aside from the stated familiar networks expanding from Benzú onto Morocco, the shared water source is another continuity in the territory, as Benzú receives its drinking water from the natural springs of Belyounech. Since it is the most remote area of the enclave, its water supply is inconvenient and is therefore forced to rely on Belyounech's springs (Bioeconomy Project 2011).

Furthermore, in spite of the border infrastructure, there are remaining courteous neighbourhood relations, as in the case of emergency medical situations, for a small practice has been established in Benzú to attend to the inhabitants and, should there be an emergency, the citizens of Belyounech too (Soria Pedraza 2017). Also, the shared cemetery is located in Belyounech and is where the residents of Benzú are buried. Finally, the previously lively urban life of Benzú, with cafés hosting Moroccans and travellers or the flea market transited by citizens from Belyounech, has been deeply altered, for currently, only one coffee shop remains open and the market no longer exists.

The public-funded society for construction, infrastructure and environment of Ceuta, (Obras, Infraestructuras y Medio Ambiente de Ceuta or OBIMASA), published a report on the distribution of the land in the district of Calamocarro-Benzú, concluding that 25.24% of the land is owned by the Assembly of the Autonomous City of Ceuta, 23.34% is privately owned, while the remaining 57.42% is owned by the Ministry of Defence (Prointec 2009). Furthermore, the border district is heavily impacted by the *security areas* law, protecting the military facilities, as it eliminates the use of the border bypass road, one of the only three roads connecting the district to the rest of the city (see Fig. 3.4). Therefore, the military plays a defining role in

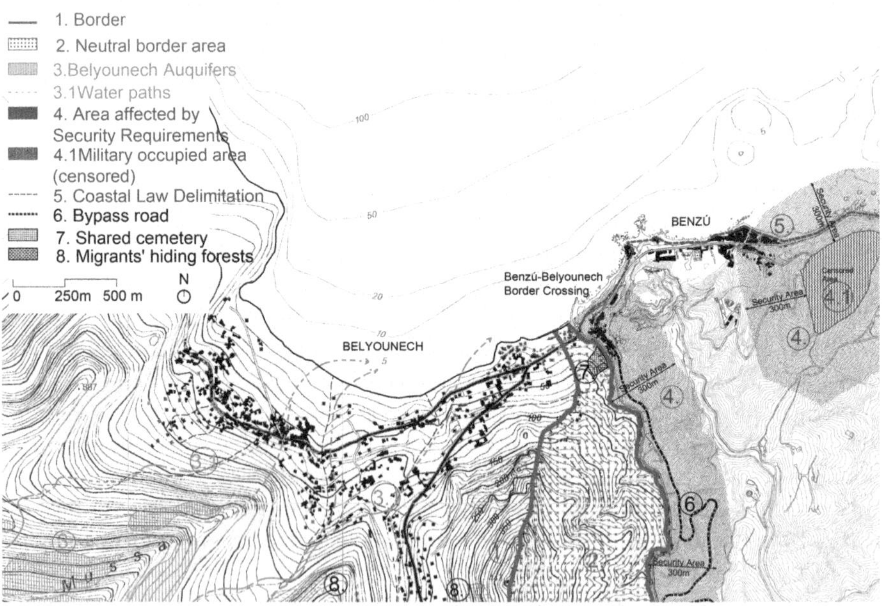

Fig. 3.4 Benzú and Belyounech articulated by the border, commonalities and impositions. (Source: The author)

the urban negotiations, depriving it of a much needed communication route and isolating it from the city centre.

To counterbalance these shortcomings, the municipality has tried to promote the district's economy and reduce its segregation from the city through architectural competitions and the implementation of development plans. Already in the 1992 master plan, Benzú's substandard housing was highlighted along with the lack of urban infrastructure and the coexisting religious groups. To improve the district's condition with quality public housing promotion, the municipality accentuated the need for a freeing up of military land (Ciudad de Ceuta 1992).

In 2005, in an attempt to consolidate the district as a proper urban centre, the municipality sponsored an international architectural competition, EUROPAN-8, to improve the urban infrastructure and environmental quality of the district (Prointec 2009). The winning project designed by the studio *DJ Arquitectura* focused on natural space and intends to create a sustainable relationship between the environment and civilisation (DJArquitectura 2005). To achieve this the coastal housing is to be removed and from the ruins and expanded territory, new housing infrastructure created, where the district's inhabitants would be relocated. The Ministry of Development, following the 1988 Coastal Law, commenced a possessory recovery file, claiming Benzú's infraction of the demarcation line, 25 m offset from the shore. Consequently, many families would be forced to vacate their homes and resettle in the projected houses (Oliva 2014). From the district of Benzú many initiatives, such as the *Caballas* political coalition or *Vecinos Benzú*, were carried out to defend their rights and claim ownership of the space. The *Caballas* appealed to the delegation of the government to review the delimitation of the urban fabric of Ceuta, and, ultimately, Ceuta's sub-delegation has elaborated a report to clarify the limits of the city, thus, defending the district of Benzú (Sakona 2018). Although these bottom-up initiatives are trying to save the slum, its future is still pending in government hands, a macro-scale agent governing on a *trans-local* level.

What is more, the 2021 master plan aims to eliminate Benzú's substandard housing and to promote the specific use of the isolated region by integrating the EUROPAN-8 proposal (Ciudad de Ceuta 2021). Despite the urgency with which the 1992 plan intended to improve the situation of Benzú, at last, none of its proposals were accomplished. Thirty years later, inadequate dwellings, lack of urban equipment and socio-spatial isolation persist. Though the problems have been acknowledged in 2021s master plan, the plausibility of a real improvement of the district remains questionable. The municipality is forced to assume the European constraints, negotiate the relinquishment of the military's territory and comply with their security requirements, demonstrating its lack of power to realise a plan for a better Benzú. The residents of the neighbourhood represent the powerless agent, for they are directly affected by this top-down management and consequently endure its lasting stagnation.

3.4.2 Uncontrolled Growth of El Príncipe

In contrast to Benzú's standstill is the uncontrolled expansion of El Príncipe. Due to the sealing of the northern checkpoint, the border-crossing of Tarajal assumed the entire volume of border-crossings and migration, for which previously two crossings had accounted. Since the dimensions of the Tarajal checkpoint proved not to offer the required space to accommodate the number of people crossing the border daily, in 2005, in the same Tarajal area, a new border passage was opened, El Biutz, in particular to host the *porteadora's* traffic. The commercial activity in the Tarajal trade area has since consistently increased, contributing to the uncontrolled urban growth of El Príncipe and the expansion of the city of Fnideq. Ferrer-Gallardo points out how, although this commercial activity is important in the region, from Ceutan's point of view, it is seen as a benefit, whereas from Morocco, it is seen as a drawback, hindering the economic progress of the country. In actuality, on the Moroccan side, the border is contributing to the creation of a middle class with a rising purchasing power (Ferrer-Gallardo 2011), while in Ceuta, it has contributed to the overwhelming development of El Príncipe, in form of an informal district of substandard housing and lacking amenities.

3.4.3 Morocco's Development Plan of the Northern Region

Morocco's desire to reduce its dependence of its northern region on commerce and trade with the Spanish exclave has culminated in an initiative defined as *Zone Spéciale de Developement* based on the construction of Tangier-Med, an extensive harbour complex in the region of Tangier, located on Morocco's northern Atlantic coast (Natter 2014). Expanding beyond the harbour infrastructure, this development project is associated with industrial zones scattered throughout the northern region, including areas of automotive production or aeronautics and textile manufacturing, focused on improving the port's hinterland and offering new employment possibilities for citizens of surrounding areas. Morocco aimed to build a seaport of international proportion that could compete with Algeciras' seaport (on the other side of the Gibraltar Strait), to overshadow Ceuta's port, and to generate an income from goods and commerce in the northern region of Morocco (see Fig. 3.5).

Morocco's endeavour to further develop the Atlantic coast and bolster the role of Tangier-Med to benefit the region stands in stark contrast to Ceuta's detrimental motion to close its northern border-crossing. The Spanish government's decision not only inhibited the potential to integrate the city into the developing northern Moroccan infrastructure, which would have facilitated the direct connection between Ceuta and Tangier-Med (a mere 30 km apart), but also has effectively isolated Ceuta in the region while denying it touristic and economic possibilities and further growth (Campos 2015). Spain's passive stance in the development of the region, while Morocco has implemented progressive means to modernise it,

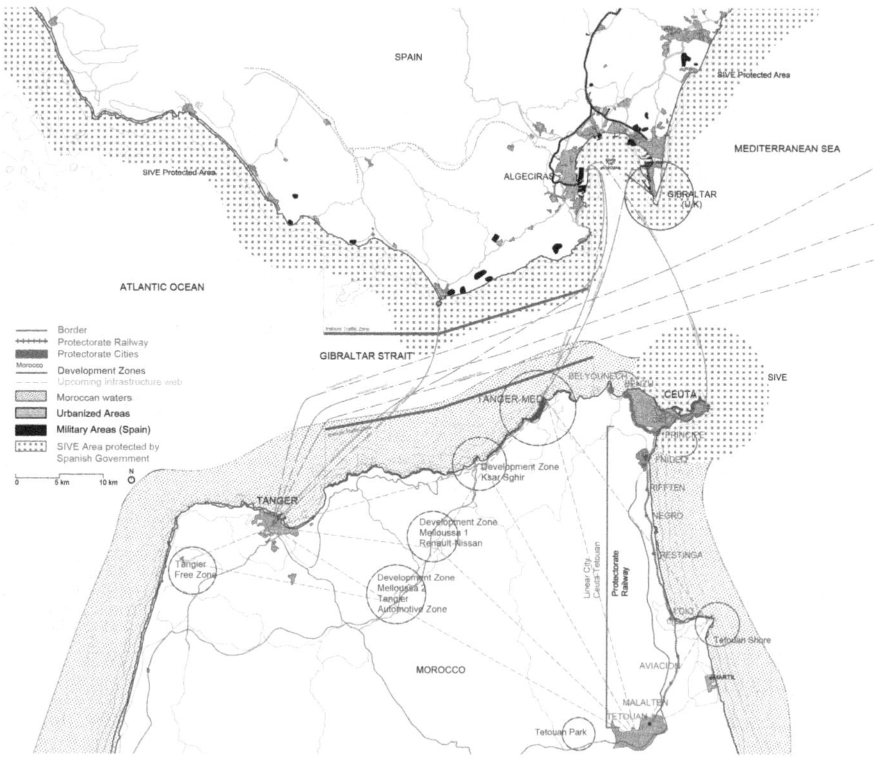

Fig. 3.5 Strait of Gibraltar, Morocco's Development plan against the Protectorate development. (Source: The author)

represents a disregard for the function and cultural importance of its exclave and, in fact, a decision against its own interest.

3.5 Conclusions

Historically, Ceuta has been subject to the restrictions set by various higher forms of governance. In the first three-quarters of the twentieth century, the city underwent extensive military occupation by the authoritarian regimes, laying claim to the northern region of Africa. The plans proposed and implemented during the regimes, limited the space for urban development in the exclave and, following a segregating agenda, defined the socio-spatial distribution of the city, which is still palpable today, for the centre is largely inhabited by a privileged class while the Campo Exterior hosts a weaker, lower-income class (see Fig. 3.1). In the last quarter of the century, under democratic rule and with Spain's entrance into the EU (1986), Ceuta has been obliged to accept the legal requirements of the EU, such as the 1986 law on foreigners, its imposed safety regulations, as well as its widespread infrastructure, constraining the city with border walls and impairing the citizens' mobility.

Throughout history and up to the present day, the macro-scale agents have disregarded the micro-scale—the municipality and locals, who live with the reality of trans-local laws. Following José María Campos' arguments concerning border spaces, a comprehensive urban plan reaching a consensus between the Spanish government, the military, the municipality, the local actors and the citizens has been missing (Campos 2015). Despite the municipality's attempt to improve the border district, its lack of power is evident, for the residents of Benzú still endure isolation from the centre, substandard housing conditions and amenities, as well as assuming a daily life impaired by the border infrastructure. It is, therefore, clear that the low-income areas, such as the case study of Benzú, which experiences the border most on a level of physical proximity, are also on the receiving end of the most negative effects it brings along, both economically and socially. To date, despite failing, every modern urban plan for the city has accordingly called for the freeing up of military-owned areas to foster the growth and inclusion of its lowest-income areas.

Moreover, the closing of the northern border and the overall dismissal of Morocco's development of the northern region has, in effect, cast Ceuta into economic isolation. The Spanish government has overlooked the possibilities Ceuta's integration into the infrastructure of the coast, bolstered by Tangier-Med, would bring. This neglect demonstrates Spain's idle stance towards the growth of the exclave, which has been largely forfeited to the military, for they have control over the territory and have subjected everyday sites to its security infrastructure. Furthermore, the function of the exclave has been reduced to merely a part of the border infrastructure, undermining the city's cultural and economic relevance and, ultimately, sacrificing its identity to that of Fortress Europe.

References

Abad R (2002) El Gobierno cierra 'definitivamente' una de las dos fronteras de Ceuta por las avalanchas de 'sin papeles.' El País. Oct 4. https://elpais.com/diario/2002/10/04/espana/1033682415_850215.html

Abad R (2003). Ceuta cierra el paso fronterizo de Benzú ocho días por incidentes. El País. Apr 11. https://elpais.com/diario/2003/04/11/espana/1050012024_850215.html

Amnesty International (2018) Morocco: relentless crackdown on thousands of sub-Saharan migrants and refugees is unlawful. Sept 7. https://www.amnesty.org/en/latest/news/2018/09/morocco-relentless-crackdown-on-thousands-of-sub-saharan-migrants-and-refugees-is-unlawful/

Bigo D (1998) Europe passoire et Europe forteresse. La sécuritisation/humanitarisation de l'immigration. In: Rea A (ed) Immigration et racisme en Europe. Éditions Complexe, Bruxelles, pp 203–241

Bioeconomy Project (2011) Diagnóstico de recursos naturales, usos y aprovechamientos. Sibe Jbel Moussa. Marruecos. España: Universidad de Córdoba. http://jbelmoussa.cicbata.org/sites/default/files/DIAGNOSTICO%20USOS%20Y%20APROVECHAMIENTO%20RRNN.pdf

Blein G (1932) Concurso para la urbanización del ensanche de Ceuta. Obras 5:30–38

BOE (1978) Boletín Oficial del Estado Reglamento de zonas e instalaciones de interés para la Defensa Nacional 89. BOE-A-1978-9612. Feb 10. https://www.boe.es/eli/es/rd/1978/02/10/689
Bonamini C (2018) Forgotten at the gates of Europe. Jesuit Refugee Service, Brussels. https://jrs.ie/wp-content/uploads/2019/04/Forgotten-at-the-Gates-of-Europe-2018.pdf
Brambilla C (2014) Exploring the critical potential of the borderscapes concept. Geopolitics 20:14–34. https://doi.org/10.1080/14650045.2014.884561
Bravo Nieto A (2004) España y Marruecos en la primera mitad del siglo XX: Arquitectura y urbanismo en un ámbito colonial. Estudios de historia de las sociedades en el mundo colonial y postcolonial. Islas e Imperios 7:45–61
Brown W (2010) Walled states, waning sovereignty, 1st edn. Zone Books, New York
Buoli A (2014) Beyond the border: exploring cross-scalar socio-spatial dynamics of conflict, resistance, and encounter at the Ceuta–Morocco borderscapes. Reg Stud Reg Sci 1(1):301–309
Buoli A (2016) BorderScaping. An explorative study on the North Moroccan/Spanish border landscapes. Doctoral Dissertation. Politecnico di Milano
Campos JM (2015) Ceuta ciudad sin rumbo. Interservicios, Ceuta
Campos JM (2020) Los cierres de la frontera con Marruecos. El Faro de Ceuta. Mar 21. https://elfarodeceuta.es/opinion-cierres-frontera-marruecos/
Casas-Cortes M, Cobarrubias S, Pickles J (2013) Re-bordering the neighbourhood: Europe's emerging geographies of non-accession integration. Eur Urban Reg Stud 20(1):37–58. https://doi.org/10.1177/0969776411434848
Castan Pinos J (2009) Building fortress Europe? Schengen and the cases of Ceuta and Melilla. Center for International Border Research, pp 1–29. http://www.qub.ac.uk/
Castan Pinos J (2014) La fortaleza europea: Schengen, Ceuta y Melilla. Instituto de Estudios Ceutíes, Ceuta
Cimadomo G (2023) Spatial transformations in Ceuta (Spain). Effects of a low-density hinterland on a border enclave. In: Zaman QM, Hall G (eds) Border urbanism. Transdisciplinary perspectives, pp 321–336. https://doi.org/10.1007/978-3-031-06604-7_19
Ciudad de Ceuta (1992) Plan General de Ordenación Urbana. Memoria Justificativa
Ciudad de Ceuta (2021) Plan General de Ordenación Urbana. Memoria Justificativa
Darias Príncipe A (2008) Una posible utopía: la Ciudad Lineal Ceuta-Tetuán. Cuadernos de arte de la Universidad de Granada 39:175–185
Darling J (2017) Forced migration and the city: irregularity, informality, and the politics of presence. Prog Hum Geogr 41(2):178–198. https://doi.org/10.1177/0309132516629004
DJArquitectura (2005) 1st Prize. Europan 8, España Benzú
Ferrer-Gallardo X (2008) The Spanish-Moroccan border complex: processes of geopolitical, functional and symbolic rebordering. Polit Geogr 27(3):301–321
Ferrer-Gallardo X (2011) La desinformalización de la frontera? Evolución de la dinámica territorial en el entramado fronterizo de Ceuta-Fnideq. In: Ribas Mateos N (ed) El río bravo mediterráneo, las regiones fronterizas en la época de la globalización. Ediciones Balterra, Barcelona, pp 365–379
Ferrer-Gallardo X, Albet-Mas A (2013) EU-Limboscapes: Ceuta and the proliferation of migrant detention spaces across the European Union. Eur Urban Reg Stud 23(3):1–4
García González MC, Guerrero S (2016) El Estatuto municipal de 1924 y la profesionalización del urbanismo en España: el caso de Ceuta. Ciudad Territ Estud Territ 48:77–95
García Mercadal F (1931) Ensanche de Ceuta. Documentos de Actividad Contemporánea, GATEPAC 2:34–35
Geddes A (2000) Immigration and European integration. Towards fortress Europe? Manchester University Press, Manchester
Graham S (2010) Cities under siege: the new military urbanism. Verso Books, London, New York
Meeus B, van Heur B, Arnaut K (2019) Migration and the infrastructural politics of urban arrival. In: Meeus B, Arnaut K, van Heur B (eds) Arrival. Infrastructures, migration and urban social mobility. Palgrave Macmillan, Cham

Mezzadra S, Neilson B (2013) Border as method, or, the multiplication of labor. Duke University Press, Durham

Misselwitz P, Rieniets T (2009) Jerusalem and the principles of conflict urbanism. J Urban Technol 16:61–78

Muguruza P (1946) Plan general de ordenación de Ceuta: memoria. Revista Nacional de arquitectura 44:281–310

Mutlu C, Leite C (2012) Dark side of the rock: borders, exceptionalism, and the precarious case of Ceuta and Melilla. Eurasia Border Rev 3(2):21–39

Natter K (2014) The formation of Morocco's policy towards irregular migration (2000–2007): political rationale and policy processes. Int Migr 52(5):15–28

Oliva J (2014) Los letrados de los vecinos de Benzú defienden que el deslinde no existe. El Faro. Feb 19. https://elfarodeceuta.es/los-letrados-de-los-vecinos-de-benzu-defienden-que-el-deslinde-no-existe/

Prointec (2009) Plan de Ordenación de Recursos Naturales y de Gestión del LIC-ZEPA Calamocarro—Benzú (Ceuta). Ciudad Autónoma de Ceuta. https://www.ceuta.es/porng/document/PORNG%20CALAMOCARRO-BENZU%20DOCUMENTO%20INTERMEDIO%20I.pdf

Remacha JR (1994) Las fronteras de Ceuta y Melilla. Anuario español de derecho internacional. Universidad de Navarra 10:195–238

Sakona J (2018) Caballas pide a Delegación que revise la Delimitación de la Trama Urbana de Ceuta para salvar Benzú y La Almadraba. Ceuta al Día. Aug 27. http://www.ceutaldia.com/articulo/politica/caballas-pide-delegacion-revise-delimitacion-trama-urbana-ceuta-salvar-benzu-almadraba/20180827141350185745.html

Sassen S (2013) When the centre no longer holds: cities as frontier zones. Cities 34:67–70

Simone AM (2004) People as infrastructure: intersecting fragments in Johannesburg. Publ Cult 16(3):407–429

Soddu P (2006) Ceuta and Melilla. Security, human rights and frontier control. In: Institut Europeu de la Mediterrània (ed) IEMED Mediterranean Yearbook Med 2006. Institut Europeu de la Mediterrània, Barcelona, pp 212–214

Solá-Morales Rubio I (1976) La arquitectura de la vivienda en los años de la Autarquía, 1939-1953. Arquitectura. Revista del Colegio Oficial de Arquitectos de Madrid 199:19–30

Soria Pedraza D (2017) Desde los límites hacia la comprensión de la ciudad contemporánea. Ceuta y la multiplicidad de la frontera. Tesis Doctoral, Dto. de historia, teoría y composición arquitectónicas, Universidad de Sevilla

Spanish Ombudsman (2016) Asylum in Spain, international protection and reception system resources. Defensor del Pueblo, Madrid. https://www.defensordelpueblo.es/wp-content/uploads/2016/07/Asylum_in-_Spain.pdf

UNHCR (2018) Desperate journeys. United Nations High commissioner for Refugees. https://www.unhcr.org/desperatejourneys/#

Walters W (2004) The Frontiers of the European Union: a geostrategic perspective. Geopolitics 9(3):674–698

Weizman E (2007) Hollow land, Israel's architecture of occupation. Verso, London/New York

Chapter 4
Fronts and Counterfronts in Aleppo: Inflicting and Resisting Violence Through the City

Omar Ferwati

Abstract The Syrian Civil War, like most modern warfare, has largely taken place in an urban environment. In the Battle of Aleppo (2012–2016), combatants contended for control of the city through multiple urban *fronts*. They did not only fight by direct armed confrontation; they also used and depended on the architecture of the city to win and hold their territory. Civilian life was thus under constant threat in attempts to control their movement, buildings and resources. These fronts formed overlapping explicit and implicit deadly borders in Aleppo that transformed its resident's relationship to it. Civilians responded to that violence by changing the architecture around them directly through spatial interventions I call 'civilian crisis architecture'. This architecture makes up a network layer of what I define in this essay as 'civilian counterfronts' that are rooted in the violence of war but aimed at survival rather than power. The artefacts of fronts and counterfronts are ephemeral and embedded in the conflict city's volatile fabric so we can only study them by looking at the city as conflict unfolded in Aleppo. I do this research by relying on social media crowdsourced footage and a forensic architectural analysis of Aleppo's built environment in it and find instances that demonstrate the intertwining of fronts and counterfronts in Aleppo.

Keywords Contested cities · Refugees · Thresholds · Borders · Aleppo · Informal urbanism

4.1 Introduction

Aleppo, Syria, March 2013

Bab Antakya street was quiet. The rubble and dust ripped from the homes and shops lining the street lay in the deserted streets where no one dared wander in the open.

O. Ferwati (✉)
Goldsmiths, University of London, London, UK
e-mail: o.ferwati@gold.ac.uk

G. Cimadomo et al. (eds.), *Urbanism at Borders*, The Urban Book Series,
https://doi.org/10.1007/978-3-032-09594-7_4

With the frontlines of the battle only a few streets away, a rebel sniper watched closely for any movement or signs of advance from government soldiers. The street separates the old city to its east and the modern city to its west. When the sniper claimed the street, residents around it were immediately trapped. As fighting raged nearby, residents of the neighbourhood who were desperate to move bypassed the sniper's vision altogether by carving out a path through Aleppo's old courtyard houses, allowing all residents nearby to pass through freely.

Stories like this one exist throughout Aleppo and illustrate how combatants and civilians depend on the architecture of the city today to both fight and survive conflict. Combatants use the city and its architecture as a weapon while civilians must use architecture to survive in response. Combatants contend for control of the city through what I call *urban fronts*. These urban fronts are more than just a line in a battlefield, they are the many forms and mediums that fighting in cities is conducted through. It includes the contest for movement, civilian space and infrastructure. Beyond military confrontations, they use these fronts to gain leverage by manipulating the urban environment. Bringing the city to the forefront of military strategy, however, puts its civilian residents increasingly at risk. Their lives are at risk when the spaces and resources they rely on are targeted and restructured by combatants. Civilians respond to those circumstances of violence by using the same urban tools as combatants, changing the architecture and spaces around them to survive the violence. I call these spatial interventions 'civilian crisis architecture' and it includes the construction of new structures or alterations to existing ones as reactions to threats against the lives of people. Their responses make up a layer of '*civilian counterfronts*' also rooted in the violence of war but aimed at survival rather than power.

For geographer James Anderson and sociologist Liam O'Dowd (1999), boundaries of space both 'shape and are shaped by what they contain'. Studying conflict urbanism requires understanding the mechanisms of control and bordering—the fronts—as well as the spatial behaviour of civilians within each territory—the counterfronts. Acts of violence through architecture and the civilian architectural responses for survival are always interconnected and their form is, as Anderson and O'Dowd (1999) put it, 'mutually formative'.

As forces that define space and agency over it, fronts and counterfronts are also practices of territoriality and *counter-territoriality*. Geographer Robert Sack (1986: 1) defines territoriality as a 'spatial strategy to affect, influence or control resources and people, by controlling area'. Territoriality in the fronts of urban warfare translates to combatants using and targeting architecture and urban elements to gain, monitor and exercise power over urban space. Counterfronts are acts of counter-territoriality because they resist that control and oppression over civil society. Counterfronts and counter-territoriality use space to reclaim or establish new agency over the public's spaces and resources. This counter-territoriality is not done as a unified political effort but is the effect of a desperation for survival. Ironically, counter-territoriality in chaos is often the only opportunity for civilian agency over their cities and exists only with crisis.

The violent fronts and civilian counterfronts in urban warfare are layers on top of one another in the transformed conflict city. Together they make the conflict city into a violent *heterotopia*, which philosopher Michel Foucault defines as a cluster of diverse worlds contained within a single intense space (Foucault 1986). Understanding the urbanism of conflict as layers of contention allows us to deconstruct and analyse the conflict's socio-political layers through tangible evidence. This understanding also allows us to learn about Aleppo and the conflict that devastated it by examining the traces it left on the urban environment. The fronts and counterfronts of conflict cities continuously redraw borders and constitute what political scientist Jacques Rupnik calls 'time written in space' (Rupnik 1994). Physical objects like civilian architectural responses that define those borders and tell of the behaviour inside them are also products and tangible manifestations of events and time.

Before the civil war, Aleppo was the largest city and the commercial centre of Syria (see Fig. 4.1). The strategic importance of Aleppo therefore made the fight for it especially vicious. The Battle of Aleppo lasted a long 4 years between 2012 and 2016 and resulted in one of the longest sieges in modern warfare, the destruction or damage of almost 36,000 buildings (Najjar 2018), and the death of more than 23,600 civilians (Center for Documentation of Violations in Syria-Killed List 2019). The extreme circumstances Aleppo went through, the duration of battle, the large size of the city, and the crucial role of the city in the events of the battle make it exemplary for studying fronts and counterfronts in urban warfare.

While field research in Aleppo is not possible, the contemporary ubiquity of smartphones and mobile networks have created a large body of media that we can easily access and interpret anywhere. Through their eyes, civilians captured every

Fig. 4.1 Map of Aleppo, Syria, with official city boundaries. (Source: The author)

event of battle but in doing so also captured the rapidly changing urban environment around them. It is in this live access to the conflict city that we can find the scars of urban fronts and the artefacts that make up the civilian counterfronts.

4.1.1 Fronts

We associate 'fronts' with the clash of foes in battle but contested urban space today is fought over in more ways than just direct armed confrontation. The urban fronts of war I refer to here are all contests for space. The fronts of contemporary warfare are layered efforts by combatants to exploit the systems and architecture of the built environment to undermine the other. Fronts are often detrimental to civilians whose lives depend on the structures war destabilised including halting movement, threatening life by targeting civil space (the places people live) and exploiting infrastructure.

Seeing urban conflict as a series of spatial fronts is predicated on understanding contemporary warfare as inseparable from architecture. Today's wars in fact *depend on* the city to be fought (Graham 2010). Urban space is militarised and continuously reshaped by appropriating, systematically targeting and creating architecture for leverage. For Eyal Weizman (2019), an architect and theorist, the city today is not just the backdrop of conflict but the 'very medium of warfare'. The urbanisation of warfare brings the weight of political crisis to the doorsteps of traditional city makers and civilians forced to take on that role. Considering architecture as persistent in chaos and as intrinsically tied to contemporary warfare and cultural production makes up an emerging field of research and theory today. Architects like Andrew Herscher (2010), Wendy Pullan (2011) and Eyal Weizman (2012) have all pioneered novel approaches to understanding the role of construction and destruction during battle, military occupation and in culture. In his analysis of the Kosovo Conflict, Herscher demonstrates how the destruction of architecture was not merely a side effect of war but an essential and intentional process in the ethno-religious violence of the conflict—a way to manipulate demographics and identity (Herscher 2010). Weizman traces the Israeli apparatus of occupation to simple architectural elements. He exposes how the separation wall, settlements, checkpoints and other border elements constitute a flexible and elastic occupation (Weizman 2012). Pullan (2011) examines how contested cities are made up of explicit and implicit urban frontiers. She analyses how the architecture of housing and settlements in Jerusalem, a checkpoint along the German-Polish border towns of Guben/Gubin and museums in Nicosia are implicit frontiers of contention and spatial extensions of conflict.

These works are valuable precedents for considering violence through architecture. They successfully expose the strategic exploitation of the built environment by state powers to grip control. They collectively overlook, however, the role of civilian agency to resist that very violence by the same means of architecture, cutting short any comprehensive account of conflict urbanism.

4.1.2 Counterfronts

We often consider civilians as helpless victims of conflict to be offered our sympathies. While life under war is of course brutal, the role of civilians and civil society in war is much more complex. They are also active participants in changing the city to adapt and survive harsh circumstances. They take part through architectural responses that change the city's form, function and system from the scale of architecture and up to the scale of the city. While fronts are made up of opposing forces fighting for power by exploiting the city and therefore threatening the lives of its citizens, 'counterfronts' are the civilian efforts to create alternative spaces and processes to resist that violence. They do not compete for political gain and take part in the clash of forces; their goal is instead survival.

Studies of conflict, especially when done by urbanists, rarely consider what we can gain by looking at the counterfronts people engage in regularly. They instead tend to focus on the relationship of power, violence and the city. While some studies of conflict cities like Belfast do call for the consideration of the effects of conflict on people, researchers often look at violence and derive its effects or look at the long-term effects on people in post-conflict cities (Leonard and McKnight 2011; Morrow et al. 2011). Looking directly at the spaces created and occupied by people *during* conflict resonates with the call to consider people but provides better insight into the immediate effects of war on their lives.

Architecture is typically a profession of the long-term future, but in this instance, architecture is created for the *immediate* future, the present. In the heat of the conflict, civilians have to make-do with what they have at hand to create rapid solutions to the threats they face. An entire typology is therefore overlooked by a field of urbanists and architects. Collectively these acts represent the wide-scale phenomenon of counterfronts.

The following sections examine three fronts and their responsive civilian counterfronts in the Battle of Aleppo: movement, civilian space and infrastructure.

4.2 Movement: Battlefronts, Barricades and Sniper Alleys

4.2.1 Front of Violence: Movement

The outbreak of Aleppo's 4-year battle in 2012 halted all normal movement in the city. Residents carefully planned their timing and routes around the dangers of battle and weighed the risks of even leaving their homes. Their movement was radically restricted as both a side effect of fighting and deliberate disruptions. The result totally restructured the city's streets and neighbourhoods.

By the turn of the year the battle divided Aleppo between rebels and government forces, with rebels controlling most of the east and south of the city. The resulting landscape was an entangled swirl of territorial claim that was fluid and changing rapidly (see Fig. 4.4). Reinforcements from both sides were spilling in from different cities and neighbourhoods to defend themselves or push on new fronts. Combatants fought over the city street by street and building by building. The frontline neighbourhoods were highly contested and so the most dangerous for civilians. The battle could arrive at the doors of residents overnight, freezing all their movement out of fear of being caught in the crossfire. Risking movement is not always a clear decision. People in contested neighbourhoods subject to shelling faced a dilemma: stay put and risk dying in your home or leave and risk dying in the streets.

Combatants intended to deter their foes but also effectively disrupted the movement of and put at risk all residents as well. They set up barricades to reduce entry points or to stop enemy soldiers and tanks from moving into their territory (see Figs. 4.2 and 4.3). Within their areas of control, opposition and government forces also set up checkpoints to regulate movement into their territory or to search and question passers-by.

Passage for Aleppines was dangerous in their own neighbourhoods even when they were away from the frontlines. Rebel and government forces stopped movement altogether by placing snipers tasked with securing their streets. 'When a sniper sets up in a building, that's it, we could be stuck for weeks', a rebel soldier told Reuters (Solomon 2012). In the process, however, they also stop all civilian movement. Together, the battlefronts, the shelling campaigns, the sniper zones, the barricades and the checkpoints created a series of overlapping and hyper-fluid inner borders that determined where people could and could not move (Fig. 4.4).

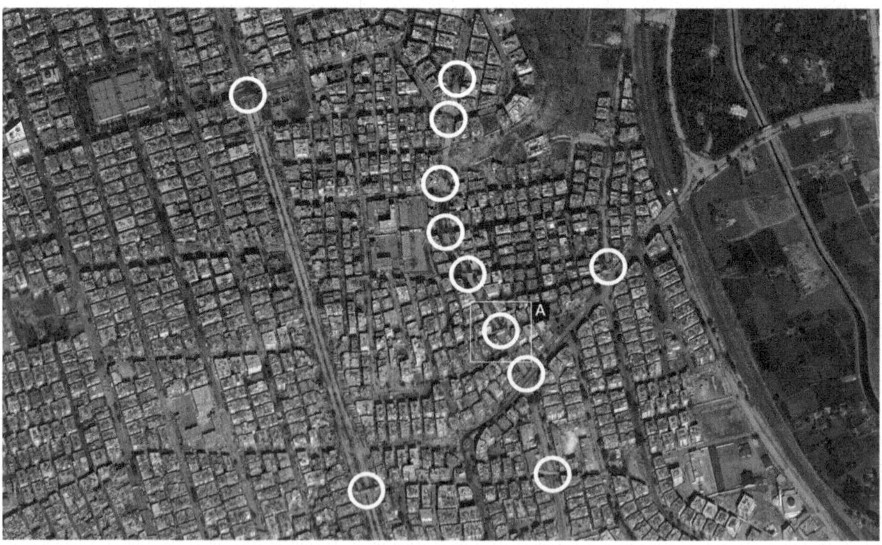

Fig. 4.2 Barricades constructed by combatants during the Battle of Aleppo between three neighbourhoods: Sayf al-Dawla, Saad al-Ansari and al-Ansari Mashad; 18 October 2016 (Maxar via Google). (Source: The author)

Fig. 4.3 Close-up of a barricade as seen in figure above; 18 October 2016. (Source: Maxar via Google)

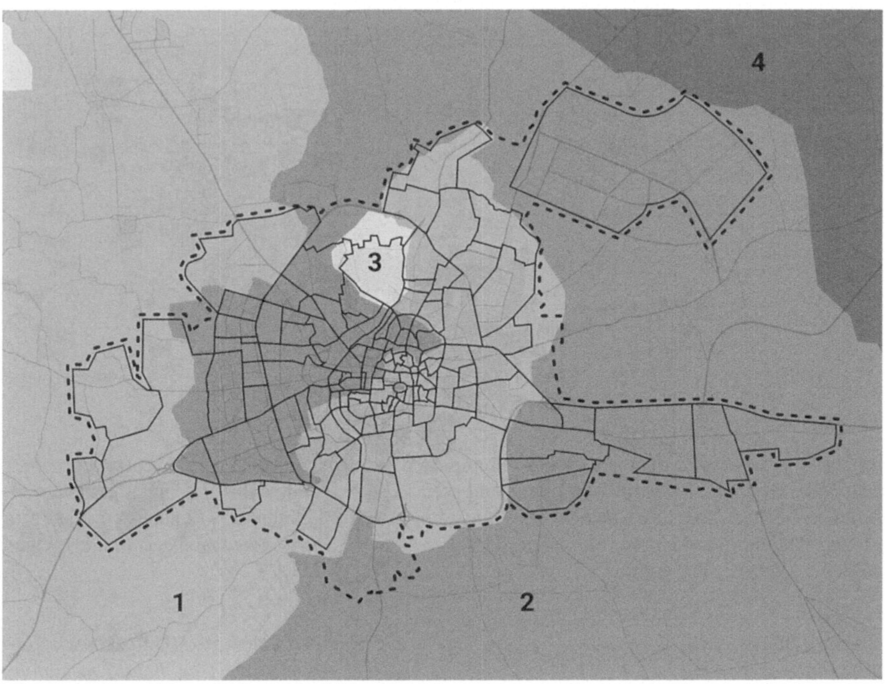

Fig. 4.4 Territorial boundaries in Aleppo, Syria, September 2016. (1) Opposition rebels; (2) Government; (3) Kurdish fighters; (4) 'Islamic State of Iraq and Syria' (ISIS). (Source: The author)

4.2.2 *Counterfront of Survival: Movement*

The front of movement in the battle restricted civilians at every level from their homes to their neighbourhoods, to their city. People had no choice but to intervene in their own homes and neighbourhoods to have any chance of navigating the conflict city. In Bustan al-Qasr neighbourhood, a small handwritten sign was tied to a fence before a street with a warning that read 'there is a sniper on this street', demarcating an urban border by the simplest intervention.

In the presence of snipers, people built visual barriers like mounds of rubble, barrels and even stitched bed-sheets hung between buildings to protect against their line of sight (see Fig. 4.5). Because these interventions exist only in images, it is difficult to know who placed every single barrier and some of these interventions do overlap with strategies that were adopted by rebel fighters. They should nonetheless be considered for their utilisation in the civilian counterfront that resists the control of movement.

Aleppines also created alternative paths of movement altogether, changing how they manoeuvre across urban space by physically reshaping it. This was the case of the story at the beginning of this chapter. In 2013, YouTube user Shams Alhorya (translating to 'freedom sun') uploaded a YouTube video titled 'Aleppo: The way to get from al-Jalloum neighbourhood to al-Kallaseh neighbourhood' (حلب : الجلوم :طريقة الإنتقال من حي الجلوم إلى حي الكلاسة 2013).[1]

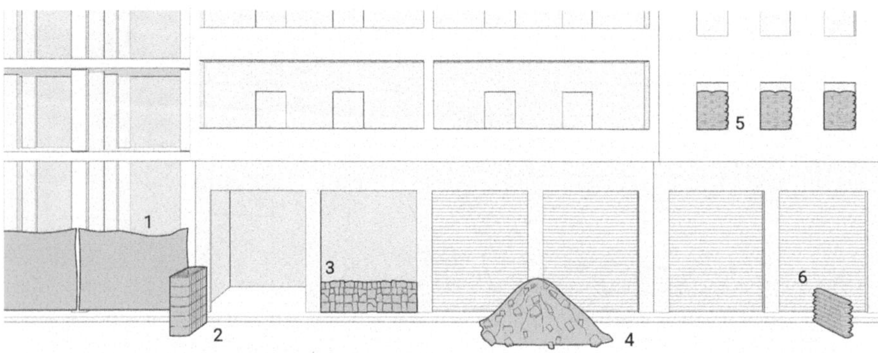

Fig. 4.5 (1) Tarps used to create shelter in unfinished construction sites. (2) Barrels used to block sight and movement. (3) Ruins scavenged and used to infill abandoned storefronts for shelter. (4) Rubble collected into a pile to block views in and protect from gunfire. (5) Sandbags used to protect against gunfire and blasts. (6) Sandbags stacked to create a protective wall used around civilian buildings. (Source: The author)

[1] Shams Alhorya, 'حلب : الجلوم :طريقة الإنتقال من حي الجلوم إلى حي الكلاسة', YouTube video, 01:50.https://www.youtube.com/watch?v=mAXxTW_BXTE.; حلب: الجلوم: طريقة الانتقال من حي الجلوم إلى حي الكلاسة["*halab: al-jaloom: tareeqat al-intiqal min hay al-jaloom eela hay al-kallaseh*"].

Fig. 4.6 Video stills from a YouTube video showing the passage through a home to avoid sniper fire on Al-Kallaseh street. (Source: Shams Alhorya; https://www.youtube.com/watch?v=mAXxTW_BXTE)

The video begins with a shot down a narrow street that connects to a larger main street. That larger street is empty (see Fig. 4.6). The videographer points to the intersecting street and introduces it as al-Kallaseh street. The recorder then explains that the street leads to Gamal Mosque, but they cannot go through the street because a sniper is watching. 'So how do we get to Gamal Mosque?' the videographer asks. 'We can go this way, no problem', an older man says. The camera turns left towards an open door of a house a few metres away. People are going in and out of the door casually as they walk. He then explains they are entering inside a house to get past the sniper. They enter and walk into a bright courtyard. This Aleppine courtyard house sits at the edge of Old Aleppo and could be centuries old. It is heavily damaged and abandoned. Across the courtyard, in the wall separating the neighbouring house, we see an artefact of the counterfront, a hole broken through by the neighbourhood's residents. The rupture is just enough for a person to pass through. Through this makeshift path they exit onto a back alley and eventually back onto a street. The camera pans to the left and zooms onto a building in the distance. 'Gamal Mosque, here it is'.

The presence of the snipers effectively privatised a public street that was no longer accessible to its own residents (see Figs. 4.7 and 4.8). The residents, desperate

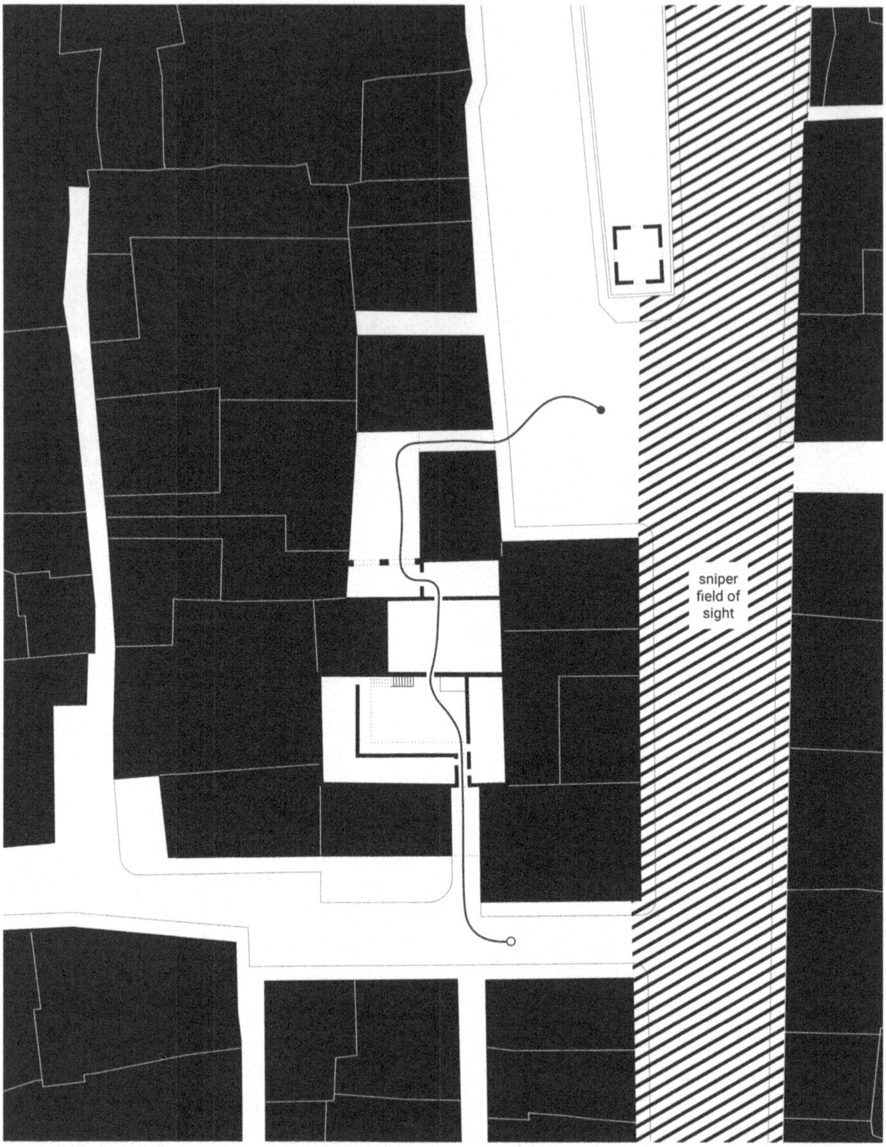

Fig. 4.7 A 'Nolli' style plan of the passage through houses showing the inversion of public and private spaces. The solid represents private space, the open represents public space and the hatched street, a semi-private space. (Source: The author)

to reclaim any freedom of movement, responded by collectively intervening in the architecture of their homes and created an alternative path. Their architectural response therefore turned the most private space, the domestic space, effectively into a public street.

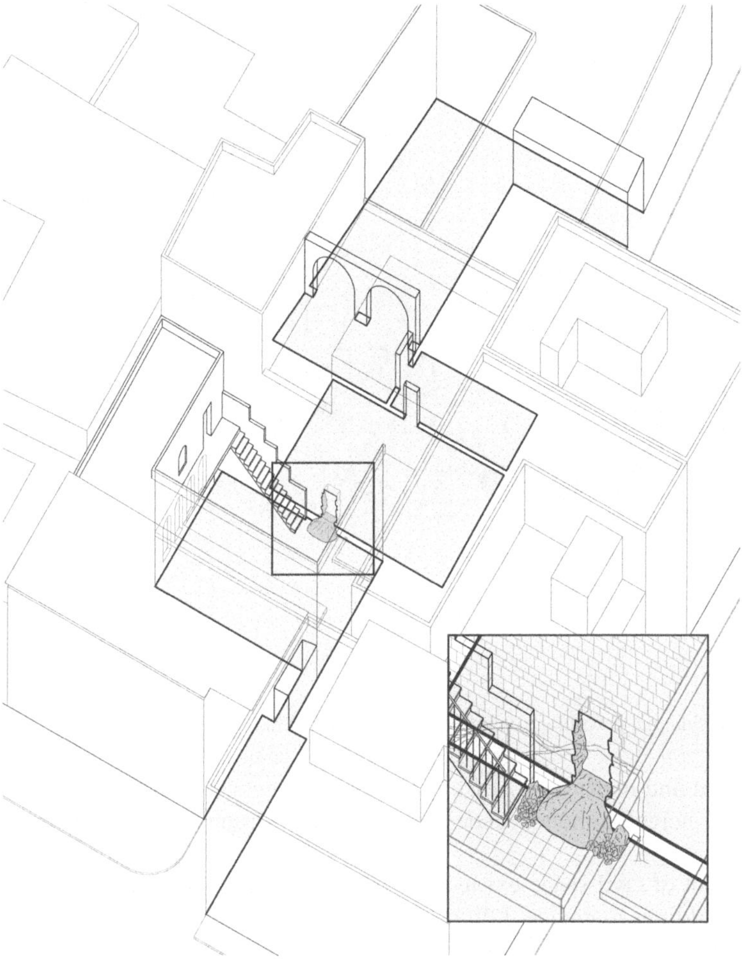

Fig. 4.8 Southeast axonometric analytical drawing of the civilians' path and their intervention. In the bottom right is a detailed drawing of the intervention. (Source: The author)

4.3 Civil Space: Homes and Hospitals

4.3.1 Front of Violence: Civil Space

In the Battle of Aleppo, civil spaces—buildings where people lived, worked and depended on for services—were under constant risk of deliberate or indiscriminate destruction. The continuous damage to the city created borders of habitability where people could and could not live. The state of architecture reflected a record of fear, pain and loss for people. Maps of damage alone tell where Aleppo's most volatile frontlines were and which neighbourhoods were most contested. The Syrian

Fig. 4.9 Residential percentage damage (data by UNOSAT). (Source: The author)

government and its Russian ally's exclusive air access is evident in the damage to residential neighbourhoods under rebel control during most of the battle (see Fig. 4.9).

The front of civil space existed in the indiscriminate or deliberate targeting of civilian buildings to coerce, kill, or weaken the resiliency of the combatants' enemies. In Aleppo, this front caused the greatest destruction of the city and took the form of indiscriminate brute force and the intentional targeting of architecture. Aleppo's typical residential typology is the French walk-up apartment building with a density that made strikes on residential neighbourhoods especially deadly.

Combatants especially targeted service buildings like hospitals to also decay will and public resilience. People in Aleppo had no safe spaces, killed even in buildings 'protected' under international law. Hurt civilians were rushed by what few ambulances were available to the few remaining and overpopulated hospitals. Of course, no one would admit to targeting hospitals because it is clearly a war crime so when evidence is presented of such attacks, combatants would simply deny any involvement (BBC News 2018; U.S. and Russia both deny bombing Syrian clinic 2019). Proof of targeting healthcare sites exists only in citizen footage from which forensic researchers could architecturally reconstruct the spaces and contextualise all footage and piece together those events. Fronts of civil space leave physical traces embedded on the urban skin of the city. Activists and non-governmental organisations (NGOs) collected this footage and created digital platforms for mapping the systematic targeting of hospitals and schools embedded with media evidence from the ground (Physicians for Human Rights 2020; Syria Direct 2020).

When conflict is fought by means of the city—through the homes and resources of people—then the very victory of that war is inseparable and depends on destroying their city and condemning them to suffering. War in the city does not kill people and destroy their homes by collateral, their decimation is therefore the inevitable means of carrying out that war.

4.3.2 Counterfront of Survival: Civil Space

The deliberate or careless destruction of the city in warfare today is an attack on the people that live there themselves. The destruction of space also destroys a sense of place, identity and culture for its inhabitants (Boano 2011). To kill the city and consequently the civilians in it, is therefore an act of urbicide (Coward 2009a). This same term can be applied to understand the systematic destruction of Aleppo and the devastating impact it had on its residents, leaving hundreds of thousands displaced and tens of thousands killed by 2016 (Templer and Shaar 2016).

Many in Aleppo were able to escape and in doing so redefined the demographic lines of the city and the implicit borders of its habitability. Many others, however, could not and had to adapt their homes to brace as much as possible for the violence to come. They filled in their windows and doors with sandbags or concrete masonry units (CMUs) to protect from gunfire and blasts. Non-governmental organisation (NGO) coordination platforms like Shelter Cluster[2] facilitated NGO efforts to distribute shelter repair kits for remedies to damaged homes as well as 'winterisation' toolkits to seal and insulate their damaged or improvised shelters (Shelter Cluster 2017). Throughout the city, people repaired damaged structures and buildings when possible or built new shelters from scavenged materials and in leftover spaces of the city. Unfinished buildings were filled, sometimes with help from NGOs like the International Committee for the Red Cross (ICRC) and the Syrian Red Crescent (Syria: Water as a weapon of war—long version 2020). People also used the distinct blue United Nations High Commissioner for Refugees (UNHCR) tarps to hang between the bare concrete columns of unfinished buildings. People displaced from their homes also created their own ad-hoc shelters. They squatted in abandoned homes, construction sites, or storefront. People scavenged rubble from ruins in the streets and buildings around them to build up a mosaic of irregular and broken stones and bricks with the little mortar they had. They retrofitted storefronts because they were easier to escape from and more secure than their destroyed or damaged

[2] Shelter Cluster is chaired by ICRC and UNHCR. During my research I spoke with researchers at both Shelter Cluster and UNHCR about their experiences in Aleppo and their roles on the ground.

apartments above. In doing so, they domesticated commercial spaces and created a new shelter urbanism on top of the old and ruined fabric.

The United Nations (UN) reported as early as 2013 that attacks on hospitals were systematic and that healthcare became a weapon of war (Cumming-Bruce 2013). The most essential civilian buildings were therefore at the centre of the front of civil space. Hospital staff and civilians dedicated to continuing to serve the vulnerable community had to find new ways to continue working and resist the persistent violence. They sometimes fortified their buildings with rubble-filled barrels or layers of sandbags. By the end of the battle, Aleppo's hospitals began to look like military barracks ready for imminent attack.

In 2017, I investigated with Forensic Architecture the repeated targeting by Syrian and Russian forces of Omar bin Abdul Aziz Hospital in Aleppo. The hospital was struck 12 times in 6 months (Czuperski et al. 2017). During this investigation I found a case of civilian crisis architecture that was part of the counterfront of civilian space, an engineered protective wall built around the hospital.[3] Understanding the danger they were under, the staff hired an engineer to design the protective wall around the hospital's perimeter (see Fig. 4.10). The wall was built with reinforced concrete columns regularly spaced along the perimeter of the hospital and CMU infill between them, leaving a gap for entrances. A horizontal concrete beam sat between the first and second level with more CMU above. The CMU wall design includes PVC pipe sections sandwiched between the blocks and mortar. The pipe section holes perforating the wall allowed the force of blasts nearby to pass through them while still protecting the building from shrapnel and gunfire (see Fig. 4.11). Despite this effort, the hospital went out of service on many occasions. They originally intended for the wall to match the five-story height of the building but could only reach two levels before it was too dangerous to continue construction. The wall stood there incomplete and with steel rebar sticking out of its concrete posts. It was an architectural response to repeated attacks on vulnerable healthcare facilities in general and M2 hospital in particular. The existence of the wall as a designed and calculated engineered structure tells of how serious and persistent the danger to the community was. The wall is a testament to the importance and vulnerability of civil services and their abuse in conflict.

The hospital stood in the Bab al-Maqam neighbourhood, which was long held by rebels during the battle including during the gruelling siege of east Aleppo. The division of the city meant that access to healthcare was restricted by location and functionality. M2 was one of the last working hospitals with a valuable paediatric department until it was itself put out of service by the end of the year. Months after the Syrian government took back the rest of Aleppo in late 2016, images surfaced on social media of the same hospital repaired and rebuilt as new (Beals 2017). The buffer wall built by residents and hospital staff is gone and the hospital now stands clad in bright new stone while buildings around it are still in ruin. It was as if nothing had happened. The architecture of civilians is the first to be erased, for their

[3] From conversation with a former nurse at M2 hospital, January 2017.

Fig. 4.10 Top: Drone footage captures Omar bin Abdul Aziz Hospital and a protective wall built around it (Ruptly, YouTube video. Illustrated by Author). Middle: M2 Hospital's protective wall under construction showing assembly (Date and author unknown). Bottom: M2 Hospital's protective wall up close. (Source: Unknown)

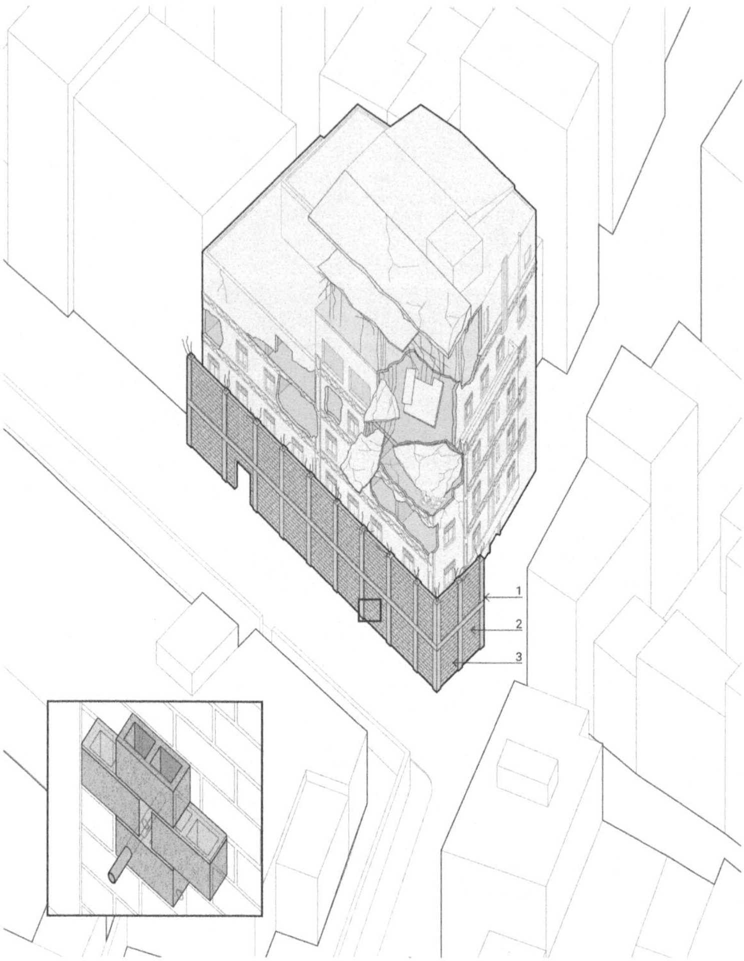

Fig. 4.11 Northeast axonometric drawing of M2 hospital and its improvised protective wall. The drawing shows its approximate state in late 2016. In the left bottom corner is a detailed drawing of the construction assembly. (1) Reinforced concrete; (2) CMU; (3) PVC pipe. (Source: The author)

destruction signals the return to an order that takes the city out of the control of civilians and returns it to the central state. The sudden disappearance of the wall demonstrates just how fragile civilian architectural responses are and reminds us why we must pay close attention to these fleeting architectural moments. With the removal of civilian counterfronts, the role of architecture to cope with violence dissolves in the victor's order.

4.4 Infrastructure: Water and Energy

4.4.1 Front of Violence: Infrastructure

The violence of urban war is not always explicitly physical. Combatants in Aleppo also undermined each other by controlling and targeting the infrastructure of the city. They cut off essential supplies and resources like water and electricity in opposing neighbourhoods. Civilians in Aleppo felt the effect of this front the most; outages were unexpected and often indefinite as far as they knew.

Infrastructure has the widest reach of all fronts because it has the power to affect residents of entire neighbourhoods or cities simultaneously and totally. The infrastructure front creates another set of implicit urban borders defined by access and reliability of essential resources like water. The methods of the infrastructure front include targeted destruction, sabotage and siege. Combatants would retaliate against one another through infrastructure while further condemning the public to precarious lives. This is not a phenomenon unique to Aleppo. Infrastructure has been increasingly weaponised since the 1990s with the continued urbanisation of war (Coward 2009b; Sowers et al. 2017). Today, infrastructure and fighting for control over it, is at the very centre of war rather than being an incidental consequence of urban war (Sowers et al. 2017: 411).

The targeting of water infrastructure has been so systematic and pervasive that in 2015, the International Committee of the Red Cross (ICRC) declared that water was being used as a weapon of war in Aleppo just as healthcare was declared 2 years earlier by the UN (ICRC 2015a). Water was used as an instrument of coercion by all sides[4]; they destroyed water pipes (Associated Press 2012), shut down water pumping stations that flowed into enemy territory, or sabotaged and damaged water towers and pumping stations altogether (BBC News 2016). Their attacks and sabotage have in turn caused water insecurity for more than two million people, with some going weeks without clean drinking water during the siege of eastern Aleppo (Serhan 2016). The damage to the pipes and pumping stations also risked the rapid spread of waterborne diseases (Guardian 2016).

What made matters worse for civilians is that the city depended almost entirely on a centralised station that pumped water from 'Lake Assad' of the Euphrates River more than 70 km away from its borders. Multiple groups struck this station as well, which only further depressed the resident's access to water. The city's electricity grid was the same, relying on a single plant to power most of the city that was also damaged and fought over. Power lines were also destroyed in bombing and shelling campaigns of the battle. The centralisation of both water and electrical infrastructure in Aleppo made them extremely desirable political elements because whoever controlled that infrastructure controlled most of the city by extension.

---

[4]Amanda Chan, Jason Danforth, and Nicolás del Valle (Students of Laura Kurgan's Conflict Urbanism Aleppo Seminar at Columbia University) have done an interesting analysis and survey of these actions here: https://bit.ly/2whayRY

4.4.2 Counterfront of Survival: Infrastructure

Infrastructure is the most difficult urban front for civilians to respond to because of the scale and complexity often required. Civilians who do engage typically respond to this front with the aid of local and international non-governmental organisations (NGOs). The counterfront of infrastructure involved prolonging the habitability of people's homes and neighbourhoods by creating alternative means to access essential resources. When the power grid failed, the community became ad hoc electricians who hooked up their homes directly to local 'temporary' generators.

The ICRC is one of the NGOs that has worked to address the lack of power and the extreme water insecurity across Aleppo. They anticipated the problem of water weaponisation and civilian accessibility as early as 2013 when they launched, in partnership with the Syrian Arab Red Crescent (SARC; an affiliate of ICRC) and local community partners, an initiative to create alternative water networks in the city. In the process they found and restored 80 disused wells throughout the city (ICRC 2015b). They often connected wells to large, red water tanks from which the community could easily drink and fill their vessels. While civilians did not exclusively carry out these responses, the ICRC implemented them in partnership with local community leaders and their success depended on community coordination. Civilian involvement in their implementation along with their architectural and urban implications qualify these tanks and infrastructural efforts as part of the civilian counterfront. They demarcated the points of water with a clear and consistent visual language of the red tanks. They place the tanks in existing community centres and visible streets. Their placement and look made the tanks urban nodes for collective resources and frequent gathering. The ICRC also created a GPS-enabled interactive map so citizens could track the location of the water sources, thus changing and defining new borders of access instantly available to all residents. One of those sites was a rehabilitated well at the Usama bin Zaid Mosque in Agyul neighbourhood, just north of the old city (see Fig. 4.12). The neighbourhood was a volatile borderland that teetered back and forth on the frontlines of battle and this volatility was directly reflected in its severe disruption of essentials like water.

By creating a secondary network, the wells and tanks acted to, even if temporarily, decentralise the infrastructure of an essential resource and therefore decreased the vulnerability and dependency on single militarised sources. The response made people more resilient and versatile as the network consisted of stand-alone sources. Resiliency empowers people because they are closer to the freedom of making their own decisions about their spaces and lives. This response works as a democratisation of infrastructure despite and simultaneously because of the violence civilians face. That extreme circumstance is the clearest urban design lesson: the vulnerability of centralisation. The civilian response shows a popular will tendency towards self-governance and autonomy over the resources they rely on. It serves as a lesson

Fig. 4.12 Southwest axonometric drawing of Usama bin Zaid Mosque and the added tanks by the ICRC in blue. In the bottom right is a detailed drawing of these tanks. (Source: The author)

for the danger of continuing single centralisation for Aleppo and other cities globally.

4.5 Conclusion

The streets, homes, hospitals and even water that Aleppines knew so well and relied on became weapons and at once sites of contention. We can no longer deal with urban conflict as merely an abstract line of battle on the ground but as one made up of molten and overlapping fronts that are formed and fought over in the civilian sphere. In their weaponisation of the city, combatants reshaped Aleppo both as a strategy of the fighting and as a consequence of sacrificing it as a battlefield.

Civilians also took an active role in constantly reshaping the city; they had no choice but to themselves intervene in their built environment, interventions that make up what I call in this essay, civilian counterfronts. The more concentrated the fronts of war are, the more counterfronts we expect to find. Individually, the interventions of counterfronts tell unique stories of their authors and what violence they responded to. A hole through a home tells us about the danger of a sniper-filled street, while a walled hospital tells us about the imminent danger felt in the places meant to be safest. When we piece them together, those stories also collectively tell another story of pervasive resistance across Aleppo, and of collaboration in a unified struggle to survive.

These artefacts of violence in the fronts of battle and the artefacts of resistance in civilian counterfronts are ephemeral traces. This fragility, together with their constant transformation during battle, has made it difficult in the past for us to track or map urban warfare comprehensively. What we are used to being left with at the end of battle is only the aftermath debris of all these fronts' movement throughout the city. That is why the open-source methods adopted in this research form the basis of a new understanding of urban conflict altogether. Social media allows us to trace those artefacts through crowdsourced footage posted by the same people shaping the conflict city—footage that captures traces of those interventions before they disappeared. This view into battle as it unfolds and through the eyes of Aleppines allows us to track where the city's implicit and explicit borders were throughout the battle and what forms they took. In turn it also allows us to trace the cause and effect in their continuous transformation; following the urban violence to acts of civilian resilience and inversely, the interventions of civilians to survive in themselves are evidence of a violent urban condition in their immediate surroundings. Neglecting the counterfronts of war therefore risks the perpetuation of its people's vulnerability in new acts of building and violence.

For a chance of justice in any post-conflict city, this new reading of the conflict city through its fronts and counterfronts must be documented in its urban history, mobilised as forensic evidence of the violence inflicted intentionally on civilians, and act as a starting point of advocacy for any just urban vision in which that agency exercised by people over their spaces in crisis can inspire an urbanism that continues to place the future of the city in its residents' hands.

References

Anderson J, O'Dowd L (1999) Borders, border regions and territoriality: contradictory meanings, changing significance. Reg Stud 33:593–604. https://doi.org/10.1080/00343409950078648
Associated Press (2012) Syria: drinking water pipe damaged in Aleppo. Hürriyet Daily News. Sept 8
BBC News (2016) Bombing leaves Aleppo "without water". Sept 24, sec. Middle East
BBC News (2018) Syria hospitals "targeted by attacks". Jan 6, sec. Middle East
Beals E (2017) Twitter

Boano C (2011) 'Violent spaces': production and reproduction of security and vulnerabilities. J Architect 16:37–55. https://doi.org/10.1080/13602365.2011.547002

Center for Documentation of Violations in Syria—Killed List (2019). http://www.vdc-sy.info/index.php/en/martyrs/1/c29ydGJ5PWEua2lsbGVkX2RhdGV8c29ydGRpcj1ERVNDfGFw-cHJvdmVkPXZpc2libGV8ZXh0cmFkaXNwbGF5PTB8cHJvdmluY2U9NnxzdGFydERhd-GU9MjAxMi0wNy0xOXxlbmREYXRlPTIwMTYtMTItMjJ8. Accessed 31 Aug 2019

Coward M (2009a) Urbicide: the politics of urban destruction. Routledge Advances in International Relations and Global Politics 66. Routledge, London/New York

Coward M (2009b) Network-centric violence, critical infrastructure and the urbanization of security. Secur Dialogue 40:399–418. https://doi.org/10.1177/0967010609342879

Cumming-Bruce N (2013) U.N. reports Syria uses hospital attacks as a 'weapon of war'. The New York Times. Sept 13, sec. World

Czuperski M, Itani F, Nimmo B, Higgins E, Beals E, Atlantic Council of the United States (2017). Breaking Aleppo

Forensic Architecture (2017) Airstrikes on M2 Hospital.

Foucault M (1986) Of other spaces. Diacritics 16:22–27. JSTOR. https://doi.org/10.2307/464648

Graham S (2010) Cities under siege: the new military urbanism. Verso, London

Guardian (2016) Syria bombings leave 1.75 million without running water in Aleppo. The Guardian. Sept 24, sec. World news

Herscher A (2010) Violence taking place: the architecture of the Kosovo Conflict. In: Cultural memory in the present. Stanford University Press, Stanford

ICRC (2015a) Syria: water used as weapon of war. News release. International Committee of the Red Cross. Sept 1

ICRC (2015b) Syria: ICRC works to avoid massive water crisis in Aleppo. International Committee of the Red Cross. Nov 10

Leonard M, McKnight M (2011) Bringing down the walls: young people's perspectives on peace-walls in Belfast. Int J Sociol Soc Policy 31:569–582. https://doi.org/10.1108/01443331111164151

Morrow R, Ciaran M, FitzGerald JD (2011) Beyond the shadow space: architecture as a professional and creative process; during and post-conflict. J Architect 16:57–70. https://doi.org/10.1080/13602365.2011.547008

Najjar A (2018) Damage caused by the Syrian Civil War: what the data say. Medium. Aug 24

Physicians for Human Rights (2020) A map of attacks on health care in Syria. http://syriamap.phr.org/#/en. Accessed 15 Feb 2019

Pullan W (2011) Frontier urbanism: the periphery at the centre of contested cities. J Architect 16:15–35. https://doi.org/10.1080/13602365.2011.546999

Rupnik J (1994) Europe's new frontiers: remapping Europe. Daedalus 123:91–114

Sack RD (1986) Human territoriality: its theory and history. Cambridge Studies in Historical Geography 7. Cambridge University Press, Cambridge [Cambridgeshire]/New York

Serhan Y (2016) UN calls for cease-fire as Aleppo's taps run dry. The Atlantic. Aug 9

Shams Alhorya (2013) حلب : الجلوم :طريقة الإنتقال من حي الجلوم إلى حي الكلاسة. YouTube video, 01:50. https://www.youtube.com/watch?v=mAXxTW_BXTE

Shelter Cluster (2017) Syria: NFI winterization support. Mar

Solomon E (2012) In Aleppo, Syrian rebels bogged down in sniper war. Reuters. Oct 18

Sowers JL, Weinthal E, Zawahri N (2017) Targeting environmental infrastructures, international law, and civilians in the new Middle Eastern wars. Secur Dialogue 48:410–430. https://doi.org/10.1177/0967010617716615

Syria Direct (2020) Mapping damaged and destroyed schools in Syria. https://www.arcgis.com/apps/MapSeries/index.html?appid=f0abac5c690d46b5be39d4f37f569d57. Accessed 23 Mar 2019.

Syria: Water as a weapon of war—long version (2020). https://www.youtube.com/watch?v=1Zzg1jHbDU8. Accessed 4 Jan 2019

Templer R, Shaar AH (2016) Urbicide or an elegy for Aleppo. Tvergastein Interdiscip J Environ:108–119

U.S. and Russia both deny bombing Syrian clinic (2019). https://www.cbsnews.com/news/russia-syria-blame-us-airstrike-msf-hospital-clinic-idlib-aleppo-civilians/. Accessed 31 Aug 2019

Weizman E (2012) Hollow land: Israel's architecture of occupation. Verso, London/New York

Weizman E (2019) Lethal theory. Online Open. https://onlineopen.org/lethal-theory. Accessed 11 Aug 2019

Chapter 5
An Interdisciplinary Approach to Make the City of the Future: A Project for the Neukölln District in Berlin, Germany

Angela Fiorelli and **Irene Tuzi**

Abstract In the past decade, Germany has been affected by socio-demographic changes that are likely to transform the future of its society. With the arrival of around one million refugees, social inclusion and social cohesion are now the main challenges for German society, which need to be addressed in the public space. Drawing on these considerations, the present research is rooted in the collaboration between architecture and sociology as an instrument of exchange. It focuses on children as the future of society and as future citizens capable of coexisting in diversity. The project is carried out through the interaction of two systems—the Schools of Cultures and the Little House of Cultures, the first being a place where social cohesion and inclusion are shaped and developed, and the second a place where they are confirmed and established. The theoretical concept at the core of this project is transculturalism, a cultural perspective broader than multiculturalism and interculturalism, which is able to overcome the boundaries of individual communities in favour of a unique, larger community, able to coexist according to common principles and values. The research behind this project was carried out in 2017 through ethnography and semi-structured interviews with humanitarian and institutional actors and local and refugee families in Berlin.

Keywords Participation · Youth · Refugee integration · Urban space · Co-design · Berlin

A. Fiorelli (✉)
Università di Roma Tor Vergata, Rome, Italy
e-mail: angela.fiorelli@uniroma1.it

I. Tuzi
Humboldt University, Berlin, Germany
e-mail: irene.tuzi@hu-berlin.de

G. Cimadomo et al. (eds.), *Urbanism at Borders*, The Urban Book Series,
https://doi.org/10.1007/978-3-032-09594-7_5

5.1 Introduction[1]

5.1.1 The Research Context

Since the so-called 'refugee crisis' that started in 2015, Germany has become the major receiving country for refugees in Europe.[2] Faced with this new challenge, Germany has adopted new measures of hospitality and asylum. Nevertheless, social cohesion and inclusion remain a challenge as many forced migrants still encounter many difficulties to access services, being included in the labour market, overcoming the linguistic barriers, and finding their own position in society. Many refugees in Germany came with their families. They flee together or reunited later with family members who were already established in the country. This led to more than 400,000 school-aged refugee children in Germany,[3] many of whom have previously experienced significant disruption of their education, having lived in unstable or precarious conditions before arriving in Germany. Integrating these children into the already decentralised education system, understaffed and unaccustomed to diversity (OECD 2018), is a challenge for Germany.

5.1.2 The Theoretical Approach

This chapter uses a transdisciplinary approach linking architecture and sociology to promote social cohesion and social inclusion among refugees and local families in Berlin. This approach builds on the concept of 'transculturalism', which is the theoretical framework underpinning the architectural design. The term 'transculturality' was firstly introduced in social science by Wolfgang Welsch (1999), who highlighted the interconnections that tie different cultures together. The author suggests that the contemporary condition of interconnectivity is a consequence of three distinct developments: the internal complexity of contemporary societies, the external networking in which they engage to build and develop links with other societies, and the tendency in all cultures today to be hybrid, as a result of increased mobility and communication technologies. As a consequence of this condition, the author

[1] The text, a synthesis of a project designed by the authors, was written in the following order: Irene Tuzi wrote Sects. 5.1 and 5.2 and Angela Fiorelli wrote Sects. 5.3 and 5.4. The section "Conclusions" has been written together by the authors.

[2] See UNHCR data: https://www.unhcr.org/data.html

[3] See The Federal Agency for Civic Education: https://www.bpb.de/die-bpb/138852/federal-agency-for-civic-education

concludes, nothing today is 'absolutely foreign' any longer (Eigeartaigh and Berg 2010: 10).

Social sciences distinguish between *multiculturalism*, *interculturalism* and *transculturalism*. A multicultural society is made by a group of people who come from different geographical, cultural and linguistic backgrounds and do not interact with each other. This is a static perspective that does not necessarily involve an interaction between social actors—who keep existing in parallel universes. When these realities connect with each other they generate interculturalism, which represents a more dynamic context, made of reciprocity, interaction and exchange. An intercultural society generates dialogue between the different cultures that live in a territory but still remain within their own borders. The next step is transculturalism, a wider perspective that embraces the will of going beyond the cultural borders and barriers of single communities, in favour of a sole and broader community capable of working together according to values and principles that are proper of the human species. The concept of transculturalism enhances the diversities of the single communities and transforms them into key assets to reshape the idea of culture and replace it with that of 'cultures', in the plural. The aim is to build a new perspective made of cohesion, dialogue and respect.

This approach of political philosophy to multicultural society has been criticised by some authors and deemed too ideological and detached from social reality, which is instead made up of differences, tensions and contradictions (Semi et al. 2009). When speaking of multicultural societies, in fact, the relevance of differences is often taken for granted and the focus is instead on the opportunity to guarantee inclusion and cooperation without erasing the uniqueness that coexists in the social space (Colombo 2010).

Cultural diversity can be approached as multicultural when there is a fusion of multiple and distinct contributions to mainstream culture, or as transcultural, when there is a fusion of cultural attributions (Grosu-Rădulescu 2012). Multiculturalism has also been regarded as the first step in the sociocultural ranking (Cuccioletta 2002) towards the construction of a cosmopolitan society. Thus, transculturalism promotes the breaking down of cultural boundaries while multiculturalism reinforces them. This does not speak of the integration of one culture into another, but of an interweaving of cultures within a nation-state, whose identity is built on the foundations of a common value system.

The architectural project presented in this chapter builds on this context. In particular, the concept of transculturalism was used as a lens to construct the architectural design and so to give an answer to the social cohesion and inclusion needs that emerged from the interactions with different actors in the German context.

5.2 'The Little House of Cultures': A Transcultural Project

The project presented in this chapter participated in the 2017 IDeA Competition[4] and arose from the interdisciplinary collaboration between Sociologist Dr. Irene Tuzi, and Architects Dr. Angela Fiorelli and Azzurra Brugiotti. The idea behind the project was to create a discourse on refugee livelihoods, which embodies the dignity and resilience of refugees working to protect their autonomy, creativity and capability in difficult conditions. The project was designed for the city of Berlin.

The project is realised through the interaction of two systems—the Schools of Cultures and the Little House of Cultures, the first being a place where social cohesion and inclusion are shaped and developed, and the second a place where these concepts are confirmed and established. The Schools of Cultures are multidisciplinary and cross-cultural after-school courses designed for children already enrolled in public schools. Children can enrol in the after-school programme spontaneously and are able to choose among a series of after-school activities ranging from music, sports, theatre, art and many others. They follow a learning path that takes them, at the end of each year, to present the outcomes of their work at the Festival of Cultures, which takes place in the Little House of Cultures. The Little House of Cultures is a travelling house that goes around the city of Berlin to bring to the streets the work done by children enrolled in the Schools of Cultures. The Little House of Cultures is a temporary object, an iconic architectural element that aims to raise curiosity. In the cyclical nature of urban events, the Little House of Cultures aims to be accepted by the community and become familiar to the population. The Little House of Cultures has a well-defined main objective: to promote social cohesion and inclusion among children.

5.2.1 *Aims and Target of the Project*

The concept of transculturalism was used in this project as an innovative method based on cooperation, interaction and exchange among actors with different histories and backgrounds. The outcome is a learning path that aims at creating inclusion and social cohesion among children through education, game and creativity.

The project presented in this chapter has three main objectives: The first objective of this project is to involve all actors in the process of social inclusion—the local and forced migrant communities. The project emphasises the need of integrating each culture with the richness of other cultures by recognising the specific

[4] IDeA—International Development in Action. (2017). Call for Entries: 'Integrating Refugee Populations within Cities'—Place & Displacement Design & Policy Competition. http://www.idevelopment.us

importance of each cultural element. Transcultural integration cannot let alone the value of cultural specificities and diversities of each group. This is done not only by acknowledging and accepting otherness but primarily by activating modes of transcultural cooperation, interaction and exchange between individuals of different cultural, perceptive, and sensitive histories and backgrounds. The second objective of the project is to support refugee and non-refugee families where both parents work. The programme provides learning and language support aside from the standard learning programme offering an opportunity also for German children to interact in languages other than German. The third objective of the project is to create a tool for new generations to become aware of public space. The concept of public space might differ from country to country or have different meanings in social contexts other than the Western one (interview with a Syrian journalist, June 2016). In authoritarian contexts, for instance, public space is conceived as a property of the government or the authority: something that does not belong to the people. The Little House of Cultures aims to bring children into public space and teach them to see the city as their own, a place to experience life as well as a place to take care of.

The target of this project is children. The target population was chosen for three main reasons: (1) Children are the future of any society and investing in their education, their well-being and their development is an opportunity that must be grasped and valued. Investing in children means projecting into the future and imagining a society capable of living together in diversity, using the differences to enrich and improve itself. (2) Children are the most permeable and flexible part of the population: they learn faster. Both cognitively and emotionally, children are able to store a large amount of information and receive a large number of stimuli, especially when they come from play activities. Childhood is the best time to form individual and collective awareness and integrity in line with the socio-demographic changes that are involving Germany. (3) Children are the most vulnerable segment of the population. In particular, refugee children suffer major traumas linked to harsh conditions in the country of origin, originating from war and violence, or by the experiences lived during the migration path.[5] This generates behavioural and learning problems or difficulties of socialisation with the new companions. However, children have a great capacity for resilience and it is on this element that the Schools of Cultures build to contribute to the creation of a society that is more aware of its opportunities.

[5] Some studies (for example, Farmakopoulou et al. 2017) have shown that children fleeing from Syria suffer from severe psychological problems related to war trauma. They are likely to develop post-traumatic stress disorder (PTSD), anxiety, conduct and sleeping problems, and other psychiatric disorders, for example, autism spectrum disorders and psychosis.

5.2.2 *The Transcultural Programme of the Little House of Cultures*

The project, as shown in Fig. 5.1, is developed in Berlin. As a multi-ethnic and cosmopolitan city, the German capital is the ideal space to start a process of childhood education that involves schools, libraries, educational spaces and neighbourhoods for the promotion of social cohesion and inclusion. Schools would respond to a statement of interest (SOI) issued by the German states, or Bundesländer and would apply to become a School of Cultures. They would receive funding and support with which they would engage in a yearly transcultural programme. These programmes will have the following requirements: (1) their activities are open to refugee and non-refugee children; (2) they follow a common thematic programme that is decided in the call for proposals and it is aimed at developing new skills, as well as at improving new languages; and (3) they involve local community entities (public libraries, civil society organisations, community hubs etc.), which participate in leading and managing the activities. At the end of each school year, Schools of Cultures prepare kids for the Festival of Cultures—a summer festival held in the travelling Little House of Cultures, where children will be able to show and bring around the city of Berlin the results of their transcultural programmes.

The learning method used by the School of Cultures is *cooperative learning*. Several studies (Leigh Smith and MacGregor 1993; Johnson 2006; Johnson et al. 2007; El-Sayedd and Mousa 2015; Le et al. 2016) have proved that through this method students achieve higher learning results, higher levels of self-confidence and self-esteem, better social competences and deeper attainment of skills and abilities. This approach also facilitates interaction and acceptance of individual

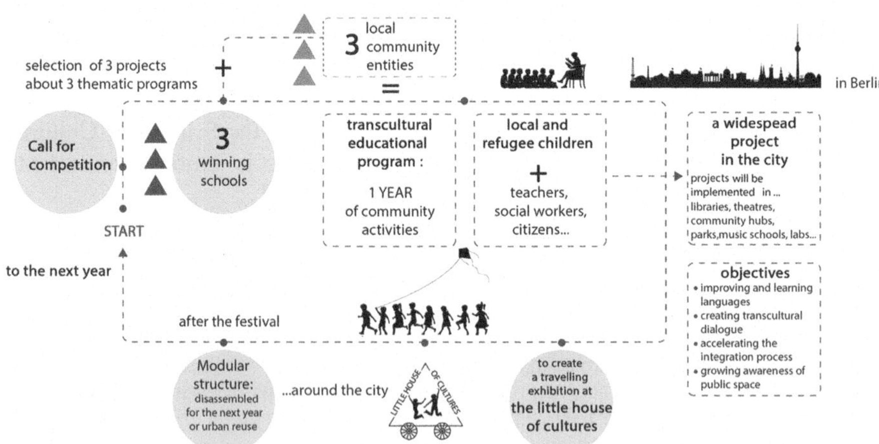

Fig. 5.1 The transcultural programme. The extracurricular programme is aimed at the collaboration between schools and local community entities. At the end of the year, final works are displayed in the Little House of Culture around the city. (Source: The authors)

differences (Bertazzi 2007). The intersubjectivity between children is the element at the base of this method, the prerequisite and the learning 'place' to develop tools for a better understanding of reality. In this sense, a cross-cultural reality appears as the place to develop multidisciplinary skills with a cooperative learning method (Yamazaki and Kayes 2004).

5.3 A Transdisciplinary Approach to Architecture

The research questions guiding this project are the following: (1) What is the role of architecture in creating a critical reflection about social inclusion and social cohesion in multicultural/postmigrant societies? (2) How can architecture promote social inclusion and cohesion in practice?

To answer these questions, the methodology was developed on a two-levelled strategy: an empirical phase was followed by a design phase. The project was carried out through the interaction of two disciplines: sociology and architecture. The reason behind this choice is firmly related to the theoretical concept of transculturalism, which in this project is expressed methodologically in terms of interdisciplinarity. The aim is to dissolve the conventional boundaries between the two disciplines to create a dimension able to (re)construct meanings and values in the context of real-world problems. Through this exchange of competencies, the project managed to address the needs of a diverse society through architectural design, in a holistic way, both on an architectural and an urban dimension.

Within this theoretical and methodological framework, some factors have been identified as central to the development and implementation of the project:

- The need for greater interaction between the local population and refugee families
- An understanding of social inclusion, which sees children as the kernel of a future society and an effective bridge for creating interactions between different ethnic groups
- The importance of the educational and training system in the process of social inclusion
- The need for a widespread project to create social cohesion and connectivity (throughout the school year and then during the Festival of Cultures)
- The value of the project aimed at a shared final outcome in which the different skills developed in the Schools of Cultures are the basis for greater awareness of diversity
- The repeatability of the project
- The sustainability of the project, in environmental, social and economic terms

5.3.1 *Empirical Phase*

The empirical investigation, which is the basis of this project, was carried out between June and September 2017 through ethnography and semi-structured interviews with humanitarian and institutional actors and local and refugee communities in Berlin. Empirical data were drawn mainly from primary sources, including in-depth individual interviews with refugee and non-refugee families in Berlin and consultation interviews with humanitarian and institutional actors. A pre-data collection phase was carried out in June 2017. The authors visited Berlin in preparation for the fieldwork, to select the pilot neighbourhood, as well as to create networks with research participants. The data-collection techniques used were mostly three: structured consultation interviews with humanitarian and institutional actors; in-depth semi-structured interviews with refugee and non-refugee families; and participant observation. Additional informal talks and conversations were carried out randomly among the population.

5.3.1.1 Structured Consultation Interviews with Humanitarian and Institutional Actors

Ten structured consultation interviews were carried out with humanitarian and institutional actors. This was needed to better understand the discourse around social cohesion and inclusion in Germany. These structured interviews were conducted face-to-face or via Skype, and each respondent was asked the same series of open-ended questions. There was little room for variation in questions since they were standardised and the phrasing of questions was kept consistent. Questions were only adapted to the different operational fields of each respondent. This method aimed at investigating various aspects of the humanitarian and institutional sectors engaged in integration.

5.3.1.2 In-Depth Semi-structured Interviews with Refugee and Non-refugee Families

The main method used to collect first-hand data for this study was the semi-structured interview. We conducted 20 in-depth, semi-structured, narrative individual interviews with refugee and non-refugee families in Berlin. Addressing humanitarian and institutional actors through consultations would have not been enough to understand the characteristics of the population subject of this study. Semi-structured interviews were helpful in this sense to deeper investigate the research questions. In-depth, semi-structured interviews were carried out in a selected neighbourhood of Berlin: Neukölln, located in the south-eastern area of the German capital, which was chosen as a pilot district. The choice fell on this district because it has a marked 'multi-ethnic' character, as well as a distinct flexibility in terms of socio-cultural interactions. It is an area adjacent to the city centre, particularly lively and with many services, where many ethnic minorities and German residents live together. Participants were chosen through existing networks following a random sampling selection. Direct interviews with families helped to deeper investigate the personal experiences of participants to collect qualitative data that would not emerge among experts.

5.3.1.3 Participant Observation

Participant observation was conducted among refugee and non-refugee families in Berlin. This method was fundamental to gaining a comprehensive understanding of the population focus of this research. In fact, the project sought to go deeper into the articulated nature of the topic and to have an insight into refugees' intimate perceptions and attitudes towards their life in a resettlement country. This strategy was useful to better understand the differences between inner and outer perceptions of the phenomenon studied. For example, it was remarkable in the phase of humanitarian actors' interrogation that the discourse about social inclusion was built around the concepts of 'integration', 'adaptation', and 'assimilation' with the local population. Less room was left for discussions related to social cohesion. In addition, because it was noticed that there was an imbalance between the humanitarian narrative and the perceptions of refugee families, the investigation meant to go deeper and explore this gap in depth (Tuzi and Fiorelli 2022).

5.3.2 Design Phase

The architectural design was carried out in parallel with the sociological investigation and it was based on three interconnected working phases (WPs; see Table 5.1). The first phase concerned the selection of the project area in Berlin: a pilot neighbourhood with a 'multi-ethnic' character was chosen for its flexibility in terms of socio-cultural interaction. Once the neighbourhood of Neukölln was identified, in a second phase the area was analysed in terms of morphological characteristics of the urban plan with the purpose of creating a widespread training programme and possible exhibition itineraries. Finally, the third phase dealt with the design of the temporary modular structure of the Little House of Cultures.

In the light of the research conducted both at an urban and sociological level, a 1-year time schedule for after-school programmes was built and possible thematic itineraries for the Festival of Cultures were identified. The main priority was to identify the possible dissemination of the project in the city. Social cohesion and inclusion call for a housing *mixité* and a more functional use of public space so as to avoid urban ghettos based on ethnicity and to raise a greater awareness of the whole population in terms of social cohesion and inclusion. The goal was to create a flexible architecture, easy to assemble and disassemble, changeable, but also

Table 5.1 Working phases

Working phases	Aims	Actions
WP 1	Identification of a pilot neighbourhood in Berlin	• Desk research about demographic characteristics based on the following indicators: – Diverse/mixed population – Presence of refugee and non-refugee families – Presence of collective services and schools for children – Presence of already existing intercultural spaces and potential for new transcultural spaces
WP 2	Analysis of the neighbourhood	• Mapping of already existing services useful for transcultural extracurricular activities • Analysis of public spaces • Identification of possible exhibition itineraries
WP 3	Design of the Little House of Cultures	• Architectural design of installation modules • Development of assembly solutions for different functionalities

easily recognisable from an iconic and figurative point of view. The project was designed to be reusable so as to allow easier repositioning within the city. Furthermore, from a formal point of view, the object responded to a clearly legible and recognisable image. The purpose was an object that, although changeable, could easily be traced back to the project.

5.4 The Architectural Design of 'The Little House of Culture'

The Little House of Cultures is a temporary object, which comes to the city once a year during the Festival of Cultures. It is also an iconic architectural element that is accepted by the community as a familiar object. It is a well-recognisable image in the urban skyline and at the same time, it is a flexible construction hosting various types of art expressions. The Little House of Cultures is a triangular module with a square base and it is designed to recall the typology of the house. Being a flexible structure, made of modular elements, the Little House of Cultures can be adapted for use and returned to the city in various forms—a theatre, a city market etc. The Little House of Cultures is an architectural object that aims at being an itinerant element for the city squares and it is the synthesis and the physical expression of transcultural learning programmes.

The structure is very simple (see Fig. 5.2). It is composed of square steel pipes connected through three-arm joints. The easy assembly and disassembly also allow the possibility of working rapidly and safely in self-building. The design offers the opportunity to use various types of panels in different materials (Medium-density Fibreboards, perforated sheet metal, canvas etc.), with rectangular or circular

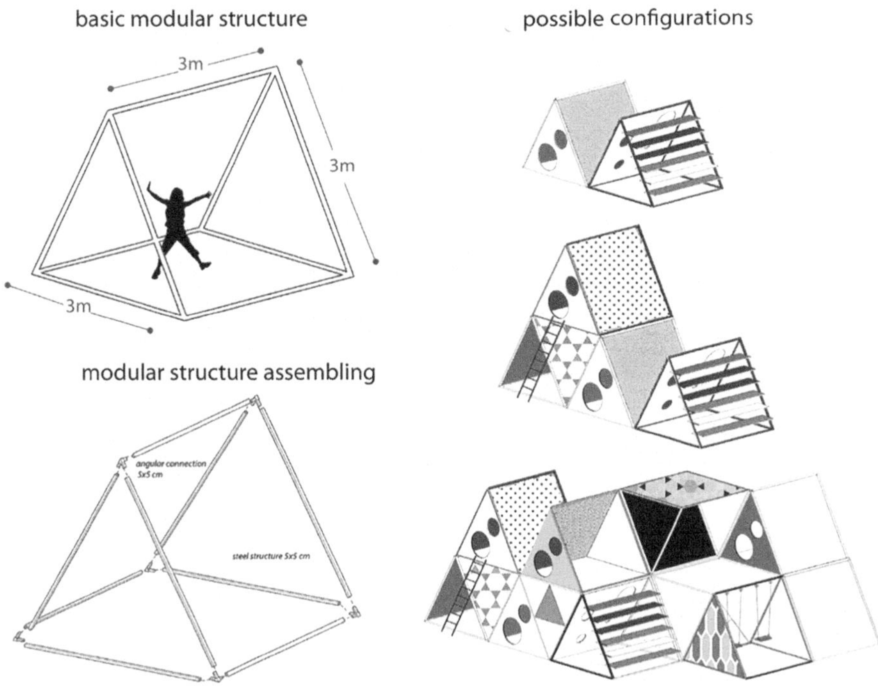

Fig. 5.2 The basic module: Structure and possible configurations aggregating several modules. (Source: The authors)

windows, or *brise soleil* elements in wood or elastic nets. The idea is to create amusing and easily transformable installations. Nevertheless, the figurative flexibility of the object responded to the need for an iconographic form.

This way, the work is always retraceable to a very precise image that recalls the Festival of Cultures and in a broad sense the abstract idea of home. For this reason, the concept of 'house' was employed to recall the form of a roof of a house, as an archetypal image of childhood. Because the final outcome of the Little House of Cultures requires different types of space depending on the thematic programme developed during the school year and based on the choice of representation and display of the contents, the modules assembled together allow maximum transformability. This constitutes a strong point in favour of the developed architectural object that, independently from the educational and didactic use, can also be reused as a temporary structure for city events of various kinds (see Figs. 5.3 and 5.4). This practice resembles existing studies (Cirugeda 2011). Some solutions studied in WP3 have produced compositional and functional types such as a theatre arena, an exhibition box, a market, or also as play spaces or street furniture, laboratories and multi-functional rooms.

The ductility of the object demonstrates the sustainability of the project in terms of recycling and reuse. The introduction of temporary installations in places of

Fig. 5.3 Modular architectures. The modular structure can be assembled in various ways and adapted to many functions. Here are some examples: a cinema, a market and canopies in a park. These objects can be reused in the city after the Festival of Cultures. (Source: The authors)

Fig. 5.4 Possible configurations in the city. (Source: The authors)

collective use (squares, parks, gardens, but also buildings or residual areas, marginal or degraded places etc.) livens up the city and stimulates creativity and dynamism (Ciaffi and Mela 2011). This approach is captivating and attractive. Temporary occupation of public space creates cohesion between people and awareness of citizens in the perception of heritage (Cicalo 2010). This urban strategy is corroborated by the recent smart city studies (Ratti 2013), where one of the six axes is precisely the smart community, a more active and aware community that participates in the development of the city, dialoguing with governance, public and private bodies.[6]

The Little House of Cultures could be defined as an experience of tactical urbanism[7] (Lydon and Garcia 2015), in which each individual is called to intervene in his own city in an active and proactive way. The whole idea is that citizens act to transform and build the environment: everybody can be a 'city builder' (Simpson 2015). Nonetheless, the experience of the Little House of Cultures takes a step forward in this sense. It aims at building new cities and new communities to raise awareness towards the increasing transcultural shift in our societies. In such a context, citizens need more exchange and interaction, no matter their ethnicity or origin. This physical and metaphysical space is therefore a space where children can find a unitary and collective dimension without forgetting their roots and strengthening their identities. In the Little House of Cultures, they can experiment with interaction and dialogue and build together an inclusive and transcultural perspective.

[6]The European Community in parallel with the definition of the criteria identified by the Massachusetts Institute of Technology (MIT), for the research project European Smart Cities, identifies six axes according to which the smart city should be developed: economy, mobility, environment, people, living, and governance.

[7]With the term 'tactical urbanism', we mean the occupation of public space by citizens, together with the consent of the public administration, with initiatives that can involve both public and private bodies. Tactical urbanism is often the driving force of urban regeneration phenomena or heritage enhancement thanks to activities promoted by citizenship.

5.5 Conclusion

In human life, social and spatial aspects are strongly connected to one another—one affects the other in terms of perception and physical interaction between individuals. Studying the space within a project of multicultural social inclusion requires careful observation of the object's figurative nature, the flexibility of its use, and the dimensional scale of reference. As argued by Simmel (2012) 'there is a strong relationship between movement in space and the differentiation of contents of social existence' (Simmel 2012: 568). Starting from the necessary relationship between architecture and sociology, the Little House of Cultures was born from the challenge of creating a city and a neighbourhood that are more inclusive. Both disciplines deal primarily with individuals and their lived experiences within a group. In fact, the *dwelling* has always been a primary collective action. In the words of Italian Architect Aldo Rossi, this project sees architecture as a creation that is inseparable from life and from the society in which it manifests itself. In fact, it is largely a collective fact (Rossi 1978: 196).

This chapter demonstrated that starting from the fruitful comparison between sociological and architectural studies, it is possible to create strategies of interaction and social cohesion within a transcultural perspective. The Little House of Cultures is the urban interface of a deep and structured path that brings all the ethnic communities of Berlin together. In this sense, the project embraces a wider meaning of social inclusion, which is the inclusion of the whole community into the public space of the German capital. The project promoted the idea that social inclusion and social cohesion in society should be promoted not only through policies but also through education, as fostering social cohesion at an early age can create harmonious interactions between children of different cultures. It is therefore important to invest in education in order to have a diversified society that coexists in harmony overcoming inequalities. In this sense, the programme of the Little House of Cultures generates new interactions at a social and urban level. The idea of a travelling festival emphasises the importance of experiencing the public space in a more dynamic, active and above all, transcultural way. The transformation of public space perception together with its fruition as a place that belongs to the whole community lies in the concept of transculturalism, and it is an expression of it. The basis of the training process is therefore essential for the dissemination of the project in the city—through a network of collective services connected to school facilities during the year programme. In conclusion, through the experience of children—the seed of future societies—urban space and culture (or cultures) together are able to generate a more integrated, cohesive and conscious community.

References

Bertazzi L (2007) Il valore del cooperative learning per la comunicazione, la solidarietà e la cittadinanza attiva. UCIIM Torino, Torino

Ciaffi D, Mela A (2011) Urbanistica partecipata. Modelli ed esperienze. Carocci ed., Rome

Cicalo, Enrico. 2010. Spazi pubblici. Progettare la dimensione pubblica della città contemporanea. Milan: Franco Angeli

Cirugeda S (2011) Reusando la ciudad, ocupando esqueletos urbanos, creando redes. Rev De Arquit 17(24):41–55

Colombo E (2010) Crossing differences: how young children of immigrants keep everyday multiculturalism alive. J Intercult Stud 31(5):455–470

Cuccioletta D (2002) Multiculturalism or transculturalism: towards a cosmopolitan citizenship. Lond J Can Stud 17:1–11

Eigeartaigh AN, Berg W (2010) Editors' introduction. In: Eigeartaigh AN, Berg W (eds) Exploring transculturalism. A biographical approach. VS Verlag fur Sozialwissenschaften, Wiesbaden, pp 7–18

El-Sayedd ASZ, Mousa MAE-GA (2015) Cooperative learning in teaching communication skills for psychiatric nursing students. Eurasia Proc Educ Soc Sci (EPESS) 3:74–77

Farmakopoulou I, Triantafyllou K, Kolaitis G (2017) Refugee children and adolescents in Greece: two case reports. Eur J Psychotraumatol 8(sup 4):1351179

Grosu-Rădulescu. (2012) Multiculturalism or transculturalism? Views on cultural diversity. Synergy 8:102–111

Johnson DW (2006) Reaching out: interpersonal effectiveness and self-actualization. Allyn & Bacon, Boston MA

Johnson DW, Johnson RT, Smith KA (2007) The state of cooperative learning in postsecondary and professional settings. Educ Psychol Rev 19(1):15–29

Le H, Janssen J, Wubbels T (2016) Collaborative learning practices: teacher and student perceived obstacles to effective student collaboration. Camb J Educ:103–122

Leigh Smith B, MacGregor JT (1993) What is collaborative learning? In: Goodsell A, Maher M, Tinto V, Smith BL, MacGregor J (eds) Collaborative learning: a sourcebook for higher education. National Center on Postsecondary Teaching, Learning, and Assessment at Pennsylvania State University, pp 10–30. http://www.cte.hawaii.edu/Webster101/docs/ED357705.pdf. Accesed 2 Aug 2019

Lydon M, Garcia A (2015) Tactical urbanism: short-term action for long-term change. Island Press, Washington, DC

OECD—Organization for Economic Co-operation and Development (2018) The resilience of students with an immigrant background factors that shape well-being. OECD Publishing, Paris

Ratti C (2013) Smart city, smart citizen. Meet the media guru. EGEA, Milan

Rossi A (1978) L'architettura per i musei. In: Canella G, Coppa M, Gregotti V, Rossi A, Samonà A, Scimemi G, Semerani L, Tafuri M (eds) Tecniche della progettazione architettonica. Dedalo libri, Bari

Semi G, Colombo E, Camozzi I, Frisina A (2009) Practices of difference: analysing multiculturalism in everyday life. In: Wise A, Velayutham S (eds) Everyday multiculturalism. Palgrave Macmillan, London

Simmel G (2012/1903) La metropoli e la vita dello spirito. Armando Editore, Rome

Simpson C (2015) An overview and analysis of tactical urbanism in Los Angeles. Urban & Environmental Policy, Occidental College

Tuzi I, Fiorelli A (2022) L'architettura come pratica inclusiva: un progetto di connettività transculturale a Berlino. DIVE-IN Int J Divers Inclusion 2(2):151–175

Welsch W (1999) Transculturality. The puzzling form of cultures today. In: Featherstone M, Lash S (eds) Spaces of culture: city, nation, world. Sage, London
Yamazaki Y, Kayes CD (2004) An experiential approach to cross-cultural learning: a review and integration of competencies for successful expatriate adaptation. Acad Manage Learn Educ 3(4):362–379

Part II
Subtle Borders of Colonial and Postcolonial Policies

Chapter 6
Inner Borders of the United States' Role in the Making of the American Landscape

Carlos Santamarina-Macho

Abstract In 1893, the historian Frederic Jackson Turner wrote a seminal essay that conceptualised 'the frontier', an idea that has been extensively used, from the westward expansion era to the futuristic space colonisation, to explain the cultural relationship between Americans and the territory. The idea of 'border', particularly the domestic ones, is also relevant to understanding how the United States' landscape has been shaped. Throughout American history, there have been some stages in which the creation of domestic borders (or their dissolution) has been decisive for the configuration of the territory. The birth of the United States as a nation was marked by the Jeffersonian land division process, which established a grid of precise borders whose footprint remains today. Afterwards, the fences in the first residential suburbs and the postwar American neighbourhood played a similar role in what John Stilgoe called 'Borderland', predecessors of the contemporary gated communities. The discourses on American cultural regionalism can also be seen as a more abstract expression of the same idea. This paper aims to go through some of these milestones, analysing how some historical domestic borders have influenced the shaping of American land, culture and society.

Keywords Property regimes · Settler colonialism · Urban form · United States · Cultural landscape

6.1 Introduction

> We guard our American borders.
> We guard the American dream.
> We guard our right to fight for democracy.
> And keep our country free.
> We guard our American spirit.
> We guard the American dream.

C. Santamarina-Macho (✉)
Independent Researcher, Valladolid, Spain

G. Cimadomo et al. (eds.), *Urbanism at Borders*, The Urban Book Series,
https://doi.org/10.1007/978-3-032-09594-7_6

These lines belong to the song *The American Dream* (Knopfler 1998), composed by Thomas Bahler and Mark Knopfler for Barry Levinson's film *Wag the Dog* (1997). The film, based on the novel *American Hero* by Larry Beinhart (1993, 2004), offered an extraordinarily acute and ironic point of view of some of the American cultural symbols and, above all, of their use by the media and the political powers (Thanouli 2013).

To achieve this, the film skilfully articulates fiction and reality through the introduction of elements and recognisable references, such as a sex scandal that refers directly to the Clinton–Lewinsky one or the idea of a war against a far and unknown enemy, depicting Albania in the movie and Iraq in the reality of that time. However, it also included references and more subtle criticism of ideas deeply rooted in the American cultural imaginary. These were joined to elements such as the song, as mentioned above, which is a part of the fictitious electoral campaign of the principal character in the story.

In the quoted lyrics, concepts such as democracy, freedom, the American Dream, or American exceptionalism can be identified. But another one, more real and material, can also be recognised: the border. Although that last reference can be understood in the context of the filmic narrative, its direct association with more abstract ideas such as those mentioned and the voluntary omission, throughout the story, of the frontier topic are interesting—and even more interesting, if we take into account how the frontier myth, closely joined to the idea of the border and deeply associated with American culture, has systematically been used by American politicians as a synonym for progress since it was conceptualised by Frederick Jackson Turner (1986) at the beginning of the nineteenth century.

Although, indeed, the borders are not alien to the most recent American policy—as it has been seen in the Donald Trump 2016 presidential campaign in which proposals about the Mexico border played a central role—the truth is that the reference made in the film goes beyond their understanding as a simple physical element of separation between countries. Their status is highlighted as a shared cultural and social American reference that had to be preserved in the same way as freedom, democracy, or the American dream.

This essay explores the role played by the idea of 'border' throughout American history and how it may have influenced the configuration of the American identity and, particularly, everything that has to shape its territory and landscape—a history in which small and imperceptible internal borders, created over the last two centuries and still shaping a large part of the territory of the United States, have been as relevant as the great borders, which seek to voluntarily isolate the Americans from the people and influence of the rest of the world.

6.2 Frontiers and Borders in the American Culture

Within American culture, the frontier and the border are two different and antagonistic ways of approaching the same phenomenon: the encounter with the other, the different one and understanding where it happens as a conflict space. But, above all, two points of view have always been present in constructing the American political space and its social and cultural interpretation. The frontier could be interpreted as the socially optimistic vision of the boundaries, which was born once American society consolidated its territory's control and stopped understanding edge spaces as a threat.

Frederick Jackson Turner himself recognises, at the beginning of *The Significance of the Frontier in American History*—the essay that gave true meaning to the mythical reading of this place—that his interpretation arises once the period of confrontation with the wilderness has ended, whether this wild condition may be understood from a social perspective or a strictly geographical one. He added that this rereading of the history and the border simultaneously as a conflictive place and a land of opportunities—a way nobody had explored yet—was essential for the American people.

> In this advance, the frontier is the the outer edge of the wave—the meeting point between savagery and civilisation. Much has been written about the frontier from the point of view of border warfare and the chase, but as a field for the serious study of the economist and the historian it has been neglected. (Turner 1986)

The cultural significance of the frontier as a space where the European settlers left behind their past and, in contact with the unknown, defined their new identity as Americans is an intelligent invention supported by specific recognisable facts. Still, it is also questionable from a strictly historical perspective. At the same time, it has also been the support of a set of extraordinarily attractive and helpful ideas not only for a newborn society such as the nineteenth-century American one, which was looking for strong elements of differentiation from European culture, but also, turned into a popular myth, for contemporary American society (Santamarina-Macho 2017).

However, Turner also pointed out that the interpretation of the border as a space of separation and conflict had been dominant until the beginning of the twentieth century. Nineteenth-century American history was a progressive overcoming of territorial and social boundaries through processes that overlapped during annexation, exploration and colonisation. This conquest of the West, real and demythologised, uncovered and consolidated a sequence of boundaries that were initially insurmountable physical barriers that can barely be interpreted, like places of encounter and progress, as the frontier myth does. The Appalachian Mountains constituted the first of these natural borders, associated with the 13 original British colonies, which were progressively being displaced in their function by others, such as the Mississippi River, which was the physical and political limit of the independence granted by Great Britain in the Treaty of Paris (1783), or the last of the great

Fig. 6.1 Col. Frémont planting the American standard on the Rocky Mountains [Baker & Godwin, c. 1856]. (Source: The Library of Congress, https://www.loc.gov/item/2003689276/ Public domain)

physical edges of North America, those Rocky Mountains firstly explored by Meriwether Lewis and William Clark in 1804.

When Colonel John C. Frémont, in 1842, managed to overcome that mountain range for the first time (see Fig. 6.1), the American natural borders, except for the two oceans, disappeared—but only to be replaced by new borders, the man-made ones, a set of limits voluntarily established by the Americans within a controlled, dominated and differentiated space. These were going to play, even more than the natural ones, a relevant role in the configuration of the American space.

6.3 The American Domestic Borders

6.3.1 Creating a Borderspace

The first genuinely American borders emerged almost immediately upon the birth of the United States as a nation, through a process that has generally been accepted as the most influential one in shaping American territory. It is, of course, the project of land parcelling and colonisation of the conquered territories, promoted by

Thomas Jefferson and supported by the *Land Ordinance Act* of 1785, although fully developed, thanks to the economic boost offered by the *Homestead Act* of 1862 signed by President Abraham Lincoln, after the American Civil War.

The Jeffersonian project was an early example of the confidence in science and technology as instruments through which it would be possible to create an isotropic and homogeneous division of the space over a complex and unknown territory, whose singularities were, until the moment of its local implementation (Corner and MacLean 1996), entirely ignored. A new egalitarian space that should be the basis for the establishment and creation of a new idealised and egalitarian society integrated by yeoman farmers. This was the first stage of the construction of American culture, in which the shaping of its social structure was connected to its territorial support (Santamarina-Macho 2016).

Denis Cosgrove (1984: 175) has pointed out that the land was the central element around which American life and ideology revolved in the first century of the United States' history. The land was, in the model raised by Jefferson and the one previously raised by J. Hector St. John de Crèvecœur (2005), the mainstay of the new settlers. But it was the means through which citizenship was obtained and, thanks to it, the rights of Americans. It would boost the development of a new kind of society, which was different (socially, economically and politically) from the one that had characterised the British colonies, but it also entirely modified the relationship between man and territory.

However, the fundamental change introduced by the new American model was neither in the relationship with the land nor in how it economically profited but in its management and tenure regime. Until the nineteenth century, traditional communities were fundamentally agrarian, but land ownership was not one of its essential features. In those communities, the land was a communal good that any of their members could use through the payment of taxes, known as 'fee simple land ownership' (Kunstler 1994: 25), or services such as the simple, productive maintenance of the land, without substantial differences between individuals or parcels. The borders had already existed within the Puritan communities but not among their members. They were just a tool to isolate the colonists from an indigenous population. The Jeffersonian agrarian ideal implied a break with this inherited system and the imposition of an alternative based on individual property since it guaranteed citizenship and rights within the new American society (see Fig. 6.2).

The progressive vanishing of 'the common' for the benefit of individual property paved the path for a non-productive and speculative economic system that was not too different from the urban and industrialised one from which the Jeffersonian ideal wanted to move away. The emphasis on individual property made it necessary to develop mechanisms not only to allow anyone to acquire, in equal opportunities and conditions, the minimum unit of land that guarantees their livelihood, but also to prove such possession before the State and the people credibly. This was the context in which the borders acquired a growing relevance, as elements that were able to support the condition of belonging, constitutionally protected since the introduction of their inalienability in the Fifth Amendment (1789), by pointing out that 'no

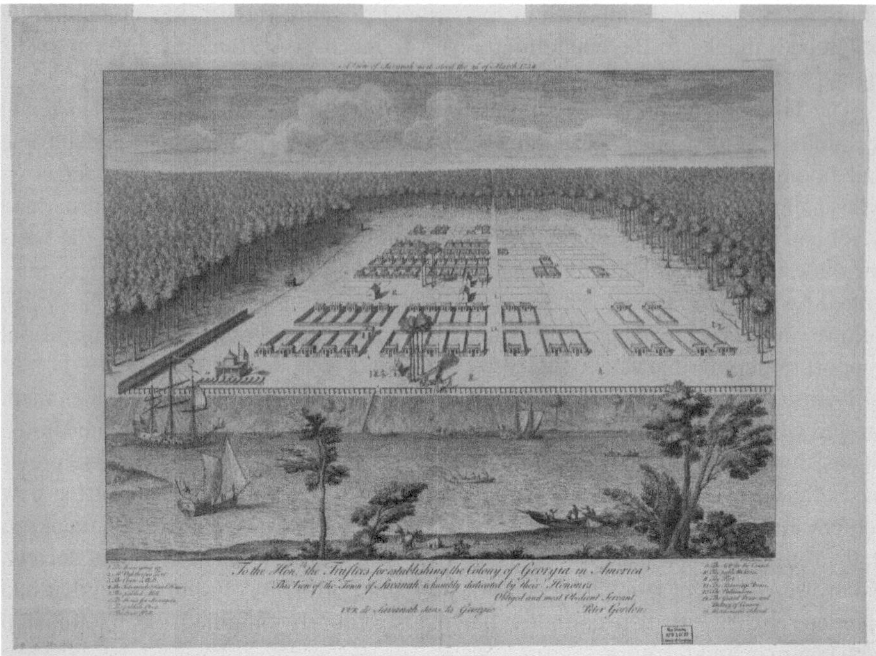

Fig. 6.2 A view of Savannah (Georgia) as it stood on the 29th of March, 1734. (Source: The Library of Congress, https://www.loc.gov/item/96514600/ Public domain)

person shall (…) deprived of life, liberty or property, without due process of law; nor shall private property be taken for public use'.

The Land Ordinance Act of 1785 was essentially the framework intended to make these new lands available to nineteenth-century settlers and, in turn, to incorporate them into the Union as States (The Northwest Ordinance 1787). Still, it was also a tool for dividing the conquered territories until they were conceived globally into small units. Therefore, the central part of its development, and in particular the *Public Land Survey System*, will be related to the implementation of tools and procedures aimed at unequivocally fixing the edges of these units, transforming a flat space into a bordered space whose physical footprints, a mesh that supports almost all territorial and administrative American subdivision systems, reach to this day.

The relevance of those limits can still be appreciated today through eloquent aerial photography (MacLean 2003: 130–147) and physical signs in the territory. Some of them are punctual, such as the landmarks that indicate the starting points of the grid. The oldest was placed in 1786 by Thomas Hutchins among the boundaries of Pennsylvania, Ohio and West Virginia. Still, there are other famous ones, such as the *Four Corners Monument*, located among the states of Arizona, Colorado, New Mexico and Utah. The most common ones are plates, poles, rocks and cairns,

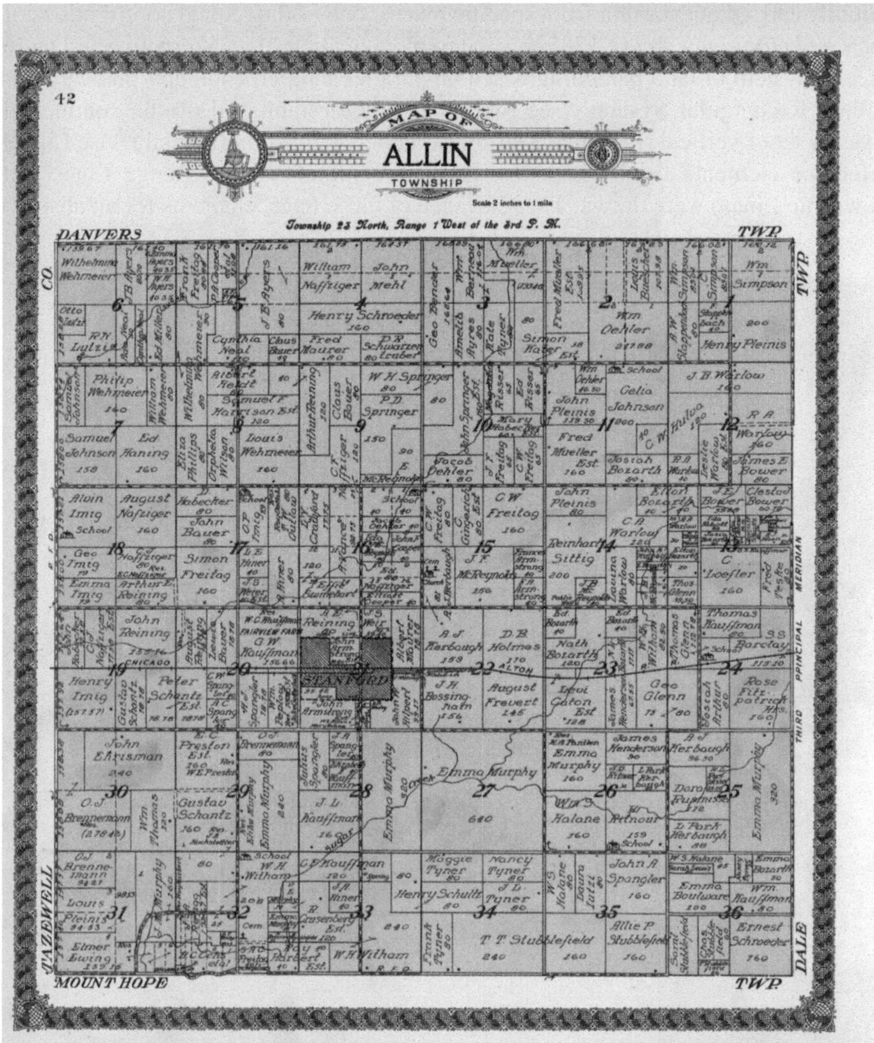

Fig. 6.3 An example of a township map showing the first division of the land into sections (numbered from 1 to 36) and their internal structure. From Standard atlas of McLean County, Illinois [Geo. A. Ogle & Co. Allin Township, 1914]. (Source: The Library of Congress, https://www.loc.gov/item/2001626356/ Public domain)

indicating the wide variety of subdivisions in American space, from the townships to the minor fraction of a section (see Fig. 6.3).[1]

However, the footprint of the Public Land Survey System is even more significant in linear divisions. Although the process of subdivision of the space was

[1] A section is 1/36 of a township (the basic unit of American Grid).

initially carried out starting from specific places, one of the critical contributions of the *Land Ordinance Act* of 1785 was the implementation of an alternative and more precise system of land parcelling based on parallel and perpendicular lines, the so-called 'Rectangular System' (see Fig. 6.4), which established on the continental space a set of vertical (Principal Meridians) and associated horizontal (Base Lines) reference elements from which the successive modifications (Range Lines and Township Lines) were drawn. The relevance of these lines, elements for an abstract

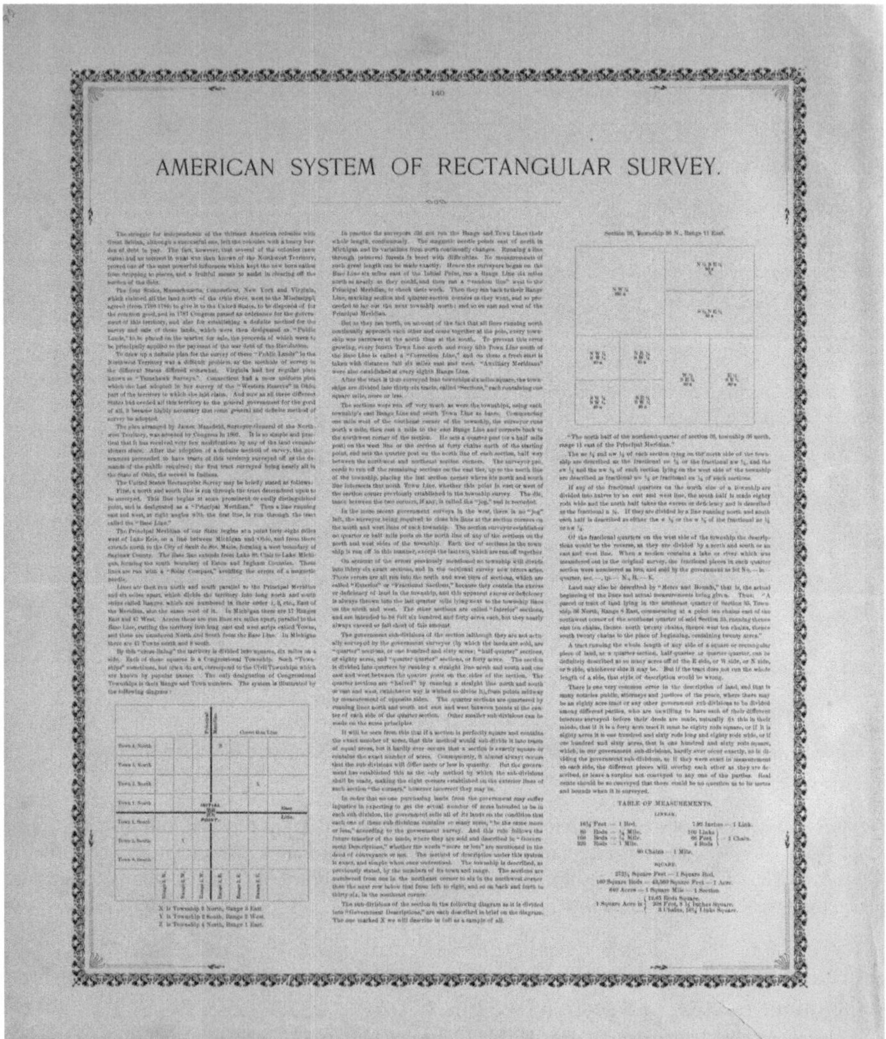

140

AMERICAN SYSTEM OF RECTANGULAR SURVEY.

Fig. 6.4 Basic rules of the American System of Rectangular Survey. From: The county of Saginaw, Michigan: topography, history, art folio. (Source: The Library of Congress, https://www.loc.gov/item/2007626770/ Public domain)

measurement of the territory, is such that they not only constituted the primary spatial references for the Geological and Geographical Surveys that took place after the American Civil War, naming expeditions such as the *Geological Explorations of the Fortieth Parallel* or the *U.S. Geological Survey West of the 100th Meridian* (Wheeler 1889), but also left real and permanent marks in the land through roads or urban streets whose names are, even today, explicit references to the system designed in the eighteenth century (the ubiquitous *Meridian Road* or *Meridian Avenues*, which divide the eastern and western parts of some American cities). They were a needed imprint if we understand that these markers, punctual or linear, continue legally determining that ownership of the land that was, in the beginning, the basis of American democracy.

It could even be affirmed that what characterised the United States, making the difference between it and the social and political territory existing before the nation's foundation, was nothing but its domestic borders, its subdivision system. In short, the transformation of America into a *borderspace*.

6.3.2 A Culturally Bordered Nation

Philip Jenkins, in *A History of the United States* (Jenkins 2009: 190–191), wrote that one of the main advances brought by the American Civil War was to transform what until then were just some united states, a multiplicity of territories weakly linked with each other, in the United States of America, in a new reality that could be understood as a single one, as a national landscape (Johnson 2010).

It is true that, at least from the political and administrative point of view, the war had definitively suppressed the lines that separated the North and the South, some as firmly drawn as the 36°30′ parallel, the spatial boundary between the free and slave states established by the Missouri Compromise (1820) (see Fig. 6.5). The *Public Land Survey System* also contributed to this dissolution of the old boundaries, at least of those established between the Atlantic states of the unoccupied western lands, defining a theoretically isotropic territory in which there seemed to be no differences among the lands to be occupied. The challenge of the conquest of the West, during which slogans such as 'In God we trust, in Kansas we busted' (Guardia 2012) became widespread, would soon show that it was not the truth: not all the territories were equally suitable to settle.

By contrast, although the war was the instrument to achieve the formal unity of the nation, it failed to do the same from a spatial, social and cultural point of view. The South itself, although politically and administratively integrated into the new nation, could not detach itself overnight from the conditions that made it intrinsically different from the North, from its geography to its climate, through its system of occupation and use of the territory, its economy, or even the social features of its population. And like the South, many other territories demanded differentiated attention that recognised their cultural singularities.

Fig. 6.5 Reynolds's political map of the United States, designed to exhibit the comparative area of the free and slave states and the territory open to slavery or freedom by repealing the Missouri Compromise. (Source: The Library of Congress, https://www.loc.gov/item/2003627003/ Public domain)

Those territorial differences had been a structural problem in the United States since the nineteenth century when the first annexation of territories to the already independent colonial domains occurred. This occurred because they were mostly incorporated into the Union, bypassing their French, Spanish, or Mexican origins, not to mention the most complex native question. The new national reality and the development of different national identity myths, such as those of the frontier, the pioneer, the self-made man, or the wilderness, were oriented to dilute these differences and build a desired and idealised unitary space. However, a few decades after the end of the American Civil War, this issue and the need to re-understand the territory as a diverse space, strongly resurfaced thanks to the rise of modern American regionalisms.

Late nineteenth-century American literature probably was the pioneer in addressing that concept. After a first half of the century was dominated by the mythical construction of America through, for example, the transcendentalist movement headed by Ralph Waldo Emerson (2010) and Henry David Thoreau, or the poetry by Walt Whitman (2009), a realistic genre, based on the local traditions, started to gain popularity, facing the ideas that came from the north-eastern cities guiding the United States towards a unitary ideal of the nation. In this context, the texts by Mark Twain, Jack London and George Washington Cable, which showed the cultural specificity of the southern states (Cuenca 2006) and some alternative literary expressions developed in the colonised territories, arose. These literary genres will have continuity in the first third of the twentieth century. They became famous thanks to authors such as John Steinbeck and William Faulkner, who generated a new iconography of southern and Midwestern America aided by the visual arts.

What is particularly relevant in this interpretation of the United States that demanded greater attention to local and regional singularities in contrast to the unifying perspective that prevailed at the end of the nineteenth century is its incorporation into a new way of seeing the American spatial reality. It was claimed as necessary to understand and act on the nation. In this sense, although the first regionalist movements, with Lewis Mumford (1925, 1938, 1961) as the primary intellectual reference, or more practical organisations such as *The Regional Planning Association of America* (RPAA) funded by Clarence Stein, promoted this idea as an alternative to the autonomous understanding of cities, this perspective will soon be extended to a more global vision.

Interpretations of regionalist ideas, such as those developed by Howard W. Odum and Harry E. Moore at the University of North Carolina, were particularly relevant due to their subsequent influence. They, firstly in *Southern Regions of the United States* (Odum 1936) and then in the classic *American Regionalism: A Cultural-Historical Approach to National Integration* (Odum and Moore 1938), called for a new interpretation of America that would once again consider the political, social and cultural differences among territories. They also helped Americans to restore 'the hope of American democracy and, according to many observers, of Western civilisation' contributing 'to better analysis and understanding of the scene and (…) to a more realistic reintegration of national diversities' (Odum and Moore 1938: 618).

The scientific analysis conducted by Odum concluded with the division of the United States into six large regions (The Middle States and Their 'Middle West'; The Northeast and Its New England; The Southeast and Its 'Old South'; The Far West and Its California; The Northwest and Its Great Plains; The Southwest and Its Texas), which had its antecedents in the census divisions devised by James D.B. DeBow in 1853. What is particularly significant is that the Jeffersonian project and the nineteenth-century idea of 'Americanism' were no longer valid in interwar America, when a more diverse approach to the nation that considered the specific conditions of each place was being boosted and socially embraced.

The idea of a national division based not only on statistical criteria, such as that of the *U.S. Census Bureau*, but also on cultural and historical ones had its impact even in the federal administration itself, which relied on the Odum system to organise and systematise one of the most relevant documentary projects of the interwar period, the photographic archive of the *Farm Security Administration* and the *Office*

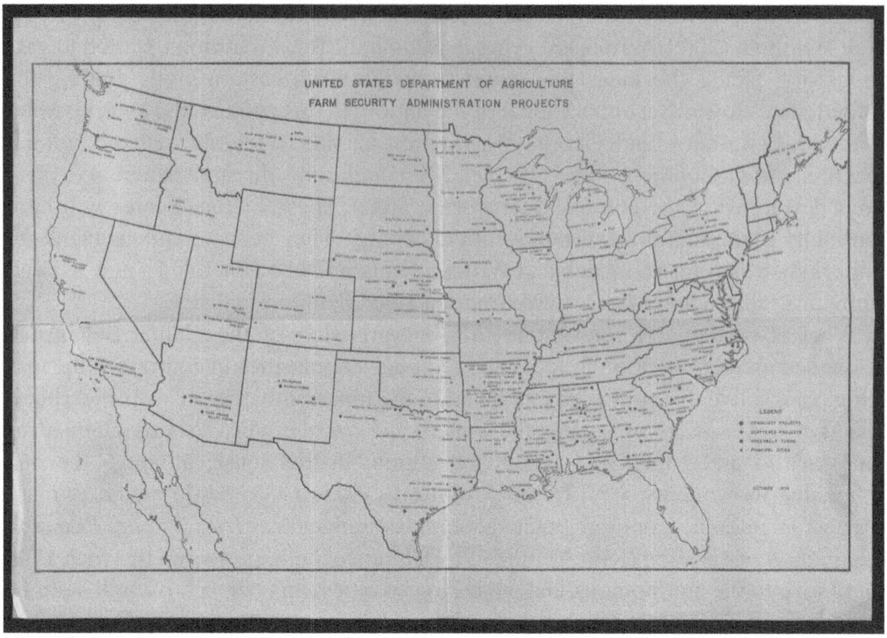

Fig. 6.6 Map of the United States showing locations of Farm Security Administration programmes. (Source: The Library of Congress; https://www.loc.gov/item/2017650358/ Public domain)

of War Information (Fleischauer and Brannan 1988; Stryker and Wood 1973) (see Fig. 6.6). But, as Odum himself pointed out, his system was just one of the 'large number of the multiple regional classifications and arrangements now being used or proposed' (Odum and Moore 1938: 618) that would not stop multiplying, defining a new tendency to read America as a union of diversities—a trend that has had its continuity to this day, both from a theoretical and a practical point of view (Wheeler 2002), seeking to discover and show us what Michael P. Conzen defined as 'the landscape regions that might be out there waiting to be recognised' (Conzen 2010: 371).

6.3.3 American Borderlands

In the mechanisms mentioned above for creating borders on American territory, the existence of specific elements that materially define the limits was irrelevant. However, this condition of invisibility in the boundaries did not imply, in any way, that their presence and effects were not palpable, whether from a material, social, cultural, or economic perspective. Nevertheless, the non-visible boundaries, elements that divide and decompose the space but are appreciated more for their effects than their physical footprints, are not the only ones in American space. American fences constitute, for example, a paradigmatic example of a physical element, an indicator of a border, which has historically fulfilled a relevant function in the configuration of the American territory and its own cultural identity. And they have also played a relevant role in the discussion on how the American landscape and society should have been.

The fences were the most common marker of division, protection and, above all, taking possession of land in the colonial period. Thanks to the extraordinarily wide variety of material solutions adopted (Stilgoe 1982: 188–192), a distinctive feature in an also diverse agricultural nineteenth-century society, they became a fundamental tool in landscape shaping. As John R. Stilgoe points out (Stilgoe 1982: 188–192; Zelinsky 1977), the building of good and solid fences was not only a source of pride for the owners of any land, devoting to it a significant amount of resources and time, but for the communities as well, providing a sense of security and stability to their members.

However, the westward expansion brought about a rupture with the material and affective linkage of the colonial society with its fences. Neither the scarcity of good resources, mainly wood, nor the accelerated expansion aided the spread of this tradition far beyond the Midwest, cutting the interest in making the old material boundaries an essential factor in the new settlement activity. The latest technologies also offered alternative ways to make divisions and take over the territory, from the markers mentioned above to the (in)famous barbed wire fences (see Fig. 6.7) that were one of the causes of some of the most relevant alterations of the ecological balance in primitive America.

The place where the changes in the attitudes regarding the role of fences in society were most easily appreciated was American housing. The expression and function of the fences, clearly defined in colonial communities, began to be a topic of debate in the American residential proposals of the late nineteenth century. They questioned their usefulness in a new model of society, especially in creating residential communities that tried to bring together the values of urban and rural life as one of its main drivers. Andrew Jackson Downing was perhaps the most influential architect from the mid-nineteenth century who was involved in this discussion.

In this new paradigm of occupation of the territory, the traditional domestic borders constituted a contradictory element, a need to guarantee the division of the property but simultaneously opposed to the aim to create a beautiful environment, connecting the house with nature—an attractive environment that was nothing but a new romantic recreation of the countryside ideal (Bunce 1994: 155–157) that Jefferson had also used to define his social ideal. At the end of the nineteenth century, a new kind of extraordinary reality was implemented, simultaneously urban and rural, called 'borderland' by John R. Stilgoe (1988).

Although in the proposals by Downing, even in those whose characters were close to large rural villas integrated into the countryside, the closure of the properties was allowed, he soon questioned the physical characteristics of these boundaries, expressing his preference for a 'low railing or hedge, instead of a formidable board fence' (Downing 2003: 174). He fostered in this way a visual disappearance

Fig. 6.7 A traditional wooden fence joining barbed wire fence in Carter County, Missouri [Vachon, J., 1942]. (Source: The Library of Congress; https://www.loc.gov/item/2017813836/ Public domain)

of the boundaries, specifically designed to create a particular type of landscape. A few years later, in 1868, Frederick Law Olmsted and Calvert Vaux entirely banned the building of fences of any kind in Riverside, the first planned community in the United States (see Fig. 6.8).

This suburban paradigm, in which the private gardens turned into what J.B. Jackson recognised as 'a national institution, as essential to every home [as] a Bible somewhere in the house' (Zube and Zube 1977: 39), was perceived as a collective public space and reproduced throughout the twentieth century. It also made it possible to keep the dissolution of the private boundaries as an essential part of the identity of the new neighbourhoods created after World War II. For instance, this was going to happen in the communities that followed the Radburn model (New Jersey) by Clarence Stein, strongly influenced by Ebenezer Howard's Garden Cities (Howard and Osborn 1946). There, only 'low and inconspicuous' fences were allowed to separate the properties, leaving a shared green space with no boundaries in the front. This model, which became an American residential ideal, would be reproduced even in the postwar speculative developments, such as the Levittowns, where the closure of front yards and backyards was initially forbidden to create a certain sense of community among its inhabitants (see Fig. 6.9). The existence of

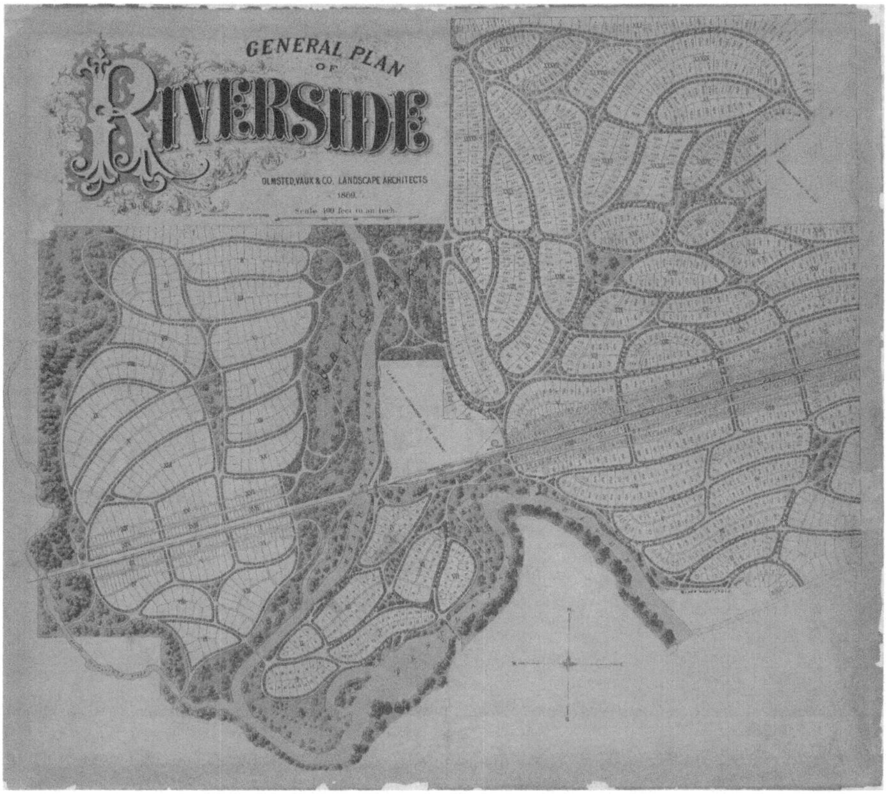

Fig. 6.8 General Plan of Riverside [Olmsted, Vaux & Co. Landscape Architects, 1869]. (Source: The United States Department of the Interior, National Park Service, Frederick Law Olmsted National Historic Site, Public domain)

fake shared and public spaces had a specific role in creating a specific social profile (Gans 1967).

This apparent lack of borders would generate, in fact, a never-ending conflict between some American social ideals, such as individualism versus collectivism, freedom versus security, or privacy versus publicity (Stilgoe 1988: 233), which, quite frequently, were solved recovering some boundary. The regulations in these urban developments, which have been extensively reproduced throughout the United States, have been progressively becoming more flexible, accepting the use of closures, boundaries, or fences. This has limited the chances of generating the desired sense of community, mainly through the more ornamental than functional use of picket fences, which changed the relationship between public and private spaces—something that has been, according to Vincent Scully, 'the basic theme (in) architecture and urbanism in the United States from 1963 to the "present"' (Scully 1996: 7).

Fig. 6.9 Levittown houses. Peg Brennan, residence at 25 Winding Lane. (Source: The Library of Congress; https://www.loc.gov/item/2018729983/ Public domain)

The culmination of this progressive and unstoppable process of destruction of the communities and public space in the United States for the benefit of privacy, security and isolation has been the spreading of the American gate communities (also known as walled communities) (Blakely and Snyder 1997) (see Fig. 6.10), a kind of urban development that seems to push the idea of 'good fences make good neighbours' by Robert Frost (1915) to the highest level. The approximately ten million housing units that have been built following this residential concept mean, according to Mike Davis (1990), the death of the American will to create a truly democratic public space for the benefit of a more exclusive, segregated and bunkered society, trapped in its countless man-made borders (Benjamin 2009).

Fig. 6.10 Entrance to a gated community (Lakehurst Drive) from Bratenahl Road in Bratenahl, Ohio, United States. (Source: https://commons.wikimedia.org/wiki/File:Lakehurst_Drive_gate_in_Bratenahl.jpg Public domain)

6.4 Domestic Borders and the American Mind

The architect and photographer Alex S. MacLean pointed out that man is a territorial being who 'marks the land with visible and invisible lines' for various purposes (MacLean 2003: 131). MacLean likely took this idea from John Brinckerhoff Jackson, one of his teachers at Harvard and, perhaps, the most lucid interpreter of the American landscape. John Brinckerhoff insistently emphasised the relevance of boundaries and borders in the configuration of territory and culture in the United States.

His interest in these elements was not exclusively formal. The importance of the borders was, in his opinion, how they helped to build and differentiate communities (Jackson 1980, 1984), not only through the segregation of social and cultural groups but also through the creation of a higher grade of internal cohesion, unity and a sense of belonging. It is meaningful that Jackson attributed the boundaries a role in the configuration of the sense of community comparable to the one given to the public roads, the public meeting places, or the monuments, elements more clearly linked to the generation of social bonds (Jackson 1970).

Although at the end of the nineteenth century, Frederic Jackson Turner conceptualised the idea of the frontier as an instrument that would be able to support a particular way of understanding the American being, the truth is that even before his positive interpretation of the border spaces, they already were playing a relevant role in the American mind, not only allowing to separate the Americans from the (still) non-Americans but also defining something close to what Odum called Americanism: the American identity itself.

The borders were an essential part of the colonial past. Still, they also support the Jeffersonian ideal, which aimed to get a democratic, egalitarian and free society through a grid 'as sameness that, overwhelmed and liberated from any sense of place, let Americans feel everywhere at home' (Jackson 1994: 15). The borders were also a helpful tool for reclaiming new ways of understanding American exceptionalism through cultural diversity and the most prominent representation of certain debates about the American way of life that remain today.

Indeed, some old barriers in the United States still have to be removed, fundamentally social and economic ones. But, on the other hand, we must not fall into the tendency, already identified by J. B. Jackson, to understand that 'all boundaries, all traditional divisions of space, whatever their original purpose (are) obstacles to beauty, to efficiency, to sociability, to change' (Zube and Zube 1977: 39), because we would be ignoring the relevant role they have played in American history. What is probably more important is the role they can continue playing if we allow them to be overcome, aiding in getting people in touch as Robert Frost's characters did while they were mending their meaningless wall.

References

Beinhart L (1993) American hero, 1st edn. Pantheon, New York

Beinhart L (2004) Wag the dog: a novel. Nation Books; Distributed by Publishers Group West, New York/Berkeley, CA

Benjamin R (2009) Searching for Whitopia: an improbable journey to the heart of white America, 1st edn. Hyperion, New York

Blakely EJ, Snyder MG (1997) Fortress America: gated communities in the United States. Brookings Institution Press and Lincoln Institute of Land Policy, Washington, DC/Cambridge, MA

Bunce MF (1994) The countryside ideal: Anglo-American images of landscape. Routledge, London/New York

Conzen MP (2010) Afterword. In: Conzen MP (ed) The making of the American landscape. Routledge, New York, pp 370–375

Corner J, MacLean AS (1996) Taking measures across the American landscape. Yale University Press, New Haven/London

Cosgrove DE (1984) Social formation and symbolic landscape. Croom Helm, London

Cuenca CM (2006) La literatura de Estados Unidos desde sus orígenes hasta la Primera Guerra Mundial. Editorial Síntesis, S.A., Madrid

Davis M (1990) City of quartz: excavating the future in Los Angeles. The Haymarket series. Verso, London/New York

de Crèvecœur HSJ (2005) Letters from an American farmer. Dover Publications, London

Downing AJ (2003) Our country villages. Ciudades 7:173–178

Emerson RW (2010) Obra ensayística. Artemisa Ediciones, Valencia

Fleischauer C, Brannan B (1988) Documenting America, 1935–1943 (Approaches to American culture). University of California Press, Berkeley, CA

Frost R (1915) North of Boston. Henry Holt and Company, New York

Gans HJ (1967) The Levittowners; ways of life and politics in a new suburban community. Pantheon Books, New York
Guardia C d l (2012) Historia de Estados Unidos. Sílex Ediciones S.L., Madrid
Homestead Act (1862) An act to secure homestead to actual settlers on the public domain
Howard E, Osborn FJ (1946) Garden cities of to-morrow. Faber and Faber Ltd, London
Jackson JB (1970) The public landscape. In: Zube EH (ed) Landscape: selected writings of J.B. Jackson. University of Massachusetts Press, Amherst, pp 153–160
Jackson JB (1980) By way of conclusion: how to study the landscape. In: Jackson JB (ed) The necessity for ruins, and other topics. The University of Massachusetts Press, Massachusetts, pp 113–126
Jackson JB (1984) A pair of ideal landscapes. In: Jackson JB (ed) Discovering the vernacular landscape. Yale University Press, New Haven, pp 8–55
Jackson JB (1994) A sense of place, a sense of time. In: A sense of place, a sense of time. Yale University Press, Binghamton, NY, pp 149–163
Jenkins P (2009) Breve historia de Estados Unidos. Alianza Editorial, S.A., Madrid
Johnson HB (2010) Towards a national landscape. In: Conzen MP (ed) The making of the American landscape. Routledge, New York, pp 127–145
Knopfler M (1998) Music from the motion picture Wag the dog. Mercury, New York
Kunstler JH (1994) The geography of nowhere: the rise and decline of America's man-made landscape. Touchtone, New York
Land Ordinance Act (1785) An ordinance for ascertaining the mode of disposing of lands in the Western Territory
MacLean AS (2003) Designs on the land: exploring America from the air. Thames & Hudson, New York
Mumford L (1925) Regions—to live in. Survey 54:151–152
Mumford L (1938) The culture of cities. Harcourt, New York
Mumford L (1961) The city in history: its origins, its transformations, and its prospects, 1st edn. Harcourt, New York
Odum HW (1936) Southern regions of the United States. The University of North Carolina Press, Chapel Hill
Odum HW, Moore HE (1938) American regionalism; a cultural-historical approach to national integration. H. Holt and Company, New York
Santamarina-Macho C (2016) Cultura y representación del man-made landscape. La construcción de la imagen de un territorio. EE.UU. 1925–1975. Universidad de Valladolid, Unpublished PhD dissertation
Santamarina-Macho C (2017) The new new Old West. A landscape without a place. Ra-Revista Arquit 19:57. https://doi.org/10.15581/014.19.57-66
Scully V (1996) The architecture of community. The University of Michigan, Ann Arbor, MI
Stilgoe JR (1982) Common landscape of America, 1580 to 1845. Yale University Press, New Haven/London
Stilgoe JR (1988) Borderland: origins of the American suburb, 1820–1939. Yale University Press, New Haven
Stryker RE, Wood N (1973) In this proud land: America 1935–1943 as seen in the FSA photographs. Graphic Society, New York
Thanouli E (2013) Wag the dog: a study on film and reality in the digital age
The Northwest Ordinance (1787) An ordinance for the government of the Territory of the United States North-West of the River Ohio
Turner FJ (1986) The significance of the Frontier in American History. In: Ridge M (ed) Frederick Jackson Turner: Wisconsin's Historian of the Frontier. State Historical Society of Wisconsin, Madison. United States Bill of Rights. 1789
Wheeler GM (1889) Report upon United States geographical surveys west of the hundredth Meridian. Government Printing Office, Washington

Wheeler SM (2002) The new regionalism: key characteristics of an emerging movement. J Am Plan Assoc 68(3):267–278

Whitman W (2009) Hojas de hierba. Visor libros, Madrid

Zelinsky W (1977) Walls and fences. In: Zube EH, Zube MJ (eds) Changing rural landscapes. The University of Massachusetts Press, Massachusetts, pp 53–63

Zube EH, Zube. MJ (1977) Changing rural landscapes. The University of Massachusetts Press, Massachusetts

Chapter 7
Identity Crisis: Social Schizophrenia and Blackness in 'Post-Racial' America

Moriah J. Snowden

Abstract 'One nation under God, indivisible, with liberty and justice for all'. The closing clause of the United States' Pledge of Allegiance, observed by all Americans, with the hand upon the chest, cites a sacred oath to a nation that asserts undivided equity. From schools to town halls to congressional sessions, this phrase is uttered with solemn veneration—yet there is an anomaly for the Black child, constituent and congressional member. Glaring *schism* somehow, simultaneously: 'American'—yet, not. National decree falling short of their melanin. A 'border within border'; those of African-American descent embodying the schizophrenic divide of American fealty severed betwixt racial reality. Echoes of illusive concepts, 'liberty' and 'justice', are left unrealised, reverberating beyond nationality into Black society, economy and geographical context. The identified and deconstructed issues presented delve into their unique crisis: The perilous lines etched between citizenry and ethnicity in a not-yet United States—the dual-life of the African-American. And the ramifications that follow when habitually forced to choose between either state. From societal borders demarcated by race versus allegiance to the economic divergences associated with Black net worth and income to the policy borders of segregation and redlining. Utilising national statistics, demographical data and the narratives of Blacks living in America from January to June 2019—as well as 29 years of personal experience—I will compile a presentation illuminating the daily antithesis of native equality for African-Americans and its effects across social, economic and policy realms with the definitive objective of revealing the divide between inherent identity and projected nationality—sense of self and sense of place—along with the physical, cultural and psychosomatic damages wrought on the African-American psyche.

Keywords Racialised borders · Spatial justice · Race · Identity · Blackness · United States

M. J. Snowden (✉)
Impact Collective, Atlanta, GA, USA

G. Cimadomo et al. (eds.), *Urbanism at Borders*, The Urban Book Series,
https://doi.org/10.1007/978-3-032-09594-7_7

7.1 Introduction

There is a condition that exists solely for those Black-born within the verges of the United States, chained—still—to sanctioned slavery and its legacy upon their ancestors: The fate of 'living border'. 'Breathing boundary'. Terms 'Black' and 'American' are ultimately unequal, left severed in a single being. This crisis state emerges from the unresolved contradiction of racial identity met with inherent nationality—the inevitable encounter of social schizophrenia is its result. For the purposes of this paper, *social schizophrenia* will be defined as the internal disruption that occurs when American birthright occupies an inferior stratum to race—and the fractured personas adopted when pledged liberties and justices assured to all US natives are habitually denied. 'American', then, becoming conditional, technical, anecdotal and wholly separate, if not thoroughly at odds with the Black psyche.

While 'when' Blacks in the United States definitively became American whether through legislation or common belief is beyond the scope of this account, the disparity of '*African-American*'—most aptly segregated by hyphen—remains a dilemma forged by centuries. W.E.B. Du Bois more than 120 years prior, also wrote of the particular plight of the US Black native. A conflict he would notably account for as '*double-consciousness*'.

> It is a peculiar sensation, this double-consciousness, this sense of always looking at one's self through the eyes of others, of measuring one's soul by the tape of a world that looks on in amused contempt and pity. One ever feels his two-ness,—an American, a Negro; two souls, two thoughts, two unreconciled strivings; two warring ideals in one dark body, whose dogged strength alone keeps it from being torn asunder. (Du Bois, 1903: 2)

This depiction of the Black citizen's inner strife concerning their estranged two selves proves applicable to this day. Indeed, more than a century has passed, and the question remains: When 'American' cannot be resolved with 'Black', what is left in its wake? Even still, there lies a more profound symptom, a more excellent issue: the Black condition spanning further—beyond *self*, alone. Their internalised 'warring' [ibid], while halted from tearing bodies asunder, manifests within the environment as ethnic isolation and national disconnect. An ever-widening chasm between the sense of selfhood and perception of place, the African-American, by all measures, wholly embodies 'border within border'. A status, which, as Du Bois (1903) so elegantly posited, is to be '*outcast and a stranger*' *in one's* '*own house*' [ibid]. In other words, active division simultaneously occurs on both the micro and macro scales within one's being and one's larger society, resulting in a schism between race and country.

Instinctively, one probes cause to effect: key factors perpetuating the African-American dual-life—as well as their lasting ramifications physically, culturally and psychosomatically when the 'house' of the United States refuses to accept the Black native as its own. While numerous boundaries exist to sustain the existential crisis of identity versus nationality for Blacks in America, the primary focus posited is on three significant borders: social, economic and policy. Research into social borders will include the investigation of race as a social construct, a comparison of social capital between Black and White spheres and the societal view of the African-American in forming the cultural basis for how Blacks self-identify—as well as how others view them. Economic borders will examine disparities in Black asset-building contrary to massive debt accumulation, the significant racial wealth gap in the United States and successful Black economic endeavours in American history and their ultimate fates due to blatant discrimination, terrorism and White supremacy. The exploration of policy borders will delve into the nationwide practice of redlining, the legal segregation of Black and White communities, and government-issued legislation, as well as government-endorsed private mandates set in place to obstruct Blacks' equal standing to White Americans. Finally: After each border analysis, damages' specific perpetuation of 'double-consciousness', utilising Erikson's *Stages of Psychosocial Development* (Cherry 2019), will be reviewed. Living borders embodying the rift between ethnicity and nationality within the United States, Black Americans—across social, economic and policy realms—remain perpetual victims of identity crisis and social schizophrenia so long as the state of their clashing inner selves remains unresolved.

7.2 Social Borders: Blackness in America

Race is a social construct, little more than nationally imposed 'fiction' employed by a dominant group in power to subject dissimilar groups deemed 'lesser' under their authority (Jones 2015: 1). This proves disproportionately prevalent within the hyper-racial confines of the United States: Whites'—the indisputable dominant race in the country—massive appropriation of Coloured races' lands, bodies and resources along with the absorption of similarly hued Cultures is all but the American modus operandi. The established concept of '*race* before all' is so pervasive within the United States that one honestly cannot know its totality until one resides outside its jurisdiction—or comes from the outside in. This social paradigm not merely affects Black natives—but also dark immigrants of African descent who naturalise to American citizens. Chimamanda Ngozi Adichie asserted upon first experiencing the severity of racial lines: 'First of all, I wasn't black until I came to America. I became black in America. [...] You have to check a box. You have to be something. And, I came here and very quickly realised to Americans I was just black' (Reese 2018: 6). Indeed, there is an ominous force, frightful coercion within the United States for identity to be made definitive—commonly under the gauge of race. The scale of a nation weighs a requisite to be bound to a hierarchical category.

So complete is the power of White in America that 'American', once, was their privilege alone. The Naturalization Act of 1790—the first statute passed in the United States concerning immigration—restricted US citizenship to free Whites while remaining voiceless on all other races born on American soil. Residing in the nation for 2 years, demonstrating good moral character and swearing an oath of allegiance to the U.S. Constitution are the three requirements for achieving such status (Statues and Stories 2018). Enslaved Black people, considered property and less than human, were strategically denied access to citizenship until they were legally and militarily freed in 1865. Consequently, the racial complication of defining *who* was 'White' followed—American rights meted out by vague notions of ancestry called into question. Replicating the common occurrence of preserving America's racial divides, White lawmakers saw it imperative to craft a clear distinction. They did so in 1921— notably after other races legally claimed liberties equivalent to their White counterparts.

> A White person has been held to include an Armenian born in Asiatic Turkey, a person of but one-sixteenth Indian blood, and a Syrian, but not to include Afghans, American Indians, Chinese, Filipinos, Hawaiians, Hindus, Japanese, Koreans, negroes; nor does white person include a person having one fourth of African blood, a person in whom Malay blood predominates, a person whose father was a German and whose mother was a Japanese, a person whose father was a white Canadian and whose mother was an Indian woman, or a person whose mother was a Chinese and whose father was the son of a Portuguese father and a Chinese mother. (Jones 2015: 2–3)

The invention of race took on societal—then political—significance. The concepts of 'White' and 'Black' in America, and on a grander scale 'White' and 'everyone *else*', were conceived in Virginia before spreading to the nation, effectively categorising ethnic differences and instilling entitlement based on skin colour—alongside the belief of racial inferiority. White Americans were allowed significantly more privileges of law than free Blacks, who—before the institution of slavery—were not categorically prohibited from claiming the same. These biases were bolstered by popular conjecture of the day, the likes of Samuel Morton, the '*father of scientific racism*', who divided humans into five primary '*races*', with Blacks occupying the lowest stratum (Kolbert 2018). The needed salve to ease America's schizophrenia; first to avow 'all men created equal' while simultaneously championing Black enslavement and exploitation. Adichie (2008: 43) ascribes the fallacy of racism and Black inferiority as the sought 'biological and social reality' to justify—and cultivate—slavery and Race is becoming the indelible implement to mould the legacy of Black subjugation in the United States.

No scientific basis for this social construct exists—despite the global adoption of race-based boundaries. According to the Human Genome Project (2003), in terms of biological traits, humans are 99% genetically identical; the scientific community maintains a consensus on the only existing 'race': 'Human'. We all originate from a single continent: Africa (Angier 2000). Instead, the carefully fabricated conception of race in America became a central pillar in the bid for White supremacy, leading to the establishment of a systemic framework for racial inequality as a whole and forever compartmentalising the country's social fabric by becoming its main perpetrator concerning disparities in African-American housing, income, education and more for centuries to come.

7.2.1 Social Capital: Black Vs. White

While broad in definition and scope, social capital pertains to relationships formed within social groups and community networks and the influence yielded by such connections in terms of economic, political or emotional support within any particular society's institutions (Brondolo et al. 2012; Yosso 2005). This also includes access to personal resources and opportunities from established social networks. Within the United States, like most social conventions, social capital remains bound by the nucleus of race—Black social capital utterly divergent from White—most noticeably in the degree of influence. White Americans, unhindered by race's impediments as the predominant group in the United States, naturally possess more significant access to betterment opportunities and extensive resource networks. Another key difference lies in *why* social capital is utilised. White utilisation of social capital, while certainly employed as a support system for its members, is already sanctioned by American society; conversely, Black social capital was engineered as a means of survival and potential foothold to prosperity within a country partitioned by segregation, racism and prejudice.

Racism primarily affects the provision of social capital by limiting social interactions between racial groups and restricting the selection of resources to a single race. Whites, already possessing the highest societal influence in the United States, allocate capital to other privileged Whites; Blacks, then, are given little opportunity to engage outside their communal networks when so often isolated by race, class or income. Brondolo et al. (2012: 372) surmise 'interracial interactions critical for access to economic and social resources and political power in the United States'. Specifically, cultural, institutional and internalised racism—individually and collaboratively—hinder allocation, induction and *reach* of social capital while marring affected groups' public and physical wellness (Brondolo et al. 2011). Cultural racism about drivers affecting how Blacks are perceived within American society via stereotypical portrayal in mainstream media; institutional racism, legislation and policies—and their legacies—deliberately set in place to regress or stall Black progress; and internalised racism as the self-fulfilling prophecy and bodily fulfilment of US society's ubiquitous views regarding Black inferiority.

While African-Americans have historically demonstrated tremendous resourcefulness as compulsory inhabitants of a nation perpetually poised to limit them, differences between White and Black social capital networks are apparent. On average, Blacks possess smaller, intimate social circles with more abundant familial ties than Whites; the latter, however, are more likely to be better educated and have more extensive communal networks (Ajrouch et al. 2001). While it is hardly surprising Blacks temper their relations given the enduring hostility in America, it proves logical Black social networks typically comprise those known and above suspect. Ajrouch, Antonucci and Janevic [ibid] infer lifetime upon lifetime of generational hostility, and discrimination determines who within Black social networks is trusted and granted membership. However, while more intimate circles are likely far more supportive, invested and demonstrative than Whites' expansive social spheres,

Black networks lack the diversity needed to obtain advancement resources outside immediate networks (Granovetter 1973). In other words, accessing new and consistent opportunities and resources becomes more complicated when all members within linked social circuits know the same people.

7.2.2 *Societal View of Blackness: How Blacks Self-Identify; How Others View Them*

Cultural racism remains a considerable determinant in how Blacks self-identify and how other racial groups perceive them. Often stereotypical and exaggerated media portrayals being the sole informant to the American public on differing ethnic spheres, manufacturing societal outlooks on Black Americans as a whole—such as the manipulative, gold-digging Black female '*Jezebel*' and the hypersexualised, menacing Black male '*Brute*' (Burrell 2010). Prejudice projected through a societal lens, cultural racism influences 'attitudes and behaviours toward others through various social-cognitive processes, including priming, modelling, schema creation and social tuning' (Brondolo et al. 2012: 373). Stereotypes, then, of minority racial groups—specifically those of Blacks—showcased in media are subconsciously and consciously adopted as ethnic avatars and perceived as the norm, if not wholly representative of the depicted racial group. This proves incredibly potent when the lack of interracial interaction between ethnicities causes different races' social networks to be estranged and homogenous, providing no significant real-life parallel to combat societal interpretations. Evidence exists that when society views mass media's representation of race outside their designated mainstream role, viewers become dismissive and agitated at the shift from conventional stereotypes (Mastro and Kopacz 2006).

The societal caricatures of race broadcasted in media not only distort racial perceptions of Blacks in America but also give undue merit to judgement of ethnic groups—and if they are deserving of total social inclusion (Dovidio 2009). Viewers begin to believe what they see on screens, allocating mass media as the '*ultimate truth*' in depicting social reality—even when aware that what they are viewing is scripted and unreal. Life imitating art and art informing life, America's media and its actors merely continue to mirror what is prescribed and normalised for each race.

In 2015, CNN and Kaiser Family Foundation distributed surveys to close to 2000 adults in an attempt to identify Americans' perception of race; one of its questions involving race portrayal and accuracy in media:

> When people of your race or ethnic background are featured in entertainment programming on TV and in the movies, how often do you think they are portrayed accurately? Most of the time, some of the time, rarely, or never? (CNN and Kaiser Family Foundation 2015: 24)

For entertainment programming, White American respondents replied 37% 'most of the time', 44% 'some of the time', 15% 'rarely' and 5% 'never'; while Black

American respondents replied 22% 'most of the time', 49% 'some of the time', 22% 'rarely' and 4% 'never'—these numbers 38%, 45%, 11%, 4% and 23%, 45%, 23%, 8%, respectively, for news portrayals [ibid] (see Table 7.1). Other studies have found that American newscasts present Whites far more than Blacks as victims of crimes compared to the actual rates either are victimised (Dixon et al. 2003). There is also the frequent demonisation of Black perpetrators in media compared to White—often granted sympathy and leniency despite committed crimes. Acceptance of misrepresented racial roles is not limited to outside races; Blacks themselves knowingly and unknowingly personify or deify what society's mirror reflects them—falling into the role of 'mythmakers' while sustaining historically contrived conventions of the Black persona (Burrell 2010: 100).

Internalised racism is the insidious inward acceptance of societal notions negatively defining any one racial group; majority opinion so pervasive one cannot help but concede it as truth. This internal prejudice is not consistently recognised within the individual, yet generally manifests as self-rejection—including distortion of self to better fit societal paradigms, imitation of socially dominant races or rejection of personal culture (Brondolo et al. 2012). Given the far-reaching effects of American media's dictate on race, brainwashed Blacks helplessly or willingly emulate—if not ultimately *believe*—how their country defines them, a permeative undercurrent of inferiority infecting the African-American psyche. A harrowing example lies in *The Clark Doll Experiment*, performed by social psychologists Mamie and Kenneth Clark in the 1930s and 1940s, attempting to prove that 'separate but equal' segregation of American schools negatively affected Black children's psychological and emotional health. Identical Black and White dolls were provided to Black youths who were subsequently asked questions ranging from individual doll attractiveness to which doll they would rather play with—overwhelmingly resulting in a preference for the White doll over the Black (Suggs 2019). Finally, when asked which doll they looked like, some children became emotionally distraught or violent when having to identify with what they rejected, proving societal racial standards of beauty and worth impacted even those who were not fully able to understand racism's complexities.

Table 7.1 Resume of CNN and Kaiser Family Foundation survey

'When people of your race or ethnic background are featured in entertainment programming on TV and in the movies, how often do you think they are portrayed accurately?'

	Entertainment programming		News portrayals	
	White American (%)	Black American (%)	White American (%)	Black American (%)
Most of the time	37	22	38	23
Some of the time	44	49	45	45
Rarely	15	22	11	23
Never	5	4	4	8

7.2.3 *Damages: Identity Vs. Role Confusion*—'Who Am I?'

Ego identity, a crucial element of Erik Erikson's psychosocial stage theory, is defined as the regard of self one continuously constructs through social interaction; it remains in flux from daily new encounters and social data received from various sources of engagement (Cherry 2019). Eight stages of ego comprise Erikson's theory—each associated with a corresponding point in lifetime development: Erikson believing each stage presents an inevitable conflict that, if surmounted, would result in growth or regression in failure. *Identity vs. Role Confusion*—apt given the state of African-American ego within the US societal prism—is linked with the stage of adolescence. The conflict: Successful development of personal identity. If left unresolved, fractured identity evolves into role confusion. Instead of internal cohesion of self and role, an endless lack of inner surety becomes standard—along with the constant questioning of place in a world that does not allow one to be anything more than its stereotype. This results in disappointment, insecurity and confusion about adequate role in society. With the United States' assault of societal undertones systematically presenting Blacks as 'lesser'—severing a sense of identity beneath the dictates of nationality—the institutionalised stigma of 'Blackness' is unable to mitigate the White-manufactured concept of 'American'.

Other damages due to exposure to racism and its corresponding stressors are the development of depression and other destructive behaviours. These symptoms are vital perpetrators in the depletion of energy needed for interracial engagement, left unaddressed, hindering the ability to cultivate healthy social relationships, and further exacerbating psychosomatic effects due to stress exposure. Both interpersonal and internalised racism have also been associated with the habit-forming practice of cigarette smoking (Brondolo et al. 2012).

7.3 Economic Borders: A Gaping Divide

A prevalent notion in America permeates the country's core: The vaunted conviction that if any American works hard enough and is dedicated enough, they can achieve financial prosperity. Indeed, this hypothesis is the principal suture binding the American Dream: A 'dream' made inaccessible to Blacks attempting to close the rift in racial wealth manufactured by centuries of discriminatory practices effectively crippling access to economic resources in times of financial emergency. The ability to establish generational wealth, inherited and later utilised, is unaided by 'hard work' like attaining a college degree or even active employment; Black college graduates, on average, possess 33% less wealth (23,400 dollars) than White dropouts of high school (78,300 dollars); even the poorest White families in the United States still possess slightly more wealth than middle-class Black families (Hamilton et al. 2015). Despite these dismal disparities, approximately 87% of African-Americans—assured education the 'great equaliser' and lacking other viable options in generating adequate wealth—are forced to take out loans for community, public *and* private colleges without a generational financial foundation to rely on; Black graduates accumulating 7400 dollars more in debt than Whites (Fay 2018). Black student borrowers also overwhelmingly default on their loans (nearly half compared to 20% of Whites) and generally owe more than the original amount borrowed as much as 12 years after attending college, regardless of graduating with a degree or dropping out (Miller 2017).

Conversely, White students owe substantially less, settling, on average, 35% of loan debt simultaneously [ibid]. Even after surpassing all hardships to obtain a degree, Blacks still generate less wealth via education than Whites with equivalent schooling (4846 vs. 55,869 dollars, respectively) (Hamilton et al. 2015)—racially charged financial pitfalls as these plunging Black Americans further into the cycle of chronic economic insecurity.

Even for those fortunate enough to gain employment after enrolment in an institution, Black employees earn $20,000 less a year than White ($50,400); Blacks only see a $4.80 wealth increase for every White dollar earned ($19.51) [ibid]. Further, racial discrimination in US hiring practices has not significantly altered or improved since 1989; White Americans still receive 36% more callbacks than Black job seekers with identical qualifications (Quillian et al. 2017). This quite contrary to the general optimistic perception of abating racism and skin-based prejudice in a rehabilitated 'post-racial' America. Discrimination in hiring practices, while unchanged in consistency, has adapted in method—biases now less explicit yet more insidious in execution. Notably, Quillian et al. (2017), utilising studies employing race-neutral procedures in applying for positions, found stability in hiring discrimination against Blacks in their 26-year analysis period (1989–2015)—instead of decline. Granted, while such studies can only trace racism at points of hire consideration and hiring offer—not during employment or compensation consideration—initial racial discrimination remains significant, as these points are the gateway to economic welfare and stability. More so when the unemployment rate for Blacks is 8.6%, almost double that of White workers (4.4%) (Asante-Muhammad et al. 2016).

7.3.1 The Ever-Growing Gap in Racial Wealth

For this account, racial discrepancies in wealth will be analysed instead of racial gaps in income. While substantial divergences in income between US races exist [median weekly earnings of Blacks 737 vs. 935 dollars for Whites (Bureau of Labor Statistics 2019)], income is short-term, mainly with a focus on day-to-day necessities. Wealth speaks to substantive economic autonomy and overall monetary well-being for racial groups. Presently, in the United States, inequality between households possessing wealth and households subsisting in poverty is so considerable that a mere 20% of Americans hold 99.4% of US wealth, while the remaining 80% are left a mere 0.6% to divvy amongst themselves; the nation's richest collective 2.34 trillion dollars net worth being *more* than the entire wealth of the Black population (Asante-Muhammad et al. 2016). By race: African-American households possess *seven times* less wealth (85,000 dollars) than White households (656,000 dollars); wealth accumulation analysis from the years 1983 to 2013 shows average White wealth increasing 84%—three times the rate of growth for Black families who only saw their wealth increase by 750 dollars [ibid]. These staggering racial and economic divides did not occur by happenstance: the acute and ever-growing chasm between Black and White wealth was human-induced, constructed by centuries of US policies—past and present—deliberately or casually obstructing Black ability to accumulate significant wealth. Subsequently, Blacks face double the rate of financial insecurity of Whites: over two-thirds of Black households (67%) lack proper financial stopgaps in the event of economic crisis, unable to persist at the US poverty level for even 3 months [ibid].

The Great Recession, occurring from 2007 to 2010, not only further exacerbated America's gaping racial wealth divide but overwhelmingly harmed minorities already assailed by economic disadvantage—Black wealth essentially reset with the nullification of three decades' worth of financial growth. While all races were affected, the median wealth for Whites never dipped beneath 1983 levels (102,300 dollars), while Black households possessing 6800 dollars median wealth in 1983 plunged to 1700 dollars in 2013—Black families losing four times more wealth than the average White household (Asante-Muhammad et al. 2017). Homeownership, one of Americans' most significant sources of wealth generation, was only realised by 41% of Blacks compared to 71% of Whites [ibid]. As a whole, the racial wealth divide—even if Black Americans were historically able to garner wealth on the same level as White—would still exist: a theoretical 123,000 dollars (38,000 more than actual) at a total of 560,000 dollars still eclipsed by White wealth at 656,000 [ibid]. Regarding retirement, Blacks—if afforded access—possess over 100,000 dollars less in savings than Whites (19,049 vs.130,472 dollars) [ibid].

It is estimated in the year 2043, Americans of Colour will outnumber White Americans, meaning the economic fates of African-American households will have a tremendous impact on America's economy—the United States effectively reaching 'a racial and economic apartheid state' lacking direct intervention (Asante-Muhammed et al. 2017: 5). If the current trend remains uninterrupted, America's median household wealth (excluding durable goods and adjusted for inflation) will decline to 52,300 dollars—32% less than the median in 1983 (Asante-Muhammad et al. 2017). All the more sobering, the US rift between Black and White wealth is so staggeringly vast that it will take at least 228 years for Black Americans to reach the level Whites hold *today*, a mere 17-year difference of the span of slavery (Asante-Muhammad et al. 2016). Breaching a certain point, the number of years will be made irrelevant; the racial wealth gap will never close.

7.3.2 Historical African-American Economic Successes and Their Fates

Disparagingly coined 'Little Africa' by neighbouring Whites, the affluent Black community of Greenwood in Tulsa, Oklahoma, infamously known today as *Black Wall Street*, boasted a self-sufficient Black economy with nearly 10,000 calling it home—before suffering an unspeakable act of racial terrorism in 1921. Claimed the worst civil disturbance since the American Civil War, from 31 May to the afternoon of 1 June, 1256 Black homes were methodically looted and then burned to the ground; over 35 square blocks of Black-owned schools, hospitals, churches, libraries and businesses were decimated while 100–300 Blacks were killed. Commonly referred to as a 'race riot'—a term offering the impression *both* groups are to Blame—Black Tulsans fought bravely to defend their homes as White invaders, both civilians and officials, besieged them with arms and planes to bomb the town from the air. To this day, not one of the criminal acts committed—from outright murder to flagrant arson—has been prosecuted or punished by any level of the US government; Oklahoma's National Guard, finally arriving on the scene, detaining 6000 Black residents, instead (Oklahoma Commission 2001). The razing of Greenwood—and its subsequent reconstruction—left entirely to its homeless Black survivors, literal refugees within their own country: the decimation of Black homes so thorough residents were forced to live in tents. In its end, four million dollars was filed in damages by affected Blacks, with only two minimal claims approved by the city commission 'for guns and ammunition taken during the racial disturbance of June 1st' [ibid: viii].

While one of the more well-publicised acts of racial terror enacted within the United States against prosperous Black-predominant communities, Tulsa's *Black Wall Street* is far from being the sole example of race-driven genocide. Similar acts of racial violence occurred in the Atlanta riots (1906), the Red Summer riots (1919) and in other Black economic meccas, such as the Black business district in Little Rock, Arkansas (1927). Post-World War I America was abundant with White veterans threatened by Black veterans they perceived as stealing their livelihoods. Today, while outright racial terror still exists, yet is debatably less overt, Blacks find themselves still beleaguered by legacies of stolen economic progress. It is not only economically more challenging for Blacks to purchase their businesses but also—even when able to invest—these businesses possess significantly lower value than those purchased by Whites; on average, White-owned businesses worth *eight* times more than Black (Asante-Muhammad et al. 2016).

7.3.3 *Damages: Generativity Vs. Stagnation*—'How Can I Contribute to the World?'

The seventh stage of Erikson's psychosocial development theory, *Generativity vs. Stagnation*, deals with the individual desire to contribute to the greater society and commit acts that benefit future generations—associated with the life stage of middle adulthood, those who complete this phase feel they have positively contributed to the world around them—and the people within it. However, stagnation results from the determination to add to society when continuously thwarted (Verywell Mind 2019). 'Stagnation', evolving into *existence*, the Fate of many African-Americans beset by generational economic dearths and the absence of financial foundations. Despondency sets further in when Black natives look to their immediate environment and find an inordinate amount of payday loan centres, liquor stores and sprawling food deserts intended to keep them fiscally trapped where they are (Charron-Chénier 2018). Hope for change is dismal when the most alarming damage to Blacks is economic *regression*—until utter loss. Median Black household wealth is projected to reach zero, with current racial wealth trends, by 2053.

7.4 Policy Borders: The 'Divided' States of America

The institution of redlining, the systemic credit evaluation process grading communities based on location, race and economic standing, in America began with the Home Owners Loan Corporation (HOLC) in 1933; examiners charged with determining 'at-risk' areas resulting in residential security maps of hundreds of US cities (Mitchell and Franco 2018). These colour-coded maps utilise green ('best'), blue ('still desirable'), yellow ('definitely declining') and red ('hazardous'), determining lending viability—and the direct result of enduring racial segregation in American communities to this day. A persistent pattern emerging of race-based residential and economic isolation coinciding to HOLC maps drafted more than 80 years ago: 91% of green-coded 'best' areas middle-to-upper income (MUI), yet 74% low-to-moderate income (LMI) families reside in redlined neighbourhoods; 85% green and 71% blue-designated areas still majority White, while 63% of red zones remain minority dominant [ibid].

Established as a New Deal programme to aid the average American in affordable housing—along with the Federal Housing Administration (FHA) in 1934—innovations such as amortised loans and government-backed mortgages were wholly restricted to Whites via race-based approvals. Redlining practices effectively bar Blacks from accessing the legitimate home-mortgage market while eliminating investment options wherever Blacks lived (Coates 2014). Interest rates on conventional loans offered in residential zones deemed 'hazardous' were considerably inflated for minority residents; Blacks, in particular, were left victim to predatory lending practices. Sellers offer houses at exorbitant sums only to evict Black families unable to meet payments, subsequently repossessing recouped homes and beginning the vicious cycle anew [ibid]. Blacks, effectively snared within HOLC-mandated red zones, were denied equal financing and residence anywhere else.

7.4.1 A Tale of Two Countries: One Black, One White

In a report to the US President, the National Advisory Commission on Civil Disorders (1968) gravely surmised the nation was 'moving toward two societies, one black, one white—separate and unequal' (1). This conclusion formed amid violent civil disorders concentrated within Black-inhabited ghettos crippled by White condoned segregation and poverty. The monopolistic FHA, embracing HOLC's racially inhibitive redlining maps, favouring newly constructed racially 'homogenous and harmonious' neighbourhoods (FHA 1938: 121)—primarily White. These are birthing the US middle class while simultaneously depriving Blacks of the benefits of home-ownership and home equity growth. Societal pressures and FHA regulations endorsing racial exclusion exacerbated 'White flight' to affordable suburbs, leaving Blacks allowed only in overcrowded inner cities languishing behind. Racial segregation was widespread despite no evidence that Black occupation of homes decreased overall house values (Rothstein 2017). Conceived in the late nineteenth and early twentieth centuries, the 'Black urban ghetto' was a direct response to the Great Migration of southern Blacks to the north from 1915 to 1970; Whites spatially segregating Blacks fleeing racial terror and seeking economic advancement to socially, economically and politically immobilise them (Massey 2015: 2). This deliberate division of a group already disadvantaged by racial inequality contributes to the concentration of Black poverty while amplifying its effects (Massey and Denton 1993). Planned 'hypersegregation' (Mitchell and Franco 2018: 4) unevenly distributing race in American cities and actively eliminating cohabitation between Blacks and Whites.

Today, a mere 10% of metropolitan areas in the United States are the Black majority; homes in neighbourhoods 50% Black or more experiencing value cuts roughly half that of homes with no Black residents (Perry et al. 2018). This occurs even with homes of equitable amenities: houses in majority Black communities worth 23% less than White; differences in home and neighbourhood quality unable to adequately justify devaluation adequately—leaving bias distorting views of Black-owned assets as the sole explanation [ibid]. Disparities also exist in the prevalence of food deserts within minority-concentrated cities: Blacks are half as likely to have sufficient access to supermarkets, yet far more likely to live near fast food chains or smaller convenience stores offering overpriced, lower quality stock (ACLU and New York Law School 2012).

7.4.2 'American-Ish': Separate and Not Equal

With the construct of race and racial inferiority flourishing in the United States, policies used to enforce segregation and reinforce Whites as the 'dominant race' were commonplace. From 1881 to 1964, Jim Crow laws, penalising every facet of interracial contact, divided Americans in 26 states. Seeping into housing, segregationist state guidelines forbidding Whites and Blacks from dwelling together—struck down in Federal court in 1917—were swiftly replaced with private racial covenants employed by suburban associations to curb Black occupation of homes in White neighbourhoods. Discrimination in the sale and rental of housing is openly practised locally and enforceable by law (Kaplan and Valls 2007). Initially enforced by violence and intimidation, the more subvert—and successful—utilisation of restrictive covenants effectively kept Blacks at bay as majorities of White homeowners signed binding contracts prohibiting sale to outside races; violators sued for breach [ibid].

Jim Crow segregation also permeated the real estate industry: The National Association of Real Estate Brokers 1924 officially adopted the prohibition of new races into established White neighbourhoods in their code of ethics, remaining in effect until 1950 (Helper 1969). Migrating Blacks—restricted to inner cities—caused Black populations to swell; Whites threatened by rapid Black growth and expansion sanctioning colossal public housing projects and rampant urban renewal—necessitating more public housing for later displaced Black residents (Massey 2015). One example of 'urban renewal'—following the Federal Aid Highway Act of 1956—highways slicing through predominantly Black communities considered crime-ridden, impoverished and dirty 'slums' (Sherman 2014). The FHA, the sole affordable mortgage lending option for Americans then, did not provide loans to urban areas where Blacks resided. Over 80% of newly constructed suburbs built between the 1930s and the 1940s contain racial covenants prohibiting Black occupation in White areas across the nation (Massey 2015). The VA loan programme, outlined within the Serviceman's Readjustment Act of 1944, was racially neutral and—while in theory, accessible to all US veterans—included provisions limiting Black benefits; the most hampering: federally funded GI Bills that were state-enforced. Discouraged by local Whiteboards, this significantly bridled Black applications for loans; boards assumed FHA-like racially restricted regulations and all but excluded Black neighbourhoods for subsidised credit (Katznelson 2005). Beleaguered by three decades of racist policies (1930–1960), impotent acts and watered down amendments doing little to impede mortgage discrimination, African-Americans overwhelmingly bear the crippling costs of exclusion from government-assisted home acquisition and property-generated wealth; the pricing out of many Blacks in the housing market directly correlated to home-ownership inequality in the United States. Lasting implications of race-based residential segregation also including major disparities in access to quality education, financial institutions and other advantageous neighbourhood assets (Perry et al. 2018).

7.4.3 Damages: Trust Vs. Mistrust—'Can I Trust the People Around Me?'

The first stage of Erikson's psychosocial development theory—echoing the infant hope African-Americans still bear for a nation that continuously fails them—*Trust vs. Mistrust* is linked to the life stage of birth to infancy (18 months). This is considered the most crucial stage with the inception of individual world-view, personality and societal trust. Individuals learn who and what they can trust in the world around them—their outlook is entirely informed by the care and support offered by parents and adult guardians (Cherry 2019). If trust is insufficient, one is stricken with mistrust and a pervading belief the world is dangerous and unpredictable. America, placed in the role of a parental figure to its African-American offspring—utterly dependent on the nation's equal and liberal care—eschewing its sacred duty, perverting inherent instinct and suffocating hope.

Erikson believed the initial trust relationship between parent and child influenced the future ability to trust relationships throughout one's life [ibid]. Unable to fully trust a country that habitually marginalises, Blacks carry a crippled sense of national trust: Native, but unable to acclimatise in a land that proves, time and again, Black interests are not its own. Thus, the Black self becomes a constant adaptation to racial discrimination, personality reflecting societal instability.

7.5 Conclusion

What *is* 'African-American' but a state of constant rejection? Amongst America's greatest ironies is people of African descent, born on US soil, are undoubtedly the most '*American*' of all welcomed racial groups, occupying the land since its inception—building it from the ground up into what it is today. No people are more deserving of its creeds—yet no race is more alienated, more conflicted when pondering the levelling of nationality and identity. Du Bois (1903: 3) further went on to write the Black citizen 'simply wishes to make it possible [...] to be both a Negro and an American without being cursed and spit upon by his fellows, without losing the opportunity of self-development'. If African-Americans are cursed with 'double-consciousness'; White Americans, as the dominant race, embody 'double-conscience'—perpetuating a societal standard of Black marginalisation in the United States while pushing freedom and equality in the same breath.

Yet, an ordinary, terrifying thought persists. With time, these social, economic and policy borders will miraculously mend, and America has already achieved racial equality. Unrealistic and unfounded, given the careful construction of policies, mindsets and barriers against Blacks, plans of recourse and remediation require no *less*. What becomes of identity when the Black citizen reaches for 'American'? When *Blackness* is so grafted onto the American psyche to be synonymous with 'inferior'. Presenting a fractured self, an incomplete citizen who knows little more of their worth than the justice of a country they call 'home'. Blacks in America are left to the unmitigated fate of border within the border.

References

ACLU and New York Law School (2012) Unshared bounty. Racial Justice Project. http://www.racialjusticeproject.com/wp-content/uploads/sites/30/2012/06/NYLS-Food-Deserts-Report.pdf. Accessed 10 May 2019

Adichie CN (2008) "Authenticity" and the Biafran experience. Transition 99:42–53

Ajrouch KJ, Antonucci TC, Janevic MR (2001) Social networks among Blacks and Whites: the interaction between race and age. J Gerontol Soc Sci 56B:S112–S118

Angier N (2000) Do races differ? Not really, genes show. The New York Times. https://www.nytimes.com/2000/08/22/science/do-races-differ-not-really-genes-show.html. Accessed 19 Mar 2019

Asante-Muhammad D, Collins C, Hoxie J, Nieves E (2016). The ever-growing gap. Prosperity Now. https://prosperitynow.org/sites/default/files/resources/The_Ever_Growing_Gap-CFED_IPS-Final.pdf. Accessed 3 Apr 2019

Asante-Muhammad D, Collins C, Hoxie J, Nieves E (2017) The road to zero wealth. Prosperity Now. https://prosperitynow.org/sites/default/files/PDFs/road_to_zero_wealth.pdf. Accessed 1 Apr 2019

Brondolo E, Love EE, Pencille M, Schoenthaler A, Ogedegbe G (2011) Racism and hyper-tension: a review of the empirical evidence and implications for clinical practice. Am J Hypertens 24:518–529

Brondolo E, Libretti M, Rivera L, Walsemann KM (2012) Racism and social capital: the implications for social and physical well-being. J Soc Issues 68:358–384

Bureau of Labor Statistics (2019) Usually weekly earnings of wage and salary workers first quarter 2019. U.S. Department of Labor. https://www.bls.gov/news.release/archives/wkyeng_04162019.pdf. Accessed 16 Apr 2019

Burrell T (2010) Brainwashed challenging the myth of Black inferiority. SmileyBooks, United States

Charron-Chénier R (2018) Payday loans and household spending: how access to payday lending shapes the racial consumption gap. Soc Sci Res 76:40–54

Cherry K (2019) Erik Erikson's stages of psychosocial development. Verywell Mind. https://www.verywellmind.com/erik-eriksons-stages-of-psychosocial-development-2795740. Accessed 2 Mar 2019

CNN and Kaiser Family Foundation (2015) Survey of Americans on race. Kaiser Family Foundation archive. http://files.kff.org/attachment/topline-methodology-survey-of-americans-on-race. Accessed 13 Mar 2019

Coates T-N (2014) The case for reparations. The Atlantic. https://www.theatlantic.com/magazine/archive/2014/06/the-case-for-reparations/361631/. Accessed 14 Mar 2019

Dixon T, Azocar CL, Casas M (2003) The portrayal of race and crime on television network news. J Broadcast Electron Media 47:498–523

Dovidio JF (2009) Racial bias: unspoken but heard. Science 326:1641–1642

Du Bois WEB (1903) The souls of Black folk. University Press John Wilson and Son, Cambridge

Fay B (2018) Demographics of debt. Debt.org. https://www.debt.org/faqs/americans-in-debt/demographics/. Accessed 21 May 2019

FHA—Federal Housing Administration (1938) Underwriting manual. U.S. Government. https://www.huduser.gov/portal/sites/default/files/pdf/Federal-Housing-Administration-Underwriting-Manual.pdf. Accessed 23 May 2019

Granovetter MS (1973) The strength of weak ties. Am J Sociol 78:1360–1380

Hamilton D, Darity W Jr, Price AE, Sridharan V, Tippett R (2015) Umbrellas don't make it rain: why studying and working hard isn't enough for Black Americans. https://www.researchgate.net/publication/293486132_Umbrellas_Don't_Make_It_Rain_Why_Studying_and_Working_Hard_Isn't_Enough_for_Black_Americans. Accessed 29 Mar 2019

Helper R (1969) Racial policies and practices of real estate brokers. University of Minnesota Press, Minneapolis

Jones B (2015) The social construction of race. Jacobin magazine. https://www.jacobinmag.com/2015/06/racecraft-racism-social-origins-reparations/. Accessed 22 Mar 2019

Kaplan J, Valls A (2007) Housing discrimination as a basis for Black reparations. Public Aff Q 21:255–273

Katznelson I (2005) When affirmative action was White: an untold history of racial inequality in twentieth century America. W. W. Norton, New York

Kolbert E (2018) There's no scientific bias for race—it's a made-up label. Natl Geogr. https://www.nationalgeographic.com/magazine/2018/04/race-genetics-science-africa/. Accessed 19 Mar 2019

Massey DS (2015) The legacy of the 1968 Fair Housing Act. Social Forum (Randolph N J) 30:571–588

Massey DS, Denton NA (1993) American apartheid: segregation and the making of the underclass. Harvard University Press, Cambridge

Mastro DE, Kopacz M (2006) Media representations of race, prototypicality, and policy reasoning: an application of self-categorisation theory. J Broadcast Electron Media 50:305–322

Miller B (2017) New federal data show a student loan crisis for African American borrowers. Center for American Progress. https://www.americanprogress.org/issues/education-postsecondary/news/2017/10/16/440711/new-federal-data-show-student-loan-crisis-african-american-borrowers/. Accessed 5 Apr 2019

Mitchell B, Franco J (2018) HOLC "redlining" maps: the persistent structure of segregation and economic inequality. NCRC. https://ncrc.org/wp-content/uploads/dlm_uploads/2018/02/NCRC-Research-HOLC-10.pdf. Accessed 3 Apr 2019

National Human Genome Research Institute (2003) The Human Genome Project. NIH. https://www.genome.gov/human-genome-project. Accessed 19 Mar 2019

Oklahoma Commission (2001) Tulsa race riot. OkHistory. https://www.okhistory.org/research/forms/freport.pdf. Accessed 14 Apr 2019

Perry A, Rothwell J, Harshbarger D (2018) The devaluation of assests in Black neighborhoods. Brookings. https://www.brookings.edu/wp-content/uploads/2018/11/2018.11_Brookings-Metro_Devaluation-Assets-Black-Neighborhoods_final.pdf. Accessed 21 May 2019

Quillian L, Pager D, Hexel O, Midtbøen AH (2017) Meta-analysis of field experiments shows no change in racial discrimination in hiring over time. PNAS 114:10870–10875

Reese H (2018) Chimamanda Ngozi Adichie: I became Black in America. https://daily.jstor.org/chimamanda-ngozi-adichie-i-became-black-in-america/. Accessed 18 Mar 2019

Rothstein R (2017) The Color of Law: a forgotten history of how our government segregated America. Liveright Publishing Corporation, New York

Sherman BP (2014) Racial bias and interstate highway planning: a mixed methods approach. University of Pennsylvania. http://repository.upenn.edu/curej/176. Accessed 3 May 2019

Statues and Stories (2018) Naturalization Act of 1790—First Immigration Act by the first Congress. Statues and Stories. http://www.statutesandstories.com/blog_html/naturalization-act-of-1790-first-immigration-act-by-the-first-congress/. Accessed 18 Mar 2019

Suggs E (2019) Mamie and Kenneth Clark used toys in a test that helped to overturn U.S. school segregation law. AJC. https://www.ajc.com/news/mamie-and-kenneth-clark-doll-test-challenged-attitudes-segregation/gGknWIaYpKa1Yh9Oqs6hjL/. Accessed 21 Mar 2019

U.S. Kerner Commission (1968) Report of the National Advisory Commission on civil disorders. In: Report of the National Advisory Commission on civil disorders. Bantam Books, New York, pp 1–29

Yosso TJ (2005) Whose culture has capital? A critical race theory discussion of community cultural wealth. Race Ethn Educ 8:69–91

Chapter 8
Reproduction and Transformation of Chinatown and 'Chinese Kampung' in Postcolonial Indonesia: A Case Study of Semarang

Kezia Eka Sari Dewi

Abstract Prior to and during the colonial era, Indonesia became a popular destination for Chinese migrants. Most of them became intermediaries in trade between local Indonesians and European colonialists. The Dutch colonialists gave the Chinese a place in the city as part of their 'divide and conquer' plan. They prohibited Chinese residents from residing in kampungs (indigenous settlements) with native Indonesians. As commercial districts, these Chinatowns expanded and formed a significant component of many Indonesian cities and towns. The Dutch altered the policy at the beginning of the twentieth century. Ethnic Chinese in the city gradually spread to the kampung nearby. The main street was home to wealthy Chinese merchants, while the low-income or newly arrived Chinese migrants who worked for them stayed inside the kampung. Chinatown remained an important commercial district even after independence. However, during the 32 years of the dictatorship (1966–1998), policies targeting Chinese culture and the Chinese Indonesian population, along with political and economic factors, shaped and altered the urban fabric of the former Chinatown. Following 1998, the new Indonesian government abolished anti-Chinese laws and promoted Chinatown's preservation. This research investigates how the economic dynamics and policies implemented following independence influenced and changed the (former) Chinese settlement within the city. Through observation, documentation and mapping of the building, neighbourhood and population exchange transformations, this chapter attempts to explain how Chinese settlements in post-colonial Indonesia reproduce their position within the city with a multilayered cultural, political and economic aspect.

Keywords Colonial legacies · Urban identity · Borderscapes · Indonesia · Heritage · Chinatown · Tionghoa settlements

---

K. E. S. Dewi (✉)
Urban Design, Urbanism, Landscape and Planning Unit, KU Leuven, Leuven, Belgium
e-mail: Keziaekasari.dewi@kuleuven.be

G. Cimadomo et al. (eds.), *Urbanism at Borders*, The Urban Book Series,
https://doi.org/10.1007/978-3-032-09594-7_8

8.1 Introduction

Indonesia is a country with hundreds of tribes. For ages, Indonesia, rich in natural resources, has drawn many traders from many countries. Before European traders arrived in Indonesia, traders from China, the Arab world and India arrived, traded and settled there. Many performed a new hybrid culture while living and assimilating into Indonesian society. The Indonesian government refers to Indonesians who are Chinese-descended (Chinese - Indonesian) as ‘Tionghoa peoples’. For this reason, the author of this work will also use the word ‘Tionghoa’.

During colonial times, the colonialist government gave these foreign eastern groups a better social status than native Indonesians. These foreign eastern groups were second-class citizens, below the European group and above the native Indonesian group. The colonial government also assigned a place for every ethnic group in the city. For the Tionghoa people, their settlement is Chinatown. Settlements for foreign eastern groups serve as a buffer between European quarters and indigenous settlements (in the case of Semarang, mostly Javanese people / tribe). This is typical of colonial cities, which have tightly packed shophouse areas for the Chinese, Arabs and Indians, spacious low-density ‘compounds’ for the Europeans and indigenous settlements (‘kampung’ in Southeast Asia) scattered around the city (King 1974). The descendants of Chinese migrants are concentrated in urban areas as a result of the policies of the colonial authority (seventeenth century–1945) and after independence, particularly the circumstances during the War of Indonesian Independence (1945–1949). They quickly took control of the industrial and economic activities that fuelled the city during the colonial era. A contributing factor is the perception of the native population as being untrained in trade and too antagonistic to live in such proximity to European settlers (Ford 1993). They dominated the nation’s economic activity and mainly occupied the commercial district.

Owing to ethnic Chinese dominance in the nascent Republic of Indonesia’s economy during the first decade of the country’s independence, Indonesia’s government implemented several laws to limit the economic activities of this ethnic minority. Such measures aimed to reduce economic disparities between the Tionghoa people and other ethnic groups, particularly the native Indonesians. The communist uprising was put down in 1965, and a new dictatorship was installed. The new regimes implemented laws that discriminated against the Tionghoa people even more than the previous one did, allegedly as a result of accusations that ethnic Chinese in Indonesia had been involved in the communist uprising. The new regime forced the Tionghoa people to assimilate into the general population of Indonesia by prohibiting Tionghoa cultural expression, prohibiting the presence of the Chinese alphabet in public spaces, prohibiting Tionghoa from celebrating Chinese festivals and forcing all Tionghoa to change their Chinese names to Indonesian sounds. All of these limitations influenced and changed the old Chinatown’s urban fabric.

Indonesians overthrew the dictatorship in 1998. Since 1998, the current administration has gradually repealed all legislation discriminating against the Tionghoa. Since then, the government and the Tionghoa community have launched many projects to restore Chinatown. However, after 32 years of prohibition, Chinatown's physical environment no longer reflects much Chinese character. Since Chinatown's primary purpose in Indonesian cities is to serve as a business district, the area's urban fabric has also been affected by the pressures of economic activity. This chapter aims to investigate how the post-independence economic dynamics and policies influenced and changed the previous Chinese settlements within the city. Semarang was chosen as a case study. The Chinese settlements in this research include Chinatown and the nearby kampung, home to a sizable Tionghoa population.

As the capital of Central Java Province, Semarang is home to one of Indonesia's largest populations, the Tionghoa people, which is why the research was done there. These days, between 5 and 10% of Semarang's population is Tionghoa. One of Indonesia's most well-known Chinatowns is located in Semarang. The revitalisation of Chinatown in Semarang is likewise seen as a success story. Every year, there are Chinese festivals to celebrate Zheng He, a famous Chinese envoy from fifteenth century among the Chinese diaspora in Southeast Asia. The festivals and the food market in Semarang Chinatown have attracted many tourists to this day. Figure 8.1 shows the area of study.

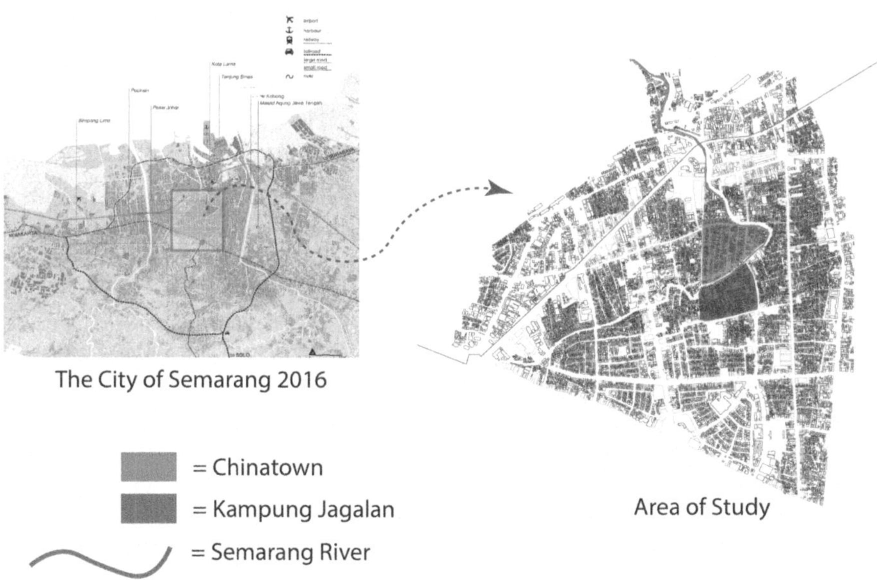

Fig. 8.1 The City of Semarang and the area of study. (Source: Department of City Planning, Sultan Agung University)

8.2 Method of Research

This chapter will give a brief history and policies towards the Tionghoa peoples from colonial times till the present, following the description of the political economy of place in the city. An explanation of Semarang's historical timeline and city planning policies will come next. The author will present a map of the neighbourhood from colonial times and the present in order to address the subject of how economic activity and policy shaped the area of study. It will also describe how the building and the neighbourhood changed in reaction to the laws, the state of the economy and the development of cities.

One issue is that no city maps from 1946 to 2000 are available. The neighbourhood residents and a local university expert were interviewed as part of the building and neighbourhood transformation documentation. These interviews are valuable resources for illustrating how the neighbourhood changed between 1946 and 2000. The elderly residents of the area who have lived there since the 1950s were the subjects of an informal interview.

Because Chinatown has been the primary location for Chinese habitation and commerce since colonial times, it is designated as the first section of the research area. This section has seen significant transformations due to economic activity and governmental restrictions. Due to its proximity to Chinatown, also primarily populated by Tionghoa, Kampung Jagalan was selected as the second section of the research area. This kampung dates back to colonial times as well.

8.3 Political Economy of Place in the City

Cities in all times were hosting the crucial functions of the respective social, economic and technological regimes: defence, organisation of significant public works, administration, astronomy and justice in ancient times; finance, commerce, culture and education afterwards; industry in modern times; headquarter, tertiary and information-intensive functions in the last century; knowledge-intensive, scientific and creative functions nowadays. Cities can be assumed to be collective (economic) actors and are not the simple result of individual actions oriented towards individual advantage (Camagni 2016). To ensure the cities function well, they must develop proper plans, known as development plans, which will govern their growth until circumstances change. Land use planning forms the most crucial part of these plans.

A city is also conceived as the authentic expression of the interests of some land-based elite. Molocht argued that any given parcel of land represents interest and that any given locality is thus an aggregate of land-based interests. That is, each landowner (or person who otherwise has some interest in the prospective use of a given piece of land) has in mind a certain future for that parcel, which is somehow linked to his or her well-being (Molotch 1976). Relative land scarcity in urban environments intensifies competition over land use and ownership (Jones et al. 2014). In the case of underdeveloped countries, empirical evidence shows that most urban residents continue to be excluded from the decision-making processes of urban governance (Obeng-Odoom 2013). Power dynamics between different social groups tend to be reproduced in the participatory sphere, for example, between landlords and tenants (Cornwall and Coelho 2007).

Government policies like land taxation based on the activity in the land parcel could affect the growth chance and even cause land use to change in the future. However, government decisions are not the only social activities that affect local growth chances; decisions made by private corporations also have a significant impact (Molotch 1976). The people who participate with their energies, particularly their fortunes, in local affairs are those who, at least in vast disproportion to their representation in the population, have the most to gain or lose in land-use decisions (Molotch 1976). The increasing competition among the parties who seek profit from land-based interest and the lack of participation from the community and vulnerable groups could lead to segregation and spatial polarisation. In turn, spatial inequalities tend to reinforce social inequalities (Skop 2006). In many post-colonial cities like in Indonesia , the political economy of place determines who benefits from urban growth and who remained excluded. Government planning and private investment often privileged land - based elites, while informal and minority communities must negotiate their survival through everyday spatial practices. The city thus becomes not only a container of economic activities but also a site where power relations and socio-political histories are inscribed into space. This perspective helps explain how Semarang Chinese settlements (Chinatown and surrounding kampungs) have evolved through shifting political - economic regimes.

8.4 A Brief History and Policies Towards the Tionghoa Peoples

Originally, Semarang was a small fishing village. The first Tionghoa settlements in the estuary date back to the fourteenth century. The coastal line was 4 km away from the present coastal line at that time. During the sixteenth century, the Dutch trading company constructed their fortress. This Dutch trading company is known as VOC (Vereenigde Oostindische Compagnie). Semarang then developed into Central Java's main harbour and trading hub. Owing to the conflict between European and Tionghoa traders in the eighteenth century, the Dutch, having vanquished the Tionghoa, designated the area on the Semarang River's riverbank as the Tionghoa habitation site. This Chinatown is a buffer between the Javanese kampungs and the European quarter near the coast. The Tionghoa people have long acted as middlemen in commerce between the natives Indonesian and the European traders. Chinatown is to the east of the traditional market (Liem 2004).

Following the 1830 Javanese (Diponegoro Wars), the colonial authorities enacted the *Wijkenstelsel* Law, which mandated that every ethnic group reside inside its settlement. Mixed settlements had vanished from existence. However, early in the twentieth century, this law was repealed. The Tionghoa people were also prohibited by colonial rule from owning or purchasing land from the indigenous people, as were Arabs, Indians and Eurasian groups. However, they might be able to purchase or rent '*partikelir* land'. *Partikelir* land has ownership status equivalent to '*eigendom*,' meaning that the owner can utilise it however they like, without interference from others. In the fieldwork conducted by the author, it was revealed that several landlords own many kampungs in the city centre. Some of the landlords are Tionghoa. They owned Chinatown and surrounding kampungs inhabited by the Javanese and Tionghoa peoples. It is not clear how they got the land.

According to a 2018 interview with Professor Atiek Suprapti of Diponegoro University, the landowners leased their little pieces of property to the newly arrived people of the city. The newcomers constructed their own homes after renting the property lots. They thereby owned the building but only paid the land's rent. The area gradually grows into a kampung. Since the nineteenth century, the Tionghoa settlements have grown outside Chinatown, the area designated for them. This new community is expanding alongside the Javanese kampungs (Liem 2004).

Multiple massacres targeted the Tionghoa people during the fight for Indonesian independence (1945–1949), primarily in rural regions. As a result, the Tionghoa people moved into the cities. Subsequently, Tionghoa people were also prohibited from running businesses and trading at the village and sub-district levels in the 1950s. That policy also drove an exodus of Tionghoa people to the cities. However, upon independence, mixed-use settlements are no longer restricted. The Tionghoa settlements in Semarang have withstood political turmoil and riots across Indonesia, spanning from the independence war and the communist uprising in 1965, to the riots that targeted the Tionghoa people in the 1980s and 1998. The two political events that transformed Chinatown and some Tionghoa kampung areas were the communist uprising in 1965 and the riots in the 1980s. From 1966 onward, the use of the Chinese language and script in publications, including the names of stores and enterprises, was outlawed in Indonesia by the People's Consultative Assembly. This regulation changed the façade of many buildings in Chinatown. While the riots in the 1980s drove the neighbourhood inhabitants to build gates (locked every night) and create a new security and circulation system to protect the neighbourhood, the biggest riots targeting the Tionghoa people occurred in 1998. In 1998; however, there were no riots in Semarang.

In recent years, there has been a concentration of Tionghoa gated communities in northwest Semarang. Although the government of Indonesia encouraged assimilation, the Tionghoa naturally lived within their community. Since the 1980s, almost all big cities in Indonesia have developed gated communities. The government provides a mortgage scheme for this, and the Tionghoa could enjoy this program, too. The policy towards the Tionghoa people changed in 1998 when Indonesia had a new government. The new regime abolished all discriminatory laws against the Tionghoa people. The government supported the renewal project and exploited Chinatown as a new tourist destination.

8.5 A Brief Historical Timeline and City Planning Policy of Semarang

Before colonisation, Semarang was just a tiny fishing community. The current coastline line was 4 km ahead of the coastline. As does the coastline, the villages expand over time, and Semarang is led by a 'Adipati', a regency leader. However, when European traders first arrived, Demak or Jepara served as Central Java's principal port.

Owing to political turmoil in the Javanese sultanate during the eighteenth century, the governing Sultan gave VOC, Semarang as a gift. VOC aids him in gaining control of the realm. VOC relocated its headquarters to Semarang from Jepara. They established the European quarter south of the fortress, the Old City of Semarang. The Tionghoa and the other foreign eastern groups (the Arabs and the Indians) were also given a place on the Semarang riverbank. However, Javanese people inhabit the kampung (Liem 2004).

Since the nineteenth century, with a good port and railway network, Semarang has become a hub of sugar trade, which has influenced the economic and administrative role of the colonial government (Novita 2015). Trade with Europe increased in Semarang, resulting in the exponential growth of their population and territory. The city's development drew newcomers from neighbouring regions and Chinese migrants. Because of the city's growing population, it was expanded beyond the historical core in 1905, bounded on one side by the West Flood Canal and the Semarang River and on the other by the East Flood Canal, with the Java Sea to the north and hilly terrain to the south. The development of European and Chinese trading houses, banks and plantation agents, as well as the ship harbour and new housing buildings in the city's south corner, are examples (Akhyat 2006). This economic growth also benefited the Tionghoa traders. Many of them became very wealthy.

In the early twentieth century, Semarang and numerous other cities were granted municipality status. This required a concrete city planning document. The municipal government of Semarang hired Thomas Karsten in 1916. He was planning a new settlement towards the south of the old city centre because the old city centre had become too crowded. The new settlements in the hilly area were for the upper class (European and wealthy Chinese). However, he also designed some settlements for the middle class, separated from those for the upper class (Cote 2014).

Semarang saw a five-day war in 1946 during the independence war (1945–1949), although the Dutch mostly ruled the city. Until the 1950s and 1960s, people from the neighbouring regency migrated to Semarang in search of employment and safety, resulting in the city becoming overcrowded. Most Europeans in Indonesia departed the country after 1945 and 1949. However, before 1957, the European population in Semarang remained significant. Nevertheless, later, the government of Indonesia expelled them. Army members or new civil servants took over their settlements (Wijono 2014). From that moment on, the population of Semarang consisted mainly the Javanese tribe as the majority, and the Tionghoa, Arabs and Indians, with other native Indonesian tribes as the minority. From the 50 s until 1976, we could say that Semarang was in the consolidation phase (Nugroho 2016). During that time, no city was planned, and the city continued attracting newcomers from neighbouring regencies. By squatting on the vacant land, they established numerous new kampungs in and around the governmental, business and city centres.

Semarang municipality expanded its borders to the neighbouring regency in 1976. To the west, territory from Kendal regency is taken; to the east, territory from Demak regency is taken; and to the south, territory from Ungaran regency is seized. This rule was implemented to comply with the dictatorship Regime's revised vision. At the time, the Indonesian government desired to stimulate economic growth through industrial activity. To accommodate this, Semarang needs a new territory (Nugroho 2016). They also built a new commercial district in the south of Chinatown (Simpang Lima), initially envisioned by Thomas Karsten.

The policy on spatial planning of every region underwent periodic modifications. Within 5–10 years, the government examined the spatial planning documents. There were undoubtedly changes in land use during the process. In Indonesia, land-use modifications in the city planning document are typically the result of a review of the previous planning document, a scientific analysis of the present state of the city, and, if any, a new vision for the city. On the other hand, investor requests occasionally result in changes to the land's usage. In several instances, the local authority decided to formally transform the land use in that area into settlements due to the new kampungs that were sprouting up on vacant lots that were initially intended for green spaces or other purposes. On the other hand, changes in land usage result in rising land taxes and shifting activity in the area. Table 8.1 presents the comparison among all the spatial planning documents from 1975–2011.

According to Table 8.1, there was a significant change in the city planning plan starting in 2004. This was affected by the regime change in 1998. After 1998, the municipal government had greater authority than in the dictatorship era.

8.6 Chinatown and the Surrounding Kampung

The description in this chapter will be limited to Chinatown and one kampung adjacent to Chinatown. Many Tionghoa people live in kampung, which is called Jagalan. The land in Chinatown and Jagalan Kampung is held by the same Tionghoa landlord, according to conversations with residents in 2018 and 2019. The clan organisation decides who has the right to rent a land tract.

8.6.1 Chinatown

Semarang's existing Chinatown dates from the eighteenth century. It has a strict grid pattern. It comprises shophouses on the main street surrounding Chinatown and row houses in the centre. There are seven Chinese temples among the housing units and

Table. 8.1 Land use comparison among city planning documents. Source: the author based on (Nugroho 2016)

No.		1975	1990	1999	2004	2011
1	Government Centre	Tugu Muda	Tugu Muda	Same, more details	Tugu Muda	Same, more details
2	Commerce Centre	Raden Patah to the harbour	Same, more details	Same,more details	City centre to the South, became mixed area (settlement, commerce, Education, service); East (settlements, industry, commerce); West (industry, settlements); North (agriculture, low density settlements, mixed area settlements)	
3	Industrial Centre	Tugu, Genuk, Pedurungan (East and West developments)	Tugu and Genuk increase, Pedurungan decreases	Same, more details		
4	Cultural Centre	Simpang Lima	Same, more details	Same, more details		
5	Education	South Semarang: Banyumanik	Same, more details	Same, more details		
6	Agriculture	South Semarang: Mijen, Gunung Pati	Same, more details	Same, more details		
7	Settlements	All areas, building coverage	Same, more details	Same, more details		

shop housing units. According to the book 'Riwayat Semarang' by Liem Thian Joe, the population first filled the main streets before eventually filling in the centre. The area in the heart of Chinatown remained vacant until the end of the nineteenth century. The Chinese tribes form the basis for the neighbourhood division since they dictate the types of trading activities. The spatial arrangement has remained unchanged since colonial times. Figure 8.2 shows the development of Chinatown in Semarang, elaborated by the author based on the book 'Riwayat Semarang' and the Master thesis 'Chinese Settlement in a Changing City: an Architectural Study of the Urban Chinese Settlement in Semarang Indonesia' by Johannes Widodo.

However, since 1965, when the dictatorial administration outlawed using the Chinese alphabet in public areas, this policy has altered Chinatown's physical characteristics, particularly the building's façade. Even though the government constructed a Chinese gate to capitalise on Chinatown's tourists after 1998, Chinese alphabets and decorations are no longer present. The business owners translate the Indonesian or Javanese name into a Latin alphabet, replacing the Chinese alphabet on their signage. Because the government at the time also prohibited Chinese festivals and religious activities, in this case, Confucianism, in public spaces, the Chinese temples camouflaged their façade. Confucian adherents practised their religion in private, away from the public view.

Fig. 8.2 Chinatown development based on the book 'Riwayat Semarang' and master thesis of Widodo (1988) 'Chinese Settlement in a Changing City: an Architectural Study of the Urban Chinese Settlement in Semarang Indonesia'. (Source: the author)

The transformation of Chinatown in Semarang—shaped by new investment and evolving functions—(see Fig. 8.3), draws on fieldwork carried out by the author in 2018.

The Tionghoa peoples' business activity grows with the city's growth. According to an interview with Prof. Jamilla K. Sharma from Sultan Agung University in 2018, there was a period in the 1980s and 1990s when Semarang Chinatown was on the decline. People started to abandon Chinatown, using their property as a warehouse for a new shop in another district or for internet selling. For this reason, they entirely transformed their shophouse into a warehouse and no longer provided lodging. The Tionghoa extended their business to the recently constructed southern commercial centre. However, this neighbourhood saw increased economic activity due to Chinatown's revitalisation project in the 2000s. Owing to Chinatown's advantageous location, many banks and offices lease space here. They all invested in the building's transformation into a modern structure, altering the neighbourhood's appearance. The former resident, who no longer resides in Chinatown, profited from the rent costs of the banks and companies. The ancient Chinese architectural features have mainly changed due to commercial activity. These days, the Chinese architectural style is limited to the roof shape. Figure 8.4 shows one of the junctions inside Chinatown with new bank offices.

Chinatown is famous for its Chinese cuisine. This is one of the businesses that could survive over many generations. In Chinatown, we can enjoy Chinese food not only in the restaurant but also in various street food stalls. In Chinatown, streets are

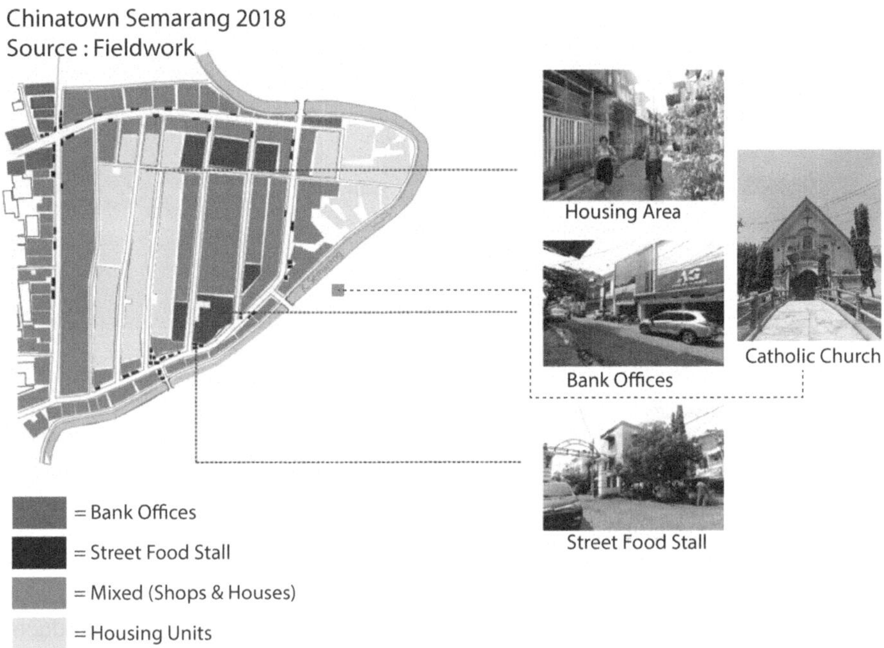

Fig. 8.3 Re-investment and changing function inside Chinatown. Source: the author

Fig. 8.4 Old houses with new bank office in the junction between Gang Pinggir Street and Gang Tengah. (Source: the author)

used for circulation and as a public space that accommodates street food stalls. The Tionghoa community saw this as a tourist magnet. There are two types of culinary tourism in Semarang Chinatown. In the first one, with the support of the government and city planning experts, they held '*pasar semawis*'. *Pasar Semawis* is a culinary festival that takes place every weekend. For over 15 years, this weekend culinary festival has survived until today. This culinary festival features a variety of cuisines, although the focus is on Semarang's '*Peranakan*' cuisine. *Peranakan* cuisine is a type of fusion cuisine that combines Chinese and indigenous cuisines.

Fig. 8.5 Day and night street food stall inside Chinatown. (Source: the author)

The second is the regular culinary businesses that have existed in Chinatown since colonial times and continue to do so now. Some restaurants date back to before Indonesia gained independence. Not all of the food vendors in Chinatown are Tionghoa. Some of the Javanese people work as food vendors. Here, the workers are served affordable meals by a street food stall not part of *Pasar Semawis*. The shop owners permit them to open their stalls before Tionghoa's shop. This collaboration encourages symbiotic partnerships. The food vendors could sell their products, while the Tionghoa received 'protection' from these Javanese neighbours, and Tionghoa employees could eat cheaply. Figure 8.5 shows different food stalls during the day and night in front of shop houses.

However, the area around the main street and along the car-wide lanes is the only one affected by this Chinatown revitalisation. The small alleys in the centre of Chinatown remain neglected, dark at night, almost empty and dangerous at night. Due to Chinatown's proximity to multiple traditional markets, some Javanese vendors from those markets lease rooms in the narrow lanes running through the heart of Chinatown. The government saw Chinatown as a potential business district. They decided to change the land use in Chinatown following the phenomenon. Previously, Chinatown belonged to the settlement area. However, the government changed the land use into a mixed area due to the revival and new activities here, such as Pasar Semawis and bank and office buildings.

8.6.2 Kampung Jagalan

Kampung Jagalan is on the Semarang River's eastern bank. This river serves as the boundary between Kampung Jagalan and Chinatown. A landowner from the Liem clan owns both Chinatown and kampung. This kampung was home to a large mansion and an alcohol manufacturing during the colonial time. The Tionghoa labourers

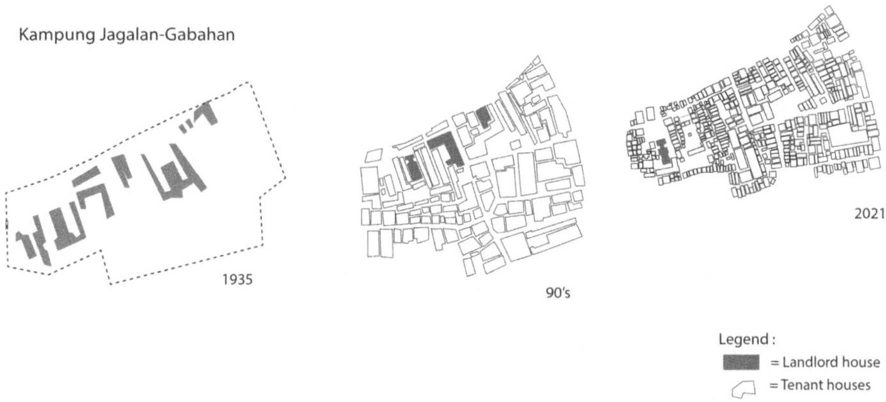

Fig. 8.6 Kampung Jagalan Development. (Source: the author)

began to settle in the vicinity of the factory. Figure 8.6 shows the development of Kampung Jagalan. The author elaborated the maps based on the conversation with the elderly inhabitants and the colonial map.

In 2018, during a conversation with an elderly resident, it was stated that the arrival of the Javanese people from the nearby regency started after 1945. She worked in a factory that produced alcohol. She arrived from the Magelang Regency circa 1946. These new residents occupied the vacant spaces in kampung. This resulted in a disorganised spatial structure distinct from Chinatown but typical of Indonesian kampung inner cities. Subsequently, the factory was shuttered and the landowner began renting out his land to the new people (Fig. 8.7).

Inside kampung, economic activity is similar to that of other inner-city kampung. Some people start a 'warung', a tiny shop that sells everyday items or handles internet trade. Nearly all the locals, Tionghoa and native Indonesian, are from the low-income class. They have a similar way of life. Most of them work as street food vendors. The Tionghoa who live here mostly could not send their children to higher education. Like their grandparents, their children are employed by the Tionghoa in Chinatown. The spatial structure in Kampung Jagalan follows the parcels apportioned up by the Liem family's landlord. However, Kampung Jagalan is a densely populated kampung. There are no vacant housing units in this area. Newcomers arrive and stay from time to time. When they could not rent a house in the kampung's main structure, they began to fill the vacant space and construct a semi-permanent building.

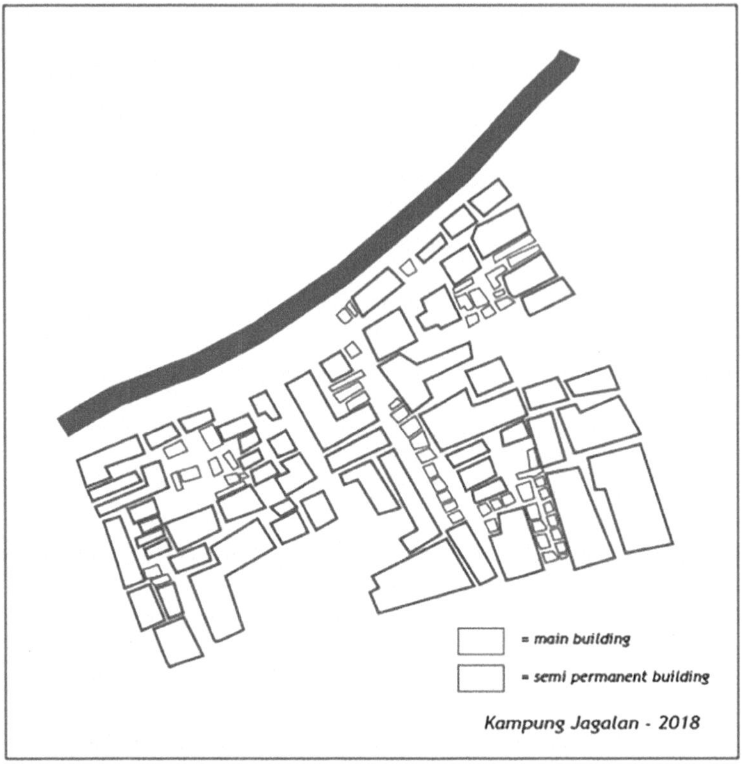

Fig. 8.7 Kampung Jagalan spatial structure. (Source: the author)

8.7 Discussion

The colonial city structure in Indonesia was divided into three sections: indigenous settlements, European settlements and foreign quarters, where migrants from China, India and the Arab world lived. Colonialism was fascinated with partitioning cities along ethnic and racial lines. Chinatown was established as a place of residence for Chinese migrants during the colonial era, allowing them to construct their own homes and showcase their cultural practices.

Since colonial times too, Chinatown has been a neighbourhood with mixed activity. It served as a place for business and also a place for residential. The occupation of the Tionghoa was mainly related to trading activities, and they served as middlemen between the Europeans and the Native Indonesian; therefore, it was very natural for Chinatown to grow as an important business district. In contrast, a kampung is a traditional village found in Southeast Asia. Nonetheless, there is economic activity within kampung. A kampung is more than just a neighbourhood. It is capable of both production and consumption. The place of employment and residence are relatively close (Funo et al. 2002). In the case of kampung in the urban areas, first and foremost, urban kampungs were the result of the colonial exclusion of native Indonesian from the formal city, expulsion to assigned places outside the town/outside of city rights where they were left to organise themselves (hence neo-traditional systems, urban-rural traditional systems). However, they were close enough to the town/workspaces to be exploited as a labour force.

During the regime from 1967 to 1998, the Tionghoa people had to obey legislation prohibiting them from displaying Chinese characters in their area. They obeyed the order and altered the signage in their shop to use the Latin alphabet. This policy had no substantial influence on their business, but it greatly impacted their Chinatown identity as a place for Tionghoa people in the city. Nonetheless, the demise of Chinatown was driven by the city's development. Selling their property inside Chinatown is also not easy since they rent the land from the landlord, and there are pretty complicated systems among the Chinese clan inside Chinatown. That is why there are so many empty buildings in Chinatown. Although the government has outlawed Chinese culture, the clans and tribes that preserve Chinatown cannot be disbanded. Without the authorisation of these organisations, no one could buy or rent property in this area. The landlord plays a significant part in this situation. However, no evidence can be found in agrarian or land agencies.

When the dictatorship period is gone, the community leader could propose a revitalisation initiative to the new government. They do not, however, intend to establish a Chinatown with a significant Chinese character. Residents have also complained about *Pasar Semawis*, the weekend food festival since it frequently interferes with Chinatown's religious and cultural activities. Within Chinatown, the community leader and clan organisation hold all the land and maintain all the temples, yet they let the firm rent and construct a new modern building here. Naturally, the construction of this new bank and office complex will preserve Chinatown's status as a business centre. The grid spatial pattern of Chinatown further supports this activity. It would seem that they value economics more than culture. They also let the street food vendors who are mostly native Indonesian to get livelihood there, because this activities support the Chinatown as business district.

However, Kampung Jagalan's facade was not significantly altered by the policies of the dictatorship regime. Due to the riots in 1998, kampung just modified the circulation and security systems. However, the city's expansion after 1945 and Chinatown's revival after 1998 rendered the kampung extremely crowded, as it is an urban settlement whose primary function is to house the labour force. Even in densely populated areas, people are drawn to them; if needed, they may even erect a temporary structure in a vacant space.

Online trading has grown in popularity in recent years. This influences Chinatown and the nearby kampungs. It is not necessary to have a shop while doing business online. They can conduct business even from the heart of kampung. To escape rising land taxes, many Tionghoas lease their property for offices and relocate their companies from Chinatown. The government rezoned Chinatown for service, business and residential purposes (mixed zone) with a higher land tax. This findings show how the political economy of place operates in Semarang. Land and space are used by both the states and private elites to maintain influence and gain profit. Clan organisations and landlords act as local growth coalition, while government policies on taxation and landuse reinforce their position. At the same time, kampung residents and small traders respond through informal agreement and everday negotiation over space. The spatial patterns that results, gated kampungs, fragmented ownership and selective revitalisation, reveal how economic and political power are materialised in the city landscape.

8.8 Conclusion

In post-colonial Indonesia and through 32 years of dictatorship, the Tionghoa in Semarang could thrive and retain their settlements as one of the city's business districts. They followed the rules that prohibited them from exhibiting their culture. They also survive by dealing with the city's economic dynamics and development. The position of the community leader, clan organisation and landlord is quite powerful in this setting. They decide what can be preserved and what can be sacrificed.

By sharing living spaces and means of subsistence, such as allowing the native Indonesian to set up a food stall in front of their store, the Tionghoa can also preserve positive relationships with the native Indonesian who are living in Chinatown and the inner city kampung. Conversely, the Tionghoa and their native Indonesian / Javanese neighbours share living space in kampungs. However, the dynamics of Tionghoa settlements continued to demonstrate that not every member of the community is entitled to participate in decision-making. The landlord, the clan organisation and the local government make the decisions regarding the Chinatown revitalisation, and they are not included in it. All parties involved in this situation prioritise the economic aspect. The case of Semarang shows that the political economy of place continues to shape post-colonial urban transformation. Spatial change in Chinatown and surrounding kampungs is not only cultural or architectural process but also the outcome of ongoing negotiation between capital, government and community interest. The persistance of landlord control, state regulation and unequal participation confirms that the city's form reflects deeper political and economic power relation that still define who can shape urban space today.

References

Akhyat A (2006) The ideology of Kampung: a preliminary research on Coastal City Semarang. Humaniora 18:15–26

Camagni R (2016) Afterthoughts on urban economic theory and its focus. Investigaciones regionales. J Reg Res 36:87–105

Cornwall A, Coelho V (2007) Spaces for change? The politics of participation in new democratic arenas. Zed Books, London

Cote J (2014) Thomas Karsten's Indonesia. Modernity and the End of Europe, 1914–1945. Bijdragen tot de taal en volkenkunde 170:66–98

Ford LR (1993) A model of Indonesian City structure. Geogr Rev 83(4):374–396

Funo S, Yamamoto N, Silas J (2002) Typology of Kampung houses and their transformation process—a study on urban tissue of an Indonesian City. J Asian Archit Build Eng 1:193–200

Jones H, Clench B, Harris D (2014) The governance of urban service delivery in developing countries. Overseas Development Institute. https://odi.org/en/publications/the-governance-of-urban-service-delivery-in-developing-countries-literature-review/

King, A.D. 1974. Colonial urban development: culture, social power and environment. London, Henley and Boston, MA: Routledge & Keagan Paul

Liem TJ (2004) Riwayat Semarang. Kata Hasta Pustaka, Jakarta

Molotch H (1976) The City as a Growth Machine: Toward a Political Economy of Place. Am J Sociol 82(2):309–332

Novita A (2015) Gerakan Sarekat Buruh Semarang Tahun 1913-1925. J Indonesian History 3(2):1–7

Nugroho MLE (2016) The spatial planning policy shifting in Semarang (from RIK 1975–2000 to RTRW 2011–2031). Diponegoro University, Semarang

Obeng-Odoom, F. 2013. Governance for pro-poor urban development: lessons from Ghana. Abingdon: Routledge

Skop E (2006) Urban space: the shape of inequality. Urban Geogr 27(5):393–396

Widodo J (1988) Chinese settlement in a changing City: an architectural study of the urban Chinese settlement in Semarang Indonesia. Master thesis. KU Leuven

Wijono RS (2014) Public housing in Semarang and the modernization of kampongs, 1930–1960. In: Colombijn F, Coté J (eds) Cars, conduits, and kampongs. The Modernization of the Indonesian City, 1920–1960. Leiden, Boston, Brill, pp 172–192

Chapter 9
Towards a Decolonial Home: Post-Partition Lahore, Pakistan (1947–1975)

Mehwish Abid

Abstract During the Partition of British India in 1947, which resulted in the creation of Pakistan and India, 12 million people were displaced, and around half a million lost their lives. Emigrants carried their objects, belongings, material assets and skills as they crossed the newly drawn borders. This chapter examines how post-Partition inhabitants of Lahore, particularly migrants, adapted to create their multigenerational homes, deviating from colonial spatial structures. The research employs the theoretical framework of decoloniality to analyse this transition. It investigates the structural change from colonial home to what I term the 'decolonial home'. This idea is presented through a comprehensive analysis of residential projects built after Partition in The Cooperative Model Town Society, Lahore, Pakistan. Methodologically, the research draws on oral histories of families and how they shaped their homes after Partition. The research led to the creation of an image-based material archive, developed as part of a teaching project in 2021. Case study documentation was conducted through a student assignment in the History of Architecture and Urbanism II course taught at the Institute for Art and Culture in Lahore. The research findings analyse how the colonial housing policies were dismantled in the case of Model Town. The research hypothesises that while the architectural influences from the global north—through foreign-trained architects and capitalist expansion—shaped a modern architectural interface, the people initiated a process toward a 'decolonial home', guided by their domestic sensibilities, sense of belonging and memory.

Keywords Decolonial home · Memory · Model Town Lahore · Oral history

M. Abid (✉)
McGill University, Montreal, QC, Canada
e-mail: mehwish.abid@mail.mcgill.ca

G. Cimadomo et al. (eds.), *Urbanism at Borders*, The Urban Book Series,
https://doi.org/10.1007/978-3-032-09594-7_9

9.1 Introduction

This chapter examines the concept of 'decolonial home' in the post-Partition city of Lahore, Pakistan. It focuses on the Model Town Society, a colonial-era settlement in Lahore established in 1917, inspired by the Garden City model proposed by Ebenezer Howard (1850–1928), reflecting its ties to British imperial planning (Anbrine and Burgess 2014). These ideas were implemented by the founders at the suggestion of Divan Khem Chand, a British-trained barrister and key figure in the initiation and execution of Model Town, Lahore (Glover 2008). Before partition, it was primarily inhabited by retired government officials only. After the Partition of British India, Model Town was declared an Evacuee property as 85% of the plots belonged to non-Muslim emigrants. At the time of Partition, Model Town had 1366 members, comprising 277 Muslims and Christians and 1089 Hindus and Sikhs (Shama 2014). After independence, migrants settled in the area, making it a socially inclusive place where people from different social segments settled, Muslims in majority.

Migrants who settled in Model Towns designed, appropriated and transformed residences according to their needs and culture (Glover 2008). This research focuses on two residential projects in Model Town that were designed after Partition, deviating from the colonial-era homes. This transition of moving away from pre-Partition policies is studies through the framework of decolonisation (Ghom and George 2023). There is a considerable gap in the human dimension aspect concerning the post-Partition cities of Pakistan and India (Talbot 2007a, b). Hence, this chapter aims to produce discourse towards analysing the traces of a decolonial home in Lahore, Model Town, which involves subaltern agency. After the creation of Pakistan, many other building typologies such as factories, schools and colleges, cinemas, hospitals and other public buildings evolved as newer building typologies—this is not the goal of research for this chapter.

9.2 Decolonisation: On a Decolonial Path

Ian Hodder stated that 'the past is present in the sense that our reconstructions of the meaning of data from the past are based on analogies with the world around us' (Hodder 2013). The collected data from this research highlights that the residential architecture produced post-partition has not perpetuated colonial nostalgia; instead, it has progressed in terms of its materiality through design interventions by local artists, designers and inhabitants. As the context around it has evolved—from a united British India to two independent states of Pakistan and India—generational memories have also evolved (Schwartz 2009). The recollections associated with the idea of home have also played a key role in post-Partition architecture, where nostalgia is to be used strategically for decolonisation purposes (Peled 2016). The evolution of post-Partition architecture in Lahore reveals a fascinating interplay between tradition, modernity and the socio-political context of a newly independent nation (Malik et al. 2019).

Drawing upon Sonya Atalay's (Atalay 2006) ideas about decolonisation in archaeology, which suggest that the majority of the knowledge base is Western, entails alternative ways of understanding and observing history, as all aspects of human life and culture are constantly evolving. This research proposes to investigate alternative histories of architecture post-Partition. Through the significant and multipronged understanding of decolonisation provided by Jan C. Jansen, Professor of Comparative History and Jürgen Osterhammel, German historian, the disassembling of colonialist structures is understood as a process that involves removing not only formal power but also the hidden aspects that continue to exert control even after colonial rule has ended (Jansen and Osterhammel 2017:12). Hence, this research undertakes a journey to uncover traces of daily life that led to a transition in residential colonial architecture to an alternative space, however, still offering colonial continuities (Jackson 2005). A critical angle is that in the postcolonial South-Asian countries, administrators resisted retaining the power reflective of Thomas Babington Macaulay's (1800–1859) infamous minutes in which Indian Education had proposed the deliberate creation in India of just such a class of 'brown, white men,' educated to value European culture above their own (Ashcroft et al. 2013:43–58). This cultural hegemony often manifested in architectural preferences, where post-Partition homes, continued to serve as models for elite housing in the newly independent nation (Anbrine, Akhtar and Qureshi 2024).

Consequently, one wonders about the existence of decolonial aspects in society. The research reveals that with the fall of the Empire, the legacy remained in the hands of different masters; however, minor yet essential steps were taken by the locals to regain their agency through appropriation of colonial policies (Khan et al., 2021). To validate this idea, the research brings forward the notion of decolonisation by Prasenjit Duara, a historian and sinologist who explains decolonisation as proposed as a more localised shift of power than a breakdown of empires. He further elaborates it as 'the process whereby colonial powers transferred institutional and legal control over their territories and dependencies to indigenously based, formally sovereign, nation-states' (Jansen and Osterhammel 2017:12–34). Such policies are also reflected in the establishment of housing policies for the colonies, as seen in the residential architecture implemented in Model Town, Lahore, Pakistan (Anbrine and Burgess 2014).

United India was the first British colony to gain independence, achieved through the division into two separate states. After World War II, there was a global urge for decolonisation (Greenberg 2005). A natural flow of decolonisation began as soon as the moment of independence and the creation of Pakistan (Hopkins 2008), whether through the new national anthem by Hafeez Jalandhari in the Persian language or the enforcement of Urdu as a national language (Rahman 2002). This chapter focuses on the structural changes that Prasenjit Duara so poignantly points out, namely the transfer of institutional and legal control of imperial power to locally based nation-states that were self-governing before the British Raj, to which Duara refers as a shift towards moral justice and political solidarity (Jansen and

Osterhammel 2017). This phenomenon has been studied in the context of two cases of residential architecture in Model Town, where the transition from colonial rule to self-governance prompted a revaluation of architectural styles and urban planning principles in Lahore.

Under the framework of decoloniality, natives must overthrow the coloniser through force, which is precisely how the coloniser established control, thereby encouraging those on the sidelines to become part of the process (Fanon 1963). This transformational process begins with understanding the works and practices of the coloniser while linking them with our own experiences (Spivak et al. 2022). Drawing upon these ideas, the case studies of residential architecture in Model Town Lahore are analysed—the intent is not to turn back time, but to question the structural changes that occur when people have the agency and control to shape their lives, in this case, their homes. Walter Mignolo puts forward Quijano's (1928–2018) notion that Eurocentric knowledge is like a 'distorting mirror', resulting in it being unable to see reality. By filtering and adapting the Eurocentric expertise with local experiences and ideas, it is possible to develop a more unique, different and personal approach that helps shape identity (Mignolo 2011). Extending these decolonial frameworks, I explore how people of Model Town Lahore took control of their homes by using local materials and craftsmanship. This negotiated design practice reflects a broader post-Partition movement to assert indigenous identities against lingering colonial influences—fotering a sense of belonging and cultural continuity in a newly independent nation (Moffat 2022).

9.3 Methodology

Drawing on Hodder's (Hodder 2013) methodology from his work in Çatalhöyük East, where he stresses that decolonisation requires community participation and engagement, this research uses oral history to analyse the agency of residents who shaped their homes in the aftermath of partition. The research method also engages with archival material, including one of the first compilations of architectural works ever published in post-partition Pakistan (Chishty 1948). Case selection focused on post-partition housing projects shaped by modernist principles, locally available materials and influenced by earlier planning models such as the Lahore Improvement Trust, established in 1936 (Talbot 2007b).

Another key methodological frame draws on Arjun Appadurai's (1986) insights into how the circulation of objects reveals information about people—in this case, homes. This research, therefore, adopts a material culture lens to examine domestic space and memory. This framework emphasises three aspects of objects: their temporality, their conceptual meaning and their spatial context. According to Appadurai, objects are not fixed in meaning; their status as commodities can shift over time depending on social context. This chapter focuses on two of these aspects—temporality and commodification—as they help reveal how objects (including the home itself) carry layered meanings within architectural and spatial contexts (Appadurai

1986). Domestic spaces are highly significant to our lives, but they are frequently undervalued, misrepresented, or overlooked in academic discourse (Buckley 2018). In this context, decolonisation refers to the ongoing, multifaceted and often subtle resistance to the lingering effects of colonial influences in everyday life (Pires 2014). To address this gap, the study integrates insights from 'architectural anthropology', championed by Supakit Yimsural, which is also used as a framework by Iqbal and Batool (2024) in their analysis of Khatija Mastur's novel, *Aangan,* to explore spatial dynamics.

The methodological work was embedded within a pedagogical context: the oral history interviews and documentation were conducted collaboratively with students enrolled in the History of Architecture and Urbanism II course at the Institute for Art and Culture in Lahore. Their participation not only enriched the empirical data but also foregrounded the pedagogical potential of teaching as a mode of collaborative knowledge production.

9.4 Modernity, Dislocation and Resettlement

Dipesh Chakrabarty notes that the conception of discipline maintained at the home and between the public and civic life reveals itself in the modernist approach to domesticity. He argues that the Indian home is a colonial construct, shaped by nationalist writings that superimposed ideals of modern domestic life. On the contrary, in the 1830s and 1840s—before the rise of nationalism—many Indians were drawn to European ways of life. Chakrabarty (2004) highlights this through the example of Madhusudan Dutt's early writings, including a poem published in *The Literary Gleaner* in 1842, where the author is shown to be naively enamoured with the coloniser. Later in life, however, Dutt returns to Bengali literature, marking a critical shift in cultural alignment (Chakrabarty 2004).

In the context of Lahore, Model Town (see Fig. 9.1)—a colonial cooperative originally designated for retired government officials—underwent a dramatic transformation after partition. Migrants from diverse occupational, class and caste backgrounds began to inhabit and appropriate its housing fabric (Glover 2008). They reshaped their homes according to their needs and cultural values. Although the Lahore Improvement Trust made efforts to address the housing crisis, its plans did not resolve the acute shortage nor the challenge of providing affordable housing (Talbot 2007a). It is essential to establish here that people settling in any locality also appropriate their spaces with their objects and varied backgrounds (Glover 2008).

Despite its significance, there is limited scholarship on dislocation, resettlement and the adaptation of domestic space in post-partition Lahore. Talbot (2007a, b) argues that many of the urban pressures that emerged in the colonial period—such as population growth and planning neglect—remained unresolved after independence. Punjab's geography and culture were permanently altered after the British annexation in 1849. Yet towns that had expanded during the colonial era were often overlooked in the postcolonial planning process (Singh and Singh 2019). While

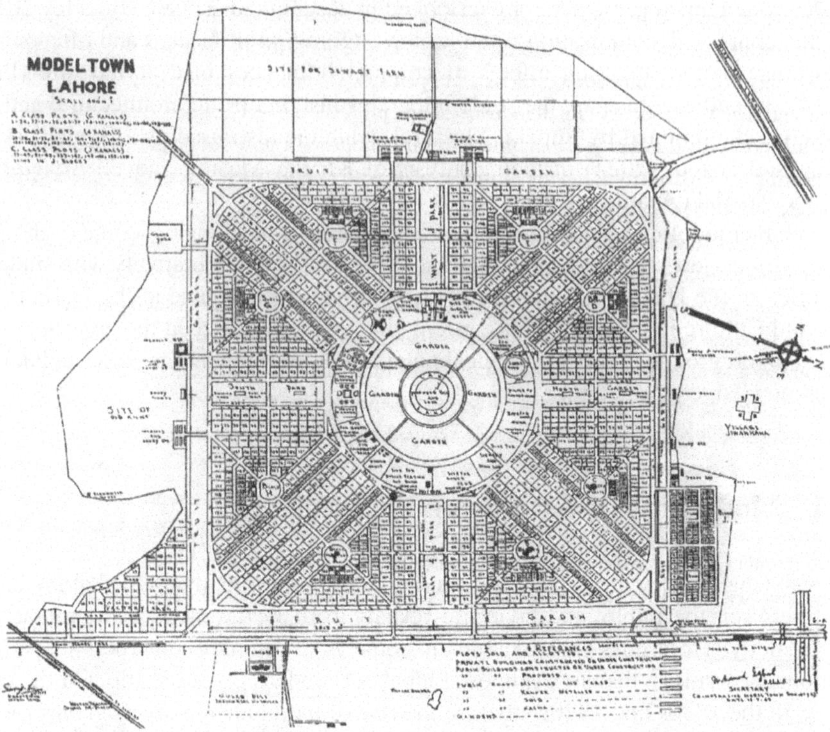

Fig. 9.1 Site plan of Model Town, Lahore, 1949. (Source: William J. Glover, Making Lahore Modern: Constructing and Imagining a Colonial City (Minneapolis: University of Minnesota Press, 2008), 152)

William Glover (2008) provides a crucial reading of colonial Lahore, this research expands the discourse by focusing on the human dimension of post-partition space-making—how people inhabited, adapted and imbued meaning into inherited structures.

9.5 General Overview of Architecture and Architectural Discourse Post-Partition

One of the earliest architectural compilations published in post-Partition Pakistan was *Homes of the Day* by S.M. Chishty (1948). The book showcases modern residential designs, emphasising affordability and efficiency. P.L. Verma, writing in the same volume, described the post-Partition chaos and administrative need for practical, rapidly buildable housing (Chishty 1948). Cities like Lahore and Amritsar turned to modernist solutions to meet urgent housing demands (see Fig. 9.2) (Talbot 2007b).

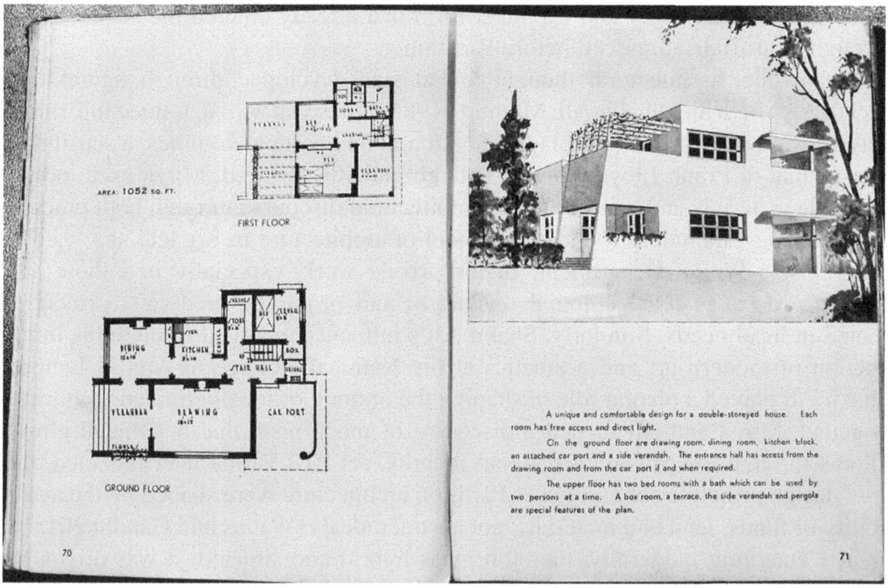

Fig. 9.2 Ground and first floor plans designed by N.B. Shroff in Lahore, with perspective view and descriptive annotation. (Source: S.M. Chishty, Homes of the Day (1948), 70–71)

Chisty's conceptual designs recommended locally available materials such as brick masonry and cement, with load-bearing walls to minimise the use of reinforced concrete. These homes were designed to serve the middle class in both rural and urban settings. Since professional architecture was still in its early stages, many designs were implemented by the Public Works Department. Chishty's work addressed colonial-era concerns such as drainage, ventilation and structural approvals—issues formalised in revised municipal laws from as early as 1862 (Chishty 1948; Vandal 2011). While layouts adhered to grid patterns, interior spaces retained traditional South Asian domestic norms, including gender-segregated drawing rooms.

An important innovation in *Homes of the Day* was the use of Urdu terminology in architectural annotations. For example, plans used the word *chulla* instead of 'stove', making them more accessible to a broader local audience, as indicated in Fig. 9.2. This linguistic shift also signalled an early form of cultural ownership and national identity formation through the built environment.

Pervez Vandal (2011) argues that Partition triggered foundational shifts in urban planning through the establishment of new bureaucratic structures and infrastructural frameworks, primary concerns being the alignment of buildings on the streets and the disposal of wastewater. Lucy Peck (2015) notes that while early twentieth-century homes retained traditional courtyard layouts, their facades increasingly reflected the influence of European stylistic trends. This hybrid approach extended beyond housing to public buildings such as railway stations, many of which were modelled on European design conventions. As Talbot (2007a) reminds us, modernity as materiality and capitalist form had already entered the subcontinent through industrialisation well before Partition.

Meanwhile, architectural thought in Pakistan developed through figures like Mehdi Ali Mirza and Shakir Ali. Mumtaz (1985) notes that Mirza, trained in Britain, was drawn to the formal interplay of horizontal and vertical volumes, a sensibility akin to that of Frank Lloyd Wright. Though globally inspired, Mirza used locally available materials and contributed to architectural discourse through both practice and pedagogy, including founding a school of architecture in Karachi and leading the Pakistan Works Department. In his private work, especially in Lahore and Karachi, Mirza rejected colonial revivalism and promoted modernist principles rooted in local needs. Similarly, Shakir Ali's influence emerged through his introduction of modern art and aesthetics at the National College of Arts in Lahore. Shakir Ali played a pivotal role in shaping the outlook of locally trained architects. Together, Mirza and Ali seeded a discourse of modernism that negotiated global idioms through the prisms of national identity. As P.L. Varma later reflected, the core design principles of early post-Partition architecture were shaped by the availability of funds, land and materials, not abstract ideals (Wattas and Gandhi 2018).

The emerging modernity, therefore, was hybrid and situated: it was driven by necessity, mediated through training abroad and realised through the resourcefulness of post-partition citizens. Architects such as N.B. Shroff, Kamil Khan Mumtaz and Pervaiz Vandal also contributed to this evolving narrative. What emerged between 1948 and 1975 was an 'indigenous modernity'—a mode of building shaped by conflict, recovery, aspiration and local adaptation. These architectural expressions of belonging played a crucial role in constructing what this study terms the 'decolonial home'.

9.6 Building After Empire: Two Stories of Home and Design

Both stories of home in this chapter have been narrated by the occupants themselves, through oral history interviews conducted during field visits, and supported by original documentation and measured drawings produced as part of the teaching project in architectural history. These accounts, drawn from memory and material evidence, offered layered insights into how lived experience, family legacy and post-Partition circumstances shaped each residence, allowing us to read domestic space not only as a built form but as an evolving archive of identity and adaptation.

9.6.1 Khalid Aftab Residence

This research examines post-Partition architecture as a blend of unfamiliar aesthetics and emergent spatial strategies, shaped by material pragmatism and shifting cultural values. One illustrative example is the residence of Dr. Khalid Aftab (former

Vice Chancellor of Government College University, Lahore, Pakistan), whose home was designed and constructed between 1975 and 1976. The plot was allotted by the university in 1972. It was developed once Dr. Aftab had accumulated the necessary funds, as described by him during an oral history interview conducted by the author and Group A of students in Lahore, March 2021 (Abid et al. 2021). The architect, Afzal Khan (late), worked within a strict budget, employing burnt brick—a cost-effective, climate-responsive material that offered both visual economy and long-term durability.

Brick Masonry with load-bearing walls enabled a relatively fast and low-cost construction process. As a further cost-saving measure, 9-inch walls were used instead of the traditional 13-inch thickness, resulting in reduced material consumption and construction time. Instead of elaborate ornamentation, the design emphasised simplicity: a rectilinear plan, flat horizontal massing and a tilted roof line. Ornamentation was consciously avoided, reflecting modernist restraint. Glass windows with screens were used instead of traditional ventilators or verandas, streamlining the structure both visually and economically.

Burnt bricks served not only aesthetic and economic goals but also offered insulation against Lahore's harsh climate, enhancing energy efficiency. The simple, linear window designs followed column-beam structural logic. The spatial organisation of the house is oriented around a central axis, facilitating visual and spatial connections between rooms. More than 50 years since its construction, the residence is now surrounded by lush local flora—a detail that, according to the resident, was deliberately integrated into the original landscape plan. The garden not only enhanced thermal comfort but also reflected the post-partition desire to reclaim environmental continuity in unfamiliar urban surroundings. The residence included a small pool in one corner, positioned to ensure visual privacy and spatial separation, reflecting the cultural emphasis on *pardah* in middle-class Muslim homes (Abid et al. 2021).

Around the same time, the house of artist and educator Shakir Ali—designed in 1964—was also under development and continued to evolve over the next 10 to 12 years. That project was led by Nayyar Ali Dada, one of Pakistan's most prolific post-partition architects. It remains unclear whether it was the late Afzal Khan or Nayyar Ali Dada who first introduced the use of burnt brick in residential design in Lahore. Interestingly, both patrons—Dr. Khalid Aftab and Shakir Ali—were associated with major educational institutions in the city, suggesting a link between academic patronage and experimental architectural practices in this period.

The house not only reflects the architectural sensibility of its time but also embodies a deeper post-Partition negotiation with memory, modesty and materiality (see Fig. 9.3). Its design decisions—from the use of burnt brick and simplified construction to spatial features like axial orientation and secluded pools—speak to the resourceful reinterpretation of modernist norms in a context shaped by displacement and reconstruction. What we see here is not only architectural adaptation, but a culturally and emotionally situated process of re-rooting: an attempt to reclaim urban space while addressing economic constraints, climatic conditions and social values such as *purdah* and domestic privacy (see Figs. 9.4, 9.5, 9.6 and 9.7). It

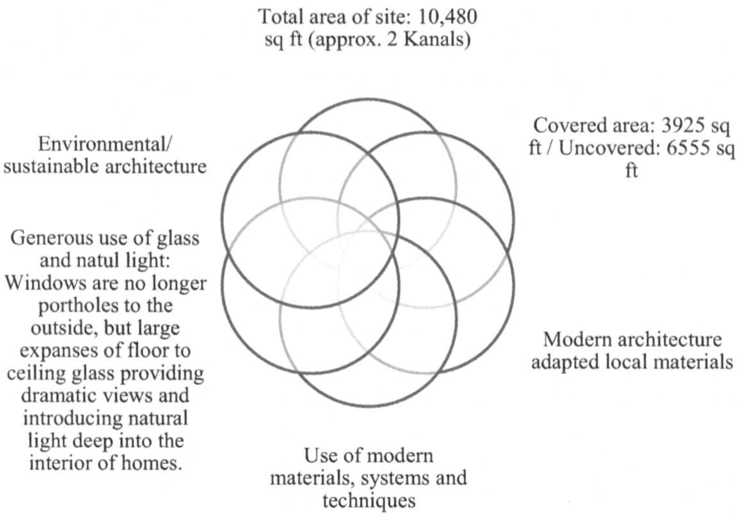

Fig. 9.3 Diagram showing total site area of 10,480 sq. ft. (approx. 2 Kanals). Student documentation by Group A on Khalid Aftab's residence, produced as part of a class project at the Institute for Art and Culture in the course *History and Theory of Architecture & Urbanism*. (Source: Group A, Institute for Art and Culture (2021))

Fig. 9.4 Photograph of Khalid Aftab's residence, produced as part of a class project at the Institute for Art and Culture in the History and Theory of Architecture & Urbanism course taught by Mehwish Abid. (Source: Zain Abbas)

Fig. 9.5 Photograph of Khalid Aftab's residence, produced as part of a class project at the Institute for Art and Culture in the History and Theory of Architecture & Urbanism course, taught by Mehwish Abid. (Source: Zain Abbas)

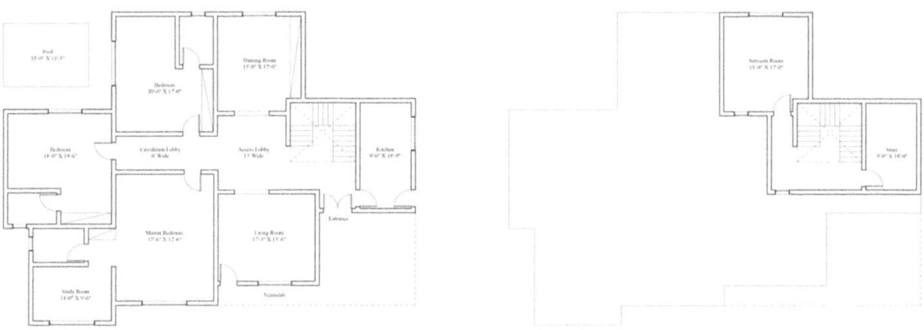

Fig. 9.6 Ground floor and first floor plans of Khalid Aftab's residence, developed as part of a class project at the Institute for Art and Culture in the History and Theory of Architecture & Urbanism course taught by Mehwish Abid. (Source: Ahmed Baasim)

stands as a living archive of how displaced communities and middle-class professionals used pragmatic, practical and culturally resonant design to construct spatial agency. This case, therefore, exemplifies how the decolonial home emerged through deliberate departures from colonial architectural inheritance, grounded not in monumental gestures but in the everyday, material and ethical choices of ordinary citizens.

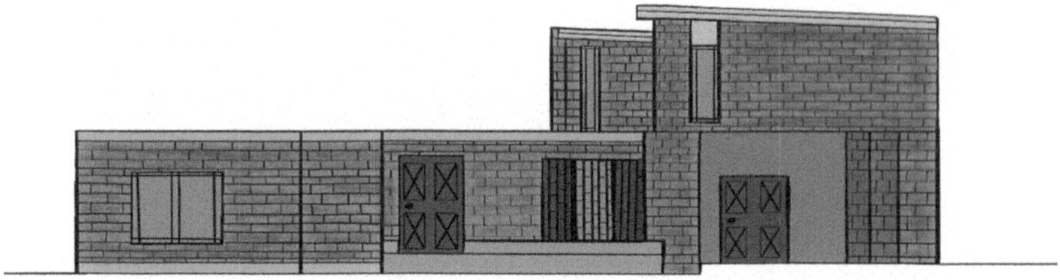

Fig. 9.7 Schematic elevation of Khalid Aftab's residence, produced as part of a class project at the Institute for Art and Culture in the History and Theory of Architecture & Urbanism course taught by Mehwish Abid. (Source: Zain Abbas)

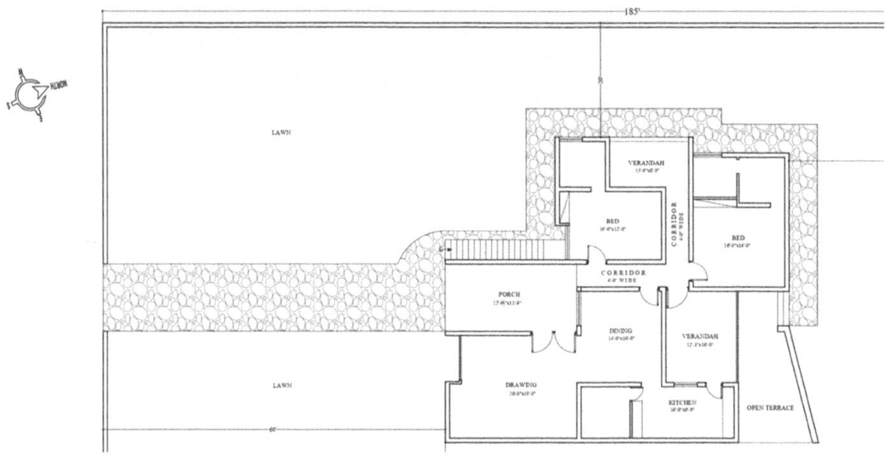

Fig. 9.8 Ground floor plan of Kashif and Hameeda Bibi's residence, produced as part of a class project at the Institute for Art and Culture in the History and Theory of Architecture & Urbanism course taught by Mehwish Abid. (Source: Maryam Tariq and Ahmed Baasim)

9.6.2 Kashif and Hameeda Bibi Residence

Architect Riaz Ahmad was strongly influenced by Charles Correa's spatial principles, particularly the use of open-to-sky spaces and integration with the landscape. The house (see Figs. 9.8 and 9.9) was situated on the plot, featuring both a front and back garden, as recounted by Hameeda Bibi and Kashif Saheb during an oral history interview with the author and student research team in Lahore, March 2021. Existing trees were retained and woven into the structure itself—one tree was allowed to pass through a cut-out beam rather than being felled, demonstrating an evident ethic of environmental continuity. The soft landscape incorporated regional species such as guavas, mangoes and plums, drawing on the model town's historical identity as an orchard zone. The walls were constructed 13.5 inches thick, with later renovations introducing thinner 9-inch masonry (Abid et al. 2021).

Fig. 9.9 Schematic drawing of layout and elevation of Kashif and Hameeda Bibi's residence, designed to evoke a sense of openness and provide comfort by fostering a natural, airy atmosphere. Produced as part of a class project at the Institute for Art and Culture in the History and Theory of Architecture & Urbanism course taught by Mehwish Abid. (Source: Maryam Tariq)

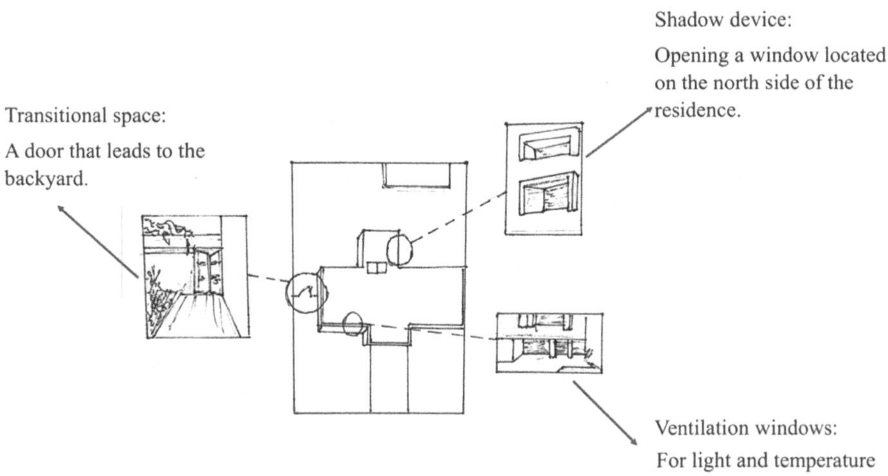

Fig. 9.10 Schematic drawing of Kashif and Hameeda Bibi's residence, produced as part of a class project at the Institute for Art and Culture in the History and Theory of Architecture & Urbanism course taught by Mehwish Abid. (Source: Maryam Tariq)

The design also embodied localised material aesthetics: the staircase balustrade was crafted with references to regional design, while the facade featured granite stone cladding. Inspired by the international style of the 1960s, it bears similarities to buildings like the University of Islamabad, designed by Edward Durrell Stone (O'Meara and Peck 2019). Window shades constructed from concrete, rather than brick, and horizontal and vertical shading devices were used on the north and west sides to mediate sun and airflow—a rare but intelligent climatic response in this region (see Fig. 9.10). Ventilation and orientation were carefully considered. Long

Fig. 9.11 Traditional architectural traces observed in the residence, including a door leading to the backyard, a functional hand pump and large vertical windows with brick shading elements. *Produced as part of a class project at the Institute for Art and Culture in the History and Theory of Architecture & Urbanism course taught by Mehwish Abid.* (Source: Maryam Tariq)

windows facilitated cross ventilation, adhering to Khem Chand's colonial-era guidelines on airflow in domestic housing (Glover 2008). East and South facades were shaded against direct sunlight. Traditional bamboo blinds offered additional filtering and ventilation, preserving indoor thermal comfort in Lahore's intense summers.

On the north side of the house, Riaz Ahmad created a transitional threshold between the front and the backyard by designing a doorway. The backyard held a still-functional hand pump and a productive kitchen garden, which the family maintained with pride (see Fig. 9.11). A servant quarter was placed at the rear, in line with the model towns bylaw that designated 1/3 of the plot for open space. A first-floor balcony was positioned to overlook this garden—a shaded, breezy corner ideal for rest and reflection.

This residence exemplifies how personal memory, family legacy and climatic logic intersect with post-partition spatial agency (see Figs. 9.12 and 9.13). Through design choices that preserved trees, shaped airflow and honoured both modernist aesthetics and local identity, this home reflects the transformation of colonial allotments into culturally embedded domestic spaces. It offers another vivid articulation of the decolonial home, one rooted in continuity of care and quiet resistance.

Fig. 9.12 Front elevation of Kashif and Hameeda Bibi's residence, produced as part of a class project at the Institute for Art and Culture in the History and Theory of Architecture & Urbanism course taught by Mehwish Abid. (Source: Maryam Tariq)

Fig. 9.13 Detail of the front elevation of *Kashif & Hameeda Bibi's residence*. Class project from the course *History and Theory of Architecture & Urbanism*, taught by Mehwish Abid at the Institute for Art and Culture. (Source: Maryam Tariq)

9.7 Concluding Remarks

This chapter has traced how post-partition residents of Lahore—through architectural adaptation, affective memory and spatial autonomy—participated in a subtle yet sustained reworking of colonial domestic forms. Moving beyond nostalgic accounts of loss or longing, such as Levi-Strauss's (Levi-Strauss 1955) romanticised evocation of Lahore or Naipaul's imperial disappointment, this study has shown that architecture in postcolonial cities does not simply inherit the melancholia of empire. Instead, it becomes a site of negotiation, resistance and cultural authorship.

By focusing on residential transformations in Model Town, this chapter repositions the home as both an archive of memory and a material terrain of decolonial practice. The concept of the *decolonial home* emerged not through monumental ruptures but through incremental acts: the retention of trees, the adaptation of ventilation systems, the reorientation of domestic space around gendered privacy and the use of indigenous and local materials within modernist forms. Through oral history, material documentation and student-led site research, this work affirms that residents—often displaced or institutionally marginalised—exercised agency in shaping built space. These interventions complicate the colonial binary of planner and subject, revealing how domestic architecture became a lived response to the fractures of partition, as well as a future-facing tool of belonging, recovery and continuity.

Ultimately, this chapter reframes Lahore's architectural history as a terrain of quiet insurgency, where homes were not passive inheritances of colonial modernity but active, situated responses to it. In doing so, it contributes to broader conversations on decolonial spatial practices, making visible the everyday politics of rebuilding and remembering after the Empire.

References

Abid M, Khurram S, Baasim A, Niazi AK, Zafar A, Gul R, Ali H, Ataina S, Ali M, Abbas Z (2021) Oral history interview with Khalid Aftab. Lahore, March 2021

Anbrine S, Burgess J (2014) The Co-operative Model Town Society: history, planning, architecture and social character of an indigenous garden suburb in colonial Lahore. The Cave Temples of India. https://architexturez.net/doc/az-cf-177373

Anbrine S, Akhtar M, Qureshi RA (2024) The evolution of Cooperative Model Town Housing Society, Lahore

Appadurai A (1986) Introduction: commodities and the politics of value. In: Appadurai A (ed) The social life of things: commodities in cultural perspective. Cambridge University Press, Cambridge, pp 3–63. https://doi.org/10.1017/CBO9780511819582.003

Ashcroft B, Griffiths G, Tiffin H (eds) (2013) Postcolonial studies: the key concepts, 3rd edn. Routledge, London. https://doi.org/10.4324/978023777855

Atalay S (2006) Indigenous archaeology as decolonizing practice. Am Indian Quarter 30(3/4):280–310

Buckley L (2018) Paint, cement, photographs: on the endurance of imagery in Western Kenya. Critical Arts 32(1):43–56. https://doi.org/10.1080/02560046.2018.1437197

Chakrabarty D (2004) Postcoloniality and the artifice of history: who speaks for Indian pasts? In: Brydon D (ed) Postcolonialism. Critical concepts. Routledge, London

Chishty, S.M. 1948. Homes of the day. Lahore

Fanon F (1963) The wretched of the Earth. Translated by Constance Farrington. Grove Press, New York

Ghom PV, George A (2023) Decolonizing South Asian architecture: sustainable and community-criented cocial housing in India. VITRUVIO Int J Architect Technol Sustain 8(1):46. https://doi.org/10.4995/vitruvio-ijats.2023.19499

Glover WJ (2008) Making Lahore modern: constructing and imagining a Colonial City. University of Minnesota Press, Minneapolis

Greenberg J (2005) Generations of memory: remembering partition in India/Pakistan and Israel/Palestine. Comp Stud South Asia Afr Middle East 25(1):89. https://doi.org/10.1215/1089201x-25-1-89

Hodder I (2013) The past and present: an introduction to anthropology for archaeologists. BT Bashford, London

Hopkins AG (2008) Rethinking decolonization. Past & Present 200(1):211–247. https://doi.org/10.1093/pastj/gtn015

Iqbal S, Batool S (2024) Embodied spaces: architectural elements as symbols of power and identity in partition-era historical fiction. Pak Lang Humanit Rev 8(3):502–514

Jackson S (2005) Cape colonial architecture, town planning, and the crafting of moden space in South Africa. Afr Today 51(4):33. https://doi.org/10.1353/at.2005.0042

Jansen JC, Osterhammel J (2017) Decolonization as moment and process. In: Decolonization: a short history. Princeton University Press, Princeton, pp 1–34. https://doi.org/10.1515/9781400884889-002

Khan R, Salim A, Baluch Z, Mehreen K, Kaleem M (2021) British colonialism and the indirect rule: a hierarchical administrative structure to control the unruly tribes. Human Soc Sci Rev 9(3):1366. https://doi.org/10.18510/hssr.2021.93136

Levi-Strauss C (1955) Tristes Tropiques. Librairie Plon, Paris

Malik S, Azhar HK, Ku. (2019) An investigation of house designs in Lahore: transformation of residential architecture from traditional to modern. J Design Built Environ 19(1):49. https://doi.org/10.22452/jdbe.vol19no1.5

Mignolo W (2011) It is our 'modernity', delinking independent thought and Decolonial freedom. In: Darker side of Western modernity. Duke University Press, Durham, London, pp 118–148

Moffat C (2022) Lahore after the modern: architecture, equality and community in Yasmeen Lari's Anguri Bagh. Glob Intell Hist 9:157. https://doi.org/10.1080/23801883.2022.2062419

Mumtaz KK (1985) Architecture in Pakistan. Concept Media Pte Ltd, Singapore

O'Meara P, Peck LK (2019) Indiana University in Pakistan. In: O'Meara P, Peck LK (eds) Indiana University and the world: a Celebration of collaboration, 1890–2018. Indiana University Press, Bloomington, pp 54–59. https://doi.org/10.2307/j.ctvkjb3z5.9

Peck L (2015) Lahore: the architectural heritage. Ferozsons (Pvt) Ltd., Lahore

Peled K (2016) Things that matter: nostalgic objects in Palestinian Arab homes in Israel. Middle East Stud 53(2):229. https://doi.org/10.1080/00263206.2016.1223055

Pires E (2014) Re-scripting colonial heritage. Cultura 11(2):131–141. https://doi.org/10.5840/cultura201411218

Rahman T (2002) Language, ideology and power: language learning among the Muslims of Pakistan and North India. http://ci.nii.ac.jp/ncid/BA9145902X

Schwartz B (2009) Rethinking the concept of collective memory. In: Tota L, Hagen T (eds) Routledge international handbook of memory studies. Routledge, London

Shama A (2014) The Cooperative Model Town Society. https://architexturez.net/doc/az-cf-177373. Accessed 20 June 2025

Singh B, Singh B (2019) Punjab under the British rule: Historicising the local transformations. Indian Hist Rev 46(2):207. https://doi.org/10.1177/0376983619889520

Spivak GC, Colpani G, Mascat JMH (2022) Epistemic daring: an interview with Gayatri Chakravorty Spivak. Postcolonial Stud 25(1):136–141. https://doi.org/10.1080/13688790.2022.2030600

Talbot I (2007a) The Punjab under colonialism: order and transformation in British India. J Punjab Stud 14:3–10

Talbot I (2007b) A tale of two cities: the aftermath of partition for Lahore and Amritsar 1947-1957. Mod Asian Stud 41(1):151–185

Vandal P (ed) (2011) Historiography of the architecture of Pakistan and the region. Thaap Publications, Lahore

Wattas R, Gandhi D (2018) Corbusier's brave New World: personal impressions/P.L. Varma. In: Wattas R, Gandhi D (eds) Le Corbusier rediscovered: Chandigarh and beyond. Niyogi Books, New Delhi

Chapter 10
Navigating Contemporary Boundaries of Johannesburg Through Toxic Landscapes

Finzi Saidi and Alona Martinez Perez

Abstract The city of Johannesburg has been a place of opportunity, since its establishment in the early 1900s. However, access to opportunities was not equitable and was controlled through racial segregation laws of colonialism and apartheid. For black people, opportunities for a better life in Johannesburg were limited, both in time and spatial location, and enforced through multiple borders and boundaries. This chapter discusses how African migrants, of little economic means, navigate new 'borders and boundaries', in the city of Johannesburg. These migrants utilise various infrastructures some of which were previously used to segregate society. This study explores migrants' experiences by adopting empathy and projective design concepts. This chapter examines two master's research by design projects of Unit15X, which tell the new stories of how migrants navigate both the visible and invisible borders of the city of Johannesburg, which have been termed Toxic Landscapes. The first project explores the visible borders, a disused train station repurposed by informal dwellers, giving it new life. The second project explores social borders and xenophobia and how they can be transcended by raising awareness and education in space. The propositional methodology of investigating the toxic landscapes encouraged the students of Unit 15X to reimagine new uses for border or boundary spaces. The *Toxic Landscape* projects speculate with socio-ecological solutions that may give rise to socio-economic justice to struggling contemporary societies within the city of Johannesburg.

Keywords Toxic landscapes · Environmental injustice · Spatial exclusion · Migrants · Johannesburg

---

F. Saidi (✉)
Department of Architecture, University of Johannesburg, Bunting Road Campus, Auckland Park, Johannesburg, South Africa
e-mail: finzis@uj.ac.za

A. Martinez Perez
De Montfort University, Leicester School of Architecture, The Gateway, Leicester, UK
e-mail: alona.martinezperez@dmu.ac.uk

G. Cimadomo et al. (eds.), *Urbanism at Borders*, The Urban Book Series,
https://doi.org/10.1007/978-3-032-09594-7_10

10.1 Introduction

Architect Rafael Moneo discusses two distinctive phases in Aldo Rossi's career as follows:

> To know and to feel. An imaginary line divides this [...] into two parts: one that shows Rossi as a slave of knowledge and one where he is a victim of feeling. Feeling prevails at the surprising conclusion that one is more capable of feeling than of knowing. In sum, in examining the architectural work of Rossi, we will witness a shift from knowledge to feeling. (Moneo 2004:150)

Taking this emphasis as a starting point, this chapter explores the city of Johannesburg in the context of new conflicts taken from the planning of the apartheid city. The metaphor in Moneo's quote is employed to delve into the theory and the projects which are at the base of this research. The theory is the one behind the creation of Johannesburg as a city, and the project is a reflection and testing of those theories as a means of exploration in Johannesburg. This project uses this theory to discuss two students projects navigating 'feelings' of desperation, as a process of knowing and understanding the inhabitants of Johannesburg who occupy two distinct positions along the city's 'border.' The first project explores the border situations that are the result of apartheid planning—the railways stations and networks, which facilitated wealth extraction through exploitation of cheap native labour, also served to separate Black and White populations within the city. The second project explores the feeling of desperation of African migrants who, having crossed South Africa's physical borders, have to navigate the psychological borders of settling in post-apartheid city of Johannesburg. Johannesburg is a city defined by gold mines, which drew thousands of migrants from across Africa and Europe. These influxes of people, attracted by the gold extraction, gave rise to a planning regime that created new borders for migrants. This chapter uses Rossi's interpretation of theory-knowing and practice-project to explore these issues in the context of the new conflicts created in South Africa. As the words of the song by the renowned South African musician Hugh Masekela (1974), it describes the situation being explored in the following sections: 'There is a train that comes from Angola and Mozambique. From Lesotho, from Botswana, from Swaziland. From all the hinterlands of Southern and Central Africa. This train carries young and old, African men. Who are conscripted to come and work on a contract. In the gold and mineral mines of Johannesburg'.

The South African architect Lindsay Bremner (2010:62) describes the urban conditions of Johannesburg: 'The first [condition] is that of 'smooth space' around which cluster ideas of looseness, deregulation, de- and re-territorialisation. The second is 'immaterial architecture', incorporating ideas of the spectacle and the imaginary. The third is 'terror' referring to new forms of political action and urban sovereignty'. In this space, we engage with urban borders that overlap yet reflect different dimensions of the African City. Johannesburg is a city of extraction, a city where workers from all over Africa arrived by train to try their luck, but at the same time, a unique example of vibrancy and culture that make the city's conditions unique. Extraction implies the accumulation of wealth, but also the legacy of a brutal and challenging period of South Africa's recent history, post-apartheid planning. Using Aldo Rossi's dualities of knowing and feeling offers us a specific order for discussing some of the borders of urban planning existing in Johannesburg.

The idea of mining and extraction is offering us the beginning of the *urbis*, of the city itself. The city, referred to as the City of Gold, became a place for migration not just of other African migrants but also white European migrants looking for a better life in Southern Africa, arriving in a country shaped by the borders of racial segregation and the urban planning that emerged from them. The system is critical of the three urban conditions that Bremner describes today in modern Johannesburg. The word extraction, according to the Cambridge Dictionary (2024), means: 'the process of removing a substance from the ground or another substance' but also 'If you say that a person is of a particular extraction, you mean that the person originally came from the stated nation or country'. The word's meaning confronts us with a painful past, where the city today is a city for all. Still, the legacy of the wealth extracted by a brutal regime from the population who was subjected to racial segregation is incredibly problematic. The legacy of the planning of that period divided the borders of the city into areas of segregation and racial separation between black, white and coloured.

> It recasts Johannesburg's enclaves, fortresses, building demolitions, building hijackings, evictions, protections, payoffs, duplicitous alliances and partnerships, as instruments of something banal, sometimes brutal war, profoundly reshaping urban space and life within it and begging understanding of how architectural agency and intelligence are deployed across it (Bremner 2010:107).

These border spaces segregated racial separation through practical planning tools and regimes, which were just a reflection of the brutality of the system of government, the Apartheid. A border where black workers lived at the edge of the Gold City in townships (such as Soweto) to extract the wealth of their own country in the mines, without accessing the down town areas only planned by the white sectors of a highly divided society. To understand the city's urban conditions, we must apply multiple lenses—revealing a richness that is uniquely and profoundly African. *Urbis* is a city of multiple layers: borders of extraction, borders of spatial planning, all cohesive of racial borders of one of the most painful and difficult regimes of racial and spatial segregation in recent history. It is also an example of a city that is now positive, resilient and welcome to all in spite of its recent past. Defining the concept of extraction here presented, Bremner writes:

> In Africa, for instance, the postcolonial nation state's incapacity to build economic underpinnings of political authority and order within its territory resulted in an enclave economy connected with the extraction of valuable resources and caught up in vast, lucrative transnational circuits. These have become privileged spaces of war and death, fed by militias, mercenaries, privateers and child armies. The new geography of resource extraction [...] which brutally attempts to immobilise and spatially fix whole categories of people (Bremner 2010:96).

> Apartheid, the official government policy of racial segregation that existed in South Africa between 1948 and 1994, conjures up in the global popular imagination of walls, fences palisades and barbed wire keeping South African of different race groups apart [...] The

> boundaries drawn by the apartheid regime were fluctuating, porous and ill-defined and its most concerted attempts to make 'borders' was the creation of 'Bantustans' (or 'homelands' as they were otherwise known) as self governing or independent black labour reserves for white industry, and of 'townships' as temporary accommodation for those working in urban areas [...] Instead, there were countless mechanisms for control and humiliation—racially discriminatory laws, administration boards, commissions of enquiry, town planning schemes, health regulations [...], These delineated South African society during the apartheid years and produced its characteristic landscapes (Bremner 2010:107).

Here we are dealing with the borders of race and spatial segregation. As an author that comes from Europe and now have worked over seven years with South African colleagues, in one of my visits to the township of Soweto, a South African Black architect and academic showed me around his neighbourhood and explained to me how his parents were segregated and how now in modern South Africa he was working and living in down town, something that his parents or grandparents were not able to do. I recall listening to him, telling me how the old regime planned ridges of land from his township so you could not see the central business district in the distance, so you could not dream or have the aspiration to even look at that kind of city to work in it. The conversation that happened in 2017 opened to me some of the brutalities and legacies of that recent period of urban planning history, but at the same time, this new city is a welcoming and positive city full of vibrancy and welcoming to all, never mind where you come from. I have always found this condition incredibly positive, a testament to the fading borders since democracy. Democratic South Africa exemplifies resilience and progressive architecture and urbanism, showcasing a city that looks ahead to a great and vibrant future as an African metropolis.

Every planning process relies on tools for its implementation, and in this case it was shaped by a single planning instrument, the Group Areas Act of 1950. Bremner writes:

> Under this legislation, each race group was consolidated into its own residential area, which was to be able to expand in case of future development [...] Areas were separated, preferably by natural barriers (ridges or rivers), or by highways, railways or buffer zones of open space. Each racial group had access to a work zone, where interaction between them was permissible. However, in moving to and from work, no race group was permitted to cross the residential area of another. A centralized production and exchange facility (a central business district) accommodated white-administered civic, commercial and financial functions (Bremner 2010:162).

These borders were created by sinister and effective planning tools of racial segregation, but also economic exploitation and segregation controlled by a racial elite that extracted and exploited the land and its resources and the inhabitants of the country.

10.2 Unit 15X: Context and Pedagogic Approach to Toxic Landscape— New Boundaries and Borders

In 2021, Unit 15X, a research unit that operates under the theme of the Unit System Africa in the Graduate School of Architecture (GSA) at the University of Johannesburg (UJ), explored the theme of 'toxicity' and 'landscape' in the mining cities and landfills. Most cities and towns in South Africa and the continent have connections that may be historic or ongoing with mining activities that include open pit or underground mining and extraction of natural resources such as coal, gold or platinum. While mines close when they are decommissioned, their toxic impact lasts many years. Similarly, landfills are an ever-growing feature of cities symbolising the urban landscape's consumption economy, and they also have a long-lasting toxic impact on the urban environment (see Fig. 10.1). This toxicity is visible in the over-burdened mine dumps, contaminated surface and ground water, polluted airscapes and 'left-over' infrastructure. Socially, too, the effect can be felt in economically unsustainable townships, derelict infrastructure that remains long after the mines have closed and working conditions that many workers are subjected to. Together with mining, other similar capitalist-driven industries (oil refineries, hazardous waste dumps), post-traumatic spaces can also be included in this array of toxic landscapes.

Unit 15X is not merely concerned with uncovering sites of environmental degradation; it is interested in exploring alternative ways to remediate the urban landscape and to create resilient public spaces. Moreover, Unit 15X aims to disrupt the ecological crisis as usual by asking what new questions may emerge by raising the term landscape [urbanism] alongside toxicity and to imagine new forms for the landscape. The Unit's interest is in re-interpreting this full and wide array of the term 'toxic' in the urban context. Unit 15X's students adopted a propositional methodology for investigating landscape architecture. They seek to imagine new ecologies for toxic landscapes through innovative infrastructure design. The Unit sought to expand the meaning of design for its students as '…a mediated way of engaging the world through situated knowing and imagining'.

A key realisation in the process of exploring toxic landscapes in Johannesburg was that the term 'toxic landscapes' could be expanded to include not only the

Fig. 10.1 Toxic landscape of Johannesburg as a result of gold-mining extraction. (Source: Leon Krige)

physically damaged landscape blighted by mining activities but also urban spaces that had been shaped by apartheid policies and their discriminatory processes. These toxic landscapes, physical or social, can be interpreted as 'borders' through which studies of new migrant realities can be explored and imagined. Unit 15X aimed to develop the students' ability to raise critical questions about new design issues in the 'toxic landscape' and their impact on cities, with a view to developing multiple and socially acceptable responses. By examining different 'border' landscape conditions and public spaces in Johannesburg, Unit 15X sought to reveal the common characteristics that migrant communities relate to and need in their new landscapes. The Unit encouraged our students to speculate on new uses for some of these toxic landscapes to contribute meaningfully to deeper understandings of them and to propose socio-economic and ecological solutions that ameliorate the damage. It also hoped to raise critical questions such as how the processes of making landscapes, through social, environmental and architectural means, can be thought of and represented? How can the experience of landscapes (including toxic ones) be constructed and represented?

The research was not merely concerned with uncovering sites of environmental degradation; it is interested in exploring alternative ways to remediate the urban landscape and to create resilient public spaces. Moreover, Unit 15X aims to disrupt the ecological crisis as usual by asking what new questions may emerge by raising the term landscape [urbanism] alongside toxicity and to imagine new forms of the landscape. Our interest is in re-interpreting this full and wide array of the term 'toxic' in the urban context, which Coquery-Vidrovitch (2005:5) loosely refers to as 'dirt-poor habitat that is poorly served and under-integrated'. Two students explored toxic landscapes as 'social spaces' navigated by contemporary migrants—both local and international—that signify new boundaries of 'extraction' of wealth by the least economically empowered inhabitants.

Borrowing from Moneo's theory of 'knowing through feeling', Unit 15X allows its students to explore the feelings of place in contemporary Johannesburg, in what we imagine to be a freer post-apartheid South Africa. The first student's project, by Linda Lesolle, explores the despondent feelings of individuals who cannot access Johannesburg's formal economy and housing opportunities. These limitations arise from their restricted capacity to participate. Lesolle critically interrogates the physical infrastructure, specifically the train station, that was historically used to segregate and deny Africans and other races fair participation in South Africa's economy. The second student, Tsaukani Masilela, explores the 'feeling of place' in the inner city of Johannesburg as experienced by southern African migrant communities as a channel through which to imagine 'spaces of justice' (Masilela 2021).

Nivesjö traces the trains' historical role and black people's lives in colonial cities like Johannesburg. Firstly, the trains transported black workers, cheap labour, from countries in the southern African region to Johannesburg to work in the mines as contract workers in the service of white capital. Secondly, the trains that shuttled to and from the city-centre to segregated townships defined the daily exploitation of black labour in the service of their masters (Nivesjö 2024). In some ways, these two projects define an in-between space which can be termed as an invisible boundary or border that is navigated by both international and internal migrant communities in Johannesburg daily. One might argue that these two student projects conjure arguments and feelings about how these ‘borders’ might be imagined.

10.3 South Africa Historical Context and the Creation of Colonial Borders

Borders can be considered transitional spaces, marking the edge of a defined space that can be identified, defended and expanded. Borders can indistinctly signify zones of peace or violence for groups of people. Since the demarcation of Africa into various small, arbitrary countries by colonial masters in 1886, it is generally true that borders have been the source of conflict to this day. The objective of the colonial masters’ demarcation of Africa was largely to extract as many economic benefits for their countries as the unity of African nations. But for the Africans in these countries, the new borders became a rallying point for their struggle for economic and cultural emancipation.

The quest for economic exploitation of colonial Africa meant the creation of infrastructures through which to extract the wealth. Colonial masters; therefore, established settlements—cities and transit routes—such as rail and roads, through which material resources could be extracted with minimal cost. Thus, two types of borders could be understood to make up the urban colonial African context. The first borders were the transportation routes that transcended the borders either to transport materials or the human resources needed in extraction and production processes. The second type of borders were formal or informal peripheral black settlements in cities and towns, designed to accommodate the black population separated from the white-only settlements, as stated above by Bremner.

Thus, the duo-legacy of the border continues to impact the lives of current residents of African cities from ‘Cape-to-Cairo’ to borrow the colonial euphemism for development. In this context, the city of Johannesburg can be explored as an ideal ‘border’ city in Africa, holding lessons of how African migrants—local and external—navigate their movement and transient existence in contemporary urban space, typified by the physical landscape or minescape barriers, the *cordon sanitaire* legislated by the Native Urban Areas Act of 1923 (Chipkin 1998:156).

10.4 Johannesburg

Johannesburg emerged from a gold prospector’s mining site in the 1880s. It can be argued that gold-mining in Johannesburg drove the modernisation of the southern African region. The industrialisation of Johannesburg required cheap labour that could only be sourced across the borders of the South African settlements, aided by natural disasters, but mainly by draconian colonial laws that forced villagers from their cultural lands to cities such as Johannesburg in search of work opportunities (Shula and Rathbone 1987). African workers began their migration through two 'borders'—first across the country, a journey eulogised by Hugh Masekela's song Stimela, as discussed above. The second border, that typified Black migrant workers' lives in Johannesburg, was that between their segregated settlements—townships—often located outside of the city centre, which was reserved for White citizens, linked to the centre by railway. Landforms, rivers and motorway infrastructure separated the black townships. Black workers required permission to enter the city centre, known as the passbook.

Officially, this forced migration, which began in the 1940s, marked the beginning of the industrialisation and modernisation of southern Africa. Upon arrival in Johannesburg, African migrants had to navigate the second 'border' between the workplace, initially the mine and their new homes. Initially, the new homes were the mine compounds that Chipkin (1998) calls 'Kampongs'. These were prison-like accommodations meant to control the mine workers' movements and to ensure that they returned to their country or place of origin at the end of their contracts, and also to maintain low production costs by keeping blacks in these pseudo-camps. Thus, we can conclude that the City of Johannesburg was never intended to be a place for black or African inhabitants to settle. Later, as the number of Africans working in the mines and town increased, coupled with growing recognised of workers' rights through mine unions, the city of Johannesburg authorities established housing for the black population, but in a segregated manner using the concept of *cordon sanitaire*, outside of settlements for the white population of the city (see Fig. 10.2). This segregation was later perfected by the racist Apartheid government of Barry Hertzog, who enacted the Native Segregation Act of 1949 that legalised separate development and established townships settlements for black only, sometimes based on tribe (Chipkin 1998).

Fig. 10.2 A way-of-life in Soweto. A segregated township for black workers (Source: Ngwato Kekana 2016)

Fig. 10.3 Toxic urban Landscape—Booysens Train stations as an abandoned train station which has been encroached by Booysens Squatter Camp, with 7000 informal settlement shacks. (Source: Linda Lesolle 2021)

The key elements of these townships were the transit routes and rail infrastructures, which ensured that the black population was removed from the city environments—the CBD—by a specific time, unless by permission in the form of the infamous Passbook. It was a small book required by Black men and women to show that they had permission to enter and work in the urban area for a stipulated time period. To black migrant workers, the passbook was known as a 'Dompas,' translated as the 'stupid pass'. From the late 1950s, every black person in South Africa required a passbook to enter the city (Alexander and Chan 2004:4).

The train stations were intended for the black population and were sites of great social activity in the city. They can be thought of as a border. Through this, the city, for black people, could be experienced. In the definition of borders, Unit 15X students reimagined ways of exploring the two 'toxic' environments: The CBD with its legacy of the 'passbook' and the train station (see Fig. 10.3) with its racial designation of a route in and out of the city for black people.

10.5 Toxic Urban Landscapes. Re-Imagining the Future Borders

Two 'border' scenarios can be drawn from the above discussion to explain the experiences of black people in Johannesburg from the historical context. The first is that Johannesburg continues to be seen as the place of 'opportunity' by migrants in the southern African region and therefore provides a site of observation of migrants' settlement process in a city that was never intended for the settlement of the black population. In the post-1994 free South Africa, migrants explore complex spatial relationships in the search for new life in Johannesburg. The 'border', migrants for migrants across Africa, marks a space of new language and socio-economic opportunities within the city. The transit points indicate the second 'border' space, the stations that initially transported black workers from the city to their townships, but have now for various reasons fallen into disuse, becoming 'sites of opportunity' that are occupied and its function 'reinvented' by migrant communities with no formal skills, except to thrive on informal economies that the city of Johannesburg offers. These two scenarios indicate the importance of studying the experiences of black people in post-independence and post-apartheid cities of Africa as representations of new 'borders' in which black African migrants search for new life experiences. In this context, this chapter discusses two speculative architectural thesis projects to provide new light on how black people, of little means of survival, navigate the city of Johannesburg. This study raises awareness of how migrants navigate the new urban challenges through multiple borders. Under the theme Toxic landscapes, these projects speculate on new spatial opportunities for migrants as they transcend the new borders.

This section discusses two Unit 15X master's students' projects describing the social and economic challenges in the toxic urban landscapes described above. In the context of what this chapter presented as 'the second concept of the border,' Linda Lesolle's project, The Urbanisation of a Ghost Town, reimagines an abandoned train station home to migrant workers working in the informal recycling business industry (Lesolle 2021). Lesolle argues that 'historically, the rail infrastructure was constructed with no cognisance of creating a good urban transportation system that would serve socio-economic advancement for African people' (Lesolle 2021:3). The project explores the abandoned Booysens Train Station located south of Johannesburg CBD (see Fig. 10.4) to reimagine a future based on multiple identities because of the diverse origins of the Booysens Train Station community. Lesolle adopts two methodologies: firstly, visual narrative is used to document and understand activities of current users of the abandoned station. Secondly, Lesolle utilises scenario generation to envision a new programme for inhabitants that includes urban farming, recycling centre, housing and trading, as a basis of new infrastructure, to reuse and repurpose Booysens Train Station (see Fig. 10.5).

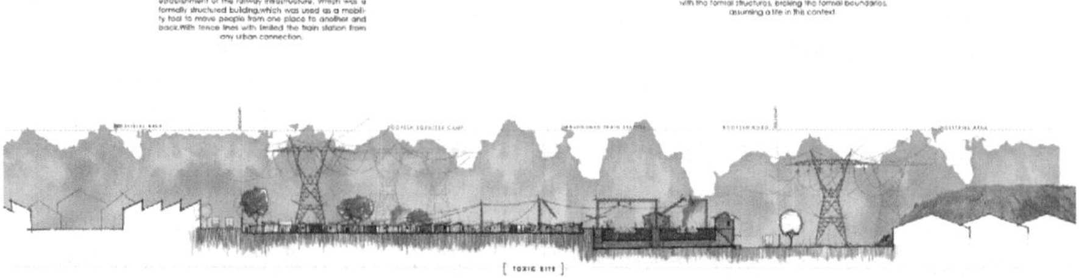

Fig. 10.4 Disused Booyesens Station showing how informal dwellers have repurposed the site. (Source: Linda Lesolle 2021)

Fig. 10.5 Plan of Booyesen Station showing its 'border' character that has been taken over by informal shack-dwellers. (Source: Linda Lesolle 2021)

Fig. 10.6 New tectonics that define new uses including recycling, urban farming and housing. (Source: Linda Lesolle 2021)

Lesolle's project attempts to re-create the former mono-functional Booysen Train Stations into a multifunction urban typology that affords communities with minimal resources and a fair chance of creating dignified urban life, as shown in Fig. 10.6 below. Thus, the former transit complex of apartheid design becomes a site of entry into a higher quality of life for the marginalised citizens of Johannesburg. In this way, the project reimagines the apartheid era 'border', into a contemporary urban space that welcomes migrants with little means to the city of Johannesburg.

The second student, Tsaukani Masilela, presents a project entitled 'The (IM)Migrant City: Re-imagining the Toxic Experiences of Migrants in Kerk Street'. This project traces the stories of xenophobia experienced by migrants in Johannesburg, focusing on Kerk Street. Masilela suggests new spatial configurations that not only recognise the needs of migrants but also educate the wider

Fig. 10.7 Masilela's Collage shows dialogic ideas of speaking about migration through knowledge and educating the public about the challenges African migrants face. (Source: Tsaukani Masilela 2021)

Johannesburg public on the issues faced by migrants, especially those coming from outside of South Africa (see Fig. 10.7).

Masilela's thesis poses a critical question: 'Is it possible for spatial practitioners to educate themselves and the general public about migrants in Kerk Street, Johannesburg through inclusive spatial design?' (Masilela 2021). Masilela documents five city blocks along Kerk Street, by telling stories of how immigrants navigate their livelihood and business on the street as a point of entry into the assimilation of their new society in South Africa. Borrowing Doug Saunders' concept of the 'Arrival City' (Saunders 2012), Masilela adopts a threefold exploration method, including the actor network theory that engaged immigrants on the street; second, Masilela used performative theory by setting up an on-site experiment to engage the broader public on migrancy; and third, a design proposal that envisions a transformed Kerk Street recognising the contribution of immigrants to contemporary Johannesburg. The author argues that through the purposeful design of public spaces and streets, dialogue that raises awareness of the challenges facing immigrants, such as xenophobia (see Fig. 10.8), can be encouraged. This may lead to improved lives for all in society, as outlined in the UN's Sustainable Development Goals.

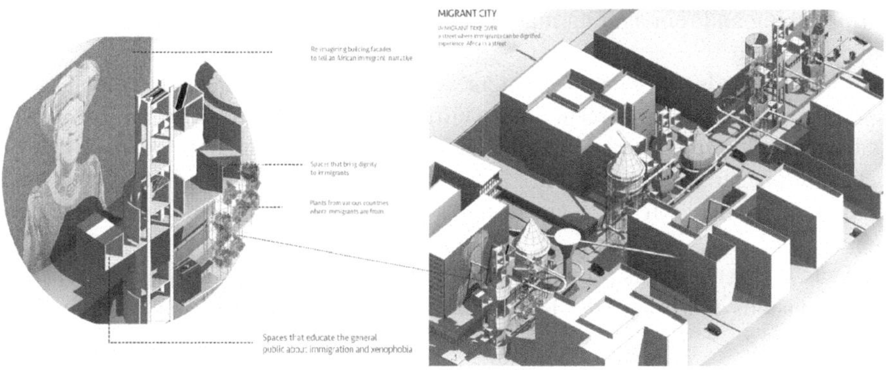

Fig. 10.8 Dialogic drawings of African identities and spatial representation in Kerk Street. (Source: Tsaukani Masilela 2021)

10.6 Conclusions

This chapter began with a theme of knowing through feeling, as advanced by Aldo Rossi's theory '...to know and to feel' (Moneo 2004). The arguments presented in the two students' work outline a methodological approach that explores their 'feelings' of empathy towards the experiences of migrant black people in the city of Johannesburg, a city that was never designed to include them. The students' explorations remind us of the feeling of pain and suffering that black people endure due to the past borders imposed by colonialism and apartheid, and of contemporary post-independence 'borders' which are socially and economically exclusive. These works engaged in a theoretical exploration to imagine places of spatial equity for disadvantaged citizens—the migrant communities—by tackling 'borders' with themes of inclusivity and reuse of existing infrastructure like abandoned train stations. In the students' projects, it is possible to see their suggestions for the adaptive re-use of infrastructures that could address some of Johannesburg's socio-economic challenges. The students explore these themes by selecting projects they identify with, a key concept and value enshrined in Unit 15X's constructivist teaching approach, a way of 'feeling'. In this approach, the projects demonstrate the 'feeling' as a legitimate process of 'knowing' as posited by Moneo. Unit 15X argues that 'feeling' provides valid motivational force to imagine multiple solutions to urban challenges in Johannesburg in a way that adds to a complex and comprehensive understanding of the city. Both students attempt to answer what Myers (2014) terms as 'spaces of dynamism and what change could look like, for African urbanism, with specific reference in this case, to the city of gold, Johannesburg'.

With this methodology of city engagement, Unit 15X fosters a new and transformative role for designers and architects to become public agents of change by getting themselves involved in policy-making and urban consultancy. Designers need to fill these public roles to broaden the view of the design disciplines and redefine architecture, landscape architecture, planning and urban design, including form-making and reimagining marginal communities' use of space and how life is supported by it.

This chapter has discussed 'borders' as they exist in Johannesburg, a city as a toxic landscape that is the result of the infrastructure of racial segregation, the train stations and the lived experiences of migrants from African countries in search of better fortunes in the 'city of gold'. In line with Aldo Rossi's theory of knowing and feeling, the students' feelings and design scenarios present opportunities for reflection on borderline communities that lie on the fringe of city planning regimes. The students' border projects provide city policy planners and designers with transformative visions that may lead to spatial and social justice in complex cities like Johannesburg.

References

Alexander P, Chan A (2004) Does China have an apartheid pass system? J Ethn Migr Stud 30(4):609–629. https://doi.org/10.1080/1369183041000169948

Bremner L (2010) Writing the City into being: essays on Johannesburg, 1998–2008. Fourthwall Books, Johannesburg

Cambridge Dictionary (2024). https://dictionary.cambridge.org/dictionary/english/extraction. Accessed 02 June 2024

Chipkin CM (1998) Preparing for apartheid. Pretoria and Johannesburg. In: Fischer RC, le Roux S, Mare E (eds) Architecture of the Transvaal. University of South Africa, Pretoria, pp 148–173

Coquery-Vidrovitch, C. 2005. The history of African cities south of the Sahara from the origins to colonization. Princeton: Markus Wiener

Kekana N (2016) The lobby. Unpublished Master of Architecture dissertation. University of Johannesburg

Lesolle L (2021) The re-urbanization of a ghost Train Station. Unpublished Master of Architecture dissertation. University of Johannesburg

Masekela H (1974) Stimela. https://genius.com/Hugh-masekela-stimela-the-coal-train-lyrics. Accessed 10 Apr 2024

Masilela T (2021) (IM)MIGRANT CITY: re-imagining the toxic experience of migrants in Kerk street, Johannesburg. Unpublished Master of Architecture dissertation. University of Johannesburg

Moneo R (2004) Theoretical anxiety and design strategies in the work of eight contemporary architects. MIT Press, Cambridge, MA

Myers G (2014) African cities: alternative visions of urban theory and practice. Zed Books, London

Nivesjö S (2024) The arrival from abroad: train travel and Mobile ideas of race in pre-apartheid south African literature. Engl Stud Afr 67(2):31–45. https://doi.org/10.1080/00138398.2024.2377516

Saunders D (2012) Arrival City. Vintage Books, New York

Shula M, Rathbone R (1987) Industrialisation and social change in South Africa. Longman, London, New York

Unit 15X (2021) Toxic urban landscapes. In: Adu-Agyei D, Mahlangu N, Makhubu J (eds) Unity yearbook. Graduate School of Architecture, University of Johannesburg, Johannesburg

Part III
Barriers to Trade

Chapter 11
The Role of Trade Centres Bordering the Public City Life: New Belgrade Case Study

Aleksandra Jevtović and Aleksandra Stamenković

Abstract During the second half of the twentieth century, New Belgrade underwent significant evolution, emerging as one of Belgrade's most substantial spatial and morphological settlements. Its construction was particularly intensive during the 1960s and 1970s. Four trade centres ('Fontana', 'Old Mercator', 'Delta City' and 'Ušće') embodied the core ideas and concepts that have guided New Belgrade's urban development, promising to transform that part of the city into a beacon of modernity that appears to be curved out as a border within the city. Their association with modernist functionalist urbanism prevailed, transforming them into significant elements in the cityscape and giving rise to numerous theories about the connection between public space and everyday life. Influenced by the prevailing ideology yet striving to align with contemporary trends, Yugoslav architects were cultivating a style that reflected a new, avant-garde, and progressive society. The article delves into the patterns of trade space utilisation for collective gatherings and their primary purpose. Their spatial forms and designs were analysed, exploring the causes behind the lack of awareness about their significance, the degradation of their forms, and their utility. This examination unveils the connections between architectural creativity and broader social and cultural changes, particularly the impact of taste commercialisation.

Keywords Trade centres · New Belgrade · Trade · Post-socialist city

A. Jevtović (✉)
Department of Business Studies Blace, Toplica Academy of Applied Studies, Blace, Serbia

A. Stamenković
Faculty of Digital Arts, Metropolitan University, Belgrade, Serbia

G. Cimadomo et al. (eds.), *Urbanism at Borders*, The Urban Book Series,
https://doi.org/10.1007/978-3-032-09594-7_11

11.1 Introduction

The capital city of Serbia, Belgrade, is situated at the confluence of the rivers Sava and Danube, which is a significant factor influencing its urban development. New Belgrade is a specific area between the old cores of Belgrade, where the Modernist paradigm was implemented in alignment with Le Corbusier's vision of a futuristic city (Jovanović and Stupar, 2022:275). The conceptual assumptions regarding New Belgrade's identity as a modern city space were established in the first decade following the Second World War. Throughout the ongoing process of its construction, there has been a continuous re-evaluation and reinterpretation of the initial modern ideas (Blagojević 2007:13). The development was initiated in subsequent years, giving rise to numerous local community places within the originally envisioned public spaces. These were referred to as 'urban hubs for meeting and gathering' (Jovanović and Stupar 2022:275). Considering their importance for strengthening communities, the article identifies four landscape topics. Their authentic settings have significantly changed the original structure since their construction in the 1960s and 1970s.

The ideological context of the construction of New Belgrade implied the aspiration to become the capital of socialist Yugoslavia as well as a respectable example for the construction of a newly formed state (Ignjatović 2012). As the architect Ivan Štraus (Štraus 1991:12) noted, sketches of New Belgrade in the years after World War II provided the basis for 'the first major architectural tasks that Yugoslav architects could never have thought of before'. The first post-war sketch of New Belgrade's urban regulation (1946), set by modern architect and urban planner Nikola Dobrović, was based on principles of modernism. He considered the concept of 'city landscape' and described it as: 'a new type of spaciousness of buildings and their plasticity, the hollow plasticity of the in-between spaces, architecture of the ground, greenery and the vistas in one organically designed composition whole' (Dobrović 1954:1–3). Dobrović's plan prioritised addressing the traffic problem, aiming to separate residential and administrative buildings from the perimeter block and position them as isolated structures within the green spaces between the main avenues (Milinković et al. 2019). During the subsequent period, numerous distinguished Yugoslav architects engaged in developing plans for New Belgrade. However, the pioneering idea from 1946, aiming to connect the new Belgrade core with the old ones, was nearly lost (Blagojević 2007:63). The urban and architectural concept of the 'City of Tomorrow' was far more obscure than its ideological premises (Bogunović 2005b:293). In late 1946, when tenders for the main administrative buildings of New Belgrade were announced, authors were also requested to submit proposals for the urban plan of New Belgrade (Blagojević 2007:73).

The area of New Belgrade began to take shape in 1948, emerging across undeveloped territory separated by the Sava River from the old town. As observed by Ljiljana Blagojević (Blagojević 2007:58), this space, nestled between older Belgrade settlements and Zemun, represented '[a] symbolically pure zone of centuries-old no-man's land between East and West'. Therefore, it is indicative that the first facility situated in that territory was the Building of the Presidency of the Federal People's Republic of Yugoslavia (FPRY), which was subsequently transformed into the Federal Executive Council building (Mišić 2011; Putnik Prica 2017).

Although New Belgrade was initially planned with the ambition to become the new administrative centre of Yugoslavia, the circumstances resulting from the political and economic crises in the early 1950s deferred the original planning idea, with a shift in focus towards housing. In the next phase of planning and construction, New Belgrade underwent a transformation into a residential neighbourhood without basic urban services (Blagojević 2007), and that paradigm has been dominant for many years. In 1957, Branko Petričić, a student of Le Corbusier and the director of the Urban Planning Institute, crafted a proposal for the General Urban Plan of New Belgrade. While this plan largely adhered to the postulates of the Athens Charter, it is worth noting that the Athens Charter itself had already faced criticism on the global stage (Frampton 1992; Perović 2009). A formal effort towards greater integration of functions was made by the General Urban Plan of New Belgrade, which was adopted the following year (1958). According to Ljiljana Blagojević (Blagojević 2007:155), the absence of a diagnostic phase and the speed at which the revision took place were identified as the causes of its failure in the realm of overlapping functions.

The commercial buildings are distinguished as functionally necessary elements that complement the residential units, but also as objects on which transformations in the domain of aesthetic, functional organisation and significance for the urban environment are particularly noticeable. Objects of that type can be seen as focal points that more or less direct the movement of residents and affect the intensity of public life in their environment. Taking into account the daily role of shopping facilities in the lives of residents, their more detailed analysis will point to the correlation between public urban space and everyday social life.

11.2 Trade Centres During the Period of Socialist Aestheticism

According to the historiography of Yugoslav architecture, the period of the sixties and seventies was recorded as a time of prosperity, ambitious projects and intensive construction (Stojanović and Martinović 1978; Manević 1986; Štraus 1991; Miletić–Abramović 2007; Ignjatović 2012). In an effort to align with contemporary global trends and under the influence of the prevailing socialist ideology, domestic architects were developing designs that reflected a new, avant-garde, progressive and free society (Ignjatović 2012). Considering aesthetic categories, sculptural forms and brutalist structures were particularly favoured. These elements conveyed a sense of monumentality and served as distinctive reference points for the territories to which they belonged.

To comprehend the emergence of modernist urban planning in the area of New Belgrade, the initial step is to examine what scholars have contributed regarding the resolution of the central zone, which extends along the axis of the Federal Executive Building. In 1960, the Urban Planning Institute's team proposed the New Belgrade Central Zone Plan. This plan envisioned the construction of three consecutive urban squares designed for events and manifestations, framed by apartment blocks and multi-storey business buildings. Besides planning, the system of industrialised construction was introduced in the practice of building New Belgrade in the same decade (Blagojević 2007:205). During the 1960s and 1970s and with the construction of the central area, the modular pattern of oversized residential blocks was greatly transformed. Despite the fact that planned squares and public buildings in the central area were not constructed as intended, the segregation of functions has largely remained unchanged (Blagojević 2007:209).

The issues of architecture and urbanism in New Belgrade engaged a considerable number of Yugoslav experts (Stojanović and Martinović 1978). Demolitions during the war, coupled with a rising number of the capital's citizens, mobilised architects, engineers, and urban planners to address the priority requirement of the time—the construction of an extensive housing stock. That should be considered when reviewing the numerous (unjustified) criticisms directed at New Belgrade during those days. The new settlement initially took form as a cluster of geometrically organised, abstract dormitories interconnected by roads. During the first decades of constructing the new city, the spontaneous insertion of stylishly personalised buildings of public content within uniform blocks played a role in breaking modernist urban dogmas. Among the earliest examples of buildings whose construction contributed to the intertwining of contents and functions were the multifunctional centre 'Fontana' in Paris Commune Street (1963–1967), by architect Uroš Martinović, and the Trade Center in Mihajlo Pupin Boulevard (1971–1972; later named 'Old Mercator'), by architect Feliks Bajlon (Bogunović 2005b:304).

The construction of dedicated trade facilities in Belgrade commenced in the early twentieth century with the advent of the secession movement. The buildings of the department store (Robni magazine, 1907) and the Army Officers' Cooperative Building (Oficirska zadruga, 1908) represented the first department stores (Škalamera 1985; Jevtović 2019) and were also considered to be among the earliest examples of the new style being transposed into the Belgrade environment. Within the history of Serbian architecture, these two buildings are regarded as examples of the reception of secession. The construction technology, metal structural elements, and large glass windows support that assertion.

During the interwar period, the most notable examples of trading premises were found in private residential and business palaces, where the ground floors were designated for the display of merchandise (Mitrović 2007b:13). After the Second World War, the socialist society reactivated the existing commercial premises, and at the end of the 1950s (1958), the first self-service shop was opened in Belgrade in the 'Flower Square' (Orig. Cvetni trg) (Bogunović 2005a:93; Bisenić 2009).

The end of the sixties in Yugoslavia was marked by the appearance of the kiosk K67 by Slovenian architect Saša Mächtig (Hubert 2017). It was one of the symbols of industrial design in Yugoslavia, which represented an approach that integrated architecture and design as common practices. With the advent of self-service shops in which customers had direct contact with products, a new type of modern consumerism was established and a hint of the coming epoch was given during which liberalisation of existing market practice occurred (Bogunović 2005a:93–94). The 'major shift in sociological view of the young socialist society' was made by the construction of buildings of that typology, while the culmination of the socialist trade was established by the construction of large department stores (Mitrović 2007b:14).

Department stores were constructed throughout the city. In the area of old Belgrade, the Fashion House on Knez Mihailova Street (1962), by architect Mihajlo Janković, and the *Belgrade Palace* (Beograđanka, 1969–1974), by architect Branko Pešić, stood out. In the basement and across four floors of the annexe of the *Belgrade Palace,* there was a salesroom of the department store *Belgrade* (Robna kuća 'Beograd'), the largest in its time. During the late socialist period, department stores in *Belgrade* were erected in numerous cities in Serbia and are thought to have played a role in their 'literal colonisation' (Ignjatović 2013:34). In terms of the cultivation of consumerism in the Serbian environment, Aleksandar Ignjatović (Ignjatović 2013:34) considered that these stores have played a significant ideological role, presenting themselves as an evident 'substitution for political and economic liberalism'.

In the territory of old Belgrade, the construction of new trade-oriented facilities usually involved the demolition of dilapidated buildings and the interpolation of new ones, transforming that way the urban matrix and original environments (Kadijević 2017), while within the boundaries of the incompletely developed New Belgrade, these insertions were designed to establish the identity of a fresh city segment, where its history was only beginning to take shape.

11.3 Multifunctional Center Fontana (1963–1967)

One of the earliest symbols of New Belgrade, the famous *Fontana,* was built by prominent architect Uroš Martinović (1918–2004) between 1963 and 1967 (Brkić 1992:190–193; Bogunović 2005c). The building was originally erected to serve as the Local Community Center (it combined administrative, economic, cultural and artistic functions, including cinema), and immediately after its construction, it was declared a cultural monument. As an integral part of New Belgrade Blocks 1 and 2 (in Paris Commune Street), Fontana was also part of the first built urban unit of New Belgrade.

The basic structure of the building was expressed by the regular geometrical shapes (see Fig. 11.1). The monumental and rigorous facade represented an attempt to liberate itself from ornamental historicism and tradition in the broad sense. The compact form with prominent frontal facade is structurally broken down into a large number of harmoniously proportioned volumes, simple in line and processing. The

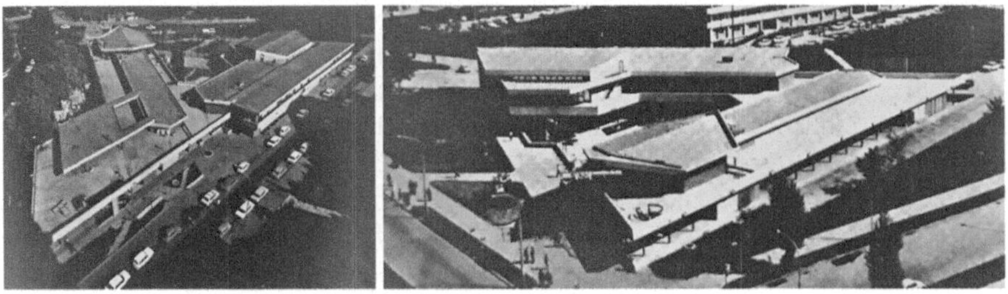

Fig. 11.1 Multifunctional Center 'Fontana'. (Source: Aleksej Brkić)

Fig. 11.2 Multifunctional Center 'Fontana'. (Source: Aleksandra Jevtović 2019)

entrance tract was accentuated by a protruding horizontal ridge block, which did not disturb the quiet flow of recessed and protruding shapes on the façade (see Fig. 11.2).

With its shape, from the appropriate perspective, it emphasised the application of a triangular mesh, which dampened the rigid orthogonal matrix of previously built blocks. Nevertheless, its architecture subtly blends into the existing architecture of nearby residential blocks. In a constructive sense, the importance of the building lies in its reliance on the geometric grid while pointing to a freely developed form characterised by a refined architectural repertoire. The avoidance of classical elements and the insistence on forms that justify their presence with their functional role were supposed to symbolically express the relatively advanced state of Serbian and Yugoslav architecture.

The composition was prevailing asymmetrical. Above the ground floor was a raised corpus, whose dimensions rested on free-standing pillars. These pillars formed a terrace that visually distinguished it from the dimensions of the ground floor. The facade canvases were filled with a large number of individually separated volumes, whose positions were creating spaces of different storey heights. All separated volumes had angular edges and ended with a flat roof, thus visually forming a cascade elevation system. These spaces, with their protrusions from the footprint of the building, further revived the composition. The triangle, as the basic element, conceptually emerged from the shape of the area and influenced the composition of the building. Starting from the triangular border of the area, an attractive structure with acute angles has been achieved. Particularly striking was the perforation of the atrium towards the main boulevard, 'allowing multiplication of visuals in the focus of the art composition and the centre of human gathering' (Mitrović 2012:60).

With its architectural qualities and by its contribution to the development of modernist forms, the Fontana is one of the distinguishing points from the socialist phase of New Belgrade. The building was renovated by virtue of the cooperation of the Municipality of New Belgrade and the Cultural Heritage Preservation Institute of Belgrade, according to the project of architects Vesna Cagić Milošević and Ksenija Bulatović (Vušović 2011; Anonym 2011). This project restored the building to its original condition, and the associated cinema was reactivated after a 10-year break in 2012, which made the facility alive in a functional sense (Anonym 2012, 2014).

11.4 Trade Center 'Old Mercator' (1971–1972)

The period from 1948 to 1980 was marked by a large amount of investments by the government in representative projects, which were shaped in the spirit of contemporary world tendencies. The slogan 'big and new' was followed by the construction of an image of a society that was advanced, free and based on contemporary beliefs. New Belgrade is considered to be the greatest achievement of socialist society and, accordingly, it is an homage to socialism. However, the interpretation of New Belgrade is much more complex than the purely ideological dimension, so buildings that unobtrusively represent the architectural tendencies of the epoch and also transformations that the state has undergone can be distinguished. The Old Mercator building, along with the Fontana, formed the corps of the socialist phase in the history of New Belgrade. It was opened as the representative venue with the cinema Yugoslavia, the hotel Putnik, a supermarket and a number of shops, in the early 1970s, soon establishing itself as an important landmark (see Fig. 11.3). The author of the original project was architect Feliks Bajlon (Mitrović 2012:26).

The architectural programme consisted of simple geometric shapes. The usage, multiplication and modification of rectangular volumes resulted in a recognisable visual record. Therefore, in the elaboration of the form, repetition of the basic element was imposed as the best solution, which gave the possibility of modular redistribution of the structure. Defined in this way, the geometric mass was prompting visitors to move along the marked street lines and to develop their visual experience from a preestablished perspective (see Fig. 11.4).

The entrance to the mall is pulled into the depths of the space and is transparent, simple and delicate in design. It led into the inner space of the mall, which was in stark contrast to the outside arrangement of enclosed cubic forms in raw concrete. Due to the narrow choice of shapes of geometric parts, emphasis was placed on spatial dynamics and movement of volumes. For the same reason, structural elements in the interior were highlighted: stairs, galleries, niches and atriums (see

Fig. 11.3 Trade and Craft Center Old Mercator. (Source: Aleksandra Jevtović 2019)

Fig. 11.4 Trade Center Old Mercator exterior details. (Source: Aleksandra Jevtović 2019)

Fig. 11.5). The structural solution of the edge of the vertical entrance amplified the visual form.

Two basic principles of modern architecture—geometry and proportion—can be recognised both in the plan and in the elevation. The programme of the building consisted of simple forms, whose aesthetic and symbolism were supposed to be amplified by the monumental form.

The building, which might have been an architectural marvel at the time of its construction, lost its distinctive appearance in 1990 due to reconstruction. The original concept and meaning, as well as the form and the harmony created by the design elements implemented, were disturbed by the upgrade and numerous changes. There are still smaller shops operating within this building, but the condition of the building is in a worryingly devastated condition. The deteriorating condition of the building can be attributed to unresolved property and legal issues. Although it still

Fig. 11.5 Trade Center Old Mercator inner space. (Source: Aleksandra Jevtović 2019)

has the status of a spatial landmark in its environment, it is mostly based on memories of its former splendour. In the last decade of this century, the Old Mercator in the media is referred to only as 'a mirror of our sloppy transition' and 'a place to avoid' (Anonym 2013; J. D. 2018a, b).

11.5 Trade Centres as Part of Globalisation

Over the following four decades, Yugoslavia, Serbia, and Belgrade underwent turbulent changes both in the field of state-building politics and the local social context, together with changes caused by the newly established global discourse. In addition to supporting the ideology of the 'free market', the new world order is based on providing protection and support to private property and the partial interests of the economic elite, which contributed to the enthronement of a few powerful and an increase in social inequalities (Tripković 2000:62; Stupar 2009:35). Globalisation has significantly affected cities around the world, conditioning transformations of urban structures that are in constant conflict between global and local powers (Stupar 2009:14). That conflict was inevitable also in Serbia, where the local elite favoured internal conflicts and isolation from the rest of the world to maintain its influence (Tripković 2000:67–68). Analysing the category of public interest, architects in Serbia perceive that today it is actually reduced to the 'structure of the sum of individual interests', which have been becoming increasingly separated and divided over time (Kovačević 2017:31).

Despite the opposing goals, in analysing globalisation in Serbia, Tripković (2000:69–70) observed that the internal ideological structures of globalists and locals were very similar in their essence. At the end of the second decade of this century, observing the transformations of urban structures, it seems that the conflict exists only at the level of media simulacrum, and that the role and influence of the local elite were reduced to that extent. While maintaining the illusion of (whilom) conflict, local authorities are assigning themselves the role of defender of the nation, while their actions say the exact opposite. The intensity and aggressiveness of these media performances grow in proportion to their adherence to globalisation flows.

When transitioning from the twentieth to the twenty-first century, the area of New Belgrade, which is the focus of this paper, became an example where the tendency of the new epoch to place the forces of the market economy at the centre of its attention is clearly visible. This is accompanied by inevitable privatisation and the logical consequence in the form of degradation of the role of urban planning (Blagojević 2007:251). Nor should the complex position of the socialist period in collective memory be diminished, coloured predominantly by critical tones, which greatly aided such a sequence of events. Thus, from the former model of socialist construction and the ideological topos of the social order at the time, New Belgrade underwent a complete transformation of its public architecture during the 2000s. In the first decade of this century, the largest concentration of so-called 'corporate' architecture was recorded in New Belgrade (Miletić–Abramović 2007:84), which is considered to be the commercialised course of globalisation in architecture that produces objects that are 'reduced in expression but not in size ratio' (Kadijević 2019:67). That distinctive spatial pattern of 'connectors of capital', represented by expansive business zones featuring luxurious glass facades and equipped with the latest technologies, according to Stupar (2009:120), highlights the current changes in the urban and world economy 'which gives primacy to tertiary activities–business, services and trade'.

According to the analysis of globalisation over time made by Malcolm Waters (2001:22–25), the process began in the field of economics and, in its early development phase (from the sixteenth to nineteenth century), causally flowed from economics through politics to culture. The crisis of capitalism caused politics to take a major role in the next development phase (nineteenth–twentieth century), balancing the economy on the one hand and the culture on the other. The crisis of the state in the last phase (twenty–first century) has resulted in the main role being played by culture, acting through politics towards the economy. As the ultimate consequence of a whole host of causes over six centuries, consumerism emerged as a lifestyle. Consequently, the city of globalisation places the consumer in its centre, and pays special attention to the purchases and spaces intended for that purpose (Stupar 2009:155).

It has been noted that the topic of shopping malls in Belgrade has been a trend since the 1980s (Miletić–Abramović 2007:82). During that period, by following the examples given in New Belgrade, a large number of shopping centres were built on peripheral settlements. The purpose of these interventions was to alleviate the pressure on the central city area and complement the new settlements in terms of content. Shopping malls which were located closer to the city centre had a more luxurious character, amongst which TC Millennium (1997–2001) on Knez Mihailova Street by architect Stevan Mićić stand out (Marković 2015).

At the end of the first decade of the 2000s, in an interval of just 2 years, two large shopping malls were built in New Belgrade. These objects were built according to the projects of foreign architects, and in addition, they are aesthetically and functionally designed in the spirit of globalisation. It is indicative that during the same decade, at the time of their creation, shopping malls were viewed 'as new universal public spaces' (Danilović–Hristić 2007), or as 'extension of public space' (Stupar 2009), while only a decade later the experts began to consequently perceive also the other side of that relationship, in which shopping malls and public urban spaces occupy opposing positions (Cvetković and Vaništa–Lazarević 2016; Đukić and Cvetković 2017). In any case, at the end of this decade, it is evident that these 'points of globalist consumerism have been imposed as priority places of social gathering' (Kadijević 2019:67).

11.5.1 Delta City (2007)

Delta City played a significant role in forming a new iconographic environment that fits the spirit of the new age. It was the first shopping mall built in accordance with international standards in Yuri Gagarin Street (Block 67) in New Belgrade, which was the most densely populated part of the city. The traffic infrastructure is designed to allow quick and easy access from all parts of the city (Jevtić 2007:23; Jevtić 2014:53). The project was authored by Israeli architect Ami Moore. The investor is Delta Real Estate, while the design and construction management was done by Delta Invest company. It was opened to the public in November 2007.

Delta City is a true 'city in small' (Danilović–Hristić 2007). It establishes new standards and advocates for a wholly different culture of purchasing compared to the one that was valid during the socialist period, but also introduces innovations in construction practices (Kadijević 2019:67; Mitrović 2007a). The facility was built on a 2.9-hectare parcel of land and covers an area of 85,000 square metres, with 30,000 squares metres designated for rental purposes. It was conceived as a complex urban form, with strong impacts on its immediate and wider environment (see Fig. 11.6). Originating from the idea of attracting and gathering, it is a generator of many powerful and diverse messages.

The aesthetic outcome of the Delta City shopping mall is predetermined due to its adherence to the criteria of commercial viability and energy efficiency in its concept (Kadijević 2019:66). The multifunctional facility has a compact structure in which the space designed for modern cinema is allocated by its volume, which is further emphasised by the oblique roof surface. In addition, conspicuous and recognisable details were erected along Jurij Gagarin Street: a tower with a logo and billboards in order to highlight information rather than the visual value of the building.

Representing a successful combination of a variety of retail establishments and leisure facilities, this multifunctional mall boasts a variety of content in stores (fashion, sports, decoration, food etc.). Many of the tenants with internationally known brands have, for the first time, 'entered' the Serbian market in this mall (Kadijević 2019:66). Unlike the complete simplicity in exterior architecture, which hides behind the advertisements of available brands, the interior is dominated by a comfortable atmosphere, besides contemporary design (see Fig. 11.7). Considering the height of the building, interior spaces were solved by the application of gallery communications. In the space above the main entrance hall, where events are often held,

Fig. 11.6 Delta City exterior. (Source: Aleksandra Jevtović 2019)

Fig. 11.7 Delta City interior details. (Source: Aleksandra Jevtović 2019)

the atrium is spherically shaped, which contributes to the ambient qualities of the area. The illumination of the building has been resolved through variously shaped skylights, which do not diminish the impression of artificially generated space.

11.5.2 Ušće Shopping Mall (2009)

The transformation of urban space in the post-socialist era has brought a new way of life and also a new value system and cultural model (Popadić 2009:15). The practice of design serving as a means of expressing a marketing strategy was further exemplified through the vision of another shopping mall in New Belgrade: Ušće The practice Belgrade: Ušće (Kadijević 2019:68–69).

The Ušće Shopping Mall was built on a vacant space next to the former Central Committee building on Mihajlo Pupin Boulevard (Block 16). The project was designed by Chapman Taylor Studio and is authored by Italian architect Gerardo Sannella. The contractor was the domestic firm MPC Properties, co-owned and partnered by Merrill Lynch Company. It was opened on 31st March 2009.

As in the aforementioned case, the idea of the Ušće mall was to intensify the public space by accommodating under one roof everything that modern man needs. Accordingly, the opening of the Ušće mall was held under the slogan '"New City Center', which clearly pointed out the desire to promote a new cultural landscape (Popadić 2009:19; Cvetković and Vaništa–Lazarević 2016).

Due to its urban position, which is recognisable in the cultural environment of Belgrade, and by the promotion of values of a global, consumer society, the Ušće Shopping Center gives a local variant of internationalism, thus fulfilling a very specific function of the city (Danilović–Hristić 2007:8–9). The architectural design incorporated a functional and compact core, adorned with vibrant and fanciful shapes which were strategically positioned on the facades. Unlike Delta City, the Ušće shopping mall is certainly a unique design achievement (see Fig. 11.8 and 11.9). This uniqueness is reflected not only in the characteristic architectural

Fig. 11.8 Ušće Shopping Mall front facade. Copyright Aleksandra Jevtović 2019

Fig. 11.9 Ušće Shopping Mall exterior details. (Source: Aleksandra Jevtović 2019)

Fig. 11.10 Ušće Shopping Mall interior details. Copyright Aleksandra Jevtović 2019

solution and application of contemporary architectural forms but also in the aesthetic vision of the ideology of consumerism.

As in the previous example, there is a lack of harmony between the interior and the exterior (see Fig. 11.10). The attractive exterior in playful forms hides the interior, which is characterised by coldness and aggressive crisp lines, defying the idea of the seductive glow of 'paradise for consumers' (Danilović–Hristić 2007:9).

Some critics considered that positioning the Ušće Mall in its current location was inappropriate in relation to the urban context of this part of the city (Jevtić 2014:71). When approaching New Belgrade from the old part of the city, this facility takes over the role of the gate which welcomes visitors. Furthermore, besides fulfilling the immediate and short-term consumer needs, the 'inappropriateness' of its location also embodies the evolving nature of this urban area and the cultural values of contemporary global society. This society does not shy away from pushing boundaries, even attempting to replace traditional cities with artificially constructed environments focused on consumptive experiences and spectacles.

11.6 Conclusion

Cities can be seen as mediators between global society and the individual (Vasiljević 2009:106). In this context, transformations of the term 'global society' over time are precisely the ones that contribute to perceiving the changed spatial relationships of the city's elements. Within the time frame of this analysis (from the mid-1940s to the first decade of the 2000s), the concept of global society encompassed notions of 'power, state, institutional space, and institutional urbanism' (Vasiljević 2009). Today, this term has broadened its scope to encompass the entirety of the world. Each element of the city plays an equally important role in connecting the global society and the individual, and the facilities intended for trade, which were selected for analysis, are particularly relevant from the present point of view, which implies the expanded dimensions of the term global society.

New Belgrade is a particularly suitable terrain for analysis because it was built on completely empty ground, without architectural remains of previous epochs. In the post-war context, the emptiness of the terrain provided a suitable ground for the realisation of a new capital 'that can be not only completely new and different but also physically separated from historic Serbia and the city of Belgrade as the capital of the former monarchy' (Blagojević 2007:248). The urban matrix of New Belgrade, which originally relied on the postulates of modern architecture and the functionalist city, has been continuously changing and upgrading with new spatial and semantic elements throughout its duration (during the second half of the last and beginning of this century). As dominant and interdependent paradigms, Blagojević (Blagojević 2007:244) singled out the concepts of a functionalist city, a governing city, a city of residence and a contemporary city.

> New Belgrade has emerged as a pivotal aspect in the establishment of a suitable identity, whether it involves crafting an image for the capital of the new Yugoslavia or creating a framework for the functioning of a contemporary democratic society. (Ignjatović 2012:693)

Considering that the concepts of functionalist and governing city were in effect in the first decade of planning and construction (from the mid-1940s to the mid-1950s), from the point of view of inhabitants (users), the paradigm shift from the city of residence to the contemporary, globalist city is of particular importance. When the idea of a city of residence was in effect, public spaces were treated as the common good of its inhabitants, and public buildings played a role in breaking the monotony of the constructed matrices. By doing so, they represented the benchmarks in their immediate environment and were contributing to better spatial orientation. In terms of function, these facilities served primarily surrounding apartment blocks.

Therefore, the radius of their attraction was directed at their immediate neighbours. The facilities of the shopping centres were an integral part of the spatial units made up of several objects, among which were those intended for cultural content (cinemas Yugoslavia and Fontana). Pedestrian traffic was achieved through open gallery communications, which continued into smaller pedestrian plateaus.

Public spaces in the context of globalisation and the idea of a 'contemporary city' are subordinated to capital, and public buildings assume the roles of 'marketing representatives of companies and brands' (Kadijević 2019:66). In New Belgrade, the private capital conquered territories of a large number of publicly and socially owned green spaces, thus contributing to making its territories no longer a field of general and common interest but 'a land divided by new frontiers' (Blagojević 2007:250). In the new structure of the city, public spaces that offer the possibility of gathering large groups of people have a new symbolic level. By bracing the myth of community and creating imagined communities, these spaces provide support for individuals in the turbulent world of globalisation (Stupar 2009:67). In that context, large shopping malls paradoxically represent some kind of oasis in modern man's daily life. Nowadays, shopping activity has the same significance as hunting activity in primordial communities. In addition to existential reasons, it is increasingly becoming an indispensable component of consumerist life and is regarded as one of the primary sources of life success and personal satisfaction. On the other side, following the possibility of dual usage of globalisation as well as Jencks' strategy of 'double coding' (Tripković 2000:63; Elin 2004:144, 193), contemporary shopping malls are also reminiscent of Jeremy Bentham's Panopticon, where the market itself performs the role of sole supervisory authority and guardian of the order.

Without a tendency to harshly judge the nature of the globalisation process itself, the resemblance between the definition of that process and the physical organisation of the observed shopping malls is indicative. Specifically, globalisation was described as a process of external unification (and acceptance of standards) and internal differentiation (between society and states) (Tripković 2000:62). Thus, the exterior of these buildings is mostly uniform, follows international architectural trends, hi-tech, reflects the latest construction technologies, does not pay attention to the context of the urban environment and the peculiarities of the existing architecture. As a result, they can be completely relocated from their environment and multiplied in different locations across the globe. The interior of these buildings is differentiated into a multitude of different shops, various showcases, billboards, stands, trends, colours and fragrances. Besides artificial simulation of urban spaces, with streets, squares and greenery, malls also have security systems. Security within these spaces is an additional asset that attracts visitors (Danilović–Hristić 2007; Cvetković and Vaništa–Lazarević 2016; Đukić and Cvetković 2017). The range of their activities has been expanded to such an extent that they are no longer just points attractive to residents of the surrounding apartment blocks, but also motives for tourist visits from other cities and countries.

Unfortunately, in terms of architectural heritage, the globalist tendency to 'favour market criteria over the artistic ones' (Kadijević 2019:67) contributed to the spatial and usable devastation of the malls inherited from the period of socialism, although they represent valuable and significant architectural monuments of their time. Undoubtedly, in terms of trends and needs of today's consumer culture, these facilities have been functionally exceeded, so it is important to constantly devise new ways of their reactivation that would prevent its further decay. In this regard, the efforts of the experts and the investment support that made possible the renovation of the Fontana Center should be welcomed, with the hope that the same awaits and other objects that have been pushed into oblivion due to globalisation. The very fact that globalisation loses its significance when it fails to operate simultaneously on both axes—local (national) and global (supranational) is reason enough.

References

Anonym (2011) Obnova MZ Fontana na Novom Beogradu (renovation of the LC Fontana in the new Belgrade). Archit Today, 16 April 2011. https://wwwyoutubecom/watch?v=KA39S7BFJNg. Accessed 21 Feb 2019

Anonym (2012) Ponovo radi bioskop "Fontana" (Cinema "Fontana" is Working Again). Ves ti onlajn, 7 March 2012. https://arhiva.vesti-online.com/Scena/Film/208536/Ponovo-radibioskop-Fontana. Accessed 17 Feb 2019

Anonym (2013) Stari Merkator: Svedok naše propasti (old Mercator: the witness of our downfall). B92, 12 July 2013 https://wwwb92net/biz/vesti/srbijaphp?yyyy=2013&mm=07&dd=12 &nav_id=731806 Accessed 7 Mar 2019

Anonym (2014) Žurka i projekcije: bioskop Fontana obeleţava dve godine rada od ponov nog otvaranja (party and screenings: cinema Fontana marks two years of reopening). Nadlanu, 6 March 2014. https://nadlanu.com/212600/zurka-i-projekcije-bioskop-fontanaobelezava-dve-godine-rada-od-ponovnog-otvaranja/. Accessed 17 Feb 2019

Bisenić D (2009) Prvoborci Jugoslovenske supermarket revolucije. Ekspanzija počela sa beogradskog Cvetnog trga Zaveštanje Milorada Jovanovića (First Fighters of the Yugoslav Supermarket Revolution The Expansion Began With Belgrade's Flower Square The Legacy of Milorad Jovanović) Danas, 18 December 2009. https://www.danas.rs/nedelja/ekspanzija-pocela-sa-beogradskog-cvetnog-trga/. Accessed 24 Feb 2019

Blagojević L (2007) Novi Beograd: osporeni modernizam (new Belgrade: contested modernism). Zavod za udžbenike—Arhitektonski fakultet Univerziteta u Beogradu—Zavod za zaštitu spomenika, Beograd

Bogunović S (2005a) Cvetni trg (Flower Square). In: Arhitektonska enciklopedija XIX i XX veka: arhitektura, arhitekti, pojmovi. (1), [Arhitektura] (architectural encyclopedia of the 19th and 20th century: architecture, architects, terms. (1), [architecture]). Beogradska knjiga, Belgrade, pp 90–94

Bogunović S (2005b) Novi Beograd (New Belgrade). In: Arhitektonska enciklopedija XIX i XX veka: arhitektura, arhitekti, pojmovi. (1), [Arhitektura] (architectural encyclopedia of the 19th and 20th century: architecture, architects, terms. (1), [architecture]). Beogradska knjiga, Belgrade, pp 290–304

Bogunović S (2005c) Martinović, Uroš (Martinović, Uroš). In: Arhitektonska enciklopedija XIX i XX veka: arhitektura, arhitekti, pojmovi. (2), [Arhitektura] (architectural encyclopedia of the 19th and 20th century: architecture, architects, terms. (2), [architects]). Beogradska knjiga, Belgrade, pp 951–954

Brkić A (1992) Znakovi u kamenu: srpska moderna arhitektura 1930–1980 (Signs in Stone: Serbian Modern Architecture 1930–1980). Savez arhitekata Srbije, Belgrade

Cvetković M, Vaništa–Lazarević E (2016) The image of the city vs. semi–public spaces of shopping malls: case studies of Belgrade. In: Vaništa-Lazarević E (ed) 3rd international academic conference on places and technologies. Faculty of Architecture, Belgrade, pp 481–487

Danilović–Hristić N (2007) Tržni centri kao novi univerzalni javni prostor (Shopping Malls as a New Universal Public Space). DaNS 60:8–9

Dobrović N (1954) Što je gradski pejzaž—Njegova uloga i prednost u savremenom urbanizmu (what is the City landscape—its role and the advantage in contemporary urban planning). Čovjek i prostor (Man Space) 20:1–3

Đukić A, Cvetković M (2017) Javni gradski prostor nasuprot tržnom centru u konzumerističkoj kulturi (public space vs. shopping mall in consumerist culture simulation of the City in order to legitimize itself as a public space). Kultura 154:228–246

Elin N (2004) Postmoderni urbanizam, dopunjeno izdanje (Postmodern Urbanism, Revised Edition). Orion art, Belgrade

Frampton K (1992) Modern architecture—a critical history, 3rd edn. Thames & Hudson Ltd, London

Hubert D (2017) The enduring lives of Sasa Machtig's modular creations. Metropolis, 23 February 2017 https://metropolismag.com/profiles/the-enduring-lives-of-sasa-machtigs-modular-creations/. Accessed 12 Jan 2019

Ignjatović A (2012) Tranzicija i reforme: arhitektura u Srbiji 1952–1980 (transition and reforms: architecture in Serbia 1952–1980). In: Šuvaković M (ed) Istorija umetnosti u Srbiji XX vek (history of art in Serbia XX century). Orion art, Belgrade, pp 689–710

Ignjatović A (2013) Dve kule u Beogradu XX veka: uspon, hibris i pad (two towers in XX c. Belgrade: ascent, hubris and decline). In: Merenik L (ed) Zbornik seminara za studije moderne umetnosti Filozofskog fakulteta Univerziteta u Beogradu 9 (the journal of modern art history Department Faculty of Philosophy University of Belgrade 9). Filozofski fakultet, Belgrade, pp 29–38

J. D (2018a) "Stari Merkator" za izbegavanje ("Old Mercator" To Avoid) Večernje novosti, 12 June 2018. http://www.novosti.rs/vesti/beograd.74.html:732528-Stari-Merkator-za izbegavanje. Accessed 7 Mar 2019

J. D (2018b) Stari Merkator" nema domaćina ("Old Mercator" Has No Hosts). Večernje novosti, 13 June 2018. http://www.novosti.rs/vesti/beograd.74.html:732702-Stari-Merkatornema-domacina. Accessed 7 Mar 2019

Jevtić MH (2007) Arhitektura prostornog kolaža (architecture of spatial collage). DaNS 60:22–23

Jevtić MH (2014) Kritički gambit—osvrti na savremenu arhitekturu Srbije (critical gambit—reviews of contemporary Serbian architecture). Stylos art, Novi Sad

Jevtović A (2019) Spomenik kulture kao resurs svakodnevice na primeru Robnog magazina u Beogradu (a cultural monument as a resource in everyday life on the example of a department store in Belgrade). Kultura 164:352–368

Jovanović PR, Stupar AB (2022) The emerging community planning in the super-blocks of new Belgrade. Urban Des Int 27:275–287

Kadijević A (2017) Tipologija arhitektonskih i urbanističkih preinačavanja Beograda (19–21. век) (the typology of architectural and urbanistic transformations of Belgrade (19th–21st century)). Kultura 154:11–23

Kadijević A (2019) Globalizacija—generator krize autorske arhitekture i potiskivanja nasleđa u Srbiji (globalization—a generator of the crisis of authorial architecture and the suppression of heritage in Serbia). In: Dimitrijević Marković S (ed) Kulturno nasleđe, rizici i perspektive: zbornik radova (Cultural Heritage, Risks and Perspectives: Proceedings). Zavod za zaštitu spomenika kulture grada Beograda, Beograd, pp 64–73

Kovačević B (2017) Preispitivanje osnovnihm kategorijalnih pojmova vezanih za preinačavanje arhitektonsko–urbanističke matrice Beograda (Rethinking Basic Categorical Concepts Connected With Revision of the Architectural–Urbanistic Matrix of Belgrade). Kultura 154:24–34

Manević Z (1986) Srpska arhitektura XX veka (Serbian architecture of XX century). In: Manević Z, Domljan Ţ, Šumi N, Štraus I, Konstantinovski G, Milić B (eds) Arhitektura XX vijeka (architecture of the XX century). Prosveta; Zagreb: Spektar; Mostar: Prva knjiiţevna komuna, Belgrade, pp 19–31
Marković IR (2015) Arhitekta Stevan Mićić: opus 40 (architect Stevan Mićić: opus 40). Dosije studio, Belgrade
Miletić–Abramović L (2007) Paralele i kontrasti, srpska arhitektura 1980–2005 (Parallels and Contrasts, Modern Serbian Architecture 1980–2005). Muzej primenjene umetnosti, Belgrade
Milinković M, Ćorović D, Vuksanović-Macura Z (2019) Historical enquiry as a critical method in urban Riverscape revisions: the case of Belgrade's confluence. Sustainability 11(4):1177
Mišić B (2011) Palata Saveznog izvršnog veća u Novom Beogradu (the building of the Federal Executive Council in new Belgrade). Zavod za zaštitu spomenika kulture grada Beograda, Belgrade
Mitrović M (2007a) "ConQUEST": Arhitektura kupovine (Conquest: shopping architecture). DaNS 60:4–7
Mitrović V (2007b) Prilog istoriji arhitekture socijalističkog šopinga: Samoposluga robna kuća, buvljak, megamarket (the supplement to the history of architecture of socialist shopping: supermarket, department store, flee market, mega market). DaNS 60:13–15
Mitrović M (2012) Arhitektura Beograda 1950–2012 (Belgrade Architecture 1950–2012). Sluţbeni glasnik, Belgrade
Perović MR (2009) Teorija arhitekture XX veka: antologija (the theory of architecture of XX century: anthology). Građevinska knjiga, Belgrade
Popadić M (2009) Znak na Ušću (the sign at Ušće). Sintezis 1:15–30
Putnik Prica V (2017) From socialist realism to socialist aestheticism: three contrasting examples of state architects in Yugoslavia. In: Preda C (ed) The state artist in Romania and Eastern Europe. The role of the creative unions. Editura Universitatii din Bucuresti, Bucharest, pp 347–373
Škalamera Ž (1985) Secesija u srpskoj arhitekturi (secession in Serbian architecture). In: Zbornik Narodnog muzeja. Istorija umetnosti XII–2. Narodni muzej, Belgrade, pp 7–12
Stojanović B, Martinović U (1978) Beograd 1945–1975: urbanizam i arhitektura (Belgrade 1945–1975: Urbanism and Architecture). Tehnička knjiga, Belgrade
Štraus I (1991) Arhitektura Jugoslavije 1945–1990 (Architecture of Yugoslavia 1945–1990). Svjetlost, Sarajevo
Stupar A (2009) Grad globalizacije: izazovi, transformacije, simboli (the City of globalization: challenges, transformations, symbols). Arhitektonski fakultet Univerziteta: Orion art, Belgrade
Tripković M (2000) Globalizacija i Srbija (Globalization and Serbia). Sociološki pregled XXXIV:61–74
Vasiljević A (2009) Kreiranje identiteta, "brendiranje grada" (identity creation, City branding). Kultura 122–123:104–119
Vušović A (2011) Biće "Fontana" kao u najboljim danima ("Fontana" Will Be As In The Best Days). Blic, 29 March 2011. https://www.blic.rs/vesti/beograd/bice-fontana-kao-u najboljim-danima/v6jkkhk. Accessed 19 Feb 2019
Waters M (2001) Globalization, 2nd edn. Routledge, London

Chapter 12
Urban Transformations in Hong Kong

Pina Giusi Ciotoli

Abstract This chapter aims to analyse the megalopolis of Hong Kong and its progressive vertical growth that has involved both the financial districts (Central and Admiralty) and the suburban areas (Sheung Wan, SoHo). As regards the vertical growth of the central districts, it should be noted that, in addition to the concentration of buildings for the wealthier classes, there is an exciting hybridisation between vertical buildings and the pedestrian streets of the Central Elevated Walkway (CEW), a unique architectural feature of Hong Kong. Therefore, this chapter, in addition to focusing on these aspects, also intends to analyse the repercussions urban policies have had in experimenting with new multifunctional aggregative typologies.

Keywords Density · Urban growth · Public infrastructure · Public housing · Hong Kong · Tall buildings

12.1 Introduction

This chapter focuses on some phenomenal aspects of urban planning in Hong Kong: the first concerns the social policy conducted on public housing from the 1950s until the present (beginning with the Shek Kip Mei fire of 1953); the second is related to the vertical growth of the city's central districts (like Admiralty and Central), where in addition to the concentration of buildings for the wealthier classes, there is an interesting hybridisation between vertical building and the pedestrian streets of the Central Elevated Walkway (CEW). Such aspects turn out to be central to understanding the architectural experiments taking place in Hong Kong. A testament to this is Hong Kong's pavilion at the 2018 Venice Biennale, titled 'Vertical Fabric: Density in Landscape', which explored the possibility of defining the existence of a 'vertical fabric' (Ciotoli 2017a) determined not only by the presence of numerous

---

P. G. Ciotoli (✉)
Independent Researcher, Rome, Italy

G. Cimadomo et al. (eds.), *Urbanism at Borders*, The Urban Book Series,
https://doi.org/10.1007/978-3-032-09594-7_12

skyscrapers and towers but by a whole range of social as well as architectural instances, some of which will be explored in this chapter.

12.2 The Public Housing Programme in Hong Kong

The city of Hong Kong is concentrated in a limited area of about 1.104 square kilometres within which are concentrated 657 skyscrapers with a total height of 150 meters or more, (Council on Tall Buildings and Urban Habitat n.d.) for a total population of about 7.23 million inhabitants. The extraordinary vertical growth of Hong Kong concerns not only the administrative and financial districts but also represents a common theme for most of the peripheral areas of the megalopolis, for the numerous housing quarters built since the 1950s. During the second half of the twentieth century, the city became a sort of *urban experiment* on densification issues, going beyond the definition of thematic and/or specialised neighbourhoods (let's think of the residential dormitory districts or to the downtown areas of American cities). In fact, in many residential districts of Hong Kong designed as public housing complexes, in addition to the functional *mixitè*, it is possible to observe the use of retail and covered streets as elements to connect (visually, functionally and architecturally) different parts of the city (Falsetti 2022:43–49). In this sense, the Central and Admiralty districts denote a certain *continuity*, even formal, with more decentralised areas of the megalopolis.

Since the second half of the twentieth century, there has been vertical experimentation that can be traced in the design of public housing estates. In fact, it is important to point out that the housing problem in Hong Kong started already during the 1930s, particularly following the Japanese invasion of China (1937), when about 750,000 refugees poured into the city.

The housing problem in Hong Kong, which became central during the 1950s, was not just a matter of numbers. It was a human crisis, with a large number of refugees and immigrants fleeing from other Asian regions following the Second World War. The population of Hong Kong, according to evaluations made by the Hong Kong Housing Society, surged from 300,000 inhabitants at the beginning of the century to around 1,600,000 after the outbreak of the Second World War. The inability to pay the rent, even just for a bed place, led to the construction of numerous slums located along the edge of the Peak and the Lion Rock, the two hills that create a physical limit for further developments of the city respectively in the island of Hong Kong and in the Kowloon area. This human aspect of the housing problem underscores the urgency and complexity of the urban planning challenges that Hong Kong faces.

The orography of the territory determined the choice of the hilly limit as the only possible place to auto-build these houses; for this reason, refugees and desperate people started to live in the squatter hills far from the business centre of Hong Kong. Following the dramatic fire that broke out on Christmas Eve 1953 in the squatter of Shek Kip Mei, the central government decided to act by trying to give more concrete answers to this problem. For this reason, the government established the Resettlement Department of the city of Hong Kong (1954) that was the first institution able to manage the construction of specific residences for refugees. Indeed, Shek Kip Mei's dramatic event strongly demonstrated the need to find better solutions for the population; more than 20,000 people who survived to the fire were again without a roof to shelter, which is why:

> They put up sheds on the streets as transitory shelters. The Government distributed relief money and household necessities, such as rice, bowls, dishes and blankets, to the victims. The Government spent up to $50,000 a day on such emergency relief. (Heritage museum Government Hong Kong archive 2010:3)

Furthermore, a few months after the Shek Kip Mei fire, the city government decided to build two-storey dwellings to house the slum refugees. These constructions were called 'Bowring Bungalows'. Subsequently, the government promoted the construction of some 6–7 storeys residential buildings in the so-called Mark I Estate. It was still an emergency solution with which to tackle the housing problem since there was no electricity or running water inside the volumes. In the Mark I estate:

> [...] most [of the buildings] were in the shape of the letter "H", while others were in the shape of the letter "I". The flats were arranged back-to-back, with access via a corridor. To improve ventilation, openings were made in the partition walls at the rear of each flat. (...) Water standpipes, communal latrines and bathrooms were provided in the crossbars of the H blocks. Most flats had a floor area of 11.15 square metres to accommodate five adults, which meant that each adult occupied an average of just 2.23 square metres. Between 1954 and 1964, the Government constructed more than 140 Mark I blocks on Hong Kong Island and Kowloon. (Heritage museum Government Hong Kong archive:6)

In 1954, the Government of Hong Kong established a further institution to deal with this problem: the Hong Kong Housing Society (that still operates in the metropolis) which—unlike the Resettlement Department that would have had to work quickly to solve the emergency of the refugees of the Shek Kip Mei fire—had the task to promote an adequate housing policy, trying to improve the living conditions of the future inhabitants. From 1954 to the present day, thanks to the activities of the society, public housing has become a central theme for the architectural and urban experimentation of the city. For this reason, we can affirm how:

> [...] its public housing policy as well as the developments of its high-rise housing typology had since been considered one of the most successful housing models worldwide. (Chung et al. 2016:12)

As witnessed by a recent interview with Donal Liao Poon-Huai, one of the architects who contributed most in affirming the priority of public housing in the city's political programme, housing is not a quantitative problem but a qualitative one. Liao, referring to those who would have lived in the Ma Tau Wai Estate (1962) and Wah Fu Estate (1968) complexes he had designed, states that:

> When Hong Kong was transforming from a sleepy town into an industrial manufacturing society, many of these people contributed to Hong Kong's prosperity; therefore, they deserved what the public housing could provide them with. (Chung et al. 2016:30)

The execution of the subsidised building programme obliged architects to move towards more modern and adequate housing types and organisms in response to the demand for higher standards requested by the population. While the 1950s were characterised by addressing the problem of displaced persons and refugees who fled wars, it was during the 1960s that a turning point occurred in the public housing programme promoted by Hong Kong. As Florian Urban as argued, during the 1960s, worldwide, there was 'A strong state and egalitarian goals, the fundamentals of mass housing, were endorsed across the political spectrum by communists, nationalists, or catholic reformers' (Urban 2012:12). The programme considered the impressive economic and industrial development experienced by the city and the fact that Hong Kong would, therefore, have attracted more inhabitants and, probably, more people in the Shutter Hills. For this reason, in 1964, the so-called 'Review of Policies for Squatter Control, Resettlement and Government Low-cost Housing' was drafted.[1]

12.3 Typological Variations Within the Residential Towers of Hong Kong

The residential public housing programme carried out by Hong Kong is extremely interesting, also as regards the typological experimentation conducted on skyscrapers. In fact, in the numerous public housing complexes built over the last fifty years, we can observe a sort of 'built abacus' of variations concerning the morphology and functionality of the skyscraper type. It is worth mentioning how the skyscraper was not born in an already clear form but is the result of successive transformations. The typological experimentation on skyscrapers was—at least until the early decades of the twentieth century—an 'almost exclusive' prerogative of the Chicago architects who tried to solve the problems related to the functional adaptation of the new architectural type in the tertiary field (Steadman 2014). Daniel Burnham, John Wellborn Root, William Le Baron Jenney, Danmark Adler, Louis Sullivan—just to mention some of the major designers who worked in Chicago—had in fact devoted their efforts to decline some recurring features of the type in office buildings; at most, as interesting architectures designed by Sullivan recall—such as the Schiller Building (formerly the Garrick Theatre destroyed in the 1970s) and the Auditorium Building, both in Chicago—the experimentation was carried out in buildings that we could define as multi-functional, in which the architect has realised, ahead of time, the

[1]After preparing the review of policies for squatter control, the Municipality of Hong Kong founded the Ten-Year Housing Program in 1972 and the Home Ownership Scheme in 1976. Subsequently, starting from the 2000s, in addition to these financial and urban planning programs for the construction of new housing quarters, there was also a further financing and intervention plan for the restructuring and sometimes demolition/reconstruction of the old quarters built starting from the 1950s and now considered inefficient from a qualitative and economic point of view (housing standards, etc.).

'mixed vertical fabrics'. These buildings were equipped with a base system with an internal square, theatre and various services.

Focusing on the typology of housing, we can affirm that the possibility of typological variations of the skyscraper has been a prerogative of Le Corbusier, who conducted his architectural research on one side by idealising the role of the skyscraper in decongesting existing cities, by implementing a process of vertical densification; on the other, 'marking, on a territorial scale, the expressive role of the vertical element, drawing with its architecture new *built landscapes* surrounded by the natural landscape' (Ciotoli 2017b:47). Le Corbusier's experimentation on the theme of *living together vertically* was adopted some decades later by the Hong Kong architects involved in housing projects. In this regard, it is interesting what Florian Urban states about the different significance and success (or failure) of the modernist apartment block:

> While on the surface the modernist apartment block appears universal, it is in fact as diverse in its significance and connotations as its many different cultural contexts. (…) These differences show that design is not to blame for mass housing's mixed record of success. The comparison of apparently similar projects suggests that triumph or disaster does not depend on a single variable but rather on a complex formula that includes not only form, but also social composition, location within a city, effective maintenance, and a variety of cultural, social, and political factors. (Urban 2012:I)

Some prominent aspects of Le Corbusier's experimentation with the theme of *living together vertically* were taken up a few decades later by Hong Kong architects engaged in housing projects. These are not so much the formal aspects of Le Corbusier's works as one of the most innovative features of his urban plans, namely the distinction between vehicular and pedestrian flows. In fact, numerous housing complexes planned in Hong Kong provide for the distinction of the street circulation network from the areas left free for pedestrians, areas that, in the qualitatively and architecturally best designs, become places to live together, gathering spaces necessary to improve the quality of life of the inhabitants.

As architectural historian Gu Daqing states, numerous housing complexes have also experimented with the distinction of flows at the pedestrian level, attributing the role of connector element to the so-called *street in the sky* (Bacon 1982; Smithson and Smithson 1957) and the sky-lobbies; after all, even Le Corbusier had experimented the new possibilities of vertical fruition, giving to what we could banally call a *corridor*, the fundamental role of structuring the composition of the building. Le Corbusier, with his more experimental projects such as the Pilotis-Cities (1915) and the Unité d'Habitation (1920–1950), had indeed hoped that the horizontal structures inside the buildings, the so-called '*pedestrian walkways* could become small boulevards, made vital by all the services which, until that time, were placed along the sidewalk' (Ciotoli 2017b:145). Both the elevated passageway and the sky-lobby are two strategic solutions designed to characterise the building (especially in the case of slab blocks) and to give a better architectural connotation to the serving spaces. These elements become, in the serial aggregative typologies (slab blocks), the architectural features intended to characterise the building, organising (acting as centralising axes) (Maretto 1993:117) its different serial compartments.

The effectiveness of this solution is applied, for the first time, in Hong Kong in the Police Married Quarters (Daqing 2016) located in Cheung Sha Wan, built in 1960 (Consultancy for Heritage 2011). The complex in question (in fact, it consisted of about 13 buildings, destroyed between 2001 and 2008 following a housing recovery programme promoted by the Government of Hong Kong) was characterised by a C-shape building and, in its inner organisation, the *street in the air* fulfilled the task of connecting each residential unit with the collective spaces placed within the building. According to Gu Daqing:

> [...] the Police Married Quarters in Cheung Sha Wan can be simply understood as a 4-storeys slab building (plus a roof terrace) stacked on top of another 7-storey slab building. The ground floor of the 4-storeys slab building becomes the street in the air. There are only two groups of lift installed along the 320 metres long horizontal circulation. The residents can first take the lift to the 7th floor and then walk up or down to each flat via the staircase. (Daqing 2016:50)

A spatial conformation of this kind has sought (as indeed typical in the developments of the slabs) to use the servant space as the distribution axis inside the building, implementing a logic of distribution of the spaces that refers, by analogy, to the spatial distribution of the English arcades (let's think, for example, to the Burlington Arcade in London). In this regard, Gu Daqing underlines how:

> [...] the one-room unit attached to an open verandah is deeply rooted in Hong Kong's traditional arcade building type and is also an extreme solution to Hong Kong's economic and social conditions at that time. The street-in-the-air concept that originated from the European architectural culture was adopted by Hong Kong architects as the most economical way to solve the vertical circulation problem in high-rise housing blocks. (Daqing 2016:52)

On the one hand, the housing programme promoted by the city of Hong Kong has allowed a vital and fervent architectural experimentation, giving rise to numerous typological variants of the tower type and skyscraper type (Chung et al. 2016:3–12); on the other, it has brought to light issues related to the urban integration (Wang 2019:76–77) of the new complexes with the rest of the city. It has also underlined issues related to the marginalisation of citizens with respect to the metropolitan life with ever greater force. Public housing has highlighted not only the problems related to social activities to be carried out within individual blocks, but the need experienced by all members of residential complexes to have places of exchange and services for recreational and commercial activities.

The case of Wah Fu Estate, designed by Donal Liao Poon-Huai, is still nowadays, in terms of typological research, an interesting example of housing policies. The complex is located along the Waterfall Bay Park, in the southern end of the island of Hong Kong, far from the central areas of the metropolis.

In the Wah Fu Estate, architect Donal Liao Poon-Huai tried to take care of the design of the interstitial spaces between the various buildings of the complex, focusing, above all, on the relationship between housing and commercial fabrics. Donal Liao Poon-Huai's goal will be to create a small city with services, places to stay and gathering spaces. Furthermore, in the complex we can find all social activities that can create cohesion among citizens, avoiding the feeling of having created a ghetto neighbourhood, not only physically but also *ideally* far from the city. For this reason, the project sought to complement the area already destined for a market with other commercial spaces, intended for the ground floors of the buildings (Chung et al. 2016:32). The Wah Fu Estate, therefore, lays the foundations for a reflection not only on the morphological and typological developments of the buildings for housing, but also on the ever-inseparable relationship (proper of the Asian city) between gathering spaces, elements of transit and commercial fabric. According to Han Man and Thomas Chung's description, since the first phase of construction (1961–1968), the Wah Fu Estate was conceived as an entirely self-sufficient housing complex (Chung and Man, 2016:34–35) and, in that sense, was the first of its kind in Hong Kong.

> Liao's vision encompassed the project's comprehensive design, from siting and landscape, communal and amenity programmes, to residential type innovations. More importantly, it was Liao's insistence to have independent flats for every family that marked out Wah Fu from previous estates that had shared facilities. (...) Besides providing self-contained dwellings which included a separate kitchen, bathroom and living room, Wah Fu also had schools, banks, post- office, market, library, restaurants as well as multi-storey carparks with children recreational areas on top. (Chung and Man 2016:34)

By the end of the second phase of construction (1970–1978), the complex consisted of a total of 23 blocks for 9100 apartments (Chung and Man 2016:34–35). The design of the Wah Fu Estate is articulated using two different types: the so-called 'double donuts' or 'twin towers' (i.e. a system of two hollow towers in the centre) and that of slab blocks (Chung and Man 2016:34–35). The twin towers, four in total, range in height from 21 to 24 stories, and are located in the northern area of the site. Since the area chosen for the complex slopes toward the coast, the architect established a different height for the buildings to follow the morphology of the site and achieving a rhythmic, undulating skyline. The intersection of the two towers leads to the creation of a common area for the buildings, in which the stairs and elevators are located. The area of the complex sloping southward consists of a labyrinthine pattern of intersecting slab blocks, creating an urban continuum that opens in the direction of the sea, forming large public courtyards.

In addition, at the northern extremes of the complex, three schools were located, while in the central area, the town centre was planned, with all the necessary services for the small town of the Wah Fu Estate. While by the 1970s the Wah Fu Estate had become a model for housing complexes in Hong Kong, functional obsolescence, due to the passing of time, led to a progressive need to improve the complex. Therefore, the Hong Kong government, in 2014, announced a redevelopment project for the Wah Fu Estate area that is still ongoing.

12.4 Commercial Fabric as a New Element of Social Aggregation. The Malling Phenomenon in Hong Kong

Nowadays, the commercial fabric (declined as an arcade—a commercial covered street—or as a mall—a nodal space intended for retail) can be considered as a connective element of the city of Hong Kong (from an urban as well as a social point of view). This phenomenon, widespread over the years, also in other cities in the world, has undergone, in Hong Kong, interesting experiments that have determined, in some cases, the alteration—and in others the mere declination—of the arcade and the mall.

In fact, if the shopping centre and the mall (two relatively contemporary meanings of commercial fabric) can be considered as *non-places* because they are isolated from the urban context and usually positioned along extra-urban roads, the specific case of Hong Kong becomes the emblem of a new configuration of the commercial function as an integral part of the metropolis, often acting as a link between road and pedestrian infrastructure. Indeed, the metropolis's vast commercial system manages to connect business districts with the wealthy residential ones located in the Peak area; this feature, as we have seen in previous cases, has also developed in the public housing districts of Hong Kong's more marginal areas.

When analysing the city of Hong Kong and the high-density process that characterises it and makes it a unique example in the world, it is fundamental to study also the phenomenon of malling, and therefore the inseparable relationship that is established between residential density and the complex infrastructure system of the Chinese megalopolis. While the relationship between commercial and residential fabric and the possibility of these to give rise to specialised buildings (like shōtengai, passage, arcade etc.) is well established (Moretti 2004), the idea that the system of relationships underlying these architectures can also extend to the network of the territory has been gaining ground in recent years, as demonstrated by the pioneering study entitled *Cities Without Ground. A Hong Kong Guidebook* by Adam Frampton, Jonathan D. Solomon, and Clara Wong, dedicated to the city of Hong Kong.

What is peculiar about the Hong Kong case is, in fact, the unique integration of different levels of paths (underground and passageways) that cross the metropolitan districts, sometimes intersecting the lobby of the skyscrapers, and give shape to urban nodes that host commercial spaces. The commercial fabric is widespread throughout the city, also due to the absence of places of aggregation and exchange; for this reason, the areas destined to trade and shopping, far from the chaos of traffic, become the *urban knots* used by the citizens as *covered squares*.

This phenomenon reaches maximum levels in Hong Kong, thanks also to its particular orography and to the *synthesis* of a system that involves horizontal connective elements placed at different heights. A morphological analysis of Hong Kong and its spaces is therefore desirable, through the study of the relationships that pedestrian infrastructures establish with these urban nodes.

Hong Kong, even compared to the Asian panorama, presents several exceptional aspects: in addition to hosting the greatest concentration of vertical architectures and urban shopping centres in the world (Moretti 2004), it has an excellent public transport system (used by about 90% of the population) where metropolitan services are integrated with a dense network of skyways.

The Central Elevated Walkway (CEW) represents one of the most efficient systems of movement in the world (Moretti 2004), connecting different infrastructural knots such as underground stations (in the Central district and also in the Admiralty and Sheung Wan) and ferry docks (Shun Tak and the Central pier).

This system guarantees, through underground connection routes, the continuity of pedestrian flows between metropolitan stations and external covered roads, facilitating the passage from one building to another. The Central Elevated Walkway extends over the entire surface of the city through the so-called 'Central Mid-Levels escalator and walkway system', which is a network of covered passages and urban elevators that crosses the island of Hong Kong in a north-south direction, intercepting numerous residential neighbourhoods. The project is strongly supported by the municipality of Hong Kong, so much so that in the last few decades, the total extension of the two integrated systems has also reached many public housing districts initially *cut off* and therefore marginalised by the life of the metropolis. It is important to underline how:

> [...] despite the harsh criticism, the passageways have become one of the most tested architectural forms of the contemporary, widely used in cities with particularly climates in North America (Calgary, Minneapolis, Saint Paul, Des Moines etc.) in those from the humid climates of Asia (Bangkok, Hong Kong etc.) guaranteeing a second pedestrian level to dispose of the large quantities of transit flows. (Ciotoli 2017b:144)

Unlike other places affected by similar phenomena—such as Minneapolis and Saint Paul in the USA—in Hong Kong, these axes can structure the urban fabric of the city, not only by guaranteeing the connections between its parts, but also giving life to real urban nodes, placed at the end of the routes and destined to a commercial function. In this way, the retail is increasingly becoming a binder of the other functions that take place within the neighbourhood and, specifically, as a physical space where large flows of people can be conveyed. The need for sheltered streets reserved for pedestrians is entirely analogous to the nineteenth-century requirement 'which stimulated the construction of modern architectural forms destined for the public, such as galleries, passages' (Ciotoli 2017b: 138).

In addition to the pioneering example of the Wah Fu Estate, the relationship between social housing, commercial fabric and public space is faced in other residential complexes in Hong Kong; among the most recent, we can recognise the Hollywood Terrace. Unlike the Wah Fu Estate, the Hollywood Terrace is located more centrally, near the Sheung Wan station, in the northern part of Hong Kong Island.

This complex, built by Rocco Design Architects Associates and completed in 1999, maturely manages the interstitial space between the two 35-storey towers. The complex seeks to go beyond the so-called *podium tower typology*, which is the tower with a base initially tested in the United States in the 1960s and then rapidly spread throughout the world. Regarding this architectural type, Professor Weijen Wang states that:

> [...] the podium structure is used for public and commercial, as well as recreational and building services, and residential floors above the podium deck. The vertical separation of the building was a simple functional resolution to accommodate high plot ratio assigned to each plot of development land that needs spaces for both domestic individuals and public commerce. This mechanism of development control shaped a building complex into a typical form of a large podium structure and several towers of residential floors above the podium deck. (Wang 2016:70)

12.5 Conclusions

The construction programme promoted by the HK Housing Society meant that the city was 'studded' with many housing complexes in the most central areas, as well as in the more peripheral ones. However, many of these complexes were disconnected from the urban system, and for this reason, they acted as individual elements towards the city. Furthermore, in each estate where architects have tried to design towers with a base podium in which to concentrate all the services useful to the inhabitants, it is possible to highlight how these complexes do not communicate with each other and, in this sense, behave like small cities.

The positive and pioneering example of the Wah Fu Estate was brought to the most extreme consequences, denying the urban need for connection between architecture, citizens and the metropolis. Therefore, the use of horizontal connective elements (underground and/or suspended ones) suggests the desire to link together the places of the city (streets, housing complexes, gathering spaces and commercial ones). The road is tied to the surrounding fabric, and, in this way, its architectural conformation is shaped by the urban context. It, therefore, acts as a specific axis for pedestrian transit, able to organise the use of the architectural complex from one point to another (Ciotoli 2017b:138).

Some of the most important commercial centres in Hong Kong (such as Pacific Place, Times Square and the Landmark) have been designed in relation to the accessibility of the Central Elevated Walkway.

In this sense, these commercial places become spaces in which to channel pedestrian urban flows. The *ambiguous* relationship between private spaces and public ones should be recorded not only negatively (as a form of disadvantage for the public) but it could also have interesting turning points. After all, both the housing programme in Hong Kong and the pedestrian system of the CEW are still in full experimentation and construction. In this case, the Hollywood Terrace proves to be an example of the new order of elements and situations that are emerging in Hong Kong; indeed:

> [...] the dense and textured fabric of old Central-Western District calls for a careful response to its urban environment, and an intricate appropriation between private and public realms. A system of public spaces is developed in the form of a series of landscaped gardens and terraces that, together with the punctuating stairs and lifts, form an elaborate twenty-four-hour pedestrian access connecting Queen's Road Central with Hollywood Road through and within the site. This public thoroughfare interweaves with the private pathways that lead to the lobbies and amenity areas for residents. (Rocco Design 2016:67)

The success of Hong Kong's infrastructure system allows citizens to move around entire blocks destined for trade. According to Leslie Lu, the CEW:

> [...] attempts to facilitate pedestrian flow, to separate pedestrian and vehicular traffic, and to provide vital pedestrian links between new development and existing fabric. A new infrastructure system, conceived as simple elevated bridges and walkways, has over time evolved into a distinct and separate pedestrian network, linking existing landmarks and creating new urban foci. Although this development signaled the virtual abandonment of the street, the network of elevated walkways created a new urban configuration of Hong Kong: it became the 'new ground' of the city, reorganizing flow with new connections, indifferent to existing urban patterns. (Lu 2005:88)

The process of 'vertical fabric' in Hong Kong is a subject of a growing number of studies (Chung et al. 2019), as also evidenced by the 2018 Hong Kong Venice Pavilion, which is entirely dedicated to this theme. The creation of a *vertical fabric* in Hong Kong is indicative of an extremely current densification phenomenon. It is also the most concrete example of how even the theme of living vertically can—and should!—be combined with better conditions for citizens, taking into consideration the possible implications in the relationship between public/private spaces and between pedestrian systems and commercial fabric.

References

Bacon CW (1982) Streets-in-the-sky: the rise and fall of modern architecural urban utopia. PhD Thesis, University of Sheffield

Chung T, Man H (2016) Wah Fu Estate. HKIA J 72:34–35

Chung T, Tsang T, Wang W (2016) A conversation with Donald Liao. Define Design Direct Hkia J 72:26–32

Chung T, Tsang T, Wang W (2019) Occupy towers. HKIA J 75:1

Chung T, Tsang T, Wang W (n.d.) Occupy housing. HKIA J 72:3–12

Ciotoli G (2017a) Arcade d'oltreoceano: analogie e differenze della strada commerciale in Gran Bretagna e in Nord America. In: Belli G, Capano F, Pascariello MI (eds) La città, il viaggio, il turismo. Percezione, produzione, trasformazione. Atti AISU 2017. CIRICE, Naples

Ciotoli G (2017b) Dal grattacielo al Tessuto Verticale. Nuovi sviluppi architettonici e urbani. Officina edizioni, Rome

Consultancy for Heritage. 2011. Impact assessment for transformation of the former police married quarters site on hollywood road into a creative industries landmark Volume 1, Built heritage impact Assessment. https://www.amo.gov.hk/form/PMQ-HIA-vol_01-BHI-A.pdf. Accessed 10 May 2019

Council on Tall Buildings and Urban Habitat (n.d.). https://www.skyscrapercenter.com/cities. Accessed 20 June 2019

Daqing G (2016) Affordability as the inspiration of design for public housing. HKIA J 72:48–52

Falsetti M (2022) Less (space) is more (money): magazzini multipiano e torri della logistica in Asia Orientale e Sud Pacifico. Trasporti e Cultura 62(63):43–49

Heritage Museum Government Hong Kong archive (2010). https://hk.heritage.museum/documents/ResourceService/History/Other/Public_Housing-E_rev.pdf. Accessed 20 June 2019

Lu L (2005) The Asian arcades project: progressive porosity. Perspecta 36:88–89

Maretto P (1993) Realtà naturale e realtà costruita. Altralinea, Florence

Moretti G (2004) Entre globalisation et réalités locales: centres commerciaux et formes urbaines à Los Angeles, Montréal et Paris. Between globalization and local realities: shopping centers and urban forms in Los Angeles, Montreal and Paris. PhD Thesis in Urban Planning, McGill University

Rocco Design (2016) Housing projects in Hong Kong. Hollywood Terrace HKIA J 72:67

Smithson A, Smithson P (1957) Cluster City: a new shape for the community. Architect Rev 122:333–336

Steadman P (2014) Building types and built forms. Kibworth Beauchamp, Matador

Urban F (2012) Tower and slab. In: Histories of Global Mass Housing. Routledge, London

Wang W (2016) Mega-housing urban form in new tows of Hong Kong. HKIA J 72:70–74

Wang W (2019) Size, type and urban fabric: Hong Kong's towers in central. HKIA J 77:76–77

Chapter 13
Current and Future Urbanisms of the United Kingdom and Republic of Ireland Border

Laura Coucill and Tom Jefferies

Abstract This chapter presents an analysis of urban spatialities contextualised by and resulting from the political crisis produced by Brexit, which has brought focus on the territorial and spatial operation of the 499 kilometres long UK/Republic of Ireland border as a microcosm of the renegotiation of the relationship between the EU and UK. The location of the UK/RoI border results from the internationalisation of former county boundaries. Its status reflects shifts in connectivity between the two sides of a border created in 1921 from what was historically a single territory. Separation produced between 1968 and 98 by a militarised yet porous hard border was radically redefined by the 1998 Good Friday Agreement, defining an open border and a seamless interdependency between settlements on either side. Occupation around the UK/RoI border reveals spatial utilisation, type and movement patterns highly particular to this place. Characterised by continuous loose or dispersed urbanism, actively embodying difference; between legal structures, tax regulations, building and urban codes, currency fluctuations and cultural values, united by freedom of movement. Unique building types and settlements where access and egress are only possible by passing through another state reflect a relationship with the border that objectively embodies the lived reality of territorial difference. This situation applies at scales from individual buildings to significant territorial areas, capable of accommodating large infrastructural elements or settlements. Through critically reviewing the current border condition, this chapter proposes possible futures for this place. Brex City/Brex Cathrach addresses challenges, absolutes and ambiguities established by Brexit. Identifying latencies and potentialities through

L. Coucill (✉)
University of Salford, School of Science, Environment and Engineering, Salford, Greater Manchester, UK
e-mail: L.S.Coucill@salford.ac.uk

T. Jefferies
School of Natural and Built Environment, Queen's University Belfast, University Road, Belfast, UK

G. Cimadomo et al. (eds.), *Urbanism at Borders*, The Urban Book Series,
https://doi.org/10.1007/978-3-032-09594-7_13

enquiry by design, this chapter will demonstrate how the border is a unique place embodying duality, a range of possible futures, from seamless to frictionless space to hard border, mirroring the inbuilt paradoxes and tensions of Brexit.

Keywords Separation · Brexit · Dispersed urbanism · Speculative urbanism · Border futures

13.1 Introduction

13.1.1 The Spatial Language of Borders

Geography is a fundamental driver in the location of borders and in driving or preventing conflict. Borders defined by difficult terrain, such as mountain ranges, have largely prevented war between countries that are neither politically nor culturally aligned. (Marshall 2016:x) Technological capability diminishes nature's physical defence lines, increasing the potential for borders to be contested and for geopolitics, rather than landscape, to define territories. This situation creates specific spatial conditions and, in many situations, adopts defined architectural typologies, such as checkpoints. These typologies can adopt a location-specific vernacular language (Martén and Boano 2022) but, in contemporary conflict, are increasingly replaced by generic security perimeter fencing, in some cases erasing natural topographical features and landscapes. (Nyers 2012) This neutralisation of contested landscapes creates indistinguishable spaces, arguably compounding geopolitical tensions when one space cannot identifiably be described as a place.

The location of the UK/RoI border is the result of the internationalisation of former county boundaries. Its status reflects shifts in connectivity between the two sides of a border created in 1921 from what was historically a single territory. Separation produced between 1968 and 1998 by a militarised, yet porous hard border, was radically redefined by the 1998 Good Friday Agreement, defining an open border and a situation of seamless interdependency between settlements on either side. Occupation around the UK/RoI border reveals spatial utilisation, type and patterns of movement highly particular to this place. Doherty and Fité Matamoros (2020) describe the particular conditions of UK/RoI border as a landscape rather than a border. They describe their work as an exploration beyond the border as a manifestation of an 'intractable political situation', instead focusing on 'the systems of interrelationships that occur around and across the border'. (Ibid: 101).

Through landscape fieldwork Doherty and Fité Matamoros (2020) provide an account of the lived experience of a border which unapologetically divides community and infrastructure, such as a petrol station in Belleek, which can be entered and exited on either side of the border. Their design-based approach highlights the limitations of mainstream data collection and analysis underpinned by independent political narratives on either side of the border, to expose the reality of communities transcending imaginary divisions to occupy a shared landscape. The work presented in this chapter builds on Doherty and Fité Matamoros's (2020) approach to imagining possible spatial implications of a 'Brexit deal' for the lived experience of an otherwise porous and largely invisible border.

13.1.2 Routes to Exploring New Urbanisms

Inhabited space, whether urban or otherwise, is made up of layers of complex systems and networks. Various forms of analysis over time have revealed patterns of activity, made manifest in the physical environment (Marshall 2009). These patterns are the product of human activity and the cultures they create. Examples of how embedding performance metrics into the work processes of designers can be controversial at best, and reductive and deterministic at worst (Coucill 2013; Imrie and Street 2011). This is a situation best avoided by establishing and testing methodologies which engage with location, culture and place. This is especially pertinent in a context framed by technological integration in all aspects of urban space, from digital communication networks to monitoring and surveillance.

Technology and data are firm partners in the production of space. Urban areas must accommodate technologies which enable continuous data production. However, there is a disconnection between the continuous data production which triggers continual responses in real-time and the comparatively static qualities of the built environment as the ephemeral demands of the digital cannot be met by the built environment. The design and planning of cities is based on long-term analysis of historical precedents and the re-evaluation of situated social, political, economic and environmental needs. It is crucial that data are reflected upon and re-evaluated in the context of space to inform long-term sustainability. A visionary approach is needed to reconsider the role of design within this discourse, adopting it as a method of testing and experimentation to reassess the characteristics of urban form in response to the transformations driven by digital infrastructure to date, and to offer an evolving framework for the ongoing analysis of spatial conditions firmly rooted in cultural context.

The importance of cultural presence as a factor in the design of infrastructures can be seen in the responses to previous 'invisible' urban phenomena including sanitation, pollution and energy use. When invisible phenomena become visible as problems, a retrospective response is demanded. Data mapping work undertaken by Infrastructure Space Research Group at the Manchester School of Architecture argues that through effectively representing the operational aspects of infrastructures, society can assess the cost benefits before a crisis situation emerges. It also enables the value of 'missing' infrastructures to be understood, not simply as absences, but as a space for the opportune introduction of capacity to facilitate societal benefit.

Mapping and data visualisation techniques support a better understanding of 'place' that is legible by a wide range of audiences. It is a working approach that, through application and testing, is under continued development. As part of this project, this method was adopted and developed to generate a shared understanding of the border-territory across a multi-disciplinary team. Through data mapping and visual representation, investigations undertaken in October–December 2018 have provided a clearer understanding of the spatial conditions and experience of the Northern Irish–Republic of Ireland border regions. Investigations have begun to analyse the effects and outcomes of proposed technical interventions along the border as a consequence of Brexit. The visual nature of the work makes the findings legible and meaningful to a broad audience and the work to date has been warmly regarded by the Irish Senate.

The overarching agenda of data mapping and visualisation has been to uncover alternative forms of urbanism. In the twentieth century, ideal models of urban space were typically characterised by visionary approaches and articulated through the drawings and representations generated through the architectural design process. During this period, some of these ideals were manifested in policy to be boldly implemented in a post-war period of wholesale modernisation. Understanding how space operates at a range of scales is critical to identifying, evaluating, developing and testing new forms of contemporary urbanism. Critical evaluation is essential to generating effective planning in a context dominated by the desire for efficiency.

Mapping and visualisation represent only one aspect of spatial investigation. Exploring existing conditions and visually representing these in meaningful ways through data geolocation and other forms of spatial representation not only affords the opportunity to re-evaluate spatial character and identify emerging spatial patterns but also provides the opportunity to generate proposals which can be placed into contemporary, evidence-based descriptions of urbanism to scenario test interventions. Scenario testing that takes place in combination with spatial analysis can offer insights at both local scales that are relational to the lived experience of individuals and provide a context for exploring broader technical and economic interests, both spatially and socially.

13.1.3 Characteristics of the UK-RoI Border

Drawn from the definition of Northern Ireland in the 1920 Government of Ireland Act, the UK-Irish border traces county boundaries. Separation produced between 1968 and 1998 by a militarised, yet porous hard border, was radically redefined by the 1998 Good Friday Agreement, enabling an open border and a situation of seamless interdependency between settlements on either side, to the extent that often the only visible indication of the presence of the border, is a difference in road signage.

The departure of the UK from the European Union produced a context which highlighted the advantages of shared standards and mobility agreements for people and trade. One unintended outcome of the 23 June 2016 EU Referendum (Electoral Commission 2023) vote was to place the status of the UK-Republic of Ireland (RoI) border into sharp focus. In the situation of the UK leaving the EU, this border, historically existing as a set of county boundaries before its formalisation with the creation of the Republic of Ireland and the UK province of Northern Ireland, will become the only international land border between the UK, now a third state, and the EU, massively increasing its international importance and status. Since the referendum in 2016, it has taken 4 Prime Ministers over seven years to propose an agreeable solution to trade across the only land border between Europe and the UK. Current proposals for the purposes of trade include situating the border in the Irish Sea.

The internationalisation of country boundaries created an irregular division between two territories: this has created spaces where the most straightforward connections between two places on one side of the border, require the international boundary to be crossed. In a social and economic context developed and defined by the overarching regulation and united agenda of the EU, spatial conditions such as this remain largely inconspicuous. However, Brexit now poses unique challenges for ease of cross-border mobility. For example, the irregular spatial conditions of the border mean that businesses occupy sites with land and facilities that span both the United Kingdom (UK) and the Republic of Ireland (RoI) territory. In some cases, residents and communities have begun to take precautions against uncertain outcomes, such as constructing new access routes within property boundaries potentially affected by the possibility of two jurisdictions.

The specific spatial characteristics of the UK-Irish border mean that the effects and outcomes of Brexit are not only intimately tied up with social and economic activity but also with the use and experience of place. In this context, existing proposals that focus heavily on operational and technical issues relating to tax, logistics and regulation have so far been limited by their detachment from an understanding of spatial use and performance.

The complexity and interconnectedness across this border reflect its historical origins and the reality of the border region acting as an interconnected whole. The historic interdependencies across the border line have been interrupted by periods of military hardening as seen during the establishment of the Irish Republic and the Troubles 1968–1998. Since the 1998 Good Friday Agreement the border has reverted to an entirely porous boundary, barely noticeable when it is crossed. The complexity of the border is often discussed in abstract through news reportage and the border has been mentioned by the UK media on an almost daily basis since the referendum result, but detailed understanding of its spatiality and use is elusive, partly the result of this single place being split between two states. This raises questions around how the border actually operated, how it has informed communities on either side of it and what is the operational and societal spatiality of the border, as seen in Fig. 13.1.

Throughout ongoing Brexit negotiations, the border has been a key topic of concern and contention for all parties involved. The UK and, therefore, Northern Ireland's exit from the EU has severe social and economic implications, particularly in the event of a no-deal scenario triggering a hard border condition. Apart from contributing to the peace process within the region, current border conditions exemplify significant economic cooperation between the UK and the Republic of Ireland

Fig. 13.1 Left: The Gates of the United Kingdom, near Crossmaglen, looking into the UK from the Republic of Ireland, 2018. Right: Gateway to the European Union, Killea, looking from the UK into the Republic of Ireland, 2018. Source: Tom Jefferies

through strong supply chain activity and trade, particularly in the agri-food sector. Therefore, finding a solution to prevent potential disruption to current frictionless cross-border trade, businesses and operations is paramount.

13.1.4 Research Drivers

The UK-RoI border has exposed many new questions about the spatial integration of technology and its social and community implications. Whilst recent history continues to inform the relationship between the two countries, this project has endeavoured to be detached from issues of identity and instead, be embedded in understanding the nature of existing frictionless border operations and spatialities. Driven by the agenda to develop and implement effective and sustainable solutions to the complex problem of the UK-Irish border, the methodology has offered the capacity to observe operational impact from a distance through approaches to mapping and data visualisation, and subsequently evaluating these observations in line with the impending possibilities posed by various forms of Brexit. Appropriate methods of applying and integrating technology in space cannot be developed without first speculating and developing design scenarios which work through possibilities for the logical operation of systems or processes.

Such work is multi-dimensional and cross-disciplinary by nature; characteristics that architects and designers continually address through the design and forward planning for long-term construction projects involving a spectrum of stakeholders. Considering the implications of technologies for space in resolving the question posed by the border situation led this research to be driven by the following overarching questions:

- What are the spatial and experiential implications of technical solutions to the question of the UK-Irish border in the context of Brexit?

- How might these affect the patterns of spatial use and existing urban characteristics?
- What are the possibilities for contemporary and future forms of urbanism to address the conditions arising from Brexit?

To begin to evaluate possible urban futures for the border regions, a better understanding of their spatial condition was needed. The study presented here was conducted with a view to examine the potential of data mapping and visualisation for studying border territories while interrogating the interface between space, identity and experience and the technical operation of political, regulatory and economic perimeters in the border context. Drawing on publicly available data, the studies consisted of desktop and fieldwork to map the full length of the UK-Irish border and to expose insights into patterns of spatial use, occupation and identity. The overarching objectives of the work were to:

(a) evaluate the existing condition and value of cross-border territories from social, cultural and technical perspectives through spatial analysis.
(b) explore multi-disciplinary approaches to appraise proposed mechanisms for seamlessness and examine the spatial implications of these for existing use patterns.
(c) advance data mapping and spatial analysis methods, in order to evaluate the relevance of the method and findings for other contexts.

13.2 Data Mapping Method

13.2.1 Previous Application

Recent work that has applied data mapping and visualisation techniques has spanned a range of scales, from neighbourhood analysis to in-depth regional investigations and has specifically focused on the key challenges of connecting data, place and people across these scales. Previous studies have supported and informed decision-making for a range of organisations and stakeholders, including NHS Highland and HIE (Highland and Islands Enterprise), Cornwall and the Isles of Scilly LEP, Catapult Satellite Applications and the Southwest Centre for Excellence in Satellite Applications (Jefferies et al. 2017). These projects provided a detailed analysis of the operation and performance of regional infrastructure for social and economic benefit by focusing on four thematic areas; Demographics and Economics, Energy and Power, Networks and Connectivity and Health and Lifestyle. Through these projects, data mapping afforded the re-evaluation of place, taking into consideration the implications and benefits to be yielded from digital technologies whilst evaluating how these reconcile with social and cultural activity.

Data mapping offers insights into spatial performance, however, the ability to cross-reference themes and data sets affords an opportunity to begin to capture an understanding of lived experience in a specific place. As a result, findings from data mapping studies provide a strategic position and evidence base for projecting future scenarios and using spatial, landscape and architectural design as a mechanism to test and evaluate proposals.

13.2.2 Methodology

The Data Mapping methodology combines data analysis and spatial representation with design as a means of testing and analysing models through three, dynamic and reciprocal stages:

1. Spatial Analysis comprises data analysis and contextual studies to understand the performance and operation of space, networks and buildings.
2. Design Exploration, uses the design process to explore and appraise solutions to specific spatial, cultural and architectural problems. This is an approach to testing and evaluating meaningful responses and interventions to the findings of Spatial Analysis and is a process which simulates spatial possibilities through a range of drawn media and captures a better understanding of the implications of digital infrastructure for physical space and lived experience.
3. Critical Study includes the dissemination of findings, reflection on the wider implications of these and the development of further research.

13.2.3 Spatial Analysis

Large volumes of data are increasingly produced and readily available. The first stage of the method involves dynamic and reciprocal stages of identifying and selecting. It is crucial to carefully identify and select appropriate data that will offer valuable insights depending on the objectives of the study. Part of this stage involves the critical evaluation of the key outcomes to identify the relevant themes. Where appropriate, data identification and selection can be undertaken through consultation with stakeholders and the public to establish key thematic areas. The overarching analysis adopted the thematic framework used for previous studies, focusing on demographics and economics, energy and power, networks and connectivity and health and lifestyle. Initial evaluations review readily available data, such as that published in the census to provide an overview of the area and how it has been and might be evolving.

Once a thematic framework has been identified, data mining is the process of assessing the data set and formatting the relevant data in preparation for data mapping and visualisation. The process provides insights into the value of the information contained within the data. It involves isolating useful data for cross-referencing with other data sets and themes in addition to preparing the data for input into the visualisation process. Compiling and geolocating data through design-based visualisation techniques provides meaningful insights about place. Data prepared and formatted in the steps above are visualised in (specialised) context or represented in inappropriate visual formats. Data are explored spatially and in ways that afford comparisons at a range of scales including regional, national, continental and global. This data is represented visually, which, where possible and appropriate, emphasises the relationship of the data to space and cultural context in order to be meaningful and engaging to an audience as wide-ranging as the citizens occupying the spaces concerned. This process reconciles visible and invisible physical, social and digital networks and systems to reveal their characteristics and interplay, to propose new spatial conditions and architectural responses to contemporary and future demands.

The process of spatialisation and visualisation is key to being accessible and meaningful to a wide range of audiences, and thereby having the capacity to engage those audiences in understanding place. Through past studies, this method afforded:

- the engagement of diverse audiences in place-specific data;
- the representation of place-specific, multi-level and cross-cutting spatial and socio-economic relationships;
- the recognition of new and emerging spatial conditions and characteristics.

13.2.4 Design Exploration

Design exploration comprises the development of design proposals as a mechanism to simulate, test and evaluate the implications of a design proposal as part of existing urban context. It offers a mechanism to apply and explore abstract and technical criteria set out in policy and informed by the integration of new technologies in urban space.

Following criticisms of urban reconfigurations in the twentieth century, such as those made by Alice Coleman (1985) against visionary approaches to social housing, it has become apparent that design-led approaches which critically evaluate the implications of their propositions are useful tools in building effective urban form. This is a logic which also applies to prevailing questions with international significance, such as issues of climate change and sustainability and matters concerning international borders. These issues need to be addressed and tested through a design-led approach rather than design-leading approach. Data mapping advocates a design-led approach to sustainability which commences by focusing on the way in which data is mined and examined in relation to location. Visible and invisible physical, social and digital networks and systems are mapped to reveal their characteristics and interplay, to propose new spatial conditions and architectural responses to contemporary and future demands.

Expanding on this method, more recent studies and realised projects have used the design studio as a laboratory to foreground investigatory work in the design process to identify design opportunities that reconcile the unique socio-economic conditions of place with global, national and local objectives. The work has demonstrated the value of considering buildings as part of an expanded network and that the method enables the identification of new spatial typologies within these networks which provide effective solutions to the problem of sustainability. In the following sections examples of proposals are provided to demonstrate how the method supports the development of ideas which address questions both critical to the context in which they emerge and are transferable at broader scales.

13.2.5 Critical Study

The critical study involves the reappraisal of design proposals. Core themes emerging from the study of space through the data mapping method include the shift towards technologically enabled smart buildings and cities. The method's critical study component is not considered a final stage, but an opportunity to engage stakeholders and open up cross-disciplinary discourse in exploring possible futures.

Critical study provides the opportunity to re-frame questions which arise from spatial investigations. In the case of the UK-RoI border, it was noted that proposals made at the policy level for maintaining open borders relied principally on technology, from digital communications capacity to surveillance and tracking devices. The presentation of these proposals in abstraction from space and lived experience highlighted the need to evaluate how such technical proposals would affect established day-to-day transactions across the border. Fitting with prevailing discourse concerning the integration of smart technologies and big data, within the scope of this work, this has led to the ongoing critical evaluation of design proposals through questioning: what do we want a smart sustainable future to be and how should this be manifest within existing place and culture?

13.3 Findings

The methodologies described above identified the unrecognisable existence of the border on the ground and consequently highlighted the physical invisibility of the cartographic line which places the border and identifies two separate territories. The cross-thematic analysis identified that this has both implications for people at a local and particular scale, in addition to things such as materials and goods crossing the border, as part of a wider logistical network informing global economies.

These findings become significant when understood in the context of a fragile and incomplete infrastructural network along the UK-RoI border is articulated through the absence of certain key infrastructures: including limited rail networks, incomplete motorway networks, no access to a domestic gas supply network, limited digital communication coverage and service delivery models that engaged with cross border cooperation on a sector by sector basis. It was also clear that the costs of developing coherent infrastructural approaches to the island of Ireland either required significant politically driven prioritisation or a longer-term perspective to justify the cost of implementation.

The following section presents examples from the method to demonstrate how the spatialisation of data-informed approaches to infrastructure that both acknowledged the border and sought technological advantage would potentially be productive.

13.3.1 TIBoS: The Irish Border State

13.3.1.1 Spatial Analysis

One key opportunity afforded by data mapping is the scope to aggregate data sets through spatial representation for cross-examination. In the case of studies of the UK-RoI border, this produced a range of mappings which aggregated infrastructural, cultural and geographic data. At a macro scale, the full length of the border is stretched, and adjacent regions are included to various extents depending on unique and specific parameters. This provided an overview of the areas and how they are represented by available data sets, such as census data. Through fieldwork which travelled the full length of the border, the lived experience of those manifestations was recorded to give an understanding of the day-to-day experience at a micro-scale. Fig. 13.2 draws on Carr's *Rule of the Land* (Carr 2017) to map cultural and historic landmarks that represent historical events along the border (see Fig. 13.2).

13.3.1.2 Design Exploration

Subsequent processes of layering cultural information with statistical demographic data and fieldwork led to the re-evaluation of the border as a distinct urban area populated with dispersed communities and economic activity. It demonstrated that rather than two distinct jurisdictions, the border regions are heavily reliant on each other for access and trade. This drove analysis to explore the possibility of a border state which would act as an independent area seeking to redefine the notion of borders and propose an alternative solution to the Brexit stalemate, one that is neither 'Hard' nor 'Soft', but both (see Fig. 13.3).

In this example, data mapping led to the proposal of a new border state, TIBoS, characterised by a new citizen-led sovereign and buffer state that acts as a mediator following Brexit, where current deals and agreements between the UK and the EU cease to exist and be valid. TIBoS was formed by the declaration of independence of neglected and generally deprived villages and smaller settlements that are located alongside the border region on either side of the border. This new state does not approach the border as a two-dimensional line map but as a complex and dynamic

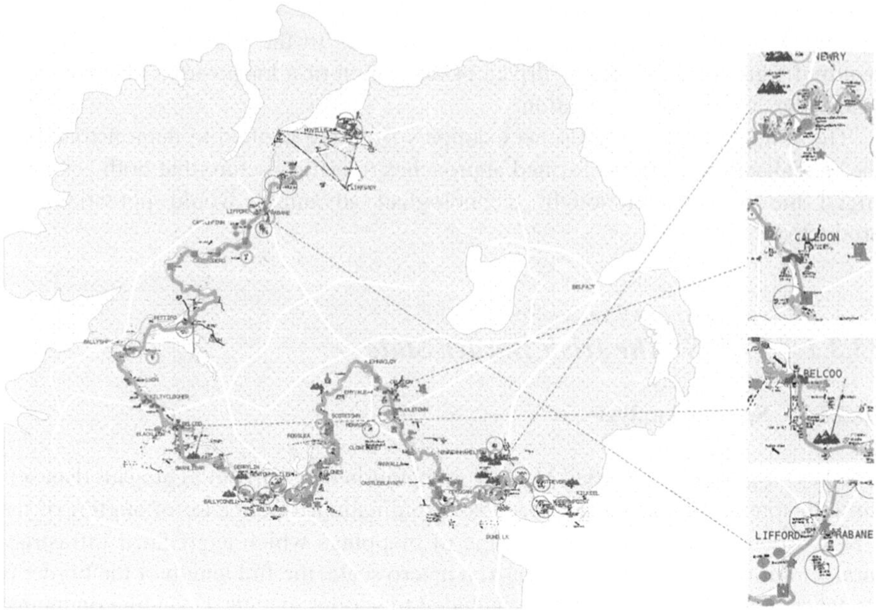

Fig. 13.2 Mapping of cultural landmarks along the border characterises the physical landscape's diversity and represents social and cultural manifestations of historical events. Developed in response to studio organisation and direction by Laura Coucill and supervised by Prof. Tom Jefferies & Richard Morton. Source: Andreas Leonidou, 2018

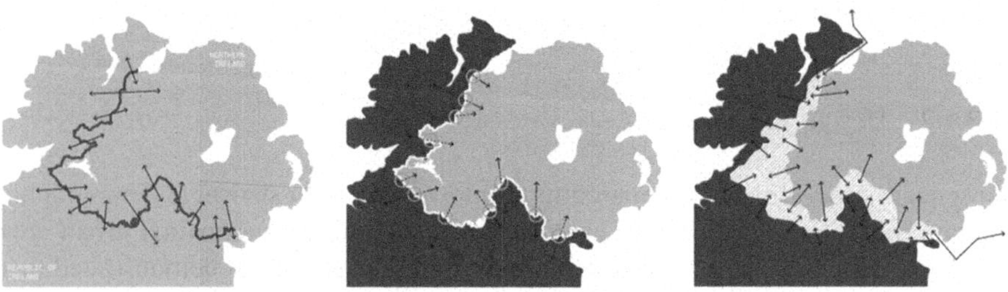

Fig. 13.3 Exploration of existing border agreements and potential scenarios which would alleviate undesirable outcomes in the event of 'Hard' or 'Soft' Brexit conditions. Developed in response to studio organisation and direction by Dr. Laura Coucill and supervised by Prof. Tom Jefferies & Richard Morton. Source: Andreas Leonidou, 2018

three-dimensional region. Each village located within the state's territory gains control of goods and people passing across the border. The project investigates what is required across the newly established territory to control and mediate all cross-country movements and flows. It embraces different levels of border control friction and proposes a situation where users have the freedom to decide the speed at which they travel through the new state and, most importantly, the level of border control friction they would like to go through.

13.3.1.3 Critical Study

The journey from the UK to the RoI and vice versa, through TIBoS, is used as a focal point. All objects, scenes, landscapes and components witnessed across this journey are compiled as spatial experiences across each village. These spatial interventions range in type, scale, timespan and level of friction, all creating a collective experience-led architectural language that belongs and is governed by its people. By challenging traditional notions of national identity where belonging derives from physical locality and proximity, the project explores and questions the meaning and existence of borders in an age of identity saturation, mass migration and large-scale political upheaval. TIBoS demonstrates that to be citizens of virtual space, we need to understand our physical space, and particularly our frontiers, challenging preconceived norms of territorial division at political and social scales (see Fig. 13.4).

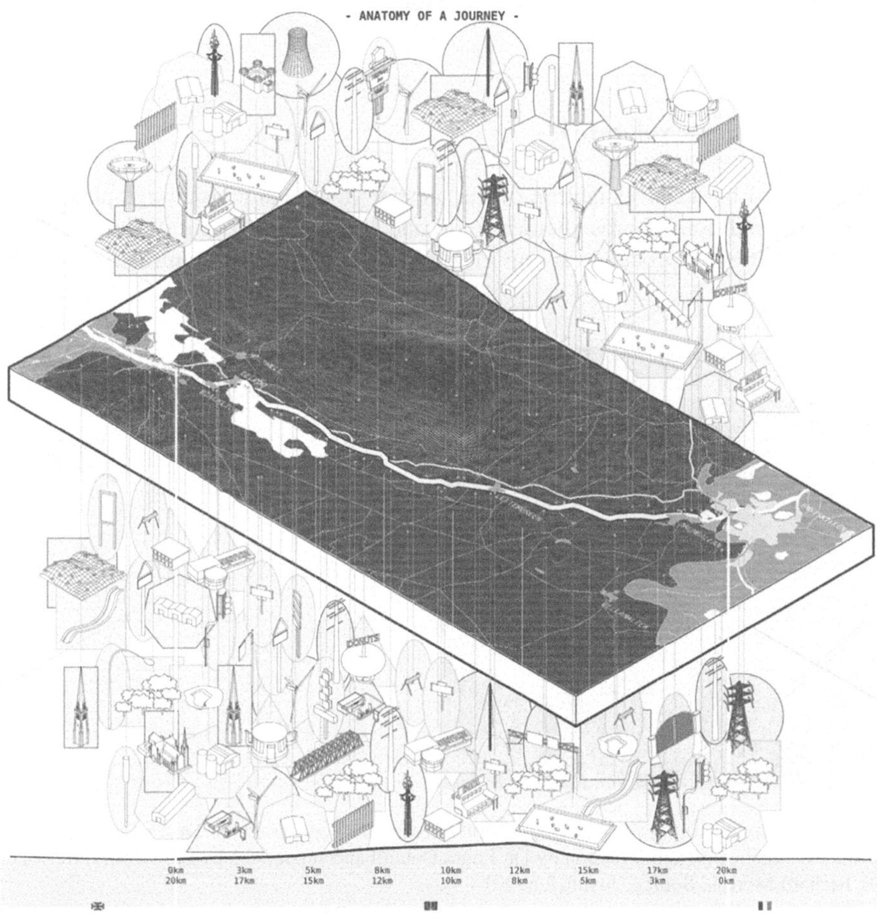

Fig. 13.4 Drawing communicating edifices and landmarks that make up layers of visual indicators marking areas of the border state in order that they can be recognised and read at various speeds. Developed in response to studio organisation and direction by Dr. Laura Coucill and supervised by Prof. Tom Jefferies & Dr. Richard Morton. Source: Andreas Leonidou, 2019

13.3.2 Space and Logistics: A Consolidated Solution

13.3.2.1 Spatial Analysis

The articulation of latent and operational infrastructural networks through data mapping was revealed. Considering trade transactions and the movement of materials and goods across the border exposed dependencies from either side of the border on trade across a range of sectors. Figure 13.5 examines road networks in relation to the placement of the border. Notably, this foregrounded the limited capacity to manage imports on the UK side of the border, as well as road networks which rely on border crossings for resilience; meaning that whilst it is possible to navigate road infrastructure without crossing the border, a road closure could force goods to make multiple crossings for a journey which begins and ends on one side. This restricted capacity for either side of the border to be able to continue to operate without the other brought the need to establish contingency for smoothness in new trade agreements.

13.3.2.2 Design Exploration

Underpinned by an investigation of the present value, nature, demographics, volume and condition of cross-border trade and logistics. Existing customs policy, regulations, border procedures, import and export arrangements, infrastructure and

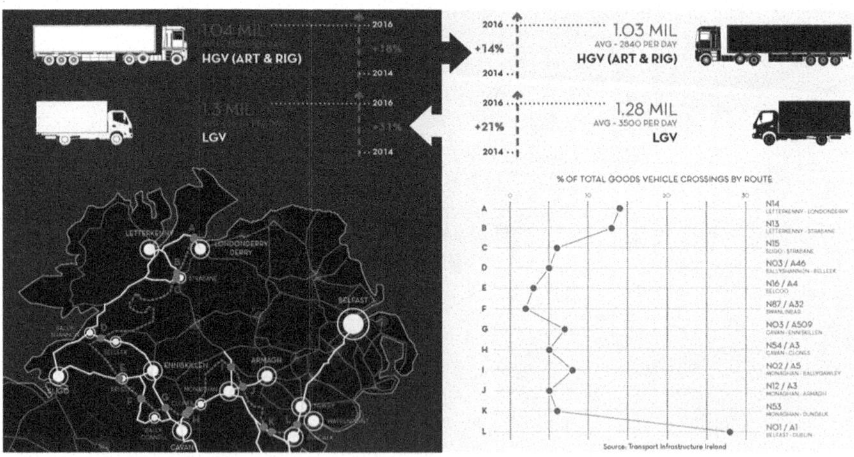

Fig. 13.5 Mapping traffic and logistics networks around the border regions. Developed in response to studio organisation and direction by Dr. Laura Coucill and supervised by Prof. Tom Jefferies & Dr. Richard Morton. Source: Justin Tan, 2018

technology adopted by the Republic of Ireland and the United Kingdom were analysed, with a particular focus on third-country trade conditions in preparation for the post-Brexit scenario. Scenario-testing reviewed the Kent Resilience Forum Contingency Plan for Dover Straits Ports as a precedent to anticipate potential disruption at the UK-RoI border. The in-depth understanding and findings informed the proposal of the All-Island Customs & Logistics Network, a series of key macro networks and strategically placed consolidation/distribution nodes that facilitate the possibility of a frictionless border condition for cross-border goods logistics. The macro spatial framework is developed around logistics demand, type and capacity, supported at the micro level by buildings and facilities that are crucial to frictionless border operation through the typology of the Customs Fulfilment Centre (CFC).

This is the architectural component enabling the transition from macro to micro scales of the Consolidated Solution. CFCs provide a network of goods consolidation facilities central to future cross-border logistics on the island of Ireland. The potential handling capacity of these CFCs is established by analysis of cross-border goods trade figures, thereby determining the infrastructural capacity and spatial requirements. This defines a typology that can be applied across the network of CFCs designed to embody scalability to different capacities (i.e. volume and type of goods and freight traffic).

Learning from the innovative retail distribution model of Amazon, the proposed establishment of a network of CFCs throughout the island streamlines customs logistic traffic through designated border crossings as well as distributing customs processes throughout the island. Legislative control of the border is expanded to these CFCs, which receive, check and inspect goods from regional traders. Goods are then consolidated and dispatched across the border via the designated crossings to fulfilment centres on the other side of the border, before being received by private traders to be distributed to the destination (see Fig. 13.6). These CFCs are equivalent to a port for ground logistics and can be nationally or privately owned as representative of the relevant customs authorities just as all seaports in the United Kingdom are currently privately owned and operated.

13.3.2.3 Critical Study

So far, the UK government in negotiation with the EU has yet to arrive at a plausible solution. Proposals include the highly controversial 'Irish Backstop', as well as the customs partnership and maximum facilitation 'frictionless' proposals. However, none of these have fully addressed the demands of the border, especially in the event of a hard Brexit. This project was framed by prevailing discussion around the 'Irish Backstop' during spatial analysis highlighted the dependence of have hinged on the

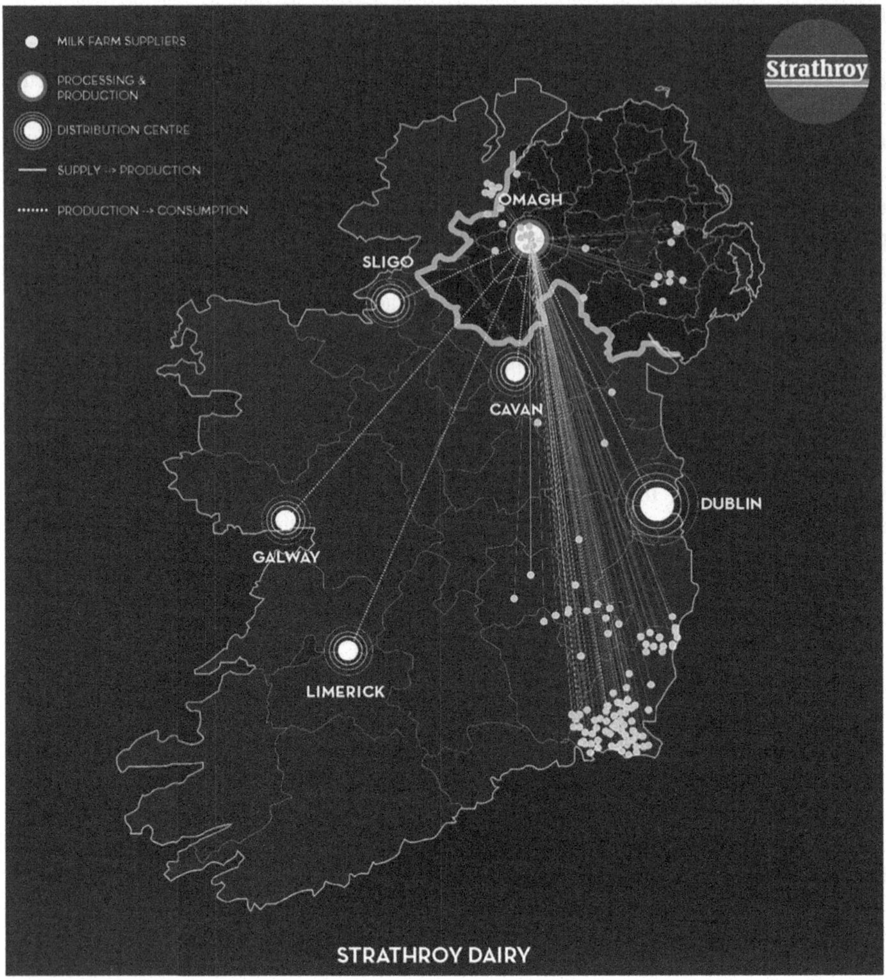

Fig. 13.6 Mapping to show the distribution of materials and goods across the border, which highlights the interdependence of the states. Developed in response to studio organisation and direction by Dr. Laura Coucill and supervised by Prof. Tom Jefferies & Richard Morton. Source: Justin Tan, 2018

possibility of developing a technological solution to frictionless cross border condition (see Fig. 13.7). The spatiality of what such a condition might entail has been left open but is currently seen as a technologically led solution. This is a flawed approach and ignores the necessary interplay between spatial scales and technological provision that could provide the basis for a frictionless border.

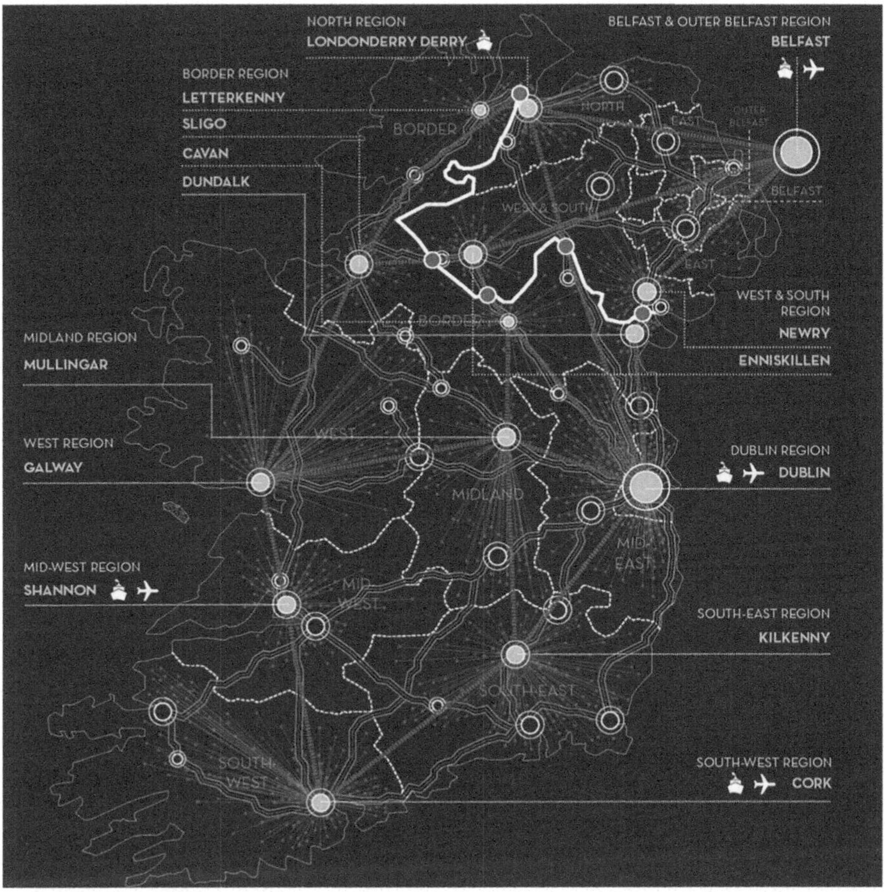

Fig. 13.7 A proposal which adopts intelligent distribution and logistics models implemented by large scale global organisations such as Amazon. Developed in response to studio organisation and direction by Dr. Laura Coucill and supervised by Prof. Tom Jefferies & Dr. Richard Morton. Source: Justin Tan, 2018

13.4 Conclusions

The key finding of this work has been the interconnectedness and interdependence of the two sides of the border. This demonstrates that a complex understanding of a place through data mapping enables the identification of latency within a situation. It may be the result of types of either realised or potential surplus or the result of two or more factors creating synergies. The examples of proposals discussed demonstrate that through a focus on maximising the potential of place, physical and systemic infrastructural solutions can act as drivers for creating inhabitable and beneficial places that are measurably good. The work has led to the recognition that technological obsolescence occurs at greater speeds than cultural change, and that society's view of what is good or bad built form also shifts over time. Consequently, new infrastructures that support the development of culture must be able to both accept redundancy as an inevitable part of their lifecycle and embody the potential to support culture as future heritage components and spaces in often radically redefined contexts.

The purpose of research through design is to critically test and extend ideas and situations that are identified through the initial data mapping phase of the work. The outcomes are proposals that can be used as mechanisms to explore the potential futures of the region and ask questions about current decisions, processes and spatial operational models.

A number of key themes emerged from the data mapping phase, ranging in scope from tightly defined built or policy-based proposals to significant and potentially radical co-created geopolitical changes, all framed by the reality of Brexit.

- Built infrastructure interventions
- Harmonised cross-border activities
- New spatial, territorial and temporal entities

13.4.1 Critical Application of the Data-Mapping Method

The ability to capture and represent data must be the precursor for developing designed outcomes that are measurably good. Data mapping has expanded our understanding of the limits of urban space through identifications of behaviours and patterns of use. This disassociates the description of space (e.g. as 'rural' or 'urban') from its appearance or density of occupation. This facilitates the designed responses to spatial and societal conditions that can be captured to provide a basis on which to make designs that respond to a measurable reality rather than an image of reality.

The emergence and adoption of smart technologies in the operation and design of contemporary urban space presents a life and development challenge to all aspects of urban design, governance and how we live. These technologies are often presented as a means to resolve urban form's dynamic challenges in relation to prevailing and emerging conditions. By placing smart responses within the context of societal needs and available spatial and infrastructural environments, it is possible to enhance the current operation and engagement of all types of space and to enable citizens to access assets by choice rather than necessity.

The work presented here demonstrates that the combination of spatial analysis and design-led exploration can offer beneficial evaluations of technical solutions as a product of visualising the experience offered by favoured proposals. The design provides a mechanism for evaluating technical proposals and understanding the spaces and building typologies required to implement those proposals. Data mapping also facilitates a shift away from a metric-based assessment context focused on demonstrating quantitative efficiency values, to a qualitative system in which value can be assigned on a place-by-place basis. In this way, the work discussed here proposes an analysis and design method that embodies interaction and complexity; allowing schemes to be understood within a wider context as components of networks, allowing policy and design decisions to be made from an evidential base that is multifaceted and user friendly.

Through the thorough interrogation and questioning of preconceived ideas of spatial efficiency, performance and character, it is proposed that a focus on effectiveness, latency and function will enable designers, policymakers and citizens to co-develop and test effective responses to the impending situation posed by Brexit in the interests of businesses and communities alike.

13.4.2 Further Work

The studies conducted as part of this project have pointed to the need to further investigate the existing data and develop a more focused understanding of space in relation to *tax jurisdictions* and *logistics*. Drawing on cross-disciplinary expertise to examine the possibilities for seamless cross-border activity will be crucial to understanding the implications of implementing technical and technological proposals to the question of the UK-RoI border.

The work presented here, however, does not reach beyond the domain of the imagination or beyond academic enquiry and critical challenge. There is a strong case to be made for live trailing and pilot studies that assess the implications of new spatial typologies and integrated proposals in the field. Opportunities to conduct live trials openly exist in technological fields to test the operationality and viability of new devices and networks. The built environment increasingly needs to keep pace with the transition to support technological capacity and capability. This requires careful and critical evaluation of the spaces we occupy, considering new technologically driven systems and processes that potentially demand new kinds of space, which simultaneously require evaluations of experience in order to achieve seamless transitions.

The next steps for this work will comprise multi-disciplinary investigations that bring together spatial studies with enterprises and organisations with the capacity to deliver technical applications. This work is currently being pursued with the support of Manchester Jean Monnet Centre of Excellence funding which will enable initial workshops to begin to establish a multi-disciplinary network to develop investigation and proposals further.

References

Carr G (2017) The rule of the land: walking Ireland's border. Faber & Faber, London

Coleman A (1985) Utopia on trial: vision and reality in planned housing. Hilary Shipman Ltd, London

Coucill L (2013) Tensions between theory and practice in architectural design. PhD. Birmingham Institute of Art and Design

Doherty G, Fité Matamoros P (2020) From line to landscape: the Irish northwest border. Archit Des 90(1):100–105

Electoral Commission (2023). https://www.electoralcommission.org.uk/who-we-are-and-what-we-do/elections-and-referendums/past-elections-and-referendums/eu-referendum. Accessed 30 Mar 2023

Imrie R, Street E (2011) Architectural regulation and design. Blackwell-Wiley, London

Jefferies T, Coucill L, Morton R, Luca Csepely-Knorr (2017) Data mapping Cornwall. Manchester School of Architecture, Manchester. https://www.msa.ac.uk/kernow/

Marshall S (2009) Cities, design & evolution. Routledge, London

Marshall T (2016) Prisoners of geography. Ten maps that tell you everything you need to know about global politics. Elliott & Thompson, London

Martén R, Boano C (2022) Checkpoint urbanism: violent infrastructures and border stigmas in the Juárez border region. Urban Stud 59(3):526–547. https://doi.org/10.1177/00420980211048414

Nyers P (2012) Moving borders: the politics of dirt. Radic Philos 174:2–6

Part IV
Nature as Intangible Border

Chapter 14
Inhabiting the Border: Between Urban and Natural

Pia Marziano, Paulina Valdivia, and Carla Zagal

Abstract Nature has always played a key role in the development and topography of cities, being a structuring and integrating part of these. The rapid urban expansion in some cases has given life to a peripheral urban fabric characterised by self--construction. The investigation compares two neighbourhoods in South America: Valparaiso, Chile and Iquitos, Peru, where the urban periphery converges with nature, linking the concept of urban ecotone. New housing typologies are born as a response to inhabiting the limit. The first case is the city of Valparaíso, in which its urban expansion into the hills unfolds towards the natural limit of the ravine, adapting to the relief of the slope. The second case is a neighbourhood in Iquitos, where the shape of the river changes with the seasons, driving the settlements to rearrange according to the temporality of fluvial cycles. In this way, the appearance of spontaneous neighbourhoods that challenge and often adjust both urban and natural limits consolidates a new border with a spatial logic and social dynamics that respond to the constant dialogue with nature.

Keywords Ecotones · Informal settlements · Urban nature · Adaptation · South America · Urban periphery · Human–nature integration

P. Marziano
La Sapienza Università di Roma, Rome, Italy

Metagora Studio, Concón, Chile
e-mail: pia.marziano@uniroma1.it

P. Valdivia
Universidad Politécnica de Cataluña, Barcelona, Spain

9CERO4 Arquitectos, Viña del Mar, Chile
e-mail: paulina.valdivia@estudiant.upc.edu

C. Zagal (✉)
Universidad Politécnica de Cataluña, Barcelona, Spain
e-mail: carla.zagal@estudiant.upc.edu

G. Cimadomo et al. (eds.), *Urbanism at Borders*, The Urban Book Series,
https://doi.org/10.1007/978-3-032-09594-7_14

14.1 The Landscape of Self-Construction as a Boundary Between the Built and Nature

In the periphery of modern towns and cities, characterised by a separation of the consolidated urban fabric from that of the unbound, the landscape of the boundaries between nature and the constructed takes centre stage. In many cases, the inhabitant appropriates the nature of this area without urban planning through the action of self-construction, like squatter settlements, but following the laws of nature, since it can never be completely dominated.

This chapter will deepen the nuances of inhabiting the limit of the urban periphery through two significant references of informal self-construction case studies: on the one hand, the city of Valparaíso, Chile, that inhabits the border of a hill responding to an interaction between the nature of the ravine and the construction of the house and, on the other hand, the Belén neighbourhood in Iquitos, Peru, which inhabits the border of the Amazon River that manifests a constant dialogue between land and water, as the shape of the river changes through the year.

In these cases, the urban limit became an opportunity for the development of spontaneous neighbourhoods in close contact with the natural environment. In this context, the concept of ecotone (from the Greek *oikos*, house, habitat, and 'tones', tension) can be applied, which arises from the ecological landscape planning, in order to define a sudden change of one ecosystem to another, giving rise to intermediate zones with great biodiversity and maximum interaction between ecological communities (Bartorila 2010). The concept was accepted by urban planning, defining urban ecotones as spaces of constant exchange and transition between two contiguous habitats, in this case, the urban and the natural, which function as potential articulators. This, in the case of the self-construction of the periphery, translates into housing typologies that can adapt to the territory.

From this point of view, the limit which allows exchange is not a separation barrier, but a space of relationship with a high value (Coppola 2004) between nature and the space of man, where the contours fade, mixing the outside with the inside. These two realities coexist in a spontaneous way, letting the inhabitants build with an intuitive knowledge about their territory, and generating different ways of interaction and encounter between the anthropic and the natural (Turri 1974). This permeability of the limit calibrates the spaces for the new urban life, which operates as an open gate in some circumstances and as a closed one in others, such as the door between two nearby enclosures (Albrecht and Benevolo 1994). In this sense, a rich space for exchange is established where elements of dialogue between nature and city can be observed. The constructive typologies of the urban ecosystem evolved through direct interaction with the natural ecosystem, thereby transforming the fundamental housing models of urban centres. These models, in the case of Valparaíso, adapt to the ravine, and in the case of Iquitos, they adapt to the river.

In the case of Valparaíso, the self-managed settlement constructions over the 100 m altimetric level are defined by the natural landform, giving life to homes that work with the concept of building from the top of the hill and down the ravine, with lighter and more flexible structures that emerge from it. On the other hand, in the case of the river in Belén neighbourhood of Iquitos, the constructions gradually adapted to the cycles of the river's rise, making it navigable or walkable depending on the season. This coexistence between the anthropic and the natural allowed the development of housing typologies that adapt to the river, being able to observe stilt houses and floating houses.

The following analysis will enable a better understanding of the complexity between humans and nature as it pertains to the urban periphery.

14.2 The Diffuse Limit of the Hills of Valparaíso

Valparaíso is a city with an unplanned origin (see Fig. 14.1). With the arrival of the Spanish to America and as a way to continue to Santiago, Chile, from the Pacific Ocean, the city of Valparaíso grew informally as a port with no urban direction, unlike the city of Santiago founded with a grid plan. It is a city formed by approximately 42 hills, which configure an amphitheatre bordering a flat extension facing the sea to the north, and to the south, west and east facing the high hills; this condition has guided the settlement by its natural edges and the growth of the population in a relationship between sea and hill. The port of Valparaíso became the entrance and exit door of Chile to the world and a way through for American and European ships entering the Pacific Ocean before the opening of the Panama Canal (1914). Due to its status as an exporting port, it experienced explosive growth from the beginning of 1835–1932, when the city extended to the nearest hills under 100 m. This part of the city is now known as the 'founding hills'. The urban perimeter grew, occupying mainly the territory on flat extensions of the planned hills (hill Playa Ancha and hill Barón), identifying the potential of the wide plateau of these two hills as the first zones of reticulated urban intervention. Parallel to this, the city expanded the flat area towards the bay, gaining territory to the sea, seeking to consolidate its flat extension (plan) from 0 to 25 m strip.

Given the remarkable growth of this enclave, a period of economic prosperity began, leading to the investment in great urban infrastructures to assemble its fabric, which connected the flat city with the nearby hills by stairs and elevators, consolidating the first terrace of residential hills (below 100 m). With the expansion of the cities in the hills, the road *Camino Cintura* (Waist road) was constructed at an elevation of 100 m, becoming the only road to traverse the central hills transversely. Over the years, an informal settlement emerged beyond this road, with *Camino Cintura* acting as a dividing line between the consolidated city below and the higher hills (from 100 to 500 m in elevation). These upper areas became populated with precarious housing, often built on increasingly complex and challenging topography as the hills rise. In Valparaíso, there is a mixed territory where the urban periphery is built autonomously by individual units that, as a row, are tied to the consolidated city. As we identify these three parts of growth, we focus on the urbanisation over 100 m where the topography of the city changes. In this part, between crests and river paths

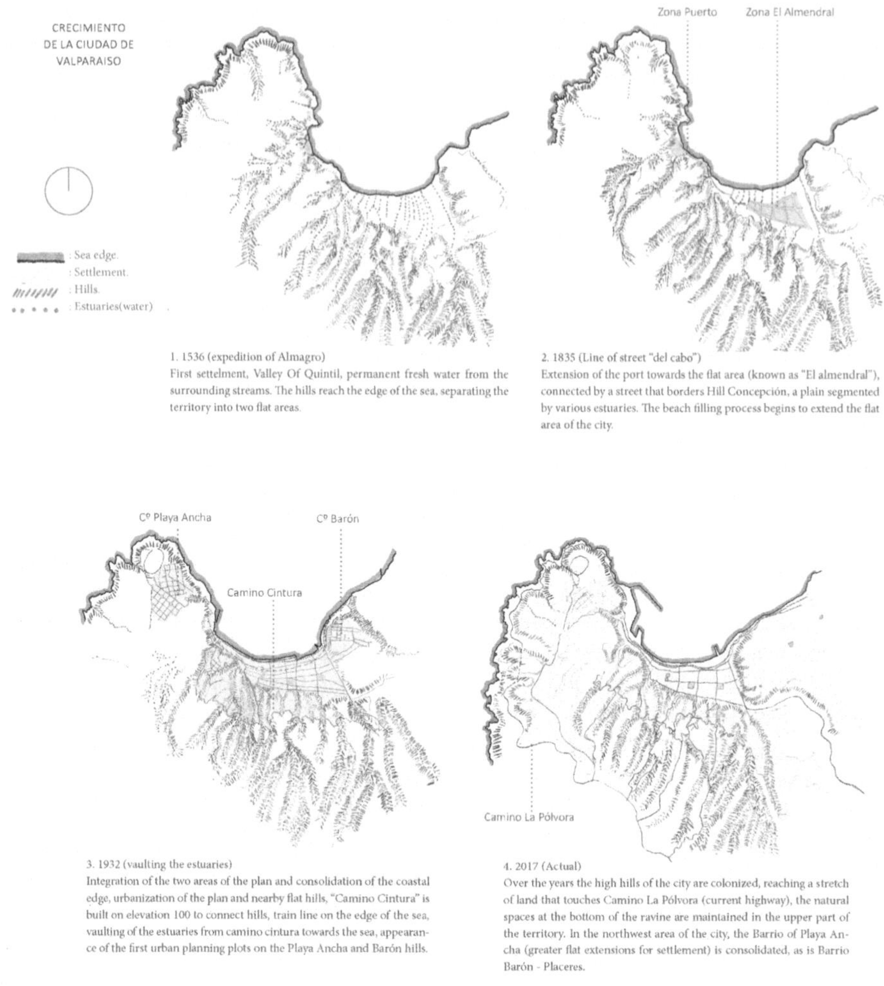

Fig. 14.1 Scheme of the growth of the city of Valparaíso. (Source: the authors)

with a rough topography, the informal neighbourhoods are built with diverse methods, as they try to attach to the consolidated city.

To understand how the informal neighbourhood continues to grow in the hills over *Camino Cintura,* the hills are sectioned at three points (see Fig. 14.2): the crest, the slope/ravine and the lower of the creek. The crest is the highest point and functions as a watershed divide. It can vary in form, being either broad or narrow. As the hill descends, the crest transitions into a slope, often marked by sporadic and ephemeral run-off channels, typically beginning with a gentle incline. These sloping areas are generally mixed in nature, starting with moderate gradients (12–40%) that may develop into steeper ravines (40–300%) or even cliffs (up to 90°). At the base lies the creek bed, the lowest point in the terrain, where surface water collects and follows the natural channel towards the sea.

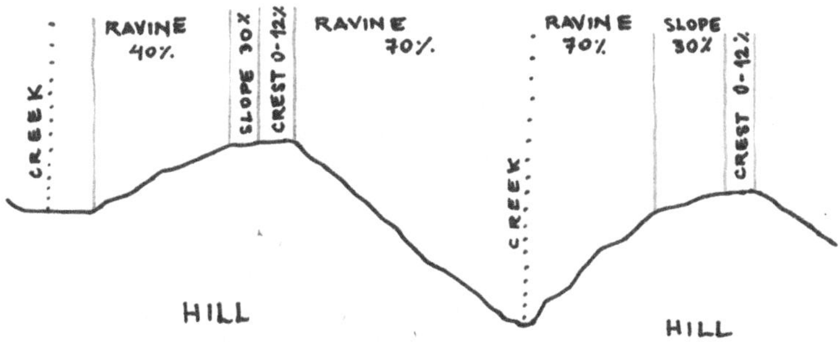

Fig. 14.2 Scheme of cross-hill composition. (Source: the authors)

Fig. 14.3 Disseminated informal constructions on the slopes and creeks around Valparaíso. (Source: the authors)

Valparaíso is a city segmented by river basins and ravines, which delineate and limit the composition and distribution of the hills. They collect the run-off of several hills to one or more river mouths (see Fig. 14.3).

The shape of the territory above level 100 is considered peripheral due to its topography. There is no connectivity between hills that allows interaction between neighbourhoods; the creek in its natural state is continuously the point of separation between similar neighbourhoods, between planned plots and self-managed occupation. The only connection between the higher hills occurs along *Camino Cintura*. In spite of the absence of a connecting route between hills above altimetric level 100, the proximity between hills is quite close. However, the steep angle of the ravine makes interaction between them impossible in its current state. Due mainly to its topography, the arrival of new inhabitants, usually of low income, forces them to intervene in the territory with simplicity. There are no major interventions and infrastructures, but rather small interventions and footprints that seek to generate access to new neighbourhoods. The landscape intermingles with self-built urbanisation, where the border between rural and urban becomes ephemeral (see Fig. 14.4). Here, self-construction follows the conditions imposed by the sloping terrain, continuing until the slope becomes a ravine or cliff, where the green landscape is a result of a complex space to inhabit autonomously and individually.

To comprehend how the hillsides and creeks are inhabited, the hill is analysed as a unit, from where the settlement starts and the limits it reaches (Solà-Morales 1997). We selected four frames of the high hill (see Fig. 14.5) in which we identify roads, plots, housing and topography. From the trace of the streets, it is possible, for example, to determine how the houses are accessed—whether from above, from below or at the same altimetric level. As a result, a tangible border is recognised: the initial streets or dirt roads along the crest, which act as limits for defining plots.

Fig. 14.4 Disseminated informal constructions on the slopes and creeks around Valparaíso. (Source: the authors)

Fig. 14.5 Frame of four spaces of urbanisation on the hills of Valparaiso. (Source: the authors)

Another clear edge is found at the bottom of the creek, where similarities in plot sizes, types of dwellings and their scattered arrangement become evident, a pattern that repeats across different areas.

The houses located in the flat part at the crest are usually of a homogeneous parcellation with rectangular sizes. As development continues down the hillside, new houses appear, adapting to the topography. In these areas, plots often extend downwards until they reach new horizontal paths, where additional families attempt to settle on the slope. These paths act as duplications of the main street at the top of the crest, and this pattern of expansion repeats as the terrain allows, replicating itself progressively down the slope until it reaches the creek, where the concept of defined land plots tends to disappear. This territory reveals how once rural spaces have been inhabited, resulting in a semi-ordered growth of new neighbourhoods (Corbóz 1983). The development of these settlements is guided by the topography of the hills, the paths of run-off, and the steepness of the ravines, which makes it impossible for individuals to inhabit them.

In the high hills, there's a form of reading that differentiates the surfaces of the upper and lower ravine. At the top, along the crest, much of the settlement is situated where traditional orthogonal plots dominate the larger flat areas. As the slope begins, the plots become deformed, adapting to the natural shape of the terrain. The slope reveals unpaved roads that continue or duplicate the line of the road of the crest. The hierarchy and organisation of the urban plots primarily respond to the steepness of the slope, which tends to follow the alignment of the main road along the crest. As urbanisation descends towards the bottom of the ravine, the number of houses and land plots decreases, making the highest areas near the crest denser and more built-up. When drawing these sketches, the 'non-habitable voids' appear between streams (see Fig. 14.6), where nature remains almost intact, with occasional houses at the bottom of the creek. These open spaces act as buffer zones within the urban area, revealing opportunities for new infrastructure connections and potential sites for urban intervention. The neighbourhood and the creek are faced with the vastness of the ravine and its depth. It is a space that we could call 'domestic', due to the shape of the relief and its interaction at the neighbourhood scale.

The hills are largely inhabited, but how far does the limit of the self-built settlement go? The peculiarities of the ways of living in different parts of the hill can be identified by the traces of the roads which lead to the houses: roads that follow the contours, roads that are structured over relatively flat areas and roads that cut the reliefs, self-managed operations as new families arrive to the city trying to attach to the new periphery of the territory. The location of a house on a hillside or a ravine seeks to cling mainly to an 'above', usually to connect with the flat surfaces at the crest. To build from the crest to the slope becomes more flexible. The support seeks to consolidate through pillars or lighter structures, leaving the house cantilevered over the sloping land. When the family grows in time, it unfolds to the ravine, towards those free spaces of the land, the houses are formed by large families, which are added to the initial nucleus, clinging to the first construction to continue growing (Pino and Ojeda 2015).

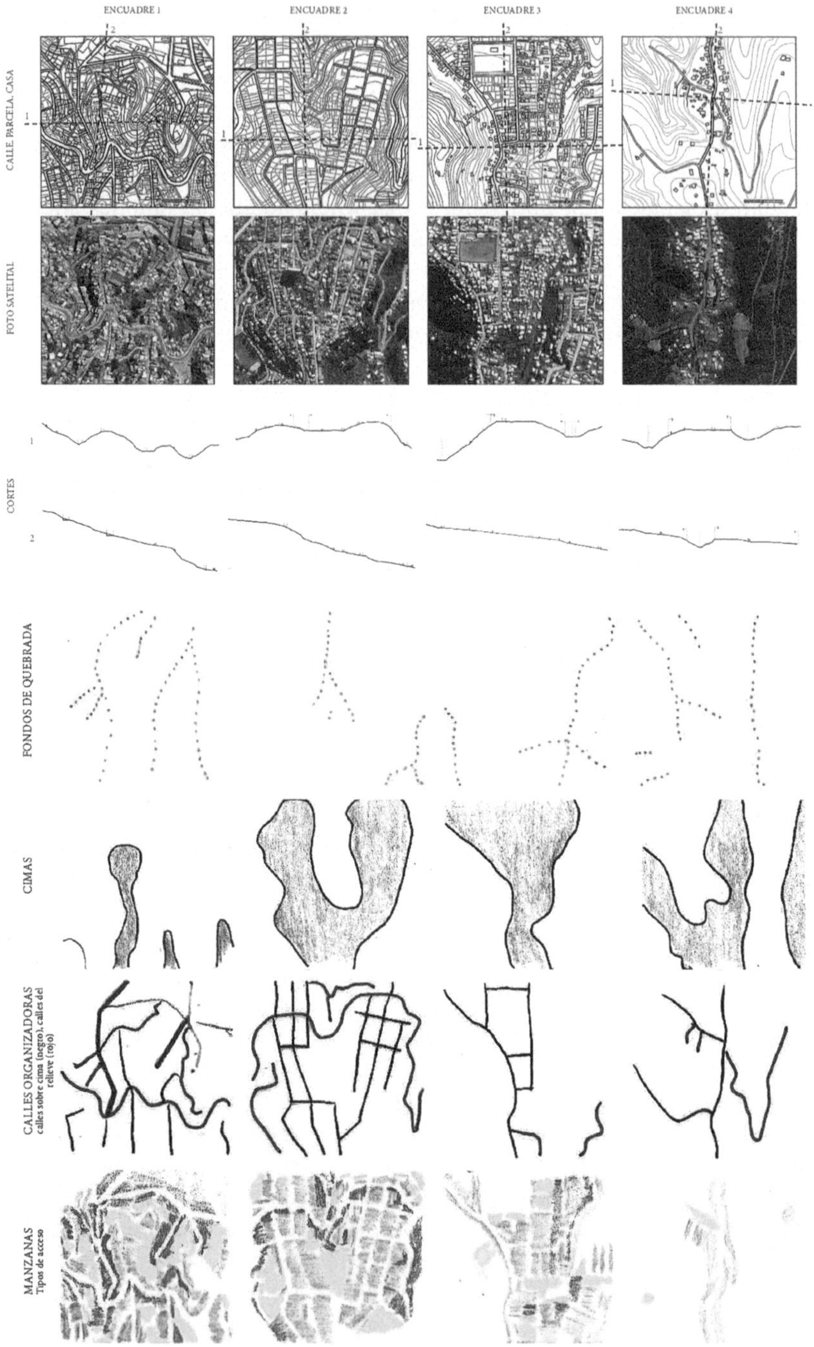

Fig. 14.6 Analysis of four settlements showing: map of the neighbourhood (row 1), aerial photographs (row 2) and topographical cross sections (row 3). (Source: the authors)

The families continue to extend the self-construction and their perimeter 'garden' to the end of the ravine without enclosure, open towards the creek, where the borders between neighbours disseminate, new routes of common use appear, being steep paths or makeshift stairs between houses (see Fig. 14.7).

The limit of the territory of the high hills, according to its topography and urban conditions, allows us to recognise a void (see Fig. 14.8), the potential of urbanisation in the ravine and the creek of the high hills that the self-construction cannot do maintaining the environmental conditions of the territory (Puentes 2009). Green spaces appear as a result of the fragmented topography, separating similar neighbourhoods from one another. These peripheral urban areas are shaped more by natural features than by planning guidelines set by the municipality or the state, responding instead to the contours of the local landscape. The appearance of self--built neighbourhoods on these hills offers the possibility to reflect on the way cities with a complex topography can be built and planned. In Valparaíso, it is said that had regulations and planning frameworks been fully enforced, this city would not have come into being, as it occupies land with slopes steeper than those allowed under national planning standards.

The city's growth is made possible only through the appropriation of the hills. In these elevated areas, the potential for valuing space, living vertically, and taking responsibility for the gorge bottom reveals both individual and collective motivations to inhabit spaces that are currently fragmented and poorly accessible. As a result, the definition of the urban landscape takes on a rural character, acknowledging that the untouched ravine and creek areas preserve the environmental quality of the territory. These spaces hold potential as sites for community infrastructure and public space (see Fig. 14.9), offering opportunities for the surrounding neighbourhoods to maintain the ecotones of the urban periphery.

Fig. 14.7 Disseminated informal constructions on the slopes and creeks around Valparaíso. (Source: the authors)

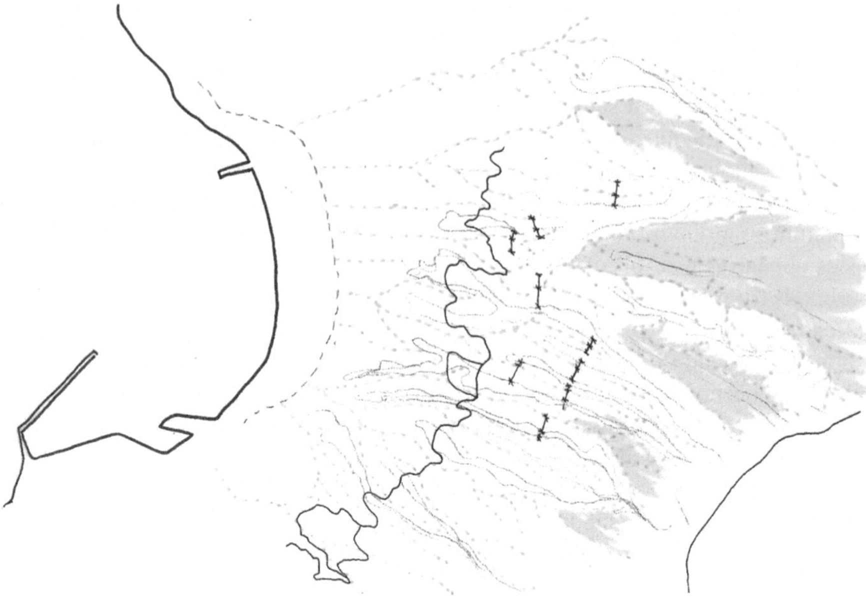

Fig. 14.8 Scheme of still natural spaces on the hills around Valparaíso. (Source: the authors)

Fig. 14.9 Urban Park in Merced Hill. (Source: https://rvl.uv.cl/noticias/6533-inauguran-kilometrico-parque-en-quebrada-santa-elena-del-cerro-merced-en-valparaiso. Source: Universidad de Valparaiso, Public domain)

14.3 The Amphibious Border of the Amazon

Today, Amazonian cities are not distanced from the phenomenon of urban expansion. In the city of Iquitos, in Peru, there is a particular case of a housing space, which responds both to urban needs and to its own territorial characteristics. It is the case of the neighbourhood of Belén, which is located on the east side of the city, on the bank that is embraced by the River Itaya, which has less flow than other rivers around. This has allowed the Amazonian neighbourhood of Belén to be a space for commercial exchange between the city and the nearby Amazonian towns, which generates a lot of social and cultural dynamism between its streets (see Fig. 14.10).

Although Belén plays an important role in urban life, the neighbourhood is seen as part of that informal city that, in many cases, is segregated by the city itself, which only recognises it as part of itself, lacking the consolidation dictated by the formal city. However, Belén is increasingly defined as a transition space and encounter between the city and nature. Belén is still one of the neighbourhoods that has stood out for increasing its growth, as it has gone beyond the opportunistic germination of a commercial-port space near the city centre, because it is a space that responds with traditional architectural patterns as well as specific variables of the Amazon territory. Since the first day that Belén exists, it has been marked by fluvial behaviour, since, according to some theories 'it was born as a fishing port on the banks of the Amazon river' (Reátegui Barra 2014: 95, see also Oliveira Valles 2010), over the years the Amazon River moved north of the city, changing its course once again. This is because in the Amazon territory great cycles of change occur that are part of the behaviour of the ecosystem, which is evident above all in the bodies of water and over time (García Sánchez and Bernex de Falen 2010).

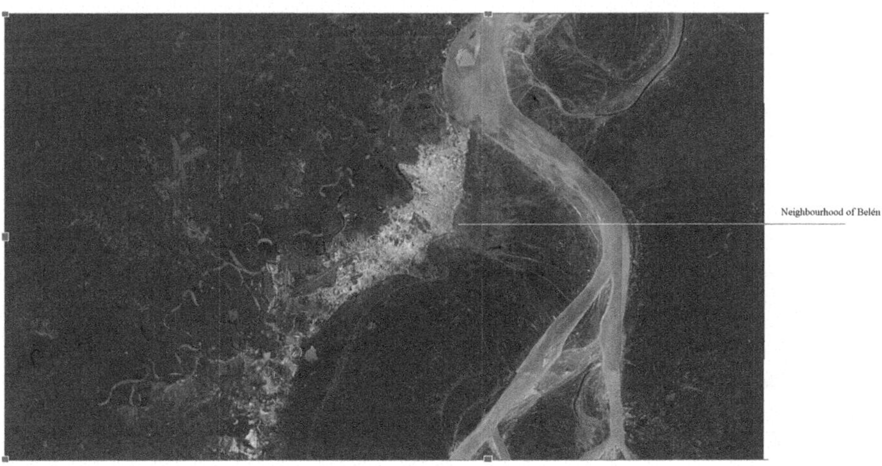

Fig. 14.10 Location of the Belén neighbourhood in Iquitos. (Source: the authors based on Google Earth)

Through the years the development of Belén has been sustained not only by keeping its urban economy active (which survives largely on a daily basis) but also by understanding the behaviour of its territory, which presents two fundamental cycles: the first cycle is characterised by the territorial transformation that is evidenced above all in the change of course of the rivers and the second cycle is characterised by seasonal changes (of growing and emptying of the flow of the rivers and streams).

When we analyse the reasons why the water cycles have an impact on the neighbourhood, we find that Belén is the result of a series of events that have been marked by two cycles, the *first cycle* is evidenced in the last gradual displacement that the Amazon River has had, which consists of moving from the east bank of the city to the north end, action that has allowed the River Itaya to surround almost entirely the east bank of Iquitos. All this has had a great effect on the river dynamics, since the Itaya River does not have as much flow and current as the Amazon River. The *second cycle* is based on Amazonian seasonal changes, which, because of its tropical climate, has only two seasons a year: the rainy season is also characterised by the increase in river waters, and the emptying season that is characterised by the decrease in rains, decrease in the flow of rivers and the formation of beaches (see Fig. 14.11*)*.

Therefore, analysing the impact of the second cycle has shown that, over the years during the emptying season, the size of the beaches is increasing, especially on the edge of the river that corresponds to the neighbourhood. And this is also partly because Belén is located on a kind of vertex that causes the sediments that the river brings to accumulate very easily (see Fig. 14.12). However, during the *rainy season*, the beaches disappear due to the increase in the level of the rivers that on average reaches between 12 and 15 m more flow. For this reason, the first and second cycles—integral to water behaviour and urban demands within the neighbourhoods of Belén—are interrelated, thereby generating a variety of

Fig. 14.11 Gradual displacement of the Amazon River parallel to the growth of Iquitos. (Source: the authors)

Fig. 14.12 Belén, a vertex located in the middle of the city which extends to the East. The Amazon River is located in the northern part of the city. (Source: Google Earth)

dynamics. For example, ports located along its entire bank move throughout the neighbourhoods as part of an adaptive process in response to the changing levels of the Itaya River.

Another of the most important dynamics of Belén is that between its streets is formed the largest market of the city, which does not have any spatial delimitation beyond the dimensions and slopes of the same streets, as well as the demand and supply of products. This shows that the neighbourhoods governs themselves, with trade being the most obvious form of organization, and that they are increasingly gaining prominence as a commercial space of regional reference. We can also find additional dynamics in urban morphology, which due to changes in level and displacement of the flow of rivers have been able to occur through the processes of territorial expansion (Solà-Morales 1997). That is, during the analysis of the morphology a subtle transition has been evidenced between the orthogonal plots located in the highest part of the neighbourhood and the linear plots located in the lower part or coast, even considering the houses located in the river, the transition of morphology is accentuated, since these houses are often separated from each other, that is, from a certain point of view the morphology of Belén could be understood as the dissolution of the block to the housing unit or vice versa depends on the position you see the neighbourhood. However, the neighbourhood of Belén has not been formed in a progressive way from the city to the river, or from the river to the city. But it is the encounter between the 'linear plot' governed by the relationship with the river and the 'orthogonal plot' established by the consolidated city. Both forms extend to the encounter with each other, through processes of occupation that contain different nuances, and this happens because the territory presents different intensities of change during its cycles of transformation (see Fig. 14.13).

Consequently, it is not the same to live in the part of the neighbourhood that presents an orthogonal plot and a more uniform urban behaviour than living in the part of the neighbourhood that presents a linear plot or in a houseboat, in the latter

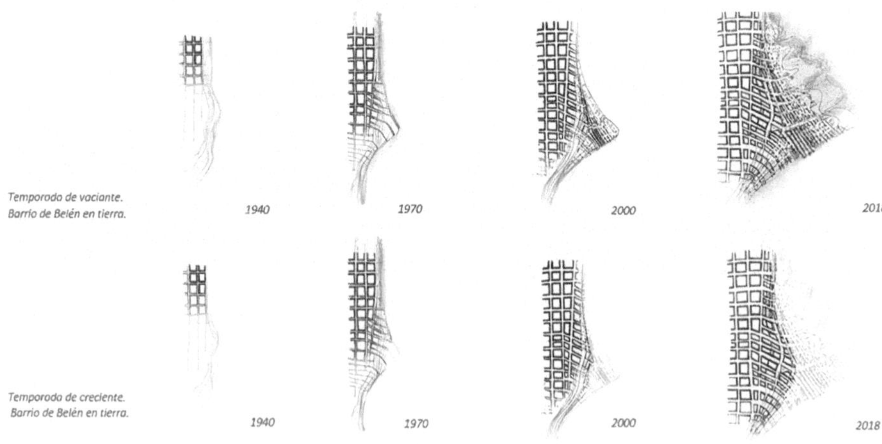

Fig. 14.13 Water process during the growing and emptying seasons. (Source: the authors)

two cases, constant adaptation to river fluctuations is needed, because, depending on the location, buildings respond formally and functionally to these patterns of change. Based on the need for adaptation, four types of housing have been detected:

Type *A:* They are the buildings located in the upper part of the neighbourhood, mostly built in armed masonry, which some have more than three levels, especially those that are mixed use (housing and commerce). They mostly have a longitudinal spatial distribution within a block (see Fig. 14.14).

Type B: This typology can be found in the area with the steepest topographic slope. Most buildings have a mixed construction system, in which the first level is built mainly in reinforced masonry and the second level is lightened in wood (see Fig. 14.14).

Type C: It is located in the beach area. They are constructions in palafitte, where the space of the housing is built between 6 and 10 m above the street level. Intermediate platforms made of wooden boards are usually built between the housing level and the street. Most of these platforms do not have perimeter enclosures and are often used as an extension of the house for public activities when water levels begin to rise (see Fig. 14.15).

Type D: They are floating houses, which are located on the River Itaya. All of them are usually moored to a trunk that they use as a reference to locate their space in the river, especially when the water level rises (see Fig. 14.15).

Fig. 14.14 Typology A and Typology B. (Source: the authors)

Fig. 14.15 Typology C and Typology D. (Source: the authors)

Fig. 14.16 The morphology of the neighbourhood. Between the orthogonal plot and the linear plot. (Source: the authors)

Considering the four typologies, it has been determined that the more adaptability buildings have, the less they are orthogonally grouped. On the other hand, they are grouped more frequently in a curved line, since in this way they can better deal with the river current. When analysing the particular shapes of the curved line, the process of displacement of the Amazon River has been evidenced, as shown in Venicia Street. Those curved lines formed by the grouping of housing are in turn the lines of past banks that the Amazon has been having along its displacement to the north of the city (see Fig. 14.16).

Thus, over time, the neighbourhood of Belén has progressively integrated the changes of the Amazonian ecosystem to such an extent that the transformations of the territory can be evidenced in its urban structure. This is largely due to the adaptive capacity of the Belén neighbourhood to natural and anthropogenic transformations. The result is a mutable urban structure, which is supported by three relational pillars: the 'river-port', 'port-market' and 'market-river' relationship. All of them are in constant synergy (see Fig. 14.17). When analysing the mutable urban structure, five areas that play an important role in the functioning of the neighbourhood have been identified. These areas are called *membranes* because the functions and activities that occur in them tend to expand and contract in relation to the cycle of increasing and emptying rivers (see Fig. 14.17).

One of these membranes contains the main port of the neighbourhood, located at the end of Venicia Street and which comes alive during the beach season. In contrast, during the season of increasing rivers, the main port fragments into several ports, located in different places of the neighbourhood that present continuity in their streets, a better connection with the orthogonal plot. Not all ports have the same dimensions; some cannot afford medium and large boats. After several months, when the river again diminishes its flow, the ports will move until they return to the place they used to occupy in the beach season (from June to September, approximately) (see Fig. 14.18).

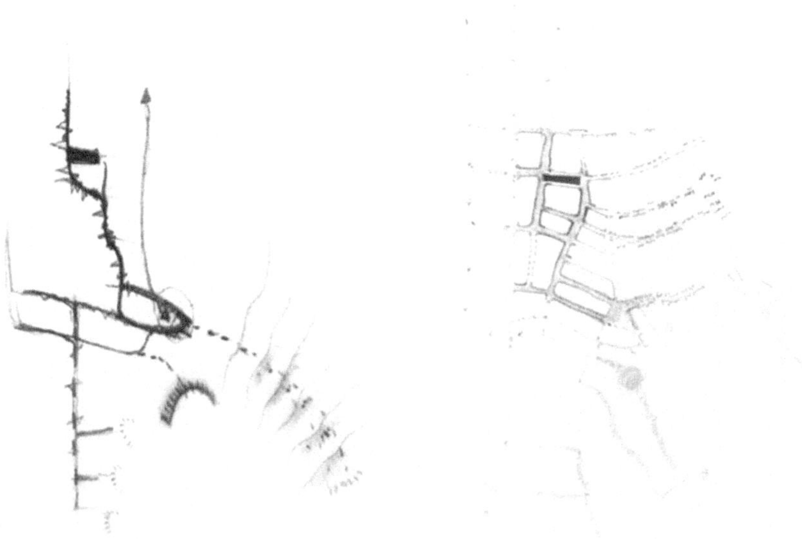

Fig. 14.17 Left: Market–river relationship. Right: Mutable urban structure. (Source: the authors)

Fig. 14.18 The mutable urban structure: displacement of one of the ports, during the growing season, which maintains a direct relationship with the market. (Source: the authors)

Fig. 14.19 Belén neighbourhood during the growing cycle of rivers. (Source: the authors)

Something similar happens with the space occupied by the market in Belén: during the beach season, the land space increases, and therefore the area occupied by the market usually expands until it reaches the main port. However, during the growing season of the rivers, the market space is compressed and linked to the different ports that are formed. In this way, the neighbourhood is adapting (Mayor Aparicio and Bodmer 2009). When the possibility of pedestrian connectivity with the rest of the neighbourhood disappears, wooden bridges are constructed to link individual houses with the surrounding urban fabric. This intervention affects the dimensions of the streets, which become channel-like spaces. Consequently, the river dynamics are altered once again, as are the modes of movement within the neighbourhoodn (see Fig. 14.19).

Ultimately, the neighbourhood of Belén exemplifies how the need of man allows living in a space of fluctuation, through the adaptability of its urban structure and its urbanity (Rykwert 1999). Recognizing the value of traditional architecture and vernacular knowledge, these should be regarded as fundamental considerations in future development plans and projects for the neighbourhood, the city, and other Amazonian settlements.

14.4 Conclusions

These two ways of inhabiting the limit between the urban and the natural allow us to visualise how self-construction spontaneously consolidates the identity of the urban periphery in the face of two different contexts of natural limits: the ravine and the river. The concept of 'urban ecotone' is confirmed, understanding that the complexity and richness of the borders in both cases are due precisely to the exchange response mechanism between two different ecosystems: the urban and the natural (Eizaguirre and Solà-Morales 2010). In this way, among the main aspects of the case studies, the transformation of urban morphology stands out as the city grows towards one of its natural limits, new housing typologies that arise from inhabiting this limit and new dynamics of spatial and social interaction.

The great growth of populated centres pushes urbanisation towards the periphery, forcing the inhabitation of more complex geographical places. The orthogonal fabric of the consolidated centres begins to mutate as it approaches the natural limit, giving life to a new urban morphology that responds to nature. In the case of Belén, Iquitos, the consolidated orthogonal grid takes on a linear and curved shape as it approaches the riverbank. It adapts to the shape of it and distributes the houses more fluidly near the water. In the case of Valparaíso, a similar phenomenon occurs, where the consolidated orthogonal grid, on elevation 100, goes up the hills, opening space between the ravines through constructions that are scattered throughout the territory. New constructive typologies emerge in response to territories defined by dominant geographical and hydrological conditions. In order to inhabit these contexts—characterized by unconventional spatialities and materials—a new mode of living is required (Viveiros de Castro 2013). In the Belén neighbourhood of Iquitos, the river has not only changed its course over the years but also as the flood cycles in which the neighbourhood follows, giving way to constructions that can coexist in the water during those periods. Stilt houses and floating houses can be seen, built mainly with light material, very different from the solid material houses located in the centre and at the top of the neighbourhood. On the other hand, in Valparaíso, it can be observed how the houses adapt to the slope in different ways depending on the point at which they are positioned on the hill. The constructions that are located at the top of the ravine stand out, characterised by downward growth where pillars and light structures work as support, leaving the construction flying over the slope.

The conditions of these natural settings, together with the adaptation of the constructions to them, give way to new public spaces where new social interactions are born. In Belén, it can be observed how it functions as a living organism that adapts its acts and public functions to the different periods of the year based on the rise of the river. As could be observed, this impacts the ports and the market, which change places and mobilise the population, giving way to various dynamics as a consequence of the flooding of its streets. On the other hand, in Valparaíso, due to the fact that the bottom of the ravine forms a physical separation that prevents connectivity between hills above elevation 100, the interaction between neighbours is mainly through voice and gestures, considering the proximity between the ravines of nearby hills. It is also interesting how the constructions, when inhabiting this natural limit, have spontaneously left a void towards the ravine that is configured as a 'back garden'.

Finally, it is important to point out that in both cases, only in the last 10 years have governments begun to develop urban plans for some sectors of these peripheral neighbourhoods. In the case of Valparaíso, the PREMVAL Metropolitan Regulatory Plan was approved in 2014, which designated part of this area as 'urban extension' and recognised the historical value of some of its hills. In the case of *Barrio Belén*, the government has generated proposals and resettlement plans, including the *Nueva Belén* project in 2017; however, they have not been successful since they move away or eliminate a large part of the labour source linked to trade. In this way, it is essential that governments develop an informed understanding of the territory and its modes of habitation, engaging in the co-production of urban plans with local inhabitants. Such plans should facilitate access to employment opportunities and public spaces. Although the government objectives for the urban periphery seek to improve the quality of life in these neighbourhoods, they must do so by valuing the unique characteristics of the buildings and the socio-spatial dynamics that have developed over time.

References

Albrecht B, Benevolo L (1994) I confini del paesaggio umano. Laterza, Roma

Bartorila M (2010) De las áreas ecológicas significativas al Plan de Ordenamiento Territorial de Montevideo. Nova Scientia 3(5):64–84

Coppola P (2004) Análisis y Diseño de los espacios que habitamos. Edit Pax, Ciudad de México

Corbóz A (1983) El territorio como Palimpsesto, traducido del original aparecido en Diogène. https://grupo1h2aboy2015.files.wordpress.com/2015/06/02-andre-corboz_el-territorio-como-palimpsesto_completo.pdf

Eizaguirre X, Solà-Morales M (2010) Las Componentes formales del territorio rural los modelos de estructuras agrarias en el espacio metropolitano de Barcelona: la masía como modelo de colonización en Torelló. Ediciones UPC, Barcelona

García Sánchez J, Bernex de Falen N (2010) El río que se aleja: cambio del curso del Amazonas, estudio Histórico-técnico. CETA—IIAP, Iquitos

Mayor Aparicio P, Bodmer R (2009) Pueblos Indígenas del Amazonía Peruana. Centro de Estudios Teológicos de la Amazonía (CETA), Iquitos

Oliveira Valles J (2010) Apuntes sobre la historia de Belén. Instituto de Investigaciones Amazónicas (IIAP), Iquitos

Pino A, Ojeda L (2015) Quebradas de Valparaíso: memoria social autoconstruida. https://www.researchgate.net/publication/282328448_Quebradas_de_Valparaiso_Memoria_social_autoconstruida

Puentes M (2009) Periferias Interiores: Un elogio a otros bordes de Valparaíso. Revista 180:24

Reátegui Barra M (2014) Iquitos. Telefónica del Perú, Varón Consultores Asociados

Rykwert J (1999) La casa de Adán en el paraíso. Gustavo Gili, Barcelona

Solà-Morales M (1997) Las Formas de crecimiento urbano. Ediciones UPC, Barcelona

Turri E (1974) Antropologia del paesaggio. Edizioni di Comunità, Milano

Viveiros de Castro E (2013) La mirada del jaguar: introducción al perspectivismo amerindio. Tinta Limón, Buenos Aires

Chapter 15
Inhabit the Borders: A Landscape Design Project for the Neighbourhood of Monteruscello (Italy)

Alessandra Como and Luisa Smeragliuolo Perrotta

Abstract The contribution focuses on the topic of urban borders and urban fragmentation within the contemporary city through the proposed transformation of the urban edges of a specific case study, Monteruscello, a district of the city of Pozzuoli, in Southern Italy. Here the urban borders affect the neighbourhood not only from an urban point of view but also within the social and economic realms. Monteruscello was born in 1983 as a new city, built on the edges of the existing urban fabric to accommodate the people of the historic centre of Pozzuoli, made uninhabitable by the phenomenon of bradyseism. The new neighbourhood, designed like an autonomous district, has become a new urban centre at the northern border of the old city. Its inhabitants never felt a sense of belonging to a place so different from their home town, with the sea, the port and related activities. Feelings of segregation and intolerance towards the local administration have grown over time; so, today Monteruscello represents a conflictual neighbourhood. Its design project is still unfinished. The neighbourhood is, therefore, characterised by a series of boundaries within its limits, which define the no man's land, creating fragmentation and lack of connections among otherwise continuous areas. Currently, these areas are undergoing several transformations, thanks to the European project MAC-Monteruscello Agro City. The project aims to transform the unused and abandoned areas into urban agricultural plots crossed by public paths, and to trigger a process of urban transformation by investing economic, social and physical aspects.

Keywords Urban agriculture · Landscape · Resilience · Peripheral urbanism

A. Como (✉) · L. Smeragliuolo Perrotta
Department of Civil Engineering, University of Salerno, Fisciano, SA, Italy
e-mail: acomo@unisa.it; lsmeragliuoloperrotta@unisa.it

G. Cimadomo et al. (eds.), *Urbanism at Borders*, The Urban Book Series,
https://doi.org/10.1007/978-3-032-09594-7_15

15.1 Introduction

Through a case study—the Monteruscello district of the city of Pozzuoli in Southern Italy—the theme of urban transformation is addressed by working on the inner urban margins with urban agriculture. Large public interventions have created places that have often become spaces of urban segregation; working on the urban edges can lead to a process of transformation that affects the whole neighbourhood from the urban, economic and social points of view.

Through the Monteruscello case study, the effects of creating a new district at the edge of the old city will be presented, together with the potentialities for its transformation starting from the analysis of its critical elements such as its urban abandoned green areas. The neighbourhood was born as a great public intervention in the 80s in order to provide homes to 20,000 displaced citizens due to the bradyseism emergency of an intense rising of the land. For the construction of the new neighbourhood an area on the edge of the city was chosen, but central to the system of the Phlegrean Fields, characterised by cultivated grounds and some widespread rural constructions. The case study is of particular interest as a significant example of interventions of great public impact that characterised many urban realities between the 1970s and 1980s. It occupies an area of approximately 245 ha with 3575 dwellings, with an urban density of about 74 inhabitants per hectare. In particular, it is interesting to observe how today a great urban transformation, such as a large district like Monteruscello, has been absorbed within the urban metabolism and what could be its transformative perspective. The main characteristics of Monteruscello are its large size—for which it can be compared to a city—and the presence of large open spaces, which together with public and residential buildings are all owned by the Municipality. If on the one hand this involves difficulties for its maintenance, on the other hand it offers the possibility to explore new public–private opportunities with the aim of increasing the quality of life.

It is in this strategic vision of transformation that the project 'MAC-Monteruscello Agro City' fits in. The MAC is funded under the European programme Urban Innovative Action (UIA) aimed at identifying and testing innovative solutions for sustainable urban development. The main aims include defeating urban poverty, understood in economic, social and urban environment terms. The project involves the transformation of 30 ha of green areas of public property, now abandoned, into a productive landscape through innovative permaculture techniques. The University of Salerno is involved in the project as responsible for the design of the Agro-Urbano landscape.

The areas that the project will transform into Agro-Urbano Landscape are urban voids that strongly characterise the district. They are areas expropriated in the years of the construction of the district but not used. Some were to be transformed into green spaces such as urban gardens and archaeological parks; others were to house public buildings such as the theatre, library and university buildings. In both cases their development was never completed and now, aside from occasional maintenance work, the areas are still in their original condition previous to urbanisation.

They are today like urban fragments that escaped urbanisation, remaining to witness the rural nature of the places before the great urbanisation of the 1980s. Their condition of areas of large dimensions enclosed by fences and inaccessible to the community, makes them like internal borders, breaks within the urban fabric. The case study of Monteruscello will show the evolution of the urban model of a large district characterised today by difficult urban conditions.

The MAC project starts from the observation of the critical aspects of the neighbourhood, mainly segregation and fragmentation, to propose a strategy of intervention for its transformation that starts from the study of its edges, identified in the large abandoned green areas. The project is offered as a contribution to the current debate on the inclusion of agriculture in urban spaces. In the case study, agriculture becomes the way to recover identity features of the landscape and at the same time to solve urban problems related to the large size, segregation and presence of abandoned and fenced areas.

15.2 Monteruscello: Urban Model, Borders, Fragments, Disconnections

The events that have decreed the urban structure of the district date back to the 80s, when a crisis of bradyseism hits the city of Pozzuoli making its historic centre unsafe. The design project was entrusted to the University Federico II of Naples, with architect Uberto Siola in charge of the project strategy and management and Agostino Renna as chief architect (Siola 1985).

The analysis of the city of Pozzuoli and its relationship with the complex system of the Phlegrean Fields convinced the architects to conceive the new district as an autonomous urban nucleus (Romano 1986). The territory is in fact interpreted as a heterogeneous system made up of different polarities, with Monteruscello as a new one, inside the municipality of Pozzuoli and linked to the whole Phlegrean territorial system (Capuano 2014; Miano et al. 2017; Maiuri 1981). The principles that guided the design of the new district were inspired by the idea of a new city, inserted in the geography of the territory (see Fig. 15.1) connected to the centre of Pozzuoli through infrastructural networks (Cerami 1988; Colpo et al. 1989; D'Antonio et al. 1989; Lirer and Pescatore 1987; Pisciotti and Miano 1988; Vitale 1989).

The Monteruscello project consists of three morphological units: the centre, the commercial area and the university area. The commercial area is located on the north side and includes the station with a large square and a sports area. In the centre of the district, connected to the commercial area by a wide pedestrian path that crosses the archaeological park and the museum, is the city centre. The city centre is at a higher altitude than the commercial area; therefore, the pedestrian paths that cross it vertically also become visual paths. The third morphological unit is the university area characterised by a large building surrounded by greenery. The three morphological units respond to the same north-south layout and to an urban grid,

Fig. 15.1 Monteruscello in the Phlegraean Fields. (Source: Unisa Mac project team based on the original Monteruscello drawing (Siola 1985: 17))

although each area has different sizes. The overall design and all the parts are held together by 'an absolutely rigid physical reference system, a modular network based on harmonic relations. This network stretches invisible threads over the territory so that each house, street and tree can find its defined place at any time and independently' (Renna 1985; Escalona and Francese 1987; Pagano 2012; Capozzi 2016).

The design project is based on specific measures that correspond to the scales of the neighbourhood, the small urban agglomeration and the territory, all of which are recognisable on a morphological, architectural and functional level. The urban layout is characterised by strong signs of mobility infrastructure, such as the Circumflegrea line that runs along the district, and the elevated highway, the state road *7 quater* Via *Domitiana*, which crosses it from north to south. The Monteruscello project is therefore conceived as a system of parts connected to each other through a dense hierarchical system of pedestrian and car routes, also defined by public buildings. Each urban unit corresponds to different types of public buildings and residential blocks. Several residential buildings of different typological origins are added to the group of single-family units: some, for example, are characterised by a central courtyard, others by the presence of commercial premises on the ground floor. Green areas, such as parks and gardens, together with the vegetation that characterises the streets' pattern, are present in all areas, even if designed differently. The elements of nature are entrusted with the dual role of preserving the character of the pre-existent landscape and marking the collective public spaces. 'Recovering the characteristics of the countryside within which the buildings are immersed, marking the large paths and collective places with pine trees, giving character to a road through the use of precious trees, making a garden welcoming and familiar with fruit trees' are some of the peculiarities indicated for green spaces (Pagano 2012: 312). A large urban park was also designed on the south side of the district.

Not all of the planned design will be realised, and not everything that has been built follows the original project. The construction process had left fragmented and uneven areas both on the edges and inside the neighbourhood because its construction had not been completed; in particular, this has affected the abandoned green areas that characterise the neighbourhood today. Not only has the project for the construction of the Renna district never been completed, but over the years an insurmountable gap has grown between the ideas of the design project and reality, which often leads to consider Monteruscello a serious design failure from an urban planning point of view.

The precarious social conditions, exacerbated by the never dormant feeling of redemption of the population combined with the desire to return to live in the city centre, sum to the bad environmental conditions characterised by degradation, repeated vandalism and lack of attention to the urban quality of the neighbourhood. Monteruscello is a district born out of emergency and political will at the administrative borders of the city. Although it has a strategic and central position with respect to the Phlegraean Fields, it has also maintained the character of a marginal place with respect to the city of Pozzuoli, both from an urban and social point of view.

The topic of the edge within the Monteruscello district has its own specificity. In particular, the district itself can be considered a margin, as it was defined by its designers; Monteruscello represents in fact the northern limit of the city of Pozzuoli, the limit of the territory with respect to the countryside and also the physical limit of the city, beyond which the next urban area begins. The boundaries of the neighbourhood towards the other territories are partly identifiable, defined by different types of elements. These are natural elements such as agricultural land, or elements of nature modified by man such as the quarries that define the northern boundary of the district and also the typical limits of modern cities represented by large road infrastructures. Towards the south, on the other hand, Monteruscello is continuously connected to the city of Pozzuoli by a residential agglomeration built in previous years in order to cope with further bradyseism crisis. Edges of the district are also the internal areas that will be transformed by the MAC project, now mostly closed to the public and only partially permeable to the pedestrian paths that mark the routes of use over time.

Fig. 15.2 The neighbourhood of Monteruscello: the green spaces. (Source: Unisa Mac project team)

These large urban areas are fenced spaces, where the imprisoned portions of nature tell of the productive areas that existed before urbanisation. Some of these areas are now unauthorised deposits of waste of various kinds, while in others there are still signs of the ancient condition of the site due to the presence of archaeological remains, farms and cisterns dating back to ancient times. Within these areas the phenomenon of illegal occupation has partially redesigned the territory and today in some of these areas, citizens have regained the territory in a non-legal and uncontrolled way creating small agricultural oases.

One of the most significant images that were found during the field works was that of a boat in a shed within one of the illegally occupied areas. This is the symbol of the inhabitants' trades before bradyseism and before being transferred to Monteruscello. The boat, covered with clothes that reveal its shape, is now inserted in the new environment, surrounded by cultivated fields and close to one of the historical farms, partially used for agricultural activities. The empty areas (see Fig. 15.2) within the Monteruscello district are therefore a further level of spatial segregation within the neighbourhood; the areas are like internal borders that characterise the district, creating interruptions and limits (Como et al. 2018; Alfano et al. 2018, 2019).

15.3 Multilayer Approach Through the Use of the Inner Borders and Urban Fragments

Complex urban issues cannot be solved by focusing solely on urban form; only a multilayered approach can address challenges such as borders and segregation, particularly in difficult neighbourhoods such as Monteruscello. The MAC project has given the opportunity for research work that aims to study the relationship between urban neighbourhoods and abandoned areas, identified as caesuras and voids within the urban fabric. The project creates new relationships between urban space and those areas transformed into productive green spaces. Until now, Monteruscello has been studied as a case study of a newly built city, focusing on its public buildings and urban layout, often overlooking the landscape value of the site. Investigating the neighbourhood today through its urban voids and its green areas represents a shift in the traditional perspective from which the area has typically been viewed.

The aim of the project is to look at the green areas of the neighbourhood, identified as inner urban margins, trying to reinterpret them as elements of the city to which it is necessary to give back value and meaning (see Fig. 15.3). The MAC project proposal aims to work with the city's internal margins through an action that involves them from the inside. This means setting aside the tools of urban morphology and design, to which these spaces have been considered in the past, but rather favouring a procedural approach to urban transformation. The edges are like internal fragments that the project intends to open and cross. The disintegration of the urban fragment is designed to facilitate the crossing of borders and to redefine the area as an urban space. This act of crossing edges also enables the formation of new relationships among parts that once constituted a single, impenetrable block. The voids, transformed into places of agriculture, are crossed by paths that unite them and stitch them to the built-up areas of the district. The project brings the neighbourhood back to a landscape dimension that was lost due to the massive urbanisation of the 1980s; it is this landscape dimension that can keep urban and rural characters together.

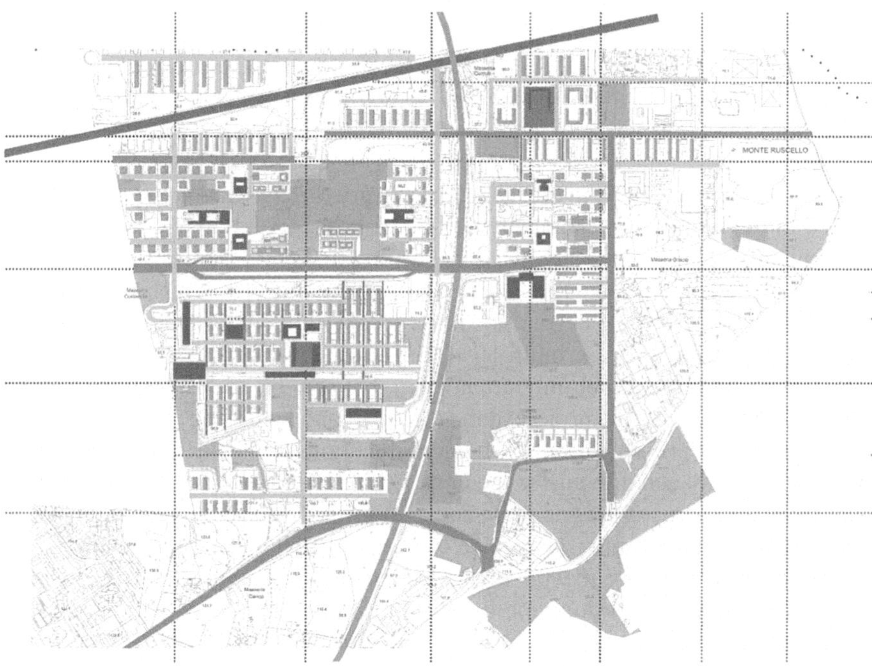

Fig. 15.3 Comparison diagram between the original project and the green spaces today. (Source: Unisa Mac project team)

Today, the coexistence of voids, infrastructure and the average quality of buildings is recognised as a defining characteristic of many urban suburbs, reflecting the forms of the contemporary urban landscape. Investigating Monteruscello involves examining all the urban elements that contribute to its classification as a degraded suburb. In particular, its voids are fragments of the interrupted landscape that characterised these places before the urbanisation of the 1980s. In this sense, the re-appropriation of the urban voids for the construction of a new Agro-Urbano landscape is an opportunity to recover a fragment of the ancient landscape—an identity that was lost.

Although the inclusion of agriculture in the contemporary urban landscape is common to many different realities, the case of Monteruscello differs from the general one. While the current proliferation of agricultural experiments in the city concerns the reappropriation of interstitial urban voids or the use of spaces such as roofs, balconies or common spaces, in the case of Monteruscello agriculture is inserted in an original way within the district. The extensive size and number of areas available for the project, which can be converted into cultivated land, make it possible to envision agriculture as a new landscape element that unifies the built urban fabric, blending urban and rural characteristics.

In general, looking at contemporary cities or rather urban conurbations, it is evident that urban transformations lead to a progressive reduction of agricultural areas. While the city was traditionally characterised by a clear and tangible separation from the agricultural territory—often emphasised by walls along its boundaries—the contemporary model of the dispersed city has extended urban development into the countryside, blurring the distinction between urban and rural areas. This has generated the model of the extended city, which today faces the challenge of re-establishing a relationship with nature and its productive spaces—connections that have been lost over time.

The main elements of Monteruscello's new Agro-Urbano landscape can be summarised through three key characteristics of the project. First, the project seeks to overcome internal urban boundaries, transforming them into elements of landscape construction. Second, it reinterprets agriculture within the city as a productive agricultural landscape, moving beyond the more recent, limited concept of urban agriculture. Third, it identifies a multi-level transformation process for the district's urban voids—addressing not only urban form, but also broader spatial, social and ecological dimensions. The project of the Agro-Urbano landscape of Monteruscello does not propose the elimination of the internal margins of the district but intends to overcome them by using them for the construction of a new identity. Boundaries, urban voids, infrastructure, agriculture and architecture are reinterpreted together as characteristic elements of the city, which is thus projected into a truly contemporary context. The proposal aims to work on the construction of a new Agro-Urbano landscape starting from overcoming the limits—those between city and countryside—present in the internal areas of the district. The aim of the MAC project is to bring agriculture back into the city through the transformation of the margins within the district.

15.4 Transformation Through a Process

A transformation strategy was then studied for the open areas of Monteruscello. More than a fixed design project, the proposal is a set of actions, which allows and triggers a process of transformation. This process-oriented approach is viewed as the most appropriate response to the contemporary condition. The design thus becomes a complex set of interconnected levels and themes, capable of adapting over time to evolving scenarios in the same location. The multi-temporality of the project is one of the central themes: urban transformations, in fact, have elastic dynamics and variable reaction times.

The proposal, more than a traditional architectural design project, takes the form of a planned process involving a variety of disciplines within a set of actions related to different scales and velocities. The main objective of the process is to reduce urban poverty in the neighbourhood, where poverty is understood not only in social and economic terms but also in relation to the physical environment. The project concept is grounded in the presence of large public properties—viewed as a resource—and their potential use for urban agriculture. Other actions addressing economic, social and urban issues are built upon this foundation. Urban agriculture—implemented through bio-intensive and organic techniques—is used to trigger the transformation process: a chain of interconnected urban policies will generate effects on the social economy, production and the physical environment. Agriculture has been linked to a production and marketing circuit where natural materials from the production processes will be transformed to be used for ecological construction and placed on the market. Both the agricultural and productive circuit will be managed through an educational programme, which will prepare young people and the unemployed for organic farming and production practices. The creation of a cooperative at the end of the educational cycles will guarantee the continuity of the cultivation, that is, the sustainability of the project itself.

A further educational path will prepare for innovative business opportunities related to agriculture. The proposal also provides links with the local market to strengthen the food production network. The involvement of schools and institutions, together with the local association, will take place through specific events. Finally, the architectural intervention will make use of abandoned public buildings, renovated to provide the necessary structures for the management of agricultural and production practices, and will define spaces for events, pedestrian and cycle paths within the agricultural areas in order to allow the integration and use of agriculture within the urban environment. All actions in the process are linked within a system of cross-relations and mutual interactions. The architectural design project itself is one of these actions, no longer holding the central role it once did within the morphological approach that defined the original plan.

The case study of Monteruscello's Agro-Urbano Landscape is a design proposal for a local productive landscape that keeps together the two souls of this contemporary suburb: the urban and the rural. Agriculture, with the unique character of its spaces, provides a solution to significant urban issues, including the challenging relationship with the critical areas of the neighbourhood. In this context, urban voids—conceived as internal edges—play a crucial role as a driving force of the transformation process that involves the neighbourhood in different ways. The overcoming of the edges and the break they create within the urban fabric thus becomes an opportunity to reconstruct the identity of the territory. The project is part of a larger process that includes the research community, local public authorities, the professional world and several associations. All together they build a network that puts the landscape at the centre of the neighbourhood development. In this context, the contribution of university research within the MAC project aims to define the character of this landscape which means, as specified by the French landscape architect Pierre Donadieu (2006, 2014; Fleury and Donadieu 1997), mobilising a set of values necessary to conceive the landscape project. Landscape cannot be something to be understood and perceived exclusively from above; it should be conquered by walking through and living it as a place of the city.

The Agro-Urbano landscape in Monteruscello is therefore designed as a public space, which can be crossed on foot or by bicycle and where it is possible to carry out leisure activities and organise small events. The perceptual character of the landscape is a further significant quality. The plant scents and the views towards the agricultural fields, towards other parts of the neighbourhood and towards the sea, build a new experience for the inhabitants. The landscape, moreover, is not a fixed but a dynamic environment that changes over time. For this reason, the project defined a spine, an infrastructural line as a single-fixed element of the landscape that determines the character and the quality of the surroundings and which builds the productive areas. The types of cultivation and the uses of space can change over time in relation to the community's needs and according to the economic opportunities of the territory. Along this extent, the landscape project is part of the main goals of the MAC project. It is a tool that powers a process of enhancement in the neighbourhood through its agricultural spaces.

The MAC partners have debated on several occasions whether the agricultural spaces should remain closed or be opened to the public. On one hand, the project aims to dissolve boundaries by integrating public pathways through these areas; on the other hand, there is a recognised need to clearly define when and how these boundaries can be opened or closed, balancing accessibility with functional and security requirements. It is highly utopian to think of being able to completely eliminate the borders in these intermediate spaces between production spaces and public paths. Indeed, it is recognised that in some parts the public crossing is limited in order to protect the agricultural areas.

The MAC project establishes a connection between agricultural spaces and public pathways by focusing on the edges of green areas, with tailored solutions developed for each specific zone. They are evaluated by observing together the opportunities that can be created within the urban environment intersecting agricultural fields and the new crossing paths. This methodology allows to build a link between public and agricultural spaces by rethinking the meaning of the edges without defining a priori a strategy but rather interpreting the space and its potential.

15.5 Inhabit the Borders

The transformation of Monteruscello's urban voids into an Agro-Urbano landscape begins with a study of the area's identity, interpreted through the historical traces embedded in the territory that reveal its agricultural vocation. Despite the extensive and forceful transformation of the land which took place to construct the neighbourhood, Monteruscello retains some important connections with its geography and ancient landscape. In fact, within some of the areas involved in the project, there are visible traces of the site's rural character. Farms, cisterns, and small water drainage channels are archaeological remains (see Fig. 15.4), witnesses of the rural use of the territory before the urbanisation of the 1980s.

In Monteruscello, the signs of the ancient agricultural landscape coexist with the complexities of the contemporary city. Today, when crossing some areas where the signs of the past stand, one cannot fail to notice the surreal urban condition that is created where archaeology, infrastructures and the contemporary city coexist in a strange balance, marked by a sense of neglect and abandonment. Some of the signs still present in the area, observed as archaeological fragments, reveal the agricultural system that existed prior to the development of the neighbourhood. These traces, which once shaped the ancient agrarian landscape, become decisive elements for the design project of the new Agro-Urbano landscape.

In order to bring the transformation of the neighbourhood's abandoned voids towards the creation of a landscape scenario for the residents, its complexity has been disaggregated into themes. These themes provided insights into the perceptual experience of moving through the neighbourhood and the use of its spaces and were subsequently integrated into the redesign of the agricultural areas. The chosen topics describe the nature of the places and the cultural and identity characters of the

Fig. 15.4 The neighbourhood of Monteruscello: some of the archaeological remains. (Source: Unisa Mac project team)

neighbourhood. The study has been carried out investigating the complexity of the existent condition, which has been disaggregated into different topics that create a set of layers, each considering one single aspect connected to the use and the crossing of the spaces. Through the overlapping of the various aspects, superimposing the grids and the main lines of the ancient landscape, it was possible to draw the new paths, thus identifying a strategy for the Agro-Urbano landscape of Monteruscello. The design project, therefore, channels all the aspects considered within the separate layers of topics. The project will transform the unused and abandoned areas into agricultural lands, while new pedestrian and cycling paths envision the possibility of traversing the neighbourhood at a human scale—an experience currently hindered by fast-moving arterial roads (Fig. 15.5).

Among the analysed layers, an essential role is given to the water system, which characterises the way the land is irrigated, and which includes extraction, distribution and irrigation for cultivation purposes. The study of the ancient traces present on the territory together with the analysis of the water opportunities in the

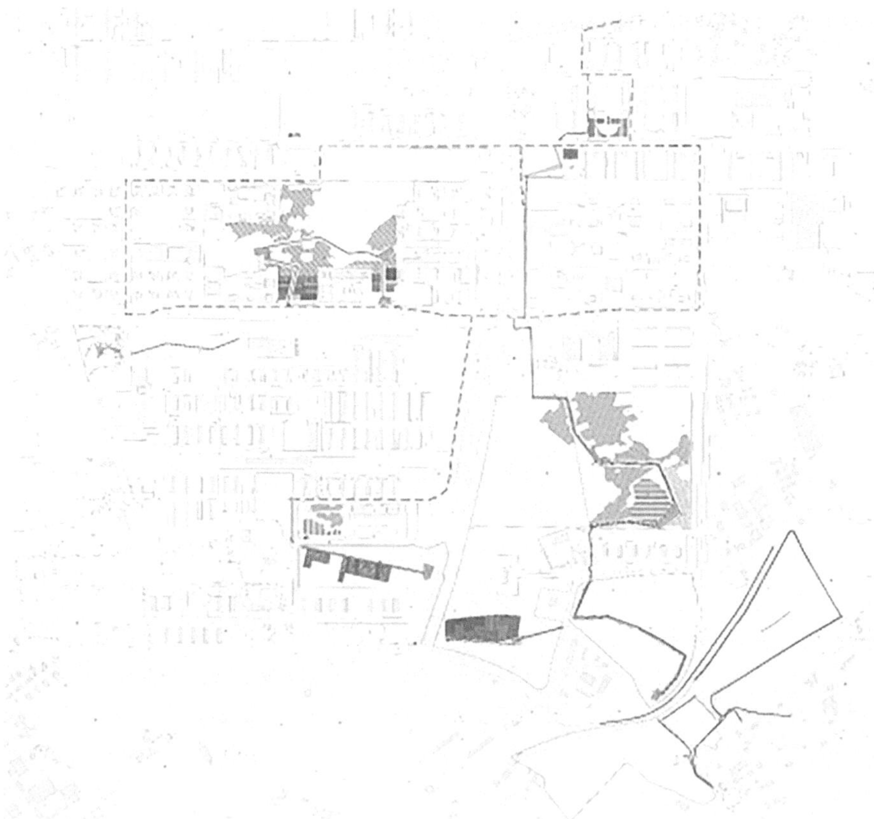

Fig. 15.5 Landscape design: plan. (Source: Unisa Mac project team)

contemporary condition has highlighted the possibility of a sustainable water system based on the sub-surface water outflows, their collection and their distribution.

The system is based on the idea of a responsible and sustainable use of resources already existing on the territory, avoiding invasive technologies; the presence of ancient cisterns shows that this system was already used in the past. Starting from the identification of the water lines and the collection tanks, the water system is defined. The main pedestrian paths follow this network and therefore the water line becomes the main trace used for the design of the new landscape. The paths are joined to the water lines realising one single infrastructure, which also contains the electric system. The water, so important for a sustainable management of the land and its agricultural use, thus becomes a visible element characteristic of the landscape. Not all technical networks are buried underground; some parts of the water system remain visible. The pipes of the water system that collect and convey the water to the various points of the territory flow below the path, while the collection tanks are visible elements that characterise paths and rest areas. The infrastructure has different arrangements, and hedges are included on one or both sides of the walkway, according to topographical and land use criteria. Its purpose is to define the path as a green infrastructure within the landscape, while also acting as a boundary for the adjacent productive areas, particularly in cases where parts of these areas may need to be made inaccessible.

Along the paths, in particular points of the territory, are arranged the so-called farmyards, which are rest areas and viewpoints (see Fig. 15.6). These elements are strategically placed in locations that offer views of significant points within the district, as well as notable landscape features such as the sea, Cuma and the Licola coast. The farmyards are designed as elements of contemporary archaeology placed

Fig. 15.6 Landscape design: sections. (Source: Unisa Mac project team)

in the landscape. They are made of a single material, whose geometry builds, through the alternation of folds and continuity of surfaces, a variety of spaces for seating and other different uses. Their location, as well as some path's directions, follow geometric lines drawn from the overlapping studies of the neighbourhood original project, the ancient routes, the directions of archaeological evidence, the historical landmarks and the historical representations of the places.

The intersecting paths through the agricultural areas and farmyards, along with the transformation of these spaces into productive agricultural land, collectively shape the Agro-Urbano landscape. This new landscape, through the construction of mainly visual relations with the enlarged territory, reconstructs the geographical and landscape dimension of Monteruscello within the Phlegrean Fields. Historical farms and archaeological sites are also incorporated into the project as spaces for outdoor activities, rest and reflection—inviting visitors to engage with the remnants of the past.

The water system, together with the paths and the farmyards for rest and contemplation of the landscape, become a single integrated element that overlaps the cultivated areas like a network. It creates a new urban system that holds together productive and public spaces. Superimposing new mobility systems onto the territory—alongside the development of an agricultural plan for the green areas—enables small-scale interventions that help mend the currently disconnected urban fragments of the neighbourhood. At the same time, building new relationships—which are routes and visual directions—is an opportunity for the neighbourhood to relocate itself within the city and the territory of the Phlegraean Fields.

15.6 Conclusions

'The Allegory of Good and Bad Government' cycle of frescoes painted by Ambrogio Lorenzetti in the Siena's *Palazzo Pubblico* (1338–1339) is often used as an example in urban studies for its contrasting representation of two cities, one well governed and the other badly governed. In particular, The Allegory and the Effects of Good Government describes a city with a perfect balance between urban and agricultural areas. The two parts are separated by a wall with a door, which is the link between two coexisting realities. Beyond the city gate there is a path that leads to the countryside. In the fresco you can see daily scenes of urban and rural life with people carrying out their activities and working. The city is described through its buildings, while the countryside is described through its natural forms and through the signs of agricultural activity.

The effects of good governance lie in the relationship between city and countryside made by cooperation and sharing. From the city one can observe the countryside and from the countryside one can get the products necessary for sustenance and to generate a domestic economy, and also to foster trade. This relationship between the city and the countryside is not only productive but is above all a cultural and landscape relationship, as shown in Lorenzetti's fresco. On the contrary, the contemporary city has lost the connection with its productive space. The countryside that feeds the city isn't around the city anymore; often it is really so far beyond its borders, that it overcomes continental limits. The blurring of strong limits brings to define the contemporary city as a city with no limits, with faraway and dispersed agricultural fields with which the city no longer has territorial relations nor landscape and cultural references.

Recently, there has been a renewed interest in sustainable agriculture, traditional crops production, and Km 0 market, thanks to the transparence in the production process where people are aware about the origin of what they eat. (Lirer and Pescatore 1987; Mougeot 2006; Zasada 2011). This situation has led to the promotion of new agricultural spaces within the city itself. The topic of urban agriculture is becoming an increasingly widespread reality in Europe. Though, if one looks at the phenomenon with a critical point of view, one realises that urban agriculture on many occasions is a slogan rather than an issue within the processes of revisiting the contemporary city. Sometimes there is even the perception that urban agriculture is supported by a nostalgic and illusory feeling towards a past that cannot be taken back in time. However, the urban agriculture topic is so deeply rooted in contemporary culture and present in the urban agendas of the main European cities that it can no longer be neglected, rather it needs to be addressed in a critical and constructive manner.

The Monteruscello case study offers a discussion on the meaning of bringing agriculture to the city. The question of the relationship between agricultural space and urban space is specifically studied within an urban condition with unsettled urban values, because of its recent urban history. Agriculture has been inserted respecting the urban condition, realising an integration between urban and agriculture values. Rather than recreating the rural landscape which existed prior to the city's foundation, it developed the idea of inserting cultivation within the voids among the buildings together with public paths crossing the areas. The result is at the same time an urban agriculture and a park, when viewed from the perspective of the pedestrians and runners through the paths. Certainly the agricultural land will be less productive because it is more parcelled out, due to the presence of the cycle and pedestrian paths crossing through the areas making it less easy to cultivate the land. The presence of the paths is though essential for an urban point of view. Productive agriculture in areas of the city without urban relations would in fact produce enclosed areas that, although aesthetically attractive from above the buildings, would not be very different from industrial areas or other types of areas creating interruption and edges within the urban fabric. The presence of the paths is therefore essential to achieve a true integration between agriculture and the city and opened the inner borders, nowadays present in the areas.

The Mac project of the Agro-Urbano landscape has therefore become an opportunity to reflect on a way of working on the edges of the city's space and on the possibility of re-discover the cultural and traditional identity of places that could involve the productive use of its most critical spaces.

References

Alfano P, Como A, Cuomo A, Guida D, Smeragliuolo Perrotta L, Terralavoro L, Vece C (2018) Multiscale, objective hydro-geomorphological mapping and modelling application: the case study of Monteruscello Agro City Project (MAC). Urb Inf 278:317–321

Alfano P, Como A, Cuomo A, Guida D, Smeragliuolo Perrotta L, Terralavoro L, Vece C (2019) Monterusciello Agro-City Landscape project: strategies for a new relationship between the built and the environment. In: Fiore P, D'Andria E (eds) Small towns...from problem to resource. Sustainable strategies for the valorization of building, landscape and cultural heritage in inland areas. Franco Angeli, Milano, pp 535–545

Capozzi R (2016) Agostino Renna: la forma della città. Clean Edizioni, Napoli

Capuano A (2014) Paesaggi di rovine. Paesaggi Rovinati. Quodlibet, Macerata

Cerami G (1988) La riqualificazione ambientale dei Campi Flegrei: gli elementi strutturali per un piano paesistico. In: Quaderni di documentazione Progetto Pozzuoli. Fratelli Giannini, Napoli

Colpo M, D'Agostino G, Previtali G, Signorelli A (1989) La valorizzazione della memoria collettiva nei Campi Flegrei. In: Quaderni di documentazione Progetto Pozzuoli. La Buona Stampa, Ercolano

Como A, Smeragliuolo Perrotta L, Vece C (2018) Agro-Urban Landscape: the case study of Monteruscello-Naples. In: Mondéjar DU, Alcácer JC, Mañanós AP, Sendra VC (eds) Proceedings 24th ISUF 2017—city and territory in the globalization age. Editorial Universitat Politècnica de València, Valencia, pp 1701–1709

D'Antonio M, Giannola A, Sicca L (1989) Elementi per la definizione di un piano di sviluppo socio-economico di Pozzuoli e dell'area flegrea. In: Quaderni di documentazione Progetto Pozzuoli. La Buona Stampa, Ercolano

Donadieu P (2006) Campagne Urbane. Donzelli Editore, Roma

Donadieu P (2014) Scienze del paesaggio. Tra teorie e pratiche. Edizioni ETS, Pisa

Escalona F, Francese D (1987) Monteruscello: l'impianto urbano e gli edifici pubblici. Fratelli Giannini, Napoli

Fleury A, Donadieu P (1997) De l'agriculture périurbaine à l'agriculture urbaine. Courrier de l'environnement de l'Inra 31:45–61

Lirer L, Pescatore TS (1987) La geologia della aree di Monteruscello e del centro storico di Pozzuoli. In: Quaderni di documentazione Progetto Pozzuoli. Fratelli Giannini, Napoli

Maiuri A (1981) I Campi Flegrei. Istituto Poligrafico dello Stato, Roma

Miano P, Izzo F, Pagano L (2017) I Campi Flegrei. In: L'architettura per i paesaggi archeologici. Quodlibet, Macerata

Mougeot LJA (2006) Growing better cities: urban agriculture for sustainable development. IDRC (International Development Research Centre), Ottawa

Pagano L (2012) Agostino Renna: rimontaggio di un pensiero sulla conoscenza dell'architettura: antologia di scritti e progetti 1964–1988. Clean Edizioni, Napoli

Pisciotti L, Miano P (1988) I Campi Flegrei: luoghi, formazione e trasformazione della città. Quaderni di documentazione Progetto Pozzuoli. Fratelli Giannini, Napoli

Renna A (1985) Il progetto di Monteruscello. In: Siola U (ed) Progetto Pozzuoli: rapporto di sintesi sul lavoro svolto al 30 giugno 1985. Quaderni di documentazione Progetto Pozzuoli. La Buona Stampa, Ercolano, pp 96–101

Romano M (1986) Agostino Renna. Monteruscello (Pozzuoli). Domus 674:25–33

Siola U (1985) Progetto Pozzuoli: rapporto di sintesi sul lavoro svolto al 30 giugno 1985. In: Quaderni di documentazione Progetto Pozzuoli. La Buona Stampa, Ercolano

Vitale A (1989) Il controllo ambientale nei Campi Flegrei. In: Quaderni di documentazione Progetto Pozzuoli. La Buona Stampa, Ercolano

Zasada I (2011) Multifunctional peri-urban agriculture—a review of societal demands and the provision of goods and services by farming. Land Use Policy 28:639–648

Chapter 16
The Binational Municipal: An Alternative to the Architecture of a Controlled Border Landscape

Felicia Anderson

Abstract *La Frontera*, the border between Mexico and the United States is the subject of overlapping myths and realities, receiving sustained attention within binational academic discourse. While many international borders present contentious issues and dynamic conditions, this chapter focuses on the distinctive malleability of this border and the possibility of reconfiguring its complex condition into a coherent territorial entity. The border is crossed daily, both legally and illegally, and remains under constant surveillance. It operates as a site of transaction and interchange, a space that negotiates the competing demands of policy and governance. This design-based research explores the potential to reduce pressure on existing cross-border networks to promote economic development between the two nations and to examine the implications of an expanded binational territory. The analysis centres on the twin cities of El Paso and Ciudad Juárez, a metropolitan oasis located within the third largest desert in the Western Hemisphere. It considers the prospects of an open border system, an emerging model that facilitates the movement of people with minimal restrictions. Despite prevailing negative perceptions, this study situates the border within recent infrastructure initiatives that enhance its habitability and promote cultural and economic integration. What distinguishes the Mexico–US border is the fragility of two contrasting nations position in proximity and subjected to 'daily political tensions', on both sides. This chapter traces a research project that examines the fragile identities shaped by maquiladora labour, the movements of workers and migrants across the border states and the deepening economic ties between El Paso (Texas) and Ciudad Juárez (Chihuahua).

Keywords US–Mexico border · Binational development · Infrastructures · Ciudad Juárez · El Paso · Transnational urbanism

F. Anderson (✉)
Structures Research Group, University of Cambridge, Cambridge, UK

G. Cimadomo et al. (eds.), *Urbanism at Borders*, The Urban Book Series,
https://doi.org/10.1007/978-3-032-09594-7_16

16.1 Introduction

The proximity of the two border towns Ciudad Juárez (Chihuahua, Mexico) and El Paso (Texas, USA) where the world's strongest economy meets a developing nation creates a uniquely complex territorial condition. The border remains active and productive, sustained by the high volume of trade and daily crossings, and exemplifies the multilayered dynamics of this international region. As El Paso and Ciudad Juárez continue to expand and their urban boundaries become increasingly indistinct, the border is strained by rising demands and the expectation to function as a cohesive binational connector.

The existing socio-political border condition reveals overlaps of culture and communities that create a zone conducive to innovation and business investment (see Fig. 16.1). The emergence of binational, bilingual architecture that extends over the border represents the next step in developing intra-national border programmes. This dual-national territory cultivates a shared culture across both sides of the border, generating binational schemes and programmes that remain unimagined. Geographer James Curtis, in his analysis of several Mexican border cities, identified the nearly 2000-mile border corridor as one of the most urbanised regions in Mexico over the past three decades (Arreola and Curtis 1993). This urbanised region,

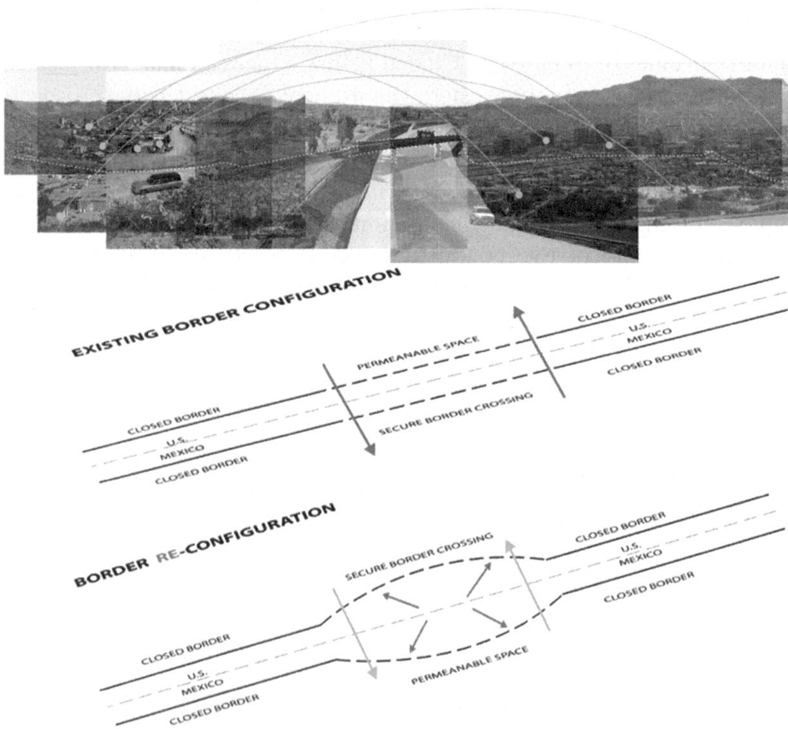

Fig. 16.1 Diagram showing expansion of the border. (Source: the author)

however, is composed of pairs of twin cities that straddle the border and are both separated by radically divergent policies and economies and vitally interconnected.

Mirrored communities such as El Paso and Ciudad Juárez have a history of dual partnership schemes that have shaped the lives of those inhabiting the region (Collins 2013). El Paso has emerged as the dominant border town (Lorey 2008), and it has leveraged trade more effectively than Ciudad Juárez. This is evident through its privileged access to trade laws, inexpensive labour across the border and increasingly strict immigration policy (Clement 2002; Massey 2001).

However, the role of border as a filter for wealth and opportunity has granted both cities political prominence within their respective national contexts, positioning them as key actors in the governance and policing of the border and contribution to the formation of an influential binational territory (Martinez 2003). Conversely, decisions made in Mexico City and Washington, D.C. penetrate the border region, affecting not only its economic regulation but also shaping its socio-cultural dynamics (Castro-Rea 2012). This form of top-down political authority often undermines binational development, as the region's nuanced independencies are overlooked and local perspectives rarely solicited in the formation of new legislation. This dynamic was clearly foreshadowed in pre-2010 critiques of border securitisation (Dunn 2010; Romero 2008) and later materialised during the first Trump administration in 2017 when wall construction and border militarisation intensified (Meissner et al. 2018).

Protests in Mexico City demonstrate a pressing need for change. This change is possible through the considered rebalancing of the relationship between national and local policy bodies. Since the United States and Mexico both operate under federal systems with vertical power sharing (Correa-Cabrera et al. 2014), the divestment of border organisation could occur not just at the state level, but become the responsibility shared by city governments and, quite possibly, distributed across a bi-national urban region. There are already examples of binational projects that start with bottom-up development involving residents and businesses to substantiate growth (Katz 2014), but this remains a rarity and development of the border still relies profoundly on cohesive border management between Mexico and the United States, which primarily focuses on privatised investment, which determines the rate at which the physical border changes (Jones and Johnson 2011; Wagner 2014).

This balance of national, bi-national, federal and local development has evolved through a complex history. Border management has adapted and diversified since the birth of the border in 1848, adhering to the modifications in policy that differ state-to-state depending on intensity and capacity (Martinez 2003; St. John 2011; Payan 2006). Consequently, the border region that constitutes the adjacent towns has become a border corridor between the two countries (St. John 2011). This continuum of spatial arrangement and the diversity of the principles and regulations that have determined its growth have created pockets of symmetry and contrasting socio-political environments (Wilson 1999; Paasi 2009). 'Twin border' cities such as El Paso and Ciudad Juárez, also commonly known as 'sister cities,' often reveal similar patterns of migration and development as well as varying degrees of integration of people and trade (Loustaunau 1999; Klein 1997). Successful twin border communities tend to address common issues collectively, which result in solving problems efficiently. Such attempts have been reinforced by overarching strategies such as that of the U.S. State Department's '*United States-Mexico Partnership: Managing our 21st Century Border Initiative (BIM)*' document which outlines management schemes to provide and care for the 13 million people who currently live at the border by improving security and developing the region without the expense of one another (Wilson and Lee 2013).

The success of BIM is vital for the cultivation of any more local bi-national programming. The potential of the twin cities El Paso-Ciudad Juárez now depends on government investment to initiate binational infrastructure projects. For instance, bi-national institutions such as the border 2012 Program focus on the environmental management of the border and future infrastructures to harvest renewable energy for use in Mexico and the United States (EPA and SEMARNAT 2003; Michel 2003). The new binational programme could expand into a free trade system, governing the flow of goods and services linked to or benefiting from energy production. However, such a step would need to reframe the terms of the North American Free Trade Agreement (NAFTA). Such a step, primarily aimed at the equitable distribution of shared energy resources, might also provide the groundwork for fairer regulation of the manufacture of goods and the provision of services. This gradual appropriation of national trade policy by the border territory, in this case, could shift the identity of the border from an enforced boundary between radically diverse levels of opportunity and prosperity, towards being a defined, shared space. El Paso and Ciudad Juárez currently maintain a controlled border system enforced by the Rio Grande. While the two cities share parts of their population and rely heavily upon one another for manufacturing, labour and trade, the sharing of resources, human and natural, is mitigated by formal immigration procedures that favour cargo over transit over more informal, cooperative exchanges. A collaborative approach to border management—one that recognises the border as a lived and shared place—could begin to address the pressures of migration and in-migration, while also tackling the issue of drug trafficking that has long shaped the public perception of the El Paso–Ciudad Juárez region.

One part of this research brings together a range of urban and architectural design interventions and provocations, academic research and primary data collection and analysis. The secondary research draws on a series of Mexican–US border geographers who have meticulously mapped the border region and studied its growth and brought together two areas of discourse concerning 'border territories' and 'bi-national development'; these categories are becoming a convention for studying border identities in regions such as Ciudad Juárez, Mexico and El Paso, Texas. While these two border cities are also characterised by much larger, global issues such as urban conflict, urban sprawl and the divisions emerging from a bilingual culture, this chapter is primarily concerned with the analysis of specific policy documents and interview material that form the bulk of the research and support the proposal for a bi-national infrastructure. To draw accurate data about these two regions, regular use of the U.S. Department of Commerce, DHS (Department of

Homeland Security) and SRE (Mexican Secretariat of Foreign Relations) has been made, as a platform to obtain current information.

16.2 Duality Mechanisms

The El Paso/Ciudad Juárez border is a site of dualities, where two nationalities and overlapping cultures coexist, often giving rise to tensions on both sides. No bi-national scheme can reverse a long history of difference, fear in inequality; however, some bi-national schemes demonstrate the ability for a shared initiative or infrastructure to allow both sides to integrate using similar agendas. Using social mechanisms to reconstruct and reorder the space to set political boundaries between two races is present in historic treaties like the Gadsden Purchase (Huffines 2005). The primary objective of this purchase was to appropriate unclaimed or truly bi-national land to the United States. The details of the purchase included a 10 million dollars awarded to Mexico, in exchange for the 26,670 square miles of land. This kind of social reconstruction was seen as necessary due to unresolved tensions left in the wake of the Mexican American War.

The strict division between El Norte and Texas was not the division of 'people' but the creation of the US–Mexico border culture and regime. Existing as two entities connected by a series of physical paths that stimulated shared development on each side. The isolation of this border region made it popular for people travelling to the United States as it became an 'in-between' space that collected people waiting for new jobs or family members in America. This transitory role of the borderlands as a liminal space has been explored in greater depth by Langewiesche (1993), who documented the psychological and spatial ambiguity of such regions. The use of bridges and eventually railways in 1880 (Martinez 1994; Richardson 2012) constituted the intensification of the border condition. The more railways built, the more people migrated to the region, settling and establishing strong nodes for urban districts. These connections allowed residents to cross over freely, which kept the two towns united. The bridges not only emerged as a means of transport, but in turn, the site of markets and street trade (Romero 2008).

Highlighted in the historical uprising of the border towns, the border began to develop as its own independent state (Huffines 2005). A strong workforce grew in correlation in the 1920s, primarily for the purpose of bridge construction, but El Paso, with its growing population and trade sectors, also required a strong skilled workforce to sustain its growth (Wilson 1992). The response was to employ Mexican citizens who would work as labourers (15%), domestic servants (40%) and artisans (30%) (Fischer 2013). El Paso relied heavily on Mexican workers, and especially women who made up most of the domestic worker percentage. By 1930, more than 3500 Mexican commuters crossed the border daily to work, forming the foundation of a resilient bilingual, binational workforce that would drive the region's industrial and infrastructural growth (Simmons and Mueller 2014). The daily commuter traffic and the dual–national workmanship created the environment for a binational landscape that set precedents for future border towns to follow. However, racial tensions still existed between the Anglo-Americans/Mexican Americans and the Mexicans, the repercussion of this leading to a dominance of Mexicans working in the unskilled workforce (Fischer 2013; Vila 2000). These tensions trace back to long-standing racial prejudice explored by León (2002), who details the deep-rooted Anglo attitudes toward Mexicans in nineteenth-century Texas.

Although the environment for binationalism had evolved, the urban districts of El Paso and Juárez developed different strategies to deal with dual-nationalities. El Paso was primarily concerned with the identity of the city and the cultivation of a skyscraper city that could challenge other industrialised states of America (Vila 2000). The ambition was to obtain a dominant capitalist future that was illustrated by the El Paso Herald (Ortiz-Gonzalez 2004), a popular magazine published by El Paso state. This particular article focused exclusively on the modernisation of El Paso—highlighting its bright electric lights, bustling pedestrian activity and the emergence of innovative businesses housed at the base of new, state-of-the-art apartment buildings. It celebrated an architecture that symbolised economic strength and portrayed the city as a place of opportunity and abundance. The advertisement, which was printed nation-wide also included the opportunity for leisure trips over the border to Mexico, enabling tourists to experience local food markets *mercados*, entertainment from the bullring—a unique Mexican/Spanish culture—and the racetrack. The advertisement was targeted at businesspeople living in America, and US tourists to encourage them to visit, invest and contribute to the sustainability of the border town (Fischer 2013). An iconic binational piece of architecture, which was intended to sit on the border, would re-enact the same advertisement for future binational projects. This type of statement architecture would represent the strong binational community that El Paso–Ciudad Juárez holds, not too distinct from the images created by the advertisement. Furthermore, in the initial stages of establishing the border between El Paso–Ciudad Juárez, the focus turned from possession of land to spaces between and next to the border. These spaces that connect and transfer across the border, like international bridges, provide an ecosystem for binationalism. However, these spaces became highly pressurised and moderated to emphasise the politics in each country (Fischer 2013; Stephen 2007). The diversion of the El Paso–Ciudad Juárez border stems from the diversion of identity and the opposition of culture and utility of space.

Juárez did not apply the same rules and regulations that were set in El Paso and made the lifestyle and Mexican identity immediately apparent in the space adjacent to the border. This was shown in several ways from the *mercados*, to the bullrings and tequila bars dotted around Juárez. Regulations on drinking and gambling were not enforced in Juárez, and at the time of prohibition in the United States in 1918, El Paso became dry and dull (Fischer 2013). At this time, a surge of Texans would indulge themselves in Mexican bars only miles away from their prohibition enforced town. As identity was the driving force for how spaces were cultivated and developed on either side of the border and manipulated to express their value, El Paso's restrictive governance caused Juárez to become known as the 'other', differences in policy eroding into what were seen as opposing traditions and standards to those held by Americans. In response to this embedded sense of difference, a bi-national, bridging territory can provide the means for a hybrid behaviour from which both cultures might draw—both socially and economically—the revenue drawn into the border towns from structures such as the international bridges ideally being shared equally. However, this does not always pan out due to a difference in opinion and strategy between municipal governments and these disparities have often led to irregular distribution of 'revenue' and therefore conflicts in state and federal governments.

The bid for a shared enterprise is naturally undermined by the ever-increasing control of the border. The US border patrol system was initiated on 28 May 1924, and administered under the Labor Appropriation Act of 1924. The Labor Appropriation Act was a branching of the 1891 Immigration Inspection Policy that was introduced to begin to control the number of Mexican migrants travelling to the United States. It also reached the Canadian–US border. By 1930, the doors to the new world were effectively closed (Dunn 2010; Timmer and Williamson 1998) the biggest change to the once 'open border'. This did not prevent all movement through the border, but the restrictions did create the paths for illegal migration and trade. As Maril (2011) argues, the construction of physical border barriers like the fence reflects a broader sociopolitical strategy of exclusion masked as security. The construction of the first low-tech border fence in 1990 at the San Diego–Tijuana crossing marked a shift toward a harsher and more sustained enforcement strategy compared to earlier efforts, such as the 1954 Operation Wetback, which resulted in the deportation of over one million undocumented migrants (Hing 2012).

16.3 The Introduction of NAFTA

While the gap in labour costs have created incentives and encouraged industrial growth on the Mexican side of the border, tax legislation remains critically biassed in favour of the United States. The manufacturing plants—the *maquiladoras*, which are subsequently located across the major border regions—create a compressed reflection of the larger cross-national situation.

Figures 16.2 and 16.3 illustrate, at a territorial scale, the connection between the two countries, highlighting how government trade policies and security legislation influence each side of the border individually, while more importantly shaping the overall economic status of the border region. The border can be understood as a tri-national governance system, which includes both US borders of Canada and Mexico. The tri-national operating system that also includes Canada CUFTA (US Free Trade Agreement of 1989) is fuelled and connected by NAFTA (the North American Free Trade Agreement) which, introduced in 1994 under President George Bush's term (Castro-Rea 2012; Kong 2015), and as the name suggests, reduced trade restrictions across the border regions. The agreement has had a contentious impact on the border regions, initially causing a massive proliferation of maquiladoras in the border region and fears in the US for the loss of American jobs. The employment statistics

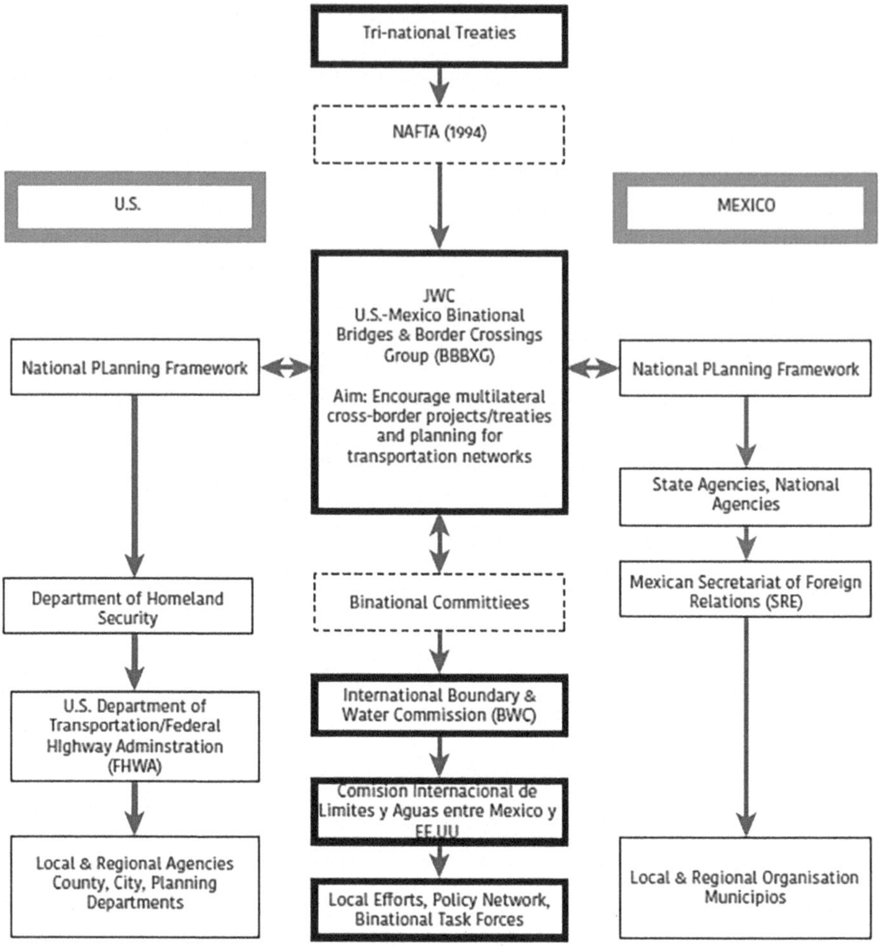

Fig. 16.2 Flow of the process in which new Cross Border Planning for Transportation Infrastructures is procured, similar to multilateral treaties and new Port of Entry. (Source: the author based on U.S. Department of Commerce 2008)

for the El Paso–Ciudad Juárez border was the lowest increase from 9.2% in 1982, to 13.3% to 1983 (Sklair 2011). By 2000 there were 450,000 Mexican workers operating in maquiladoras, half of them located in Ciudad Juárez (Stacy 2003; Hilderbrandt 2012). Certainly, the United States has profited substantially but Mexico has also seen an increase in exports and profits from agricultural production whose impact is not merely economic. Over the past two decades NAFTA has evolved and developed a complex inter-reliance on provisions, restrictions and subsidies. However, NAFTA also holds the potential to transform the border into a binational operating system—one that redefines domestic governance, creates new opportunities and reshapes international relations. The connection of NAFTA to a bi-national

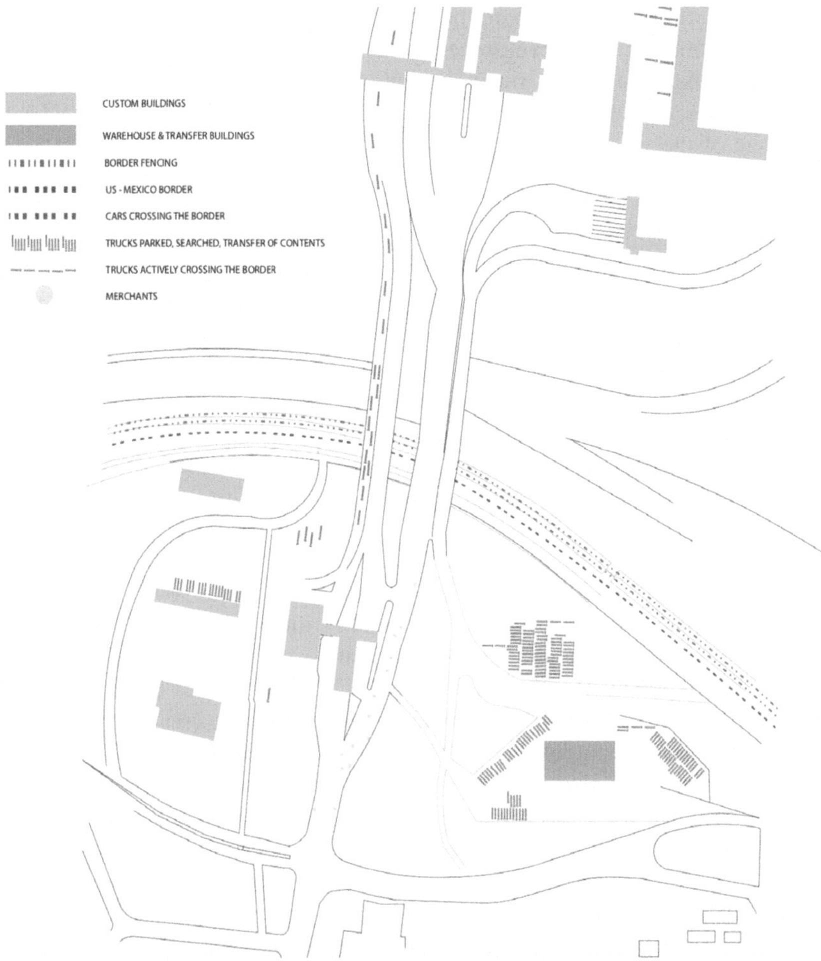

Fig. 16.3 Primary crossing at the El Paso–Ciudad Juárez border, heavily used by freight vehicles transporting goods into Mexico for assembly, often resulting in significant traffic congestion. (Source: the author)

movement may be difficult to reconcile with the agreement's tendency to favour economic freedom over cultural unity and the urgent move to tighten border movements.

Kimberly Collins authored an article for the Journal of Borderlands Studies called 'Our North America: Social and Political Issues beyond NAFTA' in which she analyses the life of residents in the US–Mexico border (Collins 2013). This in-depth research investigates the complexities of living on the border, and often diving deeper into stories, that the newspaper does not. Distinguishing between fiction and non-fiction border data and facts, Collins' conclusion revealed that, although it is extremely difficult to truly compare the quality of life on the border, a joint policy-making programme or dual development scheme would be beneficial.

The border has become an intriguing network through the implementation of NAFTA policies, which have now stood in place for over 20 years. NAFTA has become one of the world's biggest trading blocks, connecting 450 million people and the output is over 7 trillion US dollars (Kotkin 2013). A very serious analysis of the relationship between trade agreement and job creation is summarised in a report labelled *Opening Markets, Creating Jobs: Estimated U.S. Employment Effects of Trade with FTA Partners* (U.S. Department of Commerce 2013), which outlines how NAFTA has been linked to job creation since its implementation in 1994. This report, conducted by the US Chamber of Commerce in 2010, employs the 'gold standard' model, which analyses economic models known as the global tree analysis project. This model shows that five million jobs have been supported by the influx of trade created by NAFTA, also if critics claim that the agreement has led to the net loss of US jobs. However, Bureau of Labour statistics show that US manufacturers created more than 800 thousand new jobs in the first 4 years after NAFTA was implemented.

As previously discussed, the contrasts between the border cities are revealed by its open trade status, and its restriction of labour exchange. While NAFTA has appeared to increase employment rates unilaterally, the border towns united in sharing the benefit of the trade agreement, the disparities in profit from this employment are still evident (Wilson 1992). Fundamentally, NAFTA insures US control of the border and the differences in the economic standards of living created by this condition are also sharply apparent: for example, the minimum wage on the US side is almost eight times the 'salario minimo' on the Mexican side, though neither can be considered a living wage (Zartman 2010). The disparity of the scale of wages between El Paso/Ciudad Juárez and like much of Mexico-United States is enormous. The minimum wage in Mexico is between 4 and 5 dollars for 1 day of labour, in contrast to the United States, which exceeds 5 dollars per hour (Payan 2006). With such distinctions between labour costs and modern trade agreements, it is understandable that the disparities between El Paso and Ciudad Juárez continue to grow.

NAFTA in this instance has made the economic market more important than ever despite the US Chamber of Commerce not witnessing large benefits emerging from it. Arguably NAFTA should continue to play a strong role in the economic development of the bi-national bilingual border corridor and continue to grow and support the biggest bi-national bilingual workforce in the western hemisphere (U.S. Department of Commerce 2013).

16.4 Border Image

In an earlier study (see Fig. 16.4), the potential impact of reimagining the border as a series of pedestrian bridges was evaluated—proposing a model in which newly adjacent communities could connect and overlap a strategy that proved to be highly effective. The image of the border would shift and develop into a 'free zone' which would facilitate people across the border rather than trade (Yigitcanlar 2012). These provocations are partly inspired by Geenhuizen (2012), which explores urban innovation and speculative framing. But this goes beyond adding in new infrastructure, it is looking at how we flex trade policies around this new infrastructure.

Despite the trade situation, the United States is traditionally seen as a closed border, more specifically a border that is open only for trade and investment but is closed to labour (Payan 2006). Bi-national projects have commonly tackled trade and security issues or functioned as singular education partnering schemes. However, there remains scope for exploring the possibility of a physical shared infrastructure that could house further initiatives and shift the focus from the scale of the national border back to the daily interaction between locals on either side of the divide. Such an enterprise could alter the perspective of what the border image traditionally presents and changes the possibilities for how the border could be open

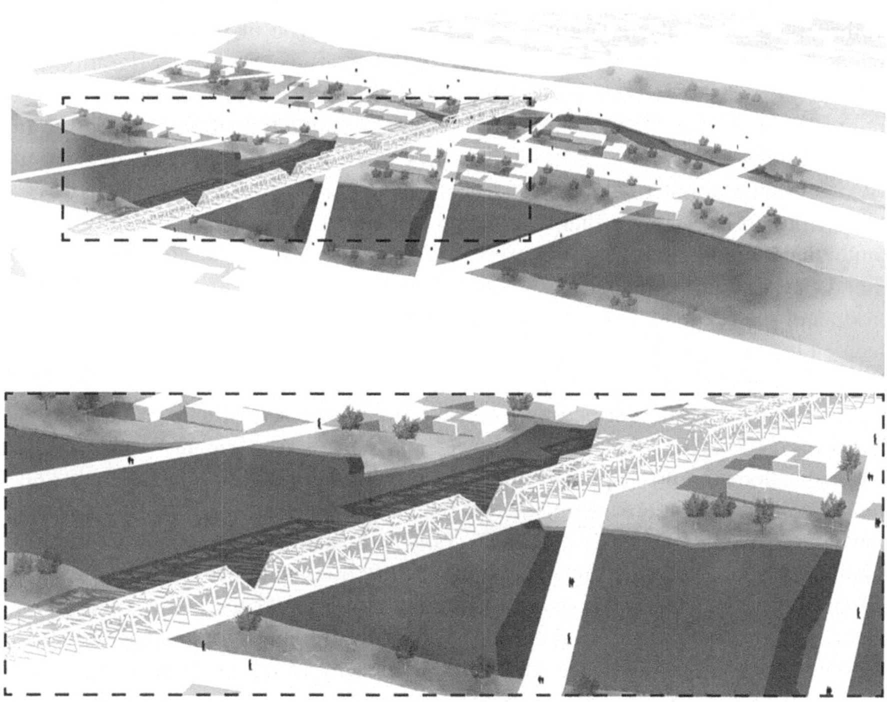

Fig. 16.4 Design proposal for a network of pedestrianised bridges across the US–Mexico border, reframing the crossing as shared public space. (Source: the author)

for people, in addition to trade. Tourism infrastructure, particularly in contested political geographies, is central to border image rebranding, as explored by Timothy (2001). Structures that house local shared institutions or physically bridge across the border to create a new shared pedestrian zone could act as a catalyst, which spreads and infiltrates the urban centres miles away from the border.

For such a project, the image of the border remains a significant problem that hinders both the tourists who want to visit the border and businesses flourishing along it. The border is not always portrayed in a good light, mainly due to cartel violence as described by Castillo (2002) reviewing women's experiences on the border. Bi-national projects, which advocate for a peaceful crossing zone, which welcomes tourists and private investments, can only help the image of the border in a positive way. A convincing argument can be made here about the presence of partnership schemes on the border, which would make a large impact on how the border is portrayed. Partnership schemes, which become physical in a binational project located in the biggest border towns El Paso-Ciudad Juárez, could become the role-model for bi-national development, creating a significant impact on the border image as it stands now.

The image of the border normally begins with a set of linguistic contrasts, drug wars versus security or illegal versus legal migration, and developed versus developing. These strong juxtapositions and contrasts begin to create a visual idea of the border, although this type of image writing does not represent the border as a whole. Anthropological border literature creates a different type of image of the border, which focuses on its identity. Roger Rouse, who wrote 'Mexican Migration and Social Space of Postmodernism', addresses identity head on through a series of interviews targeting Mexican immigrants and documenting their perspective on dual nationality and space. 'My identity now possesses multiple repertoires: I am Mexican, but I am also Chicano and Latin American' (Rouse 1991: 8). This is a quote from an interviewee that responds to the idea that identities are shifting and contextualising in multiple directions. Other social science academics, such as Gienda Tibe Bonifacio, discusses 'multidimensionality' as a theme, which is a product of interdependent relations between those who migrate and those who have decided to stay. The image of the border often does not address the ingredients of such different transnational perspectives.

The identity of a border begins with two equal societies, which are labelled and categorised by unequal relations of power and governance (Heyman 1994). A bi-national scheme could display the developing cities of El Paso-Ciudad Juárez and the radically different environments that they are rather than focusing on balancing power and livelihood. Such a project could instead bring the two communities together in a joint effort to create something truly trans-national that contends with the stark issue of largely Mexican inhabitants living under the control and governance of the border from the US side. Variables similar to this make the border region a place of distinctive contradictions and any bi-national project needs to acknowledge the diversity of the border condition, accepting that each border region has its own identity with similar variables, and are thin areas of land situated between stable places (Gupta and Ferguson 1992).

Methodologies used in anthropological research can be used to examine the power position of the border, examining the inequality of the border and its characterisation. This type of research can be used to document the condition, and the relationships that stretch across the border. Within this body of documentation, border image is something which clearly excites and intrigues people, and rightly so. The political landscape in which the border is located naturally causes attention. Well-publicised documentation about immigration laws, maquiladoras (related to NAFTA participation) (Heyman 1994) and narcotics are in the news, constantly reminding us of the 'condition' of the border. This political geography privileges one image and does not represent the entirety of the border and what it is to become; this project begins to look at the multi-layered image that represents the border more realistically and theoretically.

To better understand the context, which is needed to create a new binational border image, we have to clearly identify the broad political, economic and social factors that have set an undertone for the existing border literature. *Frontera* is a Spanish word meaning the border or the crossing between two countries. A film called *Frontera* released in 2014 depicts the journey of a Mexican man crossing the border illegally to the United States, and at the beginning of his journey, a former Arizona sheriff's wife is shot whilst patrolling the ranch she lives on.

The director Michael Berry decided to produce the film, which shows the common conceptions associated with crossing the border, focused specifically on the Arizona border with Mexico. Not only does this film show realistic events that occur when crossing, including the hiring of coyotes and the physical journey crossing the border, it also gives a universal view of migration from Mexico to the United States, a viewpoint from ranch owners who often have migrants cross through their land, and the strategies implemented to deal with it. This type of media exposure allows the public to see both views of the border. Bi-national treaties are only as successful as the people who believe and invest in the idea. Viewers of the film often transition back and forth through the border, constantly engaging with events unfolding on both sides. The idea for the film sprung from the first-hand experiences of the director who lived in Arizona, and the characters based on family members, grounding the plot to a series of true stories. This type of film exposes the different transnational perspectives of the border and the adaptation of the mainstream media outlet, with the realities of the border condition.

Understanding the movement of people across the border identifies the importance of the security and development of that border to serve a purpose. It enforces restriction and allows governments to carefully monitor the movement of people and trade. However, this movement is often imbalanced, and it is this imbalance that defines the security measures found there. Security has been evaluated and protested, since the birth of the border fence. Events of September 11th, the terrorist attack in New York, United States, affected the security measures across the entire border. The United States entered high alert mode, to prevent future terrorist attacks; this resulted in reinforcing the border with stricter immigration laws and increased patrol agents. 'Operation Blockade' also known as 'Operation Hold-the-line' mobilised over 400 border patrol agents along the border edge, to intimidate illegal crossing (Dunn 2010). This isolated event only occurred in the El Paso–Ciudad Juárez border, changing the system of border patrolling. The political statement of 'Operation Blockade' sought to apprehend illegal migrants before they crossed, rather than apprehension once in the United States. Timothy Dunn, a human rights academic, labels this strategy prevention through deterrence (Dunn 2010), which is now reoccurring through the major border cities. Trans-border dynamics are best examined in the El Paso–Ciudad Juárez border, not only because of its size but also for its history of dealing with politics, socio-economic challenges and its triumph in development. Creating binational infrastructure on this border will set a precedent for other border cities to follow.

The term 'one-sided' border has often been used to depict the easy flow of Americans to Mexico and the restriction of Mexicans to America. However, Mexican immigrants living in America constitute 15% of the population—the largest ethnic group—and this percent has been steadily increasing over the last 50 years (U.S. Department of State 2011) pointing to a flexibility of this 'one-sided' condition. This condition produces a rich and complicated border territory inhabited by several dense urban conditions that span across the national boundary. As part of this provocative design research methodology a large bi-national bridging structure was devised as a test. This research shaped the future design proposals for the border presented in this chapter.

Having a large-scale model (see Fig. 16.5) to test provocations allowed a flexibility in design and an ability to explore alternative arrangements, scales and positions in relation to the context of the twin cities. Because of the steep terrain leading down to the river, testing multiple layered bridges became an intriguing experiment. The increased number of bridges was then intended to directly increase the number of border crossings, diversify their character and offer an alternative to those specifically aimed at commercial trade vehicles.

In some border zones, public transportation has been proposed as a way of facilitating cross-border shopping (Dascher and Haupt 2011). Such projects could

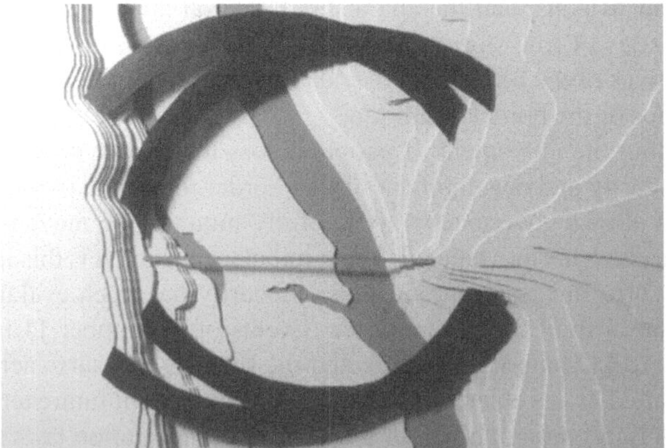

Fig. 16.5 Large-scale model of the El Paso–Ciudad border proposal. (Source: the author)

produce economic benefits in areas such as the southern border of Texas, where Mexican shoppers constitute a significant portion of retail trade. During the investigation into mobility across the border, it was apparent that pedestrian bridges had not been explored to their full potential. Although traffic is a very serious issue that needs to be addressed, so too is the need for communities to be able to access facilities over the border. This would encourage existing Mexican and American shoppers to contribute to the local economy.

16.5 Binational Project

The opportunity of creating a layered and continuously complex space allows for an experiment in elasticising and connecting resources over the border as a unique design test. We have witnessed the strong economic connections that these cities have, which are facilitated by a series of cross-trade bridges, seeing thousands of vehicles through the border daily. Existing management schemes that permit cross-border trade vehicles to travel more than 15 miles beyond the official boundary illustrate how the border is becoming increasingly elastic, with new border entities emerging as a result. This raises the question of whether an understanding of the physical border as manipulated and flexible could lead to its occupation by inhabitable zones with shared infrastructures. Furthermore, the concept of this zone also focuses on dual ownership and proprietorship. The design constraints here lie within defining the boundary of these spaces for passage, labour and importation of goods. This layering of spaces could potentially develop into its own identity, which thrives from both sides of the border. The idea itself can be extreme, but is based on the reality of the fifth bridge proposal already in place across the El Paso–Ciudad Juárez border. The appropriation and expansion of this bridge brief can prompt a new understanding of the value and regulations of intra-national space.

Objective To enclose an intra-national zone by two bridges (see Fig. 16.6).

Methodology Investigate closely the structural elements needed to support a circular bridge in order to span the Rio Grande. Inhabit the bridge with local markets, which interacted with the immediate community affected by the regime of the border. Create a ten-line cross-trade bridge, which supports the influx of trade in the region. Within this intra-national zone new communities are developed (and potentially part of a free trade zone to stabilise its growth).

Summary This design provocation was aimed to incorporate the growing population of El Paso–Ciudad Juárez and the influx of trade. Overall, this test enabled the understanding of the spaces, which might define the boundary, areas for crossing the boundary and potential set-up for proprietorship programmes. This design test continues to develop throughout this chapter.

Fig. 16.6 The provocation of intra-national zoning between the two states. (Source: the author)

A realisation of a binational infrastructure is presented in the new binational project located in San Diego (USA)–Tijuana (Mexico) border region, which began construction in 2012, for completion in summer 2015. This international project is a cross-border pedestrian bridge, which links passengers from Otay Mesa San Diego, Stateside to the Tijuana International Airport. The Cross Border Facility (CBF) will also provide a main recreational building which will provide shops and services to customers; this structure will be built on the US side because it also houses US border customs and Border Protection facilities (U.S Department of State 2011). This project began realisation in 2008, which was privately funded, and has gone through 6 years of planning and preparation. The 95,000 square foot structure has also had private investment from Chicago-based Equity Group Investments, which is owned by Sam Zell. If these local and multinational influences determine the funding and construction of such bi-national projects, one of the major issues influencing the development of the fifth bridge shared border project remains the management of the scheme. Dual management has precedents but can be difficult to control. This details the dependence on US security infrastructure that the facility must maintain and place significant restrictions on the operation of the site.

Rather, we contend that most tourists want to eat, drink and purchase selective goods and services in an idealised 'Mexico land' that is somewhat insulated from the real world and danger, but one that evokes at least a veiled sense of excitement and foreign adventure (Arreola and Curtis 1993). Understanding the value of intranational space can be explored through various avenues. The initial response was to test an inhabitable bridge in which both US and Mexican citizens could harbour.

At the outset of the design research, one of the first tests introduced an enclosed intra-national zone held within the circumference of a circular bridge that spanned the Laredo/Nuevo Laredo border. The proposal examined the possibilities of a large flexible structure that supported the bridge across the Rio Grande and housed local markets, which interacted with the immediate community affected by the regime of the border. Integrated into this was a pedestrian crossing as well as a ten-lane cross-trade bridge, which would support the increased influx of trade in the region. Within this intra-national zone, the development of new communities was projected as part of a free trade zone to stabilise its growth, incorporate the community and improve local transport links. Primarily, this design provocation was aimed at the incorporation of the growing population of the two Laredos and the increase in transnational trade. Overall, this test enabled the understanding of the spaces that might define the boundary, areas for crossing the boundary and potential set-up for proprietorship programmes.

Contemporary binational projects have become more frequent and ambitious in recent years. There are 50 official, regulated and controlled crossings between the United States and Mexico; each crossing is unique and meticulously planned to adhere to the particular site conditions. Another 17 crossings are planned by 2025 (Romero 2008). This contemporary development is demonstrated by the Cross Border Facility (CBF).

16.6 Co-operation

Such a project may become feasible through existing organisations and legislative avenues. El Paso is recognised for its efficient communication systems across and between the borders. They have developed a strategy that deals with binational agreements more effectively than other large thriving border communities such as San Diego–Tijuana. At a macro-level, the border operates coherently federal-to-federal, engaging in new binational projects on the border. However, the micro-level at which the border operates is one of the weaker elements of border systems and hinders the strengthening and implementation of binational policies. The BBC documentary 'Beyond Borders—Border Wall' shows various interviews with locals from Texas and Mexico expressing their concern for the Rio Grande.

'Baseball on the border' written by Alan Klein examines the baseball club *Los Tecolotes de los Dos Laredos* (Owls of the Two Laredos) a Mexican League team which in 1985 became the first and only binational sports team (Klein 1997). However, the team split up in 1994, but the binational enterprise gave the Texas–Mexican border a unique perception of borderlands relations, as both constituencies had collaborated effectively. This historic collaboration has been a tool to understand the larger international relationships and the value of mixed identities.

This stage of the project sought to recover this bi-national team through the integration of a new stadium complex on the border. And while the management of a programme like this would be extremely complex, the project anticipates that sponsorship could be obtained from both sides of the border, with sponsors potentially serving as the primary employers of the stadium staff. The construction of such a large programme may have to be financed initially by the Texas government on a loan basis. This type of agreement has been done before with the World Trade Bridge, and the Mexican government was able to pay off the loan in smaller amounts. This management of resources could be very beneficial for both sides of the border.

Mexico maintains its own jurisdiction over the development or modification of transport through the border, POE's (Port Of Entry), infrastructure and financing. Every border project initiated by Mexico starts in reference to The National Development Plan (NDP), outlining current and future projects in and around Mexico. The NDP is released every 6 years, following a new presidential election (Laredo International Bridge System 2013). One of the initiatives that was incorporated into the NDP was the American Recovery and Reinvestment Act, which addresses and facilitates funding for projects that modify or upgrade the borders. Its implementation amounts to a 720 million dollars average allocation for border crossings each year (Wilson and Lee 2013). These measures are directed at the ongoing prosperity of the trade relationship between the US and Mexico. However, bilateral policies and intuitions also assume responsibility for the implications of such initiatives, many of which span ecological sensitive areas, which become a management issue for both sides of the border (Herzog 1990). This is another aspect that must be considered in the design of a fourth bridge proposal. In the case of El Paso–Ciudad Juárez, the border runs through the Rio Grande and creates substantial infrastructure challenges in such an environmentally delicate area (Hamre 2003). This region is regulated by the US–Mexico Border Program Environmental Protection Agency developed in 1996, which focuses on densely populated areas, where the environment may be under heavy strain. This agency not only provides protection for the border's ecological welfare, but they also provide management schemes at new infrastructure sites. This is an important overview that calls for a longer-term understanding of impact and gradual change on site.

Beyond the environmental sensitivity of the site, the humanitarian aspect of this border is a vivid issue. The U.S. Department of State releases an annual report e on '*United States-Mexico Partnership: Managing our 21st Century Border Initiative (BIM)*'. This report discusses management schemes to provide and care for the 13 million people who currently live at the border. Collaborative border management depends on sustained participation from local stakeholders, including communities and multiple levels of government. Such environments present their own obstacles and opportunities. While larger institutional structures manage the economic arrangements at the border, the bilateral BIM is one of the institutions that aims to settle policy structures to facilitate collaborative border architecture accommodating local interest.

Apart from these direct concerns, the US–Mexico Binational Bridges & Border Crossing Group (BBBXG) is responsible for coordinating meetings to discuss border crossing projects around three times a year (U.S. Department of Commerce 2008). The aim of the meetings is to improve the efficiency of existing border projects and organise the planning for new binational plans (U.S. Department of Commerce 2013). The BBBXG meetings always include the State Department as the co-chair, for the US side of the border. The State Department has the authoritative Presidential power to permit plans, build or alter an international border crossing. It also gives an opportunity to view current/ongoing binational POEs projects on the border to review their progress. Fundamentally, these meetings determine which projects go through to planning permission and which do not.

16.7 Conclusion

Throughout this research, distinctive design provocations are assessed to show how the border could be elasticated to create a new type of trans-national landscape. Defining boundary spaces through different hierarchies demonstrates the multiple activities that could benefit not only the border but also the communities that live by it. Now, the situation of mismanagement of the border, and the dual ownership of the Rio Grande, will only continue to create areas of natural decline. Binational anchors are essential to facilitate the growth of such a binational territory. Within this, strong legal infrastructure is constituted by both sides and implemented in a manner that establishes greater equality of expectation and standard of living for the border corridor. A border which is situated through a natural resource has proven very unsustainable. But a binational natural resource is much more attainable and manageable. 'NAFTA has been a blessing for the economies of the two countries, but it has been a curse to the border' (Payan 2006). In agreement with Payan, the environmental impacts of the maquiladoras, combined with the heavy traffic flows, are detrimental to both communities. More importantly, the inaccessibility of the border crossing, caused by the lack of predominantly pedestrian-designed infrastructure, fails to support or ensure the safety of residents crossing the border. This research provides evidence that there is a need for a more local-to-local binational development programme on the border region, specifically at El Paso–Ciudad Juárez. The economic impacts, as discussed before, are a 'blessing' to both countries, bringing in billions of dollars, but the strain on the border has affected the local community. Elasticising the border has naturally blurred boundaries and made various types of boundary spaces.

The project encapsulates programmes that would become anchors for the binational development of Ciudad Juárez and El Paso; these crucial anchors are vital to develop sustainable binational programmes on the border. The project brings together new and existing binational programmes, such as the scholarship scheme between UTEP and Universidad Autónoma de Ciudad Juárez, into one centre, with the opportunity to extend that into a construction training school open to residents on both sides of the border. This binational facility will also contain a research centre that allows library access to the public. Fundamentally, it is the opportunity to connect people across the border, and provide the opportunity to learn, research and share ideas on a platform that is not biased or one-sided. A 'free zone' could incorporate niche markets and communities to share business proposals or share existing infrastructure. This would start at the border and extend deep into the border towns.

References

Arreola D, Curtis JR (1993) The Mexican border cities: landscape anatomy and place personality. The University of Arizona Press, Tucson

Castillo DA (2002) Border women: writing from La Frontera. University of Minnesota Press, Minneapolis

Castro-Rea J (2012) Our North America: social and political issues beyond NAFTA. Ashgate Publishing Limited, Farnham

Clement NC (2002) The U.S.-Mexican Border Environment: U.S.-Mexican Border Communities in the NAFTA Era. San Diego State University Press, San Diego

Collins K (2013) Life in the US-Mexican border region residents' perceptions of the place. J Borderl Stud 28(1):127–146

Correa-Cabrera G, Garrett T, Keck M (2014) Administrative surveillance and fear: implications for U.S-Mexico Border Relations and Governance. Eur Rev Latin Am Caribbean Stud 36:35–53

Dascher K, Haupt A (2011) The political economy of regional integration projects at borders where poor and rich meet: the role of the cross-border shopping and community sorting. J Urban Econ 69(1):148–164

Dunn T (2010) Blockading the border and human rights: the El Paso Operation that remade immigration enforcement. University of Texas Press, Austin

EPA - U.S. Environmental Protection Agency and SEMARNAT - Secretaría de Medio Ambiente y Recursos Naturales (2003) Border 2012: U.S.–Mexico Environmental Program. EPA-160-R-03-001. https://nepis.epa.gov/Exe/ZyPURL.cgi?Dockey=P1001XX5.TXT. Accessed 6 Dec 2013

Fischer R (2013) Mobility and morality at the border—a Lefebvrian spatio-temporal analysis in early twentieth-century Ciudad Juárez and El Paso. In: Space/time practices and the production of space and time, vol 38(3). Zentrum für Historische Sozialforschung, Cologne, pp 176–196

Geenhuizen MV (2012) Creative knowledge cities: myths, visions and realities. Edward Elgar Publishing Limited, Cheltenham

Gupta A, Ferguson J (1992) Beyond "culture": space, identity, and the politics of difference. Cult Anthropol 7(1):6–23

Hamre JJ (2003) U.S.-Mexico transboundary water management. Center for Strategic and International Studies, Washington, DC

Herzog LA (1990) Where north meets south. University of Texas Press, Austin

Heyman JM (1994) The Mexico–United States border in anthropology: a critique and reformulation. J Polit Ecol 1:1–20

Hilderbrandt K (2012) Maquiladora Industry. www.lmtonline.com/textarchives/082801/b6.htm. Accessed 6 Dec 2013

Hing BO (2012) Defining America: through immigration policy. Temple University Press, Philadelphia

Huffines AC (2005) The Texas War of Independence 1835–1836: from outbreak to the Alamo to San Jacinto. Osprey Publishing, Oxford

Jones R, Johnson C (2011) Interventions on rethinking 'the border' in border studies. Polit Geogr 30(2):61–69

Katz B (2014) Innovation districts are catalysts for urban growth. https://citiesspeak.org/2014/06/02/innovation-districts-are-catalysts-for-urban-growth/

Klein AM (1997) Baseball on the border: a tale of two Laredo. Princeton University Press, Princeton
Kong HKW (2015) NAFTA and sustainable development. Cambridge University Press, New York
Kotkin J (2013) America's true power in the NAFTA century. Forbes. https://www.forbes.com/sites/joelkotkin/2013/09/13/americas-true-power-in-the-nafta-century/. Accessed 1 Dec 2013
Langewiesche W (1993) Cutting for sign. Pantheon, New York
Laredo International Bridge System (2013) Bridge information. http://www.ci.laredo.tx.us/bridgesys/bridges.html. Accessed 3 Jan 2013
León AD (2002) They called them greasers: Anglo attitudes toward Mexicans in Texas, 1821–1900. University of Texas Press, Austin
Lorey DE (2008) The U.S.-Mexican Border in the twentieth century. Rowman & Littlefield, London
Loustaunau MO (1999) Life, death, and in-between on the U.S.–Mexico Border. Greenwood Publishing Group, Westport
Maril RL (2011) The fence. Texas Tech University Press, Lubbock
Martinez OJ (1994) Border people: life and society in the U.S.–Mexico Borderlands. University of Arizona Press, Tucson
Martinez OJ (2003) U.S.–Mexico Borderlands: historical and contemporary perspectives. Scholarly Resources Inc., Wilmington
Massey DM (2001) Beyond smoke and mirrors: Mexican immigration in an era of economic integration. Russell Sage Foundation, New York
Meissner D, Kerwin D, Chishti M, Bergeron C (2018) The U.S. Asylum System in Crisis: charting a way forward. Migration Policy Institute, Washington, DC
Michel S (2003) The U.S.–Mexican Border environment: binational water management planning. San Diego State University Press, San Diego
Ortiz-Gonzalez V (2004) El Paso: local frontiers at a global crossroads. University of Minnesota Press, Minneapolis
Paasi A (2009) Bounded spaces in a 'borderless world': border studies, power and the anatomy of territory. J Power 2(2):213–234
Payan T (2006) The three U.S.–Mexico border wars. Praeger Security International, Westport
Richardson P (2012) The informal and underground economy of the South Texas Border. University of Texas Press, Austin
Romero FD (2008) Hyper-border: the contemporary U.S.–Mexico Border and its future. Princeton Architectural Press, Princeton
Rouse R (1991) Mexican migration and the social space of postmodernism. Diaspora J Transnatl Stud 1(1):8–23
Simmons WP, Mueller CM (2014) Binational human rights: the U.S.-Mexico experience. University of Pennsylvania Press, Philadelphia
Sklair L (2011) Assembling for development: the Maquila Industry in Mexico and the United States. Routledge, Abingdon
St. John R (2011) Line in the sand: a history of the Western U.S.–Mexico Border. Princeton University Press, Princeton
Stacy L (2003) Mexico and the United States. Marshall Cavendish, Tarrytown
Stephen L (2007) Transborder lives: indigenous oaxacans in Mexico, California, and Oregon. Duke University Press, Durham
Timmer A, Williamson J (1998) Immigration policy prior to the 1930s: labor markets, policy interactions and globalisation backlash. Popul Dev Rev 4(24):739
Timothy DJ (2001) Tourism and political boundaries. University of Arizona Press, Tucson
U.S. Department of Commerce (2008) Washington, D.C. U.S.-Mexico Border crossing data and economic indicators report: U.S. Department of Commerce. https://www.census.gov/. Accessed 2 Dec 2013
U.S. Department of Commerce (2013) United States Census Bureau. https://www.census.gov/qfd/states/48/4841464.html. Accessed 2 Dec 2013

U.S. Department of State (2011) U.S.–Mexico Border Master Plan: a binational approach to transportation planning and project development. U.S. Department of State, Border Affairs Office, Washington, DC

Vila P (2000) Crossing borders, reinforcing borders: social categories, metaphors, and narrative identities on the U.S.-Mexico frontier. University of Texas, Austin

Wagner BK (2014) The rise of innovation districts: a new geography of innovation in America. Metropolitan Policy Program, pp 1–34

Wilson PA (1992) Exports and local development. University of Texas Press, Austin

Wilson HD (1999) Borders: frontiers of identity, nation and state. Biddles Ltd, Wellingborough

Wilson C, Lee E (2013) The state of the border report. Wilson Center, Washington, DC. https://www.wilsoncenter.org/publication/the-state-the-border-report. Accessed 8 Jul 2025

Yigitcanlar T (2012) Building prosperous knowledge cities: policies, plans and metrics. Edward Elgar Publishing Ltd, Cheltenham

Zartman IW (2010) Understanding life in the borderlands: boundaries in depth and in motion. University of Georgia Press, Athens

Chapter 17
Cities After Nature: Venice as a Tool for Reimagining Cities Beyond Nature and Culture

Chiara Toscani

Abstract Looking at contemporary ecological issues, the most recent philosophical thoughts are strongly affirming how the idea of nature is changing, requiring a new cultural perspective that overcomes the concept of divisions—between subjects and objects, humans and non-humans, nature and culture—and moves forward hybrid entities. As a consequence, the idea of urbanity, considered the outcome of nature and culture, is changing as well. To understand how this relationship has been constituted within architecture and to clarify future models of cohabitation, this chapter will use Venice as an exploratory example for establishing a theoretical and critical framework that could serve as a starting point for future and broader research aiming to reimagine cities beyond nature and culture.

Keywords Nature · Urban ecological thinking · Seamless boundaries · Posthumanism · Hybrid space · Urban philosophy · Venice

17.1 Introduction

The list of books and essays investigating the meaning of borders is extensive in both length and breadth, taking in a number of different disciplines. The term border has always a dual meaning, an attractive feature for a research topic (Wilson and Hastings 2012). Spatially speaking, borders are places where relations are disrupted and, on the other hand, they are unique spaces that profoundly reveal the complexity of human relations.

This dualism represents two opposing conditions: the idea of division and the contrary idea of coexistence and multiplication. The same dualism is easily identified within Western territorial, social and economic policies. On the one hand, they are firmly based on an idea of continual progress and availability of global natural

C. Toscani (✉)
AA School of Architecture, London, UK
e-mail: chiara.toscani@aaschool.ac.uk

G. Cimadomo et al. (eds.), *Urbanism at Borders*, The Urban Book Series,
https://doi.org/10.1007/978-3-032-09594-7_17

resources. On the other hand, as never before, the world experiments with an increase in the number of borders based on the idea that these are natural national divisions.

Bruno Latour (2017) explains that these principles of globalism and localism are faces of the same coin, used according to the different political and economic conveniences of modern society as it tries to survive, and identifies it as a striking paradox. This is based on a warped vision of the relationship between human beings and the Earth: an anthropocentric point of view. It imposes one culture instead of many and is deeply connected with an idea of the Earth as an infinitely exploitable natural resource. The increase in physical borders and increasing social conflicts are the litmus test to prove both the paradox and, simultaneously, its weakness, demonstrated by climate change and by a new notion of nature that demands to be understood.

In setting out this paradox and its origin, Latour invites us to reframe the complex issue of nature and culture (society).

It is the purpose of this essay to begin to explore how a new conception of nature and its relationship with culture may affect contemporary cities. In fact, the city is not just the most important outcome of the relationship between nature and culture (society), but also the place where this pairing finds its greatest and most numerous expressions. It is also the place where social conflicts and contradictions happen most clearly. If the notion of nature and, consequently the dialectic pair of culture and nature changes, the effect of that change on the urban fabric would repay investigation.

As a result, this essay will begin by examining Venice as a tool for establishing a theoretical and critical framework that could serve as a starting point for future and broader research aiming to reimagine cities beyond nature and culture. This city is a magnificent palimpsest through which all the variants of the relationship between nature and culture(s) can be found. Venice is a border city, between sea and land and between different cultures throughout the centuries.

17.2 Philosophical Background

Before delving into this theoretical and critical outline, it is necessary to clarify the philosophical framework on which this investigation is based. As briefly explained within the introduction, Latour was the first author to understand and explain the modern paradox and how it has affected historically the relationship between nature and culture. Firstly, he provides a detailed examination of how the relationship between nature and culture changed in modernity. He underlines the paradox of a systematic cultural process of separation, where quasi-objects emerge that are conceived neither as belonging to the human world (culture-society) or the non-human world (nature), nor as being entitled to their own existence and autonomy. In many respects, he shifts philosophical attention towards these hybrid objects as potentially autonomous entities, overcoming the exclusively anthropological point of view as defined by modernity and opening to the complexity and contradictions of the contemporary age.

More specifically, in his book, *We Have Never Been Modern*, (Latour 1993), Latour describes the process of modernisation through the existence of two worlds and the two related practices: the first world made of hybrid elements (networks, quasi-objects), a blending of nature and culture, the result of a translation process; the second world (criticism) composed of two separated ontological zones (human beings and non-human), defined through a purification process of all objects and thoughts of the world.

In order to be modern, the two practices—translation and purification—have to contain two dichotomies: the first is the division between the hybrid world and the two ontological zones, and the second is the internal division of the two ontological zones. The paradox of modernity lies in the proliferation of hybrids—nature and culture—and their subsequent discrimination in pure nature or pure culture. This paradox rests on two denials. The first concerns the denial of the hybrid world as an autonomous entity. Unlike in pre-Modernity, the preponderant entrance of science has started to generate and proliferate objects belonging to the reproducible and immanent nature—artificial nature—and the transcendent culture.

The second concerns the denial of the interference of the purification process with the hybrid world and the latter's framing as a world belonging to a human or non-human category. As a consequence, the status of objects has been delegitimised as a product of nature or culture. In a recent book *Facing Gaia*, Latour claims that nature does not exist as a domain (anymore); it exists only as one half of a pair pertaining to one single concept. We must thus take the Nature/Culture opposition as the topic on which to focus our attention and not at all, any longer, as the resource that would allow us to get out of our difficulties.

> […] If ecology sets off panic reactions, we now understand why: because it obliges us to experience the full force of the instability of this concept [Nature], when it is interpreted as the impossible opposition between two domains that are presumed to exist in the real world. (Latour 2017: 19)

For this reason, he introduces a new difference between the matters of fact—traditional objects or free-risk objects—and the matters of concern. Both are a blend of nature and culture. Matters of fact are distinct and defined from the outside world, keeping to themselves. Matters of concern are thoroughly blended with the world, rendering themselves indistinguishable and inseparable. As a consequence, objects can no longer be exclusively categorised within the human beings or non-human beings' world.

In light of a new awareness, many other scholars have also started to re-examine the dialectical relationship between nature and culture. Debates focus on how to deal with this instability of the dialectic couple, including the possibility of reconciling it (Tsing 2015), going beyond it (Descola 2014), embracing a formation of new ecologies out of the disturbances and getting around it (Latour 2017).

Incorporating these concerns into architecture, a few questions emerge clearly since they can be considered one of the involved disciplines in environment transformations, and cities and territories are significant outcomes of the relationship between nature and culture: can the city be analysed from Latour's perspective in light of this new philosophical investigation, using, for example, the concept of hybrid-object?

17.2.1 The Idea of the City as a Hybrid Object

My first step is to demonstrate the validity of Latour's perspective when applied to cities and territories developments throughout history, then determining how and through what theories and urban interventions the modern paradox can be observed.

As described by Leatherbarrow in his essay, *Levelling the Land* (Corner 1999) the Greek myth of Chthonia captures a primitive phase in the historic interactions between culture and nature. Chthonia represents the unlimited, deep, wet, dark, irrational, generating and creative side of the ground; it represents the vibrant side. In contrast, Gaia represents the limited light and rational side of Earth. Whilst they appear to be two contrasting concepts, they are deeply connected.

The significance of this myth lies in the process of transformation. A part of Chthonia, an uncontrolled, uncanny, irrational and inhospitable space, is transformed into Gaia, which becomes a platform, place and territory for human dwelling. Despite the considerable and continuous effort to transform the dark side of the Earth, its supposed disappearance represents one of the greatest human illusions. Underneath the platform, Chthonia steadily interferes with Gaia. The disappearance of the dark side of the Earth represents one of the greatest human illusions. Underneath the platform, Chthonia steadily interferes with Gaia. Sometimes their mutual interference has an immediate effect, and other times it is less synchronised. However, connections between the two sides always exist, even when we cannot perceive them. The awareness of this forgotten connection between the two sides—nature transformed by human action and nature as soil—underlies one of the foremost contemporary questions of ecology.

As a testament to the symbolic significance of Chthonia's myth, transformations of the ground have continuously been influenced by religious and cultural beliefs, which have shaped our understanding of nature since antiquity. According to Joseph Rykwert (1988), in ancient Rome, tracing a border to define a territory meant making a furrow in the ground. Roman settlements were founded on such furrows, representing a sacred act that defined the edge of a settlement or city as a space distinct from the countryside. Since this separation was considered a violation of nature, a ritual gesture was needed to reconcile the two opposing spaces. Marcel Poëte (1958) also highlights how a continuity in the combination of ground (nature) and human transformation (culture) has always existed, supporting agricultural activities and the building of early human dwellings.

In modernity, such rituals are no longer deemed necessary in order to sanctify these actions that violate the grounds of nature. Interventions and cultural transformations have come to signify human's unlimited power over nature, which is conceived as their rightful space to conquer.

In his essay, *Reason's Adventures: Naturalism and the City in the Century of the Enlightenment,* Manfredo Tafuri (1976) gives a clear explanation of when and where this paradox began, analysing Laugier's architectural theory and its effect on the Enlightenment city. In *Essay on Architecture,* Marc-Antoine Laugier (1755) redefines the main principles of architecture through the concept of the primitive hut. He subsequently extends them, through the metaphor of the forest, to describe the principles on which a new idea of the city is based. Influenced by philosophers such as David Hume, Laugier returns to nature as a source of universal principles. But, unlike the pre-modern idea of nature as a model to imitate, so that artifice and nature would be in balance, Laugier conceives nature in terms of its rational laws, to which human beings and their corresponding artefacts also belong.

Natural processes, rationally interpreted, become the reference point for architecture and the device through which to define the essence of its constructive logic. In consequence, the hut becomes the architectural archetype of all kinds of architectural buildings, and the forest its urban counterpart. The metaphor of the forest, which slowly becomes similar to the idea of the garden as the essay develops, is the author's answer. Through the concept of nature in the form of the forest, the guarantee of a new design process, Laugier reconciles all the different urban forces, solving the increasing loss of form of the city. On the one hand, he considers the city as a natural phenomenon regulated by the rational laws of nature. The purity and universality of nature guarantee the validity and the truth of the transformations, even if they are contradictory. On the other hand, nature, through new ideas related to the picturesque aesthetic and its breakdown of a single perspectival point of view, becomes the main figurative reference to manage the urban variety and fragmentation within the urban projects. Therefore, Laugier's urban theory expresses the attempt to find an impossible synthesis between rational laws and the uncontrollable aspects of reality, especially within the city.

As a consequence, this theory allowed to overcome traditional physical borders; such walls, revealing an interesting anticipation of the idea of spatial continuity, theoretically only developed two centuries later; but on the other hand, it definitely swept away any form of reconciliation regarding nature's transformations, leading to the modern paradox, as described by Latour. In addition, Tafuri (1980) explains how this urban evolution led to the crisis of the architectural object as an autonomous entity. The architectural object was dipped in the urban fabric, slowly disappearing in favour of a typological uniformity.

From the eighteenth century onwards, the role of architecture became increasingly related to social function and this is crystallised in the concept of architectural typologies corresponding to specific functions. From Tafuri's point of view, this alteration intensifies the crisis of the architectural object, its artistic dimension and its role within the city. The functional purpose becomes so fundamental to supporting the contradictory political and economic dynamics that the human experience becomes even more critical in giving a meaning to architecture and defining a figurative language. This interpretation leads to a second and more fundamental question that informs this essay: how can changing understandings and functions of nature, as well as their relationship with culture, influence cities? Therefore, it is necessary to examine what theoretical principles and practices are involved in answering this question. As such, Venice can be considered an effective tool for navigating these complex components.

17.3 Venice as a Tool for Reimagining Cities Beyond Nature and Culture

Every city is undoubtedly the result of the encounter between nature and the human project, but Venice can be considered to be the supreme manifestation of that encounter. It marks a most magnificent encounter between nature and humanity, mediated through centuries of struggle and negotiation between opposing forces. There, the interweaving of nature and culture does not just generate a city, but has involved all humanistic disciplines over time, creating its vast cultural heritage.

Reading the description of Aschenbach's arrival by sea in Thomas Mann's novel, *Death in Venice*, (Mann 1924), everyone immediately understands the true essence of the city. Venice is not just the story's backdrop, but the place where the encounter of two domains, the water and the land, mirrors human events and emotions and vice-versa. These two domains continually build the city, metaphorically and practically, throughout its history. This is apparent when looking at its material and cultural aspects, revealed in its stones, its buildings—internal spaces and facades—its streets, its small and broader open spaces, its people and food. It is also evident in its immaterial aspects, in its atmosphere, its light, its colour, its words and its sounds. It has happened without a true reconciliation. Manfredo Tafuri demonstrated this in his study of Venice in the Renaissance (Tafuri 1989). Here, he detects a rupture and the beginning of modernity. He identifies in this period the growth of anxieties and contrasting forces that subsequently lead to the creation of the contemporary city, and to capitalist dynamics.

However, he has claimed Venice tried to endure within her origin (Tafuri 1989: 11) assuming a further peculiar approach which consisted of integrating elements belonging to its historical legacy into its dealings with the environment, such as empirical and collective practices, with others based on new scientific discoveries that seemed more responsive to the new needs. Tafuri has described this ability as Venice's *prudentia.* This latter guaranteed Venice's persistence within her origin; prudence constituted the measure of good government, founded criteria of justice, enabled tradition to resist within the new and enabled the new to live in an uninterrupted Crouzet-Pavan cosmic time (Tafuri 1989: 11). Venice still represents a clot today: it is a magnificent microcosm of man's relation to the environment (Crouzet-Pavan 2000: 51), where, however, contemporary ecological crises are condensed. In fact, Venice is currently one of the cities most affected by the rising sea level caused by climate change. An interesting map, designed by Martin Vargic (2018) and based on the latest scientific predictions, shows that the entire Italian *Pianura Padana*, like many other parts of the world, could be submerged in the next 150 years. Venice could disappear entirely, and Vicenza could become an island, as could Naples. Milan could become a seaside town.

Moreover, it is deteriorating faster than ever before; its depopulation is leading to the loss of its social space, identity and memory. For this reason, Salvatore Settis, in his book *Se Venezia Muore* (Settis 2014), defines it as a 'tool to think', a laboratory for the destiny of historic and contemporary cities with an alarming perspective. As a result, this catastrophic scenario is the perfect vehicle for investigating the current instability of the couple's nature/culture in accordance with recent ecological thinking (Harman 2018). Venice can easily function as a tool for establishing a theoretical and critical framework that could serve as a starting point for future and broader research aiming to reimagine cities beyond nature and culture.

Based on four projects related to Venice, this initial framework aims to highlight various perspectives/outcomes of the nature–culture relationship within the city: nature/culture as a critical and thinking device; nature/culture as a vibrant matter; nature/culture as a social environment and nature/culture as a spatial transformation. The first deals with the concept of nature as urban criticism. The second deals with the concept of the city as a raw transformed matter. The third deals with the idea of the city as an event and spatial stage. The fourth and last one deals with the idea of the city and its continuity as a historical, environmental and morphological place.

17.3.1 Superstudio, Critical Device

In 1972, Superstudio presented a radical project, *Salvataggio dei centri storici italiani*, in which six city centres—Milan, Florence, Pisa, Rome, Naples and Venice—are radically reimagined (Mastrigli 2016). These projects are the result of a new awareness that grew from a natural catastrophe in Florence: in 1966 the River Arno flooded the city completely, covering a vast part of its territory and ruining most of its historic buildings, houses, libraries and museums.

More than the content of the proposals themselves, what is fascinating about these reinventions is Superstudio's way of looking at nature and its cultural transformation. Nature is not considered simply in terms of a rural landscape but in terms of its catastrophic effects, mainly caused by human interventions: the matters of concern as defined by Latour. Devastating floods and high levels of pollution become the base-line for each Superstudio proposal: Florence remains flooded, Milan is enclosed within a glass box full of polluted air and Venice's canals are covered with a glass floor to solve the problem of rising sea levels. In this sense, Superstudio critiques the concept of the city through these six projects.

The first element of this criticism prioritises social reform through new behavioural models, only afterwards making their radical architectural proposals. The second element is based on different design ideas, objects and projects, also fictional, to open up a new and architectural perspective not affected by the capitalist system.

17.3.2 John Ruskin, Vibrant Matter

Nature can be also interpreted by focusing on its materiality. The ways in which human beings transform natural resources correspond exactly with their cultural perspective. This can be affected by historical events, social conditions, economic strengths and different environments. Ruskin's book on Venice represents a perfect example of this approach. *In The Stones of Venice,* John Ruskin (1851), looks at the stones through different lenses, not just as a raw material but reading the meaning and the history that each stone evokes. Through a detailed study of the stones and their features, Ruskin gives an account of the history of Venice and how the city's sculptural–architectural features have changed, focusing on three main historical periods.

Through the text and drawings, he demonstrates the extent to which the Gothic represents an elaborate style, full of vitality and vibrant quality, in comparison with Greek and Roman buildings. This vital character is created by six main features, of which savageness and changefulness (variety) are the most meaningful. Savageness represents the rude and wild aspect of the Gothic style that resonates with its Nordic origin and with the intrinsic truth of the material. For Ruskin, this quality includes a changing variety made by the mineral component of the stone and by the imperfections of the details.

Changefulness expresses the idea of encompassing previous styles within the Gothic, modifying their architectural elements. In this way, Ruskin gives an accurate description of the Venetian arch as a result of the transition from the Byzantine to the Gothic style. Furthermore, changefulness is also the result of craftsmen's contributions; they brought their own lives into the Gothic building. The architectural elements are never equal; details are not replicable; so the idea of symmetry does not exist. As a result, each building becomes unique and rich in a vital energy that derives from this union between human activity, place and matter. Through this description, Venice's landscape resonates even more with these vital features. Unfortunately, Ruskin does not discuss another important aspect that is equally relevant in Venice. Stone is a vibrant matter not just for its quality and processing, but also by its reflection on the sea green water. He mentions this aspect briefly, describing the Renaissance buildings and how their static volumes are made vibrant by the water. In contrast, he attributes to the gothic an inner vitality independent of the presence of water.

In this regard, his drawings are more effective. Most of them are watercolours, left intentionally incomplete. Their paper support and pencil strokes are still visible. The watercolours cover only some parts of the drawings, leaving others naked.

Moreover, the use of the watercolour is the result of the union of matter and water that speaks of Venice. As a consequence, they transmit a sort of transparency and blurriness of the Venetian building. They become vibrant through the water of the Lagoon and the reflected light on it. These drawings differ from the architectural details, columns and capitals, which are often drawn pencil, zooming in on details such as the flowers and plants, defining a literal connection to the world of nature without mediation.

Looking at Ruskin's drawings, nature is a constant reference for his interest and curiosity. He used nature as a direct and indirect source of inspiration and at the same time as the entrance to understanding the world (see Fig. 17.1).

Fig. 17.1 John Ruskin, Saint Mark, Venice, 1846. (Source: John Ruskin, Public domain)

17.3.3 Ralph Rumney, Vibrant Dwelling Spaces

The urban space can be considered a vibrant dwelling matter as well: a sensual space. In fact, the void between urban fabric, shaped by the borderlines of facades and by their proximity and proportions, has a physical consistency that can become the stage for urban events.

The first example is Ralph Rumney's unplanned journey in Venice in 1957. Following the Situationist approach (Andreotti 1996), he takes a sequence of pictures trying to catch the real spirit of Venice. The images, composed as a movie, attempt to include not just visual fragments but the emotions and the behaviours of Venice's inhabitants. The urban space is discovered walking into the streets, bumping into new places and people, looking for a new way to look at the world and consequently cohabit on it. This experience refers to the concept of *derive*, defined by Debord and the Situationists. The existing city and especially urban streets are privileged spaces where unpredictable encounters can take place. They are unknown and obscure places, emphasising the contrasts and the mutable and experiential aspects of the urban space. Once these places are identified, the city is represented by a new psychogeographical map, composed by what they called the *unité d'ambience*. In doing so, the Situationists overcome the neutral, undefined, hygienic and idealised public space of the modern city, in order to get at the true spatial and social reality of the existing city.

In Venice, this approach seems to be particularly beneficial in revealing the city's peculiar history and its deep connection with its citizens since its first settlement. Its challenging environmental conditions have historically created a community that is deeply interconnected with the city and its surroundings. Both citizens and institutions have always understood the close relationship between their health and the integrity of the lagoon (Vanzan Marchini 2009). For instance, as Dennis Romano (1987) pointed out, maintaining the cleanliness of canals, streets and other common areas was vital for the city's survival and proper functioning. These practices were historically rooted in the habits of the first distinct island communities responsible for daily upkeep. Further, beyond defining citizens' behaviours, this peculiar relationship has profoundly shaped Venice's urban spaces and material culture over the centuries.

17.3.4 Le Corbusier, the Concept of Continuity

The urban fabric is not a static structure. It is the result of a continuous overlapping of architectural artefacts, affected by countless political, social and economic factors. For this reason, each new building is defined by this existing context and, consequently, will itself affect subsequent projects. This process defines an idea of continuity that has to be understood in terms of both time and space. Nowadays, this concept is commonly accepted, but it has only been developed as a theory in the last century.

In 1965, Le Corbusier finished his unbuilt hospital project in Venice. It represents both his legacy and his record of the osmosis principle between architecture and the existing city. In this sense, Venice Hospital is his last urban project profoundly developed from the site, in which nature is the design resource. This differs from the Ville Radieuse, where nature is just a way to guarantee the healthiness of new urban projects. Its architectural section demonstrates this contextual approach; the floating hospital platform is drawn like a ground, in continuity with the city that extends further towards the sea. It has the same level as the surrounding buildings, the same porosity and the same way of connecting the sky and the sea, through small courtyards and dense pilotis respectively. Equally, its blueprint demonstrates the same characteristics; a sequence of streets and connected courtyards recall the morphological structure of Venice with its *calle* and narrow passages. As a result, the new hospital becomes an in-between platform that mediates the city and the sea landscape.

This project is considered one of the most important examples of *mat-building* in the urban debate at the time. Alison Smithson's utilisation of this project as the primary frame of reference in her 1974 essay *How to recognize and read Mat-building* (Sarkis 2001) serves as a clear demonstration of this. Here she explains how *mat-building*, as a new approach to architecture and the city, could affect its inhabitants and their environment, establishing a deep human and social cohesion based on a strong phenomenological experience. In her essay, Alison Smithson described several other projects, including Van Eyck's explanation of the orphanage building, which was presented at the IX CIAM in Otterlo in 1959. Here, the Dutch architect clarifies some theoretical aspects crucial to understanding his work and architectural continuity ideas. Instead of focusing on objects, Van Eyck claims that social and spatial relationships should become the core issues of post-war architecture and urban design. For this reason, he defines the concept of in-between space as an architectural and urban device to reconcile conflicting polarities, such as inside/outside, which are frequently adopted by Modern architecture.

In-between space is the place where different interactions are mediated: through the doorstep at the architectural level or through the streets and squares at the urban scale. In this way, his orphanage defines a vital buffer against the city, created by a sequence of different connected squares. Some of them are open towards the outside, others are more closed and protected. In-between is not just a place but is also a time-space, where encounters could take place and dialogues can happen.

This essay also contains Smithsons' Urban Re-identification Grid, which was originally presented at the 1953 CIAM meeting in Aix en Provence. In this project, the town structure is defined as a hierarchical pattern-association of house, street, district and city. This pattern is described as a cluster, a word used to replace historical terms and to emphasise the relationship between its constituent parts. The idea of continuity is not achieved with historical forms but through the hierarchy of human associations through urban elements.

Looking at the participants' list of the Otterlo CIAM and their projects, it is evident that the main subject they have in common is the concept of continuity, albeit differently constructed. Smithsons and the other Dutch architects focus more on the idea of spatial continuity in order to create stronger social networks. The Italian architects, such as BBPR, with Ernesto N. Rogers, propose a concept of spatial continuity based on the history and morphology of the urban fabric. Showing the Velasca Tower project, Rogers exposes his first theoretical assumptions that will be later developed (Rogers 2006).

17.4 Conclusion

In this short exploration, I have sought to provide a preliminary critical framework that could help to inform the effects of recent critical ecological thoughts affirming the instability of the dialectic of nature/culture and the introduction of a new notion of nature on urban imaginaries. To illustrate these possible urban and architectural implications, I have proposed four Venice-related projects corresponding to four levels of understanding that paradigmatically represent the current instability of the couple nature/culture in accordance with recent ecological thought (Harman 2018).

Despite their differences, these levels share two common characteristics. They all belong to the history of architecture, reaffirming the validity of the historical discourse in contributing to ecological studies that are often limited to emergencies, future needs and technological solutions. More importantly, they highlight how the new relationship between nature and culture also subverts the traditional concept of boundary established by the modern paradox (Latour 1993). In this sense, through the lens of ecological thought, the concept of boundary should be completely revised and turned into a device to overcome divisions and increase equality between human and non-human entities, while simultaneously engaging concepts such as interobjectivity, symbiosis and coexistence.

To conclude, I would like to briefly discuss these four projects/levels to illustrate how within them a different understanding of the concept of boundary within the ecological framework discussed above can be defined. In this regard, the fourth is considered the most understandable. It illustrates well how boundary spaces can be transformed into in-between spaces, as theorised by Van Eyck, into places where people can share experiences, to welcome other users, including non-human ones (Latour 1993; Harman 2018). In this vein, the concept of boundaries can be explored and reframed at other levels. The first level of understanding pertains to the relevance of critical and theoretical thinking in activating awareness, sharing and the application of new philosophical perspectives to other disciplines. Superstudio's example has the peculiarity of uniting critical theoretical ideas of nascent ecological thought through ideas, words, images and radical projects within the architectural discipline to build new imaginaries. Here, the concept of boundary is not physical or spatial but rather it may be meant as a place, a capillary network of ideas/images that can be used to test and exercise the art of describing (Tsing 2015).

In a second level of understanding, Ruskin's work highlights the relevance of geological stratifications that raw matter and objects might contain, which deeply resound of human and non-human stories, ideas and practices. Even at this level, the concept of boundary is not physical or spatial, but may serve as a bridge to discover connections between humans and non-humans, especially those whose voices have been unheard (Nixon 2023).

The third level of understanding emphasises the deep bonds between urban spaces and citizens, which have weakened over the last century (Guattari 2000). Here, the concept of boundary could acquire a new meaning and become a place of care (Puig de la Bellacasa 2017) as a principle that may rejuvenate and strengthen these tides.

References

Andreotti L (1996) Theory of the dérive and other situationist writings on the city. ACTAR, Barcelona

Corner J (1999) Recovering landscape: essays in contemporary landscape architecture. Princeton Architectural Press, New York

Crouzet-Pavan E (2000) Toward an ecological understanding of the myth of Venice. In: Martin JJ, Romano D (eds) Venice reconsidered: the history and civilization of an Italian City-state, 1297–1797. Johns Hopkins University Press, Baltimore, pp 39–64

Descola P (2014) Beyond nature and culture. University of Chicago Press, Chicago

Guattari F (2000) The three ecologies. The Athlone Press, London

Harman G (2018) Object-oriented-ontology. Pelican, London

Latour B (1993) We have never been modern. Harvard University Press, Cambridge

Latour B (2017) Facing Gaia: eight lectures on the new climatic regime. Polity Press, Malden

Laugier M-A (1755) An essay on architecture. T. Osborne and Shipton, London

Mann T (1924) Death in Venice. Alfred Knopf, New York

Mastrigli G (2016) Superstudio: opere: 1966–1978. Quodlibet, Macerata

Nixon R (2023) Slow violence and the environmentalism of the poor. Harvard University Press, Cambridge

Poëte M (1958) La città antica. Introduzione all' urbanistica. Einaudi, Turin

Puig de la Bellacasa M (2017) Matters of care: speculative ethics in more than human worlds. University of Minnesota Press, Minneapolis

Rogers EN (2006) Gli elementi del fenomeno architettonico. Marinotti, Milan

Romano D (1987) Patricians and Popolani: the social foundations of the Venetian Renaissance State. Johns Hopkins University Press, Baltimore

Ruskin J (1851) The stones of Venice. Smith, Elder & Co, London

Rykwert J (1988) The anthropology of urban form in Rome, Italy and the ancient world. MIT Press, Cambridge

Sarkis H (ed) (2001) Le Corbusier's Venice Hospital and the mat building revival. Prestel, London

Settis S (2014) Se Venezia muore. Einaudi, Turin

Tafuri M (1976) Architecture and utopia: design and capitalist development. MIT Press, Cambridge

Tafuri M (1980) Theories and history of architecture. Granada, London

Tafuri M (1989) Venice and the Renaissance. MIT Press, Cambridge

Tsing A (2015) Aura's openings: unintentional design in the Anthropocene. In: AURA. More than human. Aura working papers. https://anthropocene.au.dk/fileadmin/Anthropocene/Vol_1_updated.pdf

Vanzan Marchini N-E (2009) Venezia civiltà anfibia. Cierre Edizioni, Verona
Vargic M (2018) The world–rising sea level. Detail of Italy. https://www.halcyonmaps.com/#/the-world-climate-change/
Wilson TM, Hastings D (eds) (2012) A companion to border studies. Blackwell Publishing Ltd, Hoboken

Chapter 18
Borderlands as Urban Systems: A Case Study of the Border Between Nicaragua and Costa Rica

Luis Alonso Pérez Monge

Abstract A border is an imaginary line defined by physical–political conditions as well as natural geographical conditions, some of which are perceived as conflict-ridden and problematic. The intention of this paper is to understand and read the borders as an urban and infrastructural system, analysed from its territorial, geographical, natural and connectivity potential. The border between Nicaragua and Costa Rica represents an ideal case for this analysis, as it presents the political and geographic characteristics to be interpreted and intervened as if it were an urban project. This border has great potential due to its strategic location, it contains a large amount of ecological resources and it is located next to the Panama Canal. However, it is an area with serious political, migratory and environmental conflicts. For this reason, the reading scale in which this border can be best understood is in the territorial-geographical one, the trans-municipal scale. Based on photographic and cartographic tools, the systems that make up the border system under study are identified, also evaluated from urban and architectural theoretical concepts to detect which operational logics should be worked on in order to identify guidelines and intervention strategies. Both countries intend to develop large-scale infrastructural connectivity projects individually, which would directly impact border dynamics; these projects contain duplicated infrastructure and incompatible management and logistics systems, whose inconsistencies may cause failures throughout the region. Through a cartographic analysis, the territorial operational logistics systems can be intervened to suggest an integrated urban–architectural intervention and solution, called 'Integral Cross-Border City'.

Keywords Borderlands · Planning · Central America · Bioregionalism · Territorial integration · Infrastructural system

L. A. Pérez Monge (✉)
Universidad Politécnica de Madrid, Madrid, Spain

Escuela de Arquitectura, Universidad de Costa Rica, San Pedro, Costa Rica
e-mail: luis.perez.monge@alumnos.upm.es, luisalpe@ucm.es, luis.perez_m@ucr.acr.cr

G. Cimadomo et al. (eds.), *Urbanism at Borders*, The Urban Book Series,
https://doi.org/10.1007/978-3-032-09594-7_18

18.1 Borders Understood as an 'Urban Project'

Addressing and intervening conflicting borders is a current global issue of interest, which has integrated academic interests, especially from politics and geography, but increasingly from urbanism, sociology and architecture (Cruz and Forman 2018). This objective has led to the formation and reform of relevant international organizations, such as the UN itself, the EU and UNESCO (European Commission 2020), in order to improve the quality of life between nations. Border studies reproduce relevant research topics today, through the use of methods of analysis, exploration and intervention that were not usual before. Relations between states require a trans-scalar reading, understanding the impact and implications of border conflict from a global to a local scale, seeking to understand how to intervene in their systems and improve relations.

For this reason, I question whether it is possible not only to conceive of borders as an urban project, but also to intervene in them as such—understood as multi-layered systems that shape territories through connectivity, encounter, and exchange to understand borders as an *urban project* but to intervene the territory. Each layer or system contains different levels of information, all of which converge and interact with each other to determine their function. Therefore, at the borders where systems have been disarticulated, they can be identified and intervened to transform them into platforms of encounter and exchange, understood as a macrostructure of polycentric networks whose relationships transcend boundaries and geometric conditions. In this sense, the Dutch architect Rem Koolhaas defines *new urbanism* (Koolhaas 1994: 961) as a new discipline with an interpretation that transcends the concept and paradigm of what we understand by city, understanding the concept of city as the space where exchange relationships develop.

> If there is to be a '*new urbanism*' it will not be based on the twin fantasies of order and omnipotence; it will be the staging of uncertainty; it will no longer be concerned with the arrangement of more or less permanent objects but with the irrigation of territories with potential; it will no longer aim for stable configurations but for the creation of enabling fields that accommodate processes that refuse to be crystallized into definitive form; it will no longer be about meticulous definition, the imposition of limits, but about expanding notions, denying boundaries, not about separating and identifying entities, but about discovering unnamable hybrids. (Koolhaas 1994: 969)

The spatial study of borders from architecture, landscape and urbanism has been a trend in the last 15 years. Proof of this is the attention they have received as a topic of discussion in international exhibitions such as the Chicago Biennial or the Venice Biennial (Cruz and Forman 2018). Focused on research projects for the mediation of territories and urban networks from regional policies, addressing spatial conflicts through spaces or infrastructures that promote exchange and multi-systemic relationships. These projects propose spatial interventions with social, economic and cultural impacts in specific areas, modifying the functioning of territorial systems through the strategic arrangement of *infrastructural bridges* for territorial articulation in natural areas and social and economic zones.

For a clear reading of all the systems that interact on the border territory, it is necessary to use graphic tools of urban analysis and architectural design, adequate for the manipulation of layers of information and data that reflect the actors and dynamics involved. Therefore, cartography as a graphic resource for research in observation and functional experimentation is indispensable. Cartography offers a control and registration of qualitative and quantitative variables of the information contained in border areas, ultimately generating a study process that superimposes information at all levels and scales (McHarg 1995); in order to intervene in the dynamics of functioning of border actors in a more objective and comprehensive manner.

In the middle of the twentieth century, attempts to approach architecture and urbanism projects on a large scale were not very successful, due to the great challenge and change in the academic and disciplinary paradigm that was involved. The territorial intention was to intervene in the logic of operation based on connectivity infrastructures and specific modifications, altering the dynamics of human relations at a trans-scale level. With this idea, some modifications were made in the configuration of the metropolitan areas, seeking to understand the interaction of the architecture and the infrastructures of the city. Two good examples of this process were the *Network Cities* proposal (Hilberseimer 1944) and the *Potteries Thinkbelt* project (Price 1964). In the same way, the geographer Edward Soja redefines what we understand today as contemporary urban space under the concept of 'Postmetropolis' (Soja 2000). Soja proposes to approach the city from a more complex perspective, understood from its multiple operations, and it is through this concept that socioeconomic issues and urban historiography are applied on a large scale to the territory from a holistic and global perspective. In this way, as the world population increases and contemporary social, economic and political challenges become more acute, they can and need to be addressed from a more holistic and multidisciplinary perspective, which requires a vision of the problem at various scales.

At the end of the twentieth century and the beginning of the twenty-first century, in addition to the concept of *Postmetropolis* (Soja 2000), the English urban planner Peter Hall proposed the concept of 'World City' (Hall 1997). This urban concept is a complex but balanced system that functions as a global network linked to the idea of territory. The characteristics of the previous concept are associated with what the geographer Claude Raffestins will determine as a contemporary model for urban–territorial analysis on a geographical scale (Raffestin 2012), noting that contemporary geography is the discipline that reveals the relationships between formal systems. From this, the concept of *infrastructural geography* (Herreros 2004) has been raised, understood as a new science that would explain the functioning of the physical support of our planet and all its relationships, both physical and virtual. When we say that our planet works or that it has to work, we are referring to a series of processes and simultaneous transformations of various scales and consequences that permanently strive to achieve a balance, and these relationships are driven and optimized by connectivity infrastructures.

This theoretical–conceptual evolution of the understanding of the city and its growth is directly related to the concept of *Global City* (Sassen 2001). This concept is a more precise and updated one, as it includes the dynamics that characterize the changes and behaviours of cities. It suggests a more contemporary look at how cities work through global economies from value and supply chains, understanding global functional pathologies from the influence that some cities have on others and the sense of competing and adapting their infrastructures to improve the connectivity between them. As a matter of fact, it has led to the blurring of some conflicting historical borders. With this, the previous theoretical discussion offers the opportunity to transfer this logic and operations to cross-border territories at a practical level, enabling interventions grounded in their territorial dynamics to rearticulate systems that have been historically fragmented by political boundaries.

18.2 The Border Between Nicaragua and Costa Rica as an 'Urban and Infrastructural Project'

In the American continent, there are several conflicting borders. These conflicts range from political–military, ecological or socio-economic due to uncontrolled migration, human trafficking and illegal trade. Currently, there are groups of professionals in geography, sociology and policy dedicated to these issues, making an impact through informal settlements and housing. One of these groups works in the border area between San Diego and Tijuana. The Teddy Cruz + Fonna Forman studio is working from the University of California in San Diego; the firm has analysed the development of border neighbourhoods and their cultural production, from which they rethink urban policy, the affordability of housing and the accessibility to public space, all addressed from a macro scale while taking account of the territorial system (Cruz and Forman 2018).

In the same way that this border strip in North America, Central America presents great opportunities for its strategic location on the continent and its narrow geographical configuration, which serves as a bridge between North and South America and between the Pacific and Atlantic oceans. It stands out for its strong agricultural development thanks to the fertility of the land, with gradual growth in tourism and socio-economic sectors. However, the area has serious conflicts that impede its social progress, such as the lack of security, drug trafficking, corruption, deteriorated transportation and communications infrastructure. In spite of the great potential presented by its proximity to the Panama Canal, there is great regional commercial competition. This issue could be considered an opportunity to generate infrastructure, policies and developments that work as a network to complement, replace and reinforce this activity where necessary.

The border between Nicaragua and Costa Rica is a 309 km linear border, of which 160 km correspond to a regular geometrical–political span that intersects forests, farms and small towns. The other 149 km correspond to the natural and irregular south edge of the San Juan River, whose territory belongs to Nicaragua. Though Costa Rica has international right of transit over the river waters, it is currently a very conflictive strip. The river is also a social and economic meeting point, representing an international node of passage for a constant number of people and exchange of goods within a region rich in agricultural, tourist, industrial and ecological resources, but very abandoned by both governments.

This border experiences a high rate of illegal and uncontrolled immigration, the most critical in all of Central America. The 80% of immigrants entering Costa Rica come from Nicaragua that represent around 470,367 of habitants, the 9% of the total population of Costa Rica in 2022 (INEC 2022). This phenomenon is due to the difficult economic and political situation facing Nicaragua and Honduras. In 2018, migration to Costa Rica increased by 15% due to political and social conflicts in Nicaragua. The border works as an open door for the passage of immigrants in both directions. Although the border has two official border control points, these do not have the adequate infrastructure conditions for their proper operation as the remaining part of the border is an open strip that is mostly land dedicated to agriculture, livestock or protected forests. Nevertheless, it is conceived as a bridge of opportunities, with the potential to develop infrastructure systems that can be part of a comprehensive multinational development plan that incorporates natural resources, thus contributing to the improvement of socioeconomic conditions in Central America.

Nicaragua and Costa Rica have much in common, both share traditions, customs and influences. However, due to political issues, they have different infrastructure development plans that do not take each other into account. These projects would affect the commercial exchange and industrial dynamics of the region, especially because the projects are very close to each other. This would lead one to think that, as they are countries with not very large territorial surfaces and having a population of 5.1 million inhabitants in the case of Costa Rica and 6.5 million in Nicaragua, developing a unique infrastructure project that improves connections between both countries, would avoid the risk of environmental impact, excess industrial demand and would avoid double construction investment.

Nicaragua intends to build a new interoceanic canal, taking advantage of the San Juan River and Lake Nicaragua, while Costa Rica in turn proposes a dry canal, both infrastructures are just separated by 40 km between them and with the border line in the middle. These projects involve similar approaches and investments in supporting infrastructure such as ports, airports and logistics hubs, which need to work together. The construction of both infrastructures together would not be sustainable since it would affect industrial logistics and the economic dynamics of the region. As they are projects with common approaches, they require high energy consumption with a risk of environmental impact; their parallel development would ultimately lead to the unsustainability of the region. The status of both projects is uncertain, and they have many uncertainties. This potential condition of a common infrastructural development due to its geographical condition is what characterizes this territorial border as the ideal case study to verify the hypothesis raised in this research project (see Fig. 18.1).

This borderland has two official border crossings, the main one is in Peñas Blancas, at the western end of the border near the Pacific Ocean; the second was established in 2015, in Las Tablillas, at the central point of the border strip, near the

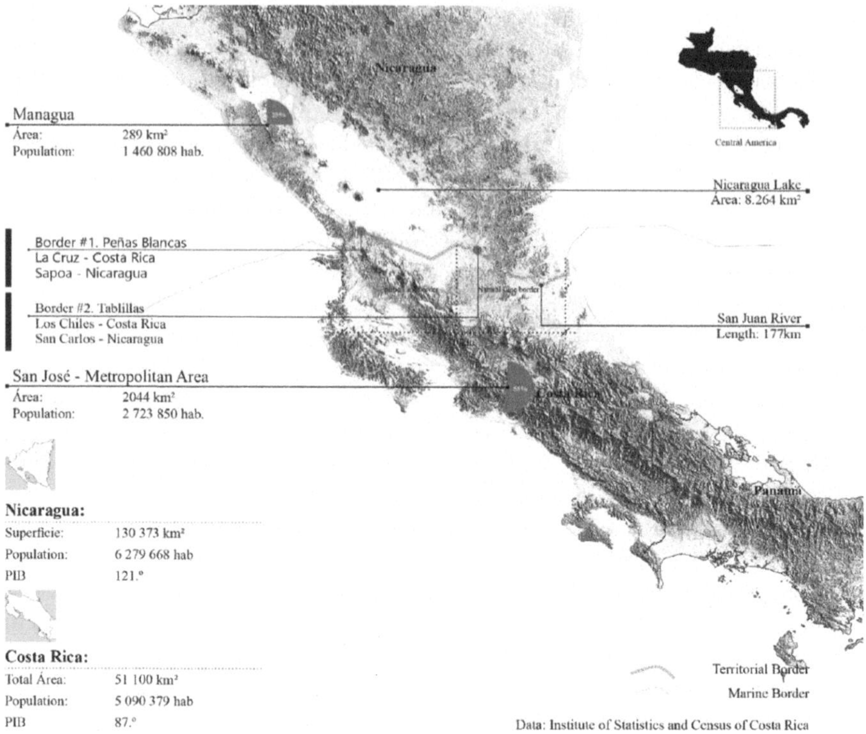

Fig. 18.1 Border between Costa Rica and Nicaragua. General data. (Source: the author, 2020)

sector that connects the San Juan River with the Lake of Nicaragua. The Peñas Blancas border cross is the oldest and has an infrastructure that reflects the economic stagnation in this area. The conditions of the current border infrastructures in both stations do not favour an adequate commercial development, nor do they have the adequate control and security platforms to prevent human trafficking and the passage of illegal goods. Therefore, it is necessary to address the dynamics of migration management in these points.

Both countries have experienced several conflictive encounters on the border since 2000 to the present. The most prominent was the case of '*Calero Island*', which lasted 5 years (2010–2015) and was originated by the invasion of the Nicaraguan army to Portillos Island, which is Costa Rican land. This event forced the Costa Rican government to denounce this fact before the International Court of Justice in The Hague, which ruled in favour of Costa Rica. Nicaragua had to remove the military camp from the Costa Rican island. As a result, Nicaragua was fined for the environmental damage caused and was ordered to suspend the dredging and cleaning process in this sector of the San Juan River, but after this event, the conflicts continued. This event is an example of how critical and frayed the relationship between these two countries is.

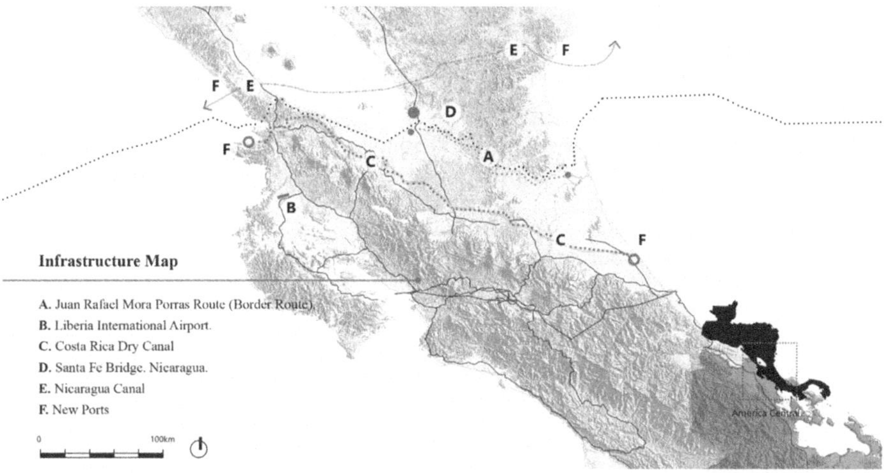

Fig. 18.2 Map of infrastructures proposed by Nicaragua and Costa Rica in the last 10 years. (Source: the author)

Although both governments have turned their backs on improving infrastructure at border posts, in the last 10 years they have planned and invested in large-scale infrastructure projects that have not only modified migratory flows and behaviours, but have also improved aspects such as accessibility and communication, energy services, the local economy and agricultural production. The projects with the greatest impact raised by Nicaragua in the last 10 years are as follows (see Fig. 18.2):

Santa Fe Bridge: It is located north of the San Juan River, just 10 km from the border, built thanks to a financial collaboration with the Japanese International Cooperation Agency. It is the longest bridge in Nicaragua measuring 365 m long and 12 m wide. The bridge was completed in August 2014 and crosses the San Juan River in the San Carlos municipality, at the centre of the border strip, with the intention that it be consolidated as a Central American corridor linked to the Dry Canal proposed by Costa Rica.

The Nicaragua Interoceanic Canal: It is the most ambitious and controversial project in the Central American Region. Currently, information on its status has not been made available by the Nicaraguan government. The costs of the construction and the administration would be in the hands of a Chinese private company specialized in the construction of infrastructures that support large-scale transportation projects. This canal would connect the Caribbean to the Pacific Ocean starting from the Punta Gorda River to Lake Nicaragua, then crossing the Rivas isthmus to finally reach the Pacific Ocean, covering an area of 270 square km. The Canal plan includes five floodgates of a larger dimension and with greater capacity than the new Panama Canal floodgates.

Two Mega-Ports: The first port will be located in Brito on the Pacific coast, 24 km from Rivas, a Nicaraguan city that is close to the border. The second port will be located in Punta Águila, in the Autonomous Region of the South Caribbean

Coast. Additionally, a free trade zone is proposed, incorporating a technological logistics centre and a financial and urban hub for 140,000 inhabitants, projected to generate more than 130,000 jobs and an annual output of $27,000 million by 2030. In addition, an international airport is planned in Rivas, with the capacity for one million passengers and 22,000 tonnes of cargo per year. This airport would be of similar scale to that of the existing airport in Liberia, 90 km away.

The infrastructure projects with the greatest impact, proposed by the Costa Rican government near the border sector during the last 10 years, are (see Fig. 18.2) as follows:

Juan Rafael Mora Porras National Route 1856: This border road consists of two phases and aims to create a road parallel to the entire border, providing access and opportunities to communities that are in this northern Costa Rican sector of difficult access and communication. Only the first section has been elaborated and inaugurated in 2012, running 160 km along the Costa Rican margin of the San Juan River. The Costa Rican government accelerated the construction process near the San Juan River sector, due to the tensions that arose from the conflict with the Nicaraguan government between 2000 and 2018.

Expansion of Daniel Oduber International Airport: Due to its capacity, this is the second most important international airport in Costa Rica. It is in Guanacaste, 90 km from the Peñas Blancas border cross point, counting with the greatest number of tourists and merchandise passes. The expansion project aimed to increase the terminal and its capacity to serve 1.5 million passengers per year.

Costa Rica Dry Canal: It is a project seen as very necessary due to the amount of heavy transport that passes through the centre of the metropolitan area. The creation of a dry canal would improve and provide the opportunity to expand the commercial and industrial sector of the Central America region. The northern sector of Costa Rica has the ideal geographical conditions to develop this infrastructure whose scope would include 325 km with 6 highway lines and 2 railway trains. This proposed system would connect the Costa Rican Caribbean sector from Puerto Limón to Santa Elena Bay in the North Pacific where there is consideration to develop another port 30 km from the Nicaraguan border. Environmental and financial studies are currently being carried out to commence the work.

18.3 Re-Reading the Border Through Cartography

Cartographies, maps, diagrams and graphics are a necessary resource for the representation of our ideas; additionally, it is an analytical method that helps to identify and recognize territorial behaviours, functional systems and the various layers of information that converge in the studied field, in this case, a fragmented territory, a border. This method, with its ability to identify scales, proportions and valuations, enables both qualitative and quantitative applications.

This research has as reference the works and methods of Scottish architect and landscape architect Ian McHarg explained in his publication *Design with the Nature* (McHarg 1995) that show the ideal use of cartography as a method of scientific analysis, known as geographic information systems, which through the intervention of the necessary territorial functional layers, relations and exchange can be optimized, reducing risks and negative impacts. Maps are not only developed to help identify and understand the multiple territorial systems that make up and interact on the border but also to intervene in them.

Another methodological reference is the book, *Taking Measures Across the American Landscape* (Corner and MacLean 2000). The author not only uses cartographic representation techniques along with aerial photography but also achieves an adequate graphic management of the information. He processes data on the territory and he perfectly explains its pathologies which are based on understanding acts on the territory. With the above, the three levels of analysis developed for this study are detailed below.

Urban and infrastructural conditions: It is a support layer that works on different scales, which manages and irrigates the territory to improve the human relationships, accessibility to resources, services, consumer products and waste management. It is related to terms such as scale, logistics, field, network and system. It shows the activity and production nodes and how they are linked through connectivity networks.

Geographical and territorial conditions: It represents the field conditions (Allen 1999), the tabula rasa (Koolhaas 1994: 1008–1089). This concept corresponds to the study and understanding of modern methods and theories that were developed from the 70s to the present, which have been applied or adapted for the intervention of the systems involved in the border. They are theoretical methods to conceive and understand the city, associated with models directly related to economic exchange, political relations, mobilization and transportation.

Border and political conditions: This manages the relationship, the encounter and the exchange capacity. Borders have been understood as a geopolitical line that contains a series of dynamics, forming part of a network of relationships and exchanges of different dimensions and directions. These are natural relationships linked to territory and geography, as well as physical and political relationships that spatially order the territory.

18.3.1 Analysis: Identification of the Main Layers on the Border

The theoretical concepts represented by each map manifest a system, a layer of information or certain dynamics that interact at the border, all of them are protagonists. These are concepts that are also referenced to theoretical analysis. The base information of the layer is represented in grey scale; the existing systems,

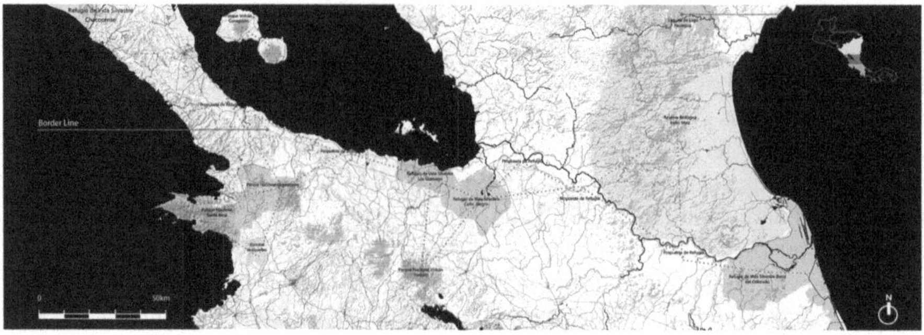

Fig. 18.3 Map 1: Support layer. (Source: the author, 2020)

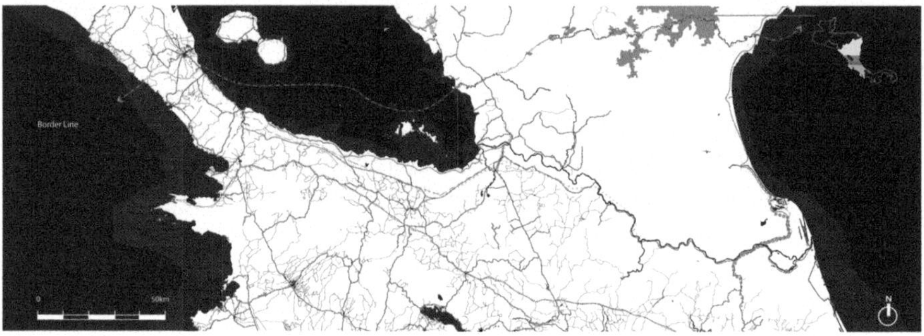

Fig. 18.4 Map 2: Networks layer. (Source: Author, 2020)

relationships or networks are shown in green and in purple those that are intended to be generated or proposed.

Map 1: *Support layer* (see Fig. 18.3). Support is an understanding of the dynamics of movement and the flexibility of relationships (Allen 1999). Support is also a tool and method for intervention through which a territory can be modified. Support takes into account all the relationships that may be possible in a topological field, where the work and infrastructure are essentially the organizational platform of this field.

Map 2: *Networks layer* (see Fig. 18.4). Technologies and mobility networks are the systems that serve as the basis for design in contemporary urbanism (Varnelis 2011). Relationships and connections are the origin of these links, owing to their necessity for communication and connectivity that is tied to the life and structure of cities.

Map 3: *Topology layer* (see Fig. 18.5). Interactions, relationships and continuous transformations (Woods 1970). It manifests the intensity and hierarchy of the influence of relationships. The ramifications exemplify the influence that the border receives, where it comes from and with what force it follows the behaviour of the territory, taking into account the size, shape, geometry, scale and its logic.

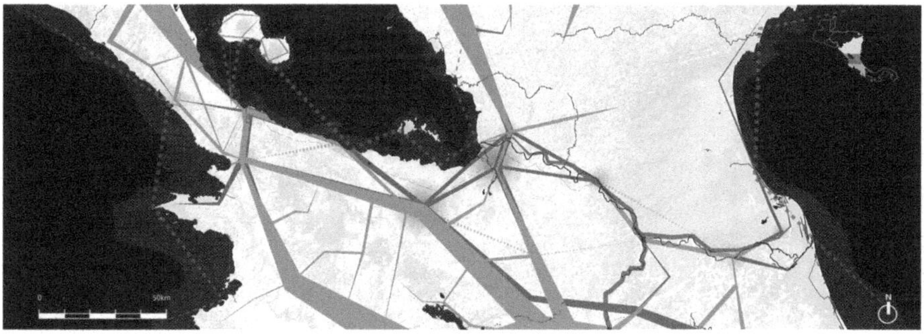

Fig. 18.5 Map 3: Topology layer. (Source: the author, 2020)

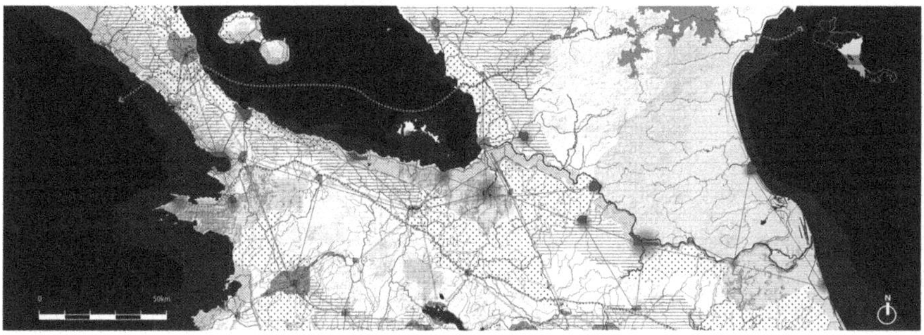

Fig. 18.6 Map 4: Polycentric system. (Source: the author, 2020)

Map 4: *Polycentric system* (see Fig. 18.6). Polycentric systems as a set of intertwined action nodes that are related to each other under a specific function (Hall 1997). These nodes have particular activities, which define the vocations of each sector. It links the concepts of convergence, connections, interactions, reciprocity, complementary interdependence, reciprocity and exchanges, a network within a macro network.

Map 5: *Porosity capacity* (see Fig. 18.7). Porosity is the relationship between the volume dedicated to the flows and the total area of intervention (Viganò 2009). To know and interpret porosity and design it means to pay attention to the practices, changes, fractures in space, urban materials and the availability and possibilities of new flows.

Map 6: *Continuity relationship* (see Fig. 18.8). Continuity is associated with ties, connections, links and relationships (Price 1964). This is the expression through which one can address the paradigm that is the border between two nations. Continuity also refers to the natural territorial codependency and geographical continuity since natural conditions and systems do not conceive boundaries. Continuity aims to find the common value to weave the space.

Map 7: *Exchanges layer* (see Fig. 18.9). Interchange adapts to new situations, to various action scenarios, to complex inclusive systems without prejudice (Cruz and

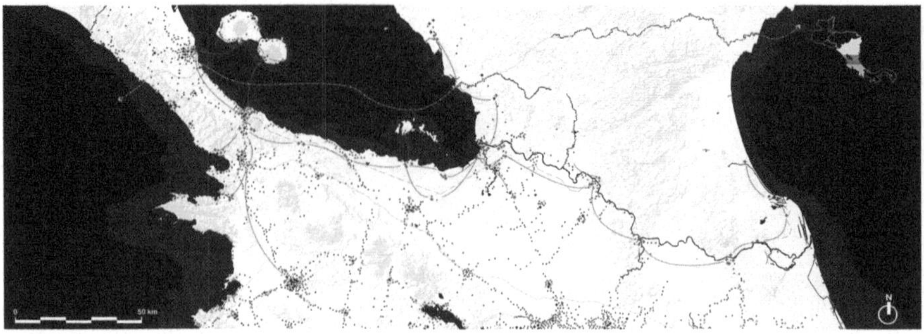

Fig. 18.7 Map 5: Porosity capacity. (Source: the author, 2020)

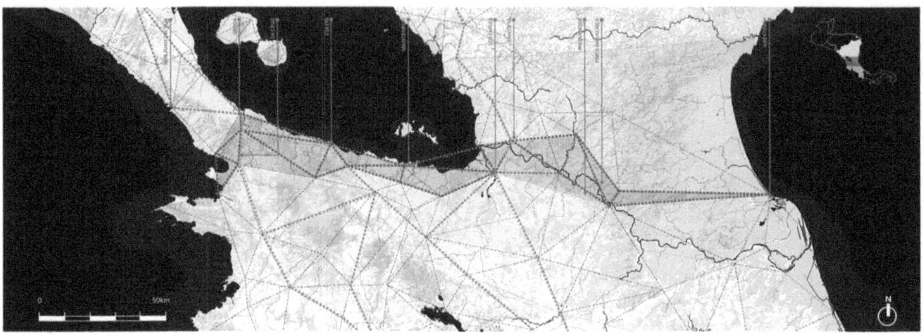

Fig. 18.8 Map 6: Continuity relationship. (Source: the author, 2020)

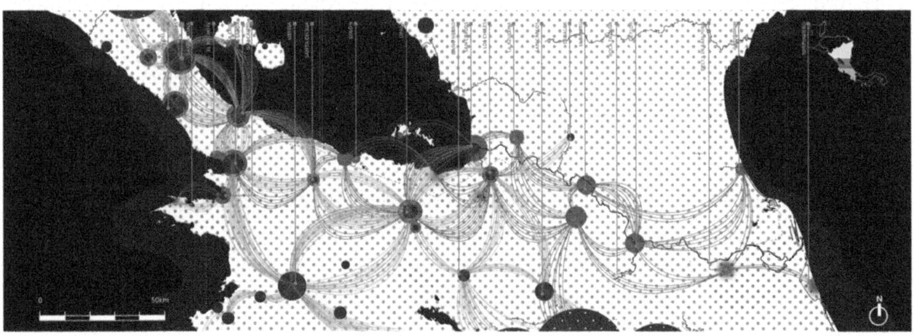

Fig. 18.9 Map 7: Exchanges layer. (Source: the author, 2020)

Forman 2018). The dynamic of these systems is inclusive, and they seek to improve connections. There is a mutual benefit between its parts that allows the flexibility of its form in space; they occur spontaneously, with no specific scale, and they adapt to the intensity or size presented.

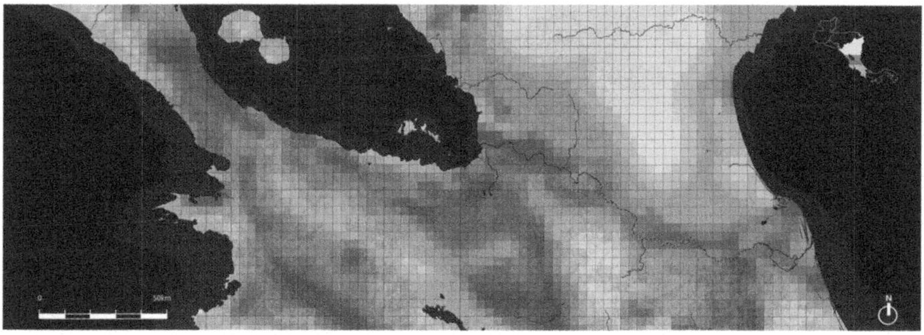

Fig. 18.10 Map 8: Perceptual notion. (Source: the author, 2020)

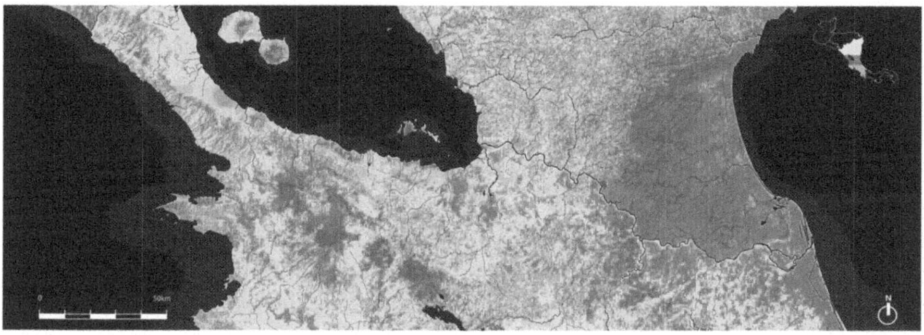

Fig. 18.11 Map 9: Environmental systems. (Source: the author, 2020)

Map 8: *Perceptual notion* (see Fig. 18.10). Ties and links are also perceptual; they create a common connection factor based on the lived experiences by those that inhabit or go through it, making and imagining the site through diverse and independent criteria (Foucault and Moore 2007). It is linked to the way the brain detects and perceives sensations.

Map 9: *Environmental systems* (see Fig. 18.11). The natural structures interact with the territorial interventions and the infrastructures, and together they are part of the most complex mega-ecosystem that humans have created (Bélanger 2016). The results are part of a long and collective evolutionary process that gathers all the necessary so that the humans can inhabit in balance with nature and its conditions.

Map 10 Economic exchanges (see Fig. 18.12). The economies in the territory are linked to agricultural production and the administrative capacity for it (Herreros 2004). These are subject to the availability and dynamics of labour and the supporting transportation logistics. Understanding how these are arranged on the territory helps us understand distribution and how it could generate new dynamics and vocations.

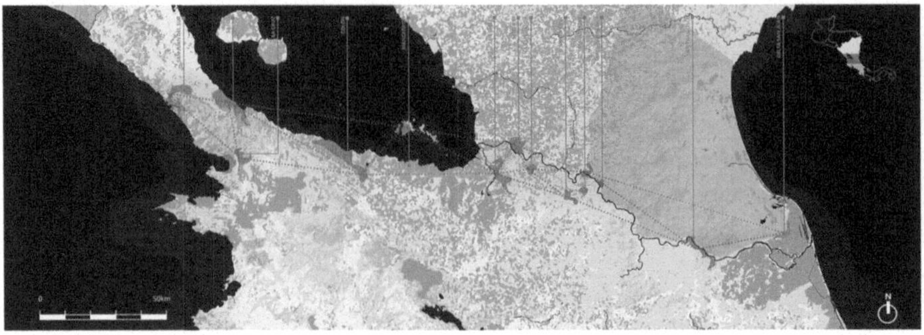

Fig. 18.12 Map 10: Economic exchanges. (Source: the author, 2020)

18.4 A Blurry Borderline

The border between Nicaragua and Costa Rica is the subject of this study not only because of its strategic location in the American continent but also because it has the geographical, political, economic, social, ecological and infrastructural conditions to be understood and intervened as an urban system. Its territorial scale allows it to be subjected to a process of observation and study in its entirety through the use of cartographic drawing, considering the global economic and political systems that affect it, and the more urban scales where human exchanges take place. Therefore, a multiscale analysis of border systems allows the control of their entire longitudinal dimension to control and superimpose factors and layers of information (see Fig. 18.13).

This allows the development of a series of maps created with concepts and methodologies used for urban analysis. These manage to reflect and decipher the different dynamics that intervene in the border, including the discovery of disconnected microsystems, potential points of intervention, new vocations and potential connectors that would improve the conditions of mobility and logistics throughout the territory. In order to achieve an integral dynamic that complements and includes the interests of both countries, these will be introduced through contemporary international legal practices that allow the creation of new political and geographic figures. These will help to blur the boundary, distorting the notion of border, promoting a sense of belonging of its inhabitants. The ten cartographic analysis carried out, adapted and applied concepts from the study of the city to the border through its three conditions. They also manage to confirm the geographical capacity of the border as a support structure for the improvement of logistics along the strip, and potential networks associated with infrastructures are discovered.

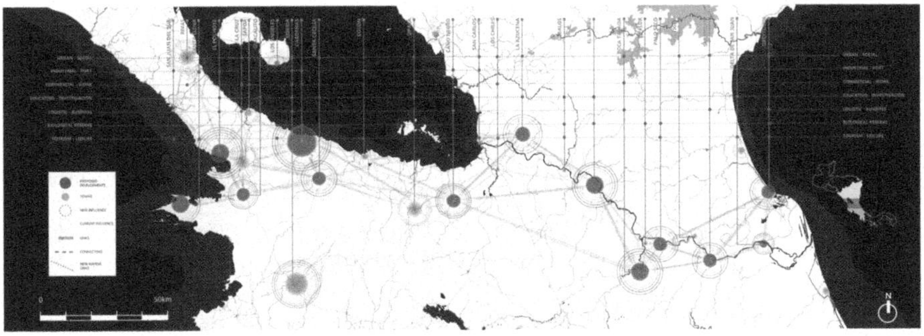

Fig. 18.13 Intervention plan, first intention. (Source: the author, 2020)

18.5 Reflections and Conclusions

An excellent reference from urbanism is the 'Delta Metropool' (Koolhaas and Harvard Project on the City 2001), a project proposed since 1960, which intervenes on an international, regional and urban scale, modifying the dynamics of some borders near the Netherlands. This project connects cities such as Amsterdam, Haarlem, Utrecht, Delft, The Hague, Leiden, Dordrecht together with the port of Rotterdam and Schiphol airport, and other cities outside the Netherlands. It proposes the development of better logistical connections with adequate infrastructure to control urban growth while maintaining a natural or green centre. Although the Netherlands is strategically located in Europe, it lacks land, and this project provided an urgent territorial and geographical strategy to provide adequate land for housing, workspaces, leisure, infrastructure, natural landscape and land for agriculture in order to avoid conflicts in the common space and also environmental conflicts. This successful project is also a reference for having a similar scale to the study border.

Some conclusive reflections are that the strategies of planning and analysis of urban projects, used in more metropolitan scales, can be adapted and used to analyse the territorial dynamics that are developed on a large scale. For this, the cartographic resource remains an indispensable research tool, and is even taken further to be used both qualitatively and quantitatively. The various systems or leading layers that interfere in a territory must be analysed separately, in order to understand which are the disconnections or those sectors that can be better focused, in order to promote appropriate networks and ensure that a fragmented territory such as the border can be articulated.

It is necessary to face the spatial paradigms and challenges of the twenty-first century. It is urgent to propose integral and cooperative solutions between states, with this, Nicaragua and Costa Rica are presented with a great platform and opportunity to carry out a joint investment process, generating economic and environmental synergies, proposing a unique cross-border integration project. The idea of blurring and erasing the boundaries may seem utopian, but it is a challenging idea, where innovative solutions emerge and exemplify the collaborative power of humanity.

References

Allen S (1999) Points + lines. Diagrams and projects for the city. Princeton Architectural Press, New York

Bélanger P (2016) Landscape as infrastructure: a base primer, 1st edn. Routledge, London

Corner J, MacLean A (2000) Taking measures across the American landscape. Yale University Press, New Haven

Cruz T, Forman F (2018) Mexus, a geography of interdependence. In: Dimensions of citizenship. United States Pavilion. 16th Biennale di Venezia. Inventory Press LLC, Los Angeles, pp 106–111

European Commission (2020) Interreg Europe Program. EU Regional Council Nord-Pas de Calais, Lille

Foucault M, Moore G (2007) The force of flight in space, knowledge and power. In: Crapton JW (ed) Foucault and geography, 1st edn. Routledge, London

Hall P (1997) Megacities, world cities and global cities. The mega cities foundation. The First Megacities Lecture, Rotterdam

Herreros J (2004) Geografía Infraestructural. In: Cambio + Energía + Información: Palacios de la diversión en la Isla San Miguel. Departamento de Proyectos Arquitectónicos. ETSAM, Madrid

Hilberseimer L (1944) The new city. Principles of planning. Paul Theobald, Chicago

INEC (2022) Costa Rica en cifras. Instituto Nacional de Estadística y Censo, San Jose

Koolhaas R (1994) S, M, L, XL, 2nd edn. The Monacelli Press, New York

Koolhaas R, Harvard Project on the City (2001) Mutations. Actar, Barcelona, pp 280–337

McHarg I (1995) Design with nature (25th Anniversary Edition). Wiley, New York

Price C (1964) Potteries thinkbelt. Conceived for North Staffordshire, England

Raffestin C (2012) Could Foucault have revolutionized geography? In: Crampton JW, Elden S (eds) Space, knowledge and power: Foucault and geography. Ashgate Publishing, London, p 129

Sassen S (2001) The global city. New York, London, Tokyo. Princeton University Press, New Jersey

Soja E (2000) Postmetropolis: critical studies of cities and regions. Wiley, New York

Varnelis K (2011) Campos Infraestructurales. Quaderns d'Arquitectura i Urbanisme, vol 261

Viganò P (2009) The metropolis of the twenty–first century. OASE, vol 80

Woods S (1970) What U can do. Architecture at Rice, Houston

Part V
Inequities of Neoliberal Policies

Chapter 19
Unchartered Boundaries: The Territoriality of Everyday Life

Luca Gaeta

Abstract A misunderstanding permeates current discussions on boundaries. This is the deeply rooted idea according to which boundaries are a hindrance to movement. Often portrayed as barriers and walls or as material artefacts purposely designed to impede the free movement of people and goods, boundaries are instead intrinsic to movement. They can be better understood by considering them as the settings of everyday life or even as tracks left by the comings and goings of people. By introducing the concept of *practice on the move*, within the framework of Bourdieu's theory of practice, this chapter argues that boundaries are integral to social motion. The latter statement applies particularly to administrative boundaries. Territories are divided into districts to secure general welfare or enforce social control. However, the administrative layers of spatial ordering are built upon the customary territoriality of people. As life passes by, administrative boundaries and daily motions inevitably diverge. Unchartered boundaries result from unpredictable variations in motion patterns and routines. The main challenge for experts is first recognising the new patterns, and then reconciling boundaries with everyday life's mobilities. This task is more easily achieved with the reuse and resignification of existing boundaries rather than the forging of new ones. Boundary makers should work as bricoleurs to take advantage of social and behavioural resources.

Keywords Boundaries · Everyday life · Thresholds · Governance · Territoriality · Socio-spatial constructs

L. Gaeta (✉)
Department of Architecture and Urban Studies, Politecnico di Milano, Milan, Italy
e-mail: luca.gaeta@polimi.it

G. Cimadomo et al. (eds.), *Urbanism at Borders*, The Urban Book Series,
https://doi.org/10.1007/978-3-032-09594-7_19

19.1 Introduction

Boundaries are not destined to disappear in the near future. The theorists of a borderless world (Ohmae 1990; Miyoshi 1993; Clark 2000) have rejoiced too soon after the demise of the Iron Curtain and the subsequent triumph of globalism. The same holds with the heralds of an Orwellian world in which boundaries and surveillance would be everywhere (and nowhere) in the body of nation-states.

Based on geographical information collected since the unification of Germany and the collapse of the Soviet Union, Foucher (1991, 2016) argues that borders are proliferating worldwide both in number and length. Indeed, new treaties are signed to delimit the sovereignty of adjoining states; borders already delimited by a treaty are demarcated on the ground; contested borders and truce lines are fought over; walls and barriers are built to reinforce international frontiers.

While international boundaries are intensively researched by political geographers and cultural anthropologists (Wastl-Walter 2011; Wilson and Donnan 2012), significantly less attention has been given to so-called internal boundaries (Prescott 1987) and their effects on local communities. Although much closer to the settings of everyday life, internal boundaries are eclipsed by the emphasis on massive migrations and the international division of labour (Mezzadra and Neilson 2013).

This chapter addresses how internal boundaries relate to everyday practices and mobilities of citizens. Everyday practices on the move can provide a promising point of departure towards a renovated theory of boundaries. This chapter builds on Bourdieu's theory of practice and recent theories of social motion to argue that boundaries are intrinsic to motion rather than detrimental to it. This approach challenges the common belief that boundaries hinder motion as far as everyday life settings are concerned.

This chapter moves progressively from international to internal administrative space, chasing the boundaries of everyday life. In this pursuit, people's motion is viewed as a primary source of territoriality, and social motion is taken as a principle of boundary-making. The administrative division of space, namely its regionalisation, can be viewed accordingly as based on the customary territoriality of citizens. As an endless source of conflict and reconciliation, the interplay between the customary and the institutional territorialities actively shapes boundary landscapes at various scales ranging from neighbourhood to region.

The concept of unchartered boundaries puts forward the spatial agency of citizens as the driver of territorial arrangements that administrative power redoubles. This dynamic is the opposite of viewing administrative boundaries as straitjackets, that is, arbitrarily posed to constrain social behaviour.

19.2 Migration, Motion and Borders

In a well-known article published 1 year after the collapse of the Berlin Wall, almost certainly without a direct relation to this historical event, Gupta and Ferguson wrote:

> Representations of space in the social sciences are remarkably dependent on images of break, rupture, and disjunction. The distinctiveness of societies, nations, and cultures is based upon a seemingly unproblematic division of space. (Gupta and Ferguson 1992: 6)

The two ethnographers reject the opinion that contact and conflict between societies presuppose spatial discontinuity. Instead, their primary purpose is 'rethinking difference through connection' (Gupta and Ferguson 1992: 8). By positing connection and exchange as the basis of identity and otherness, the neat concept of border turns into the much more nuanced concept of borderland, defined as 'an interstitial zone of displacement and deterritorialisation that shapes the identity of the hybridised subject' (Gupta and Ferguson 1992: 18). Images of melting and nuance replace traditional images of breakage and rupture to the point that the concept of diaspora comes to the fore to express a deterritorialisation process:

> In a world of diaspora, transnational culture flows, and mass movements of populations, old-fashioned attempts to map the globe as a set of culture regions or homelands are bewildered by a dazzling array of postcolonial simulacra. (Gupta and Ferguson 1992: 10)

A migration-led process of deterritorialisation (Papastergiadis 2000) is envisaged which would undermine existing spatial divisions. Areas of spatial discontinuity cease to be thought of as empty and are instead observed as generating new cultural identities. According to Balibar (1998, 2009), Europe, the birthplace of the modern territorial State, is transforming into a borderland where citizenship no longer relates to national space as closely as in the past. On the other hand, border policing is organised to chase undocumented migrants wherever they are, inland as well as overseas.

According to recent estimates, international migrants worldwide account for 3.6% of the world's population (United Nations. Department of Economic and Social Affairs 2020). When read inversely, this percentage shows that the overwhelming majority of people on Earth stay in the country where they are born, often in the same administrative district. Boundaries are part of people's identity in most cases. As the World Migration Report 2022 makes clear:

> The great majority of people do not migrate across borders; much larger numbers migrate within countries, although we have seen this slow over the past two years as COVID-19 related immobility has gripped communities everywhere. (McAuliffe and Triandafyllidou 2021: xii)

Discourses of diaspora in a borderland world are clearly political auspices based on biassed views of borders as detrimental to free movement. In a self-contradictory vein, as Massey (2005: 86) remarks, unbound space is invoked for people on the one hand, while stronger borders are invoked for capital flows on the other. The result is 'a geography of borderlessness and mobility, and a geography of border discipline; two completely antinomic geographical imaginations of global space, which are called upon in turn'. The issue of mobility is unquestionably at stake in debates over boundaries, but my argument here is that there is no need to conflate mobility with borderlessness. Exploring the nexus between bordering practices and mobilities, Paasi et al. (2019) rightly ask for whom borderless worlds are intended or claimed. Perhaps, not for those who spend most of their lives in the same region or country.

In his theory of the border, derived from parallel studies of migration across Mexico and United States, Nail (2016) recognises control of social motion as the main reason for the existence of borders. Nail's theory links social motion to borders through the concept of bifurcation. Rather than being a physical barrier, the border is 'an active process of bifurcation that does not simply divide once and for all, but continuously redirects flows of people and things across or away from itself' (Nail 2016: 4). By this bifurcation process, some of the flows freely pass through the border, while others are diverted into loops, which cyclically return upon themselves. Accordingly, 'the border cannot be properly understood in terms of inclusion and exclusion, but only by circulation' (Nail 2016: 7). Nail's theory of the border is a social theory based on the principle of motion, since 'society and space do not pre-exist the delimitation of mobile flows' (Nail 2016: 9).

Nail's promising ideas, however, are not fulfilled by his representation of the border as a material artefact: that is, something extrinsic to social motion and purposely designed to bifurcate it. Social motion is divided without being itself the generative principle of such a division. But this is contrary to the initial assumption of mobility as pre-existing society and space. A possible way to prevent this contradiction is relating motion to the concept of practice more consistently.

19.3 Practices on the Move

To address the relationship between boundaries and social motion I refer to what political geographer Anssi Paasi calls boundary-producing practices, drawing from authors like Ashley (1987) writing in the field of international relations. By this phrase, it is first assumed that boundaries are 'social, political and discursive constructs, not just static naturalised categories located between states' (Newman and Paasi 1998: 187). Far from being natural entities, boundaries are produced through a wide variety of practices, local and national, institutional and popular, discursive and non-discursive. Second, it is implied that boundaries are always in the making as practices are intrinsically recurring. Third, boundaries are not only or exclusively material artefacts located at the edges of territories but also social constructs that exist 'all over territories, in innumerable societal practices and discourses' (Paasi 2009: 215).

The concept of practice is well established nowadays in the lexicon of border studies, which are structured around the cooperation of geographers and anthropologists in the research on international borders, control of the flows that cross them and the hybrid cultures that flourish in their proximity. Practice, though, is a concept initially developed by Bourdieu (1977) in his ethnographic research. Since then, the concept has become common in the field of both natural and human sciences. Despite its various meanings, the concept exhibits an unquestionable transdisciplinary relevance, which has led some scholars to speculate about a practice turn in contemporary theory (Schatzki and Knorr Cetina 2001).

Practice is a common way of acting, a response pattern corresponding to recognisable situations in everyday life. According to Bourdieu, the correspondence of practices to situations is far from mechanical as the instinctive reply to stimuli can be. Thus, a practice is never limited to performing a role under a standard rule but requires the latter to adapt to the situation. Agents are not entirely supine to practices. Furthermore, practices have neither an objective nor a subjective character. One cannot say objectively what a practice is because it always progresses. It is not even possible to say subjectively what a practice is because this would be done retroactively. The agent who is immersed in a practice reports on things already done. Accordingly, his description of his practice is just another kind of practice. As Rouse (2007: 500) writes, this fact explains why 'some theorists [...] make central to their discussion of practices those aspects of human activity they regard as tacit and perhaps even inexpressible in language'. Within such a broad concept that holds for discursive and non-discursive practices, this chapter pays specific attention to routine practices incorporating motion. I shall call them practices on the move.

Hereafter, this chapter's focus shifts from international to internal boundaries. The problem is not that international boundaries lack routine for those who live and work in their proximity (Andersen et al. 2012), but the border, however, is remote from the everyday practices of most citizens. More often, people experience administrative boundaries. According to Newman and Paasi (1998: 197), 'administrative boundaries have a far greater impact on the daily behavioural patterns of most individuals than do national and international boundaries'.

In studying internal boundaries, the routine practices to be chosen as a reference are those which lead people away from home. Not simply practices repeated on a regular basis, but those which compel people to go somewhere to do everyday chores. The back-and-forth movement is integral to practices such as going to school. With little children, for example, it is customary to ask if they already go to school when one wants to know if they have reached school age. In asking this, the act of going to school conflates with the practice of learning to read and write. To do what they usually do, either daily or weekly, people go somewhere and return home. This practice on the move doubles back on itself many times, thus shaping a circuit or horizon.

I contend that internal boundaries originate from routine practices on the move, that is, not from lines drawn by politicians and bureaucrats. Boundaries are everyday life's horizons intrinsically fused with practices on the move: not something that intervenes from the outside to bifurcate motion, but something integral to people's regular motion patterns. Therefore, we can understand a boundary as a field of routine action when we leave its social and material determinations in a state of non-distinction. This chapter then moves away from viewing boundaries simply as material artefacts or social processes. One advantage, among others, when setting aside the binary separation of the material from the social world is the opportunity to recognise that boundaries are indistinctly material and social, embedded as they are in routine practices. This concept of boundary is adequate to Nail's basic principle of motion. It means that boundaries without practices make no sense. Once abandoned by lively routine practices, boundaries regress to the state of material artefacts.

The following section connects practices on the move to administrative boundaries through the geographical concept of human territoriality.

19.4 Disjointed Territorialities

The fields of practices on the move are boundaries of daily life, generating their own spatiality and influencing the spatiality of others. The motion patterns of people form an indistinctly material and social field day by day. This field is more than a passive backdrop: in fact, it is intrinsic to practices on the move and is intersubjective insofar as we act in the presence of others who observe, act and react. This spatiality may be intended as a tactic to control everyday life's environment, whereby control provides people with a sense of ontological security (Giddens 1984). As such, it can be viewed as a form of customary territoriality.

This territoriality of the common people differs from deliberate institutional territoriality, although both can be defined as 'the attempt to affect, influence, or control actions and interactions (of people, things, and relationships) by asserting and attempting to enforce control over a geographic area' (Sack 1983: 55). Based on the above-mentioned distinction, I shall contend that administrative boundaries can be viewed as the interplay between two territorialities, the customary and the institutional, each attempting to enforce control over geographic areas, albeit for different reasons and with different means.

The customary territoriality of people by no means lacks boundaries and disjunctions. Instead, it continuously produces boundaries as the horizons of routinised social practices. I contend that administrative power builds on customary territorialities to lay out its intentional territoriality. Administrative boundaries, therefore, can be reinterpreted as doubling and retracing of routine motions sedimented in space. Instead of viewing boundaries as a hindrance, I propose to view them as encompassing motion in ways which differ from the spatial relationship between the container and the contained. These boundaries in a nascent state are human and peaceful and do not harm anyone, but rather provide each one with a sense of stability in their daily living environment.

The intentional political strategies that draw administrative districts to enforce social control can hardly disregard the customary uses of space. Administrative districts tend to shadow the patterns of everyday life as the most effective way to enforce control. Arbitrary boundary lines are disruptive of local uses and easily arouse conflicts. Smooth administrative action requires some degree of correspondence between the two types of human territoriality to which I refer.

Quite obviously, administrative districts can be designed arbitrarily due to sheer ignorance of local routines or to intentionally disrupt them. Institutional territoriality is always a deliberate strategy. Its effect nevertheless depends on the habituation of citizens to newly shaped districts. The latter is a slow and incremental process based on regular motion practice continued to the point that it becomes unreflecting. Arbitrary administrative lines become effective boundaries as a result of social appropriation and emotional attachment.

In any case, the expected effects of social control eventually depend on people's practices. More than authoritarian, social control is mimetic and precarious when it operates by means of territoriality. A gap between the customary and institutional territoriality closes and reopens ceaselessly. Practices on the move are regular but by no means immutable. As patterns of response, practices adapt to circumstances and allow innovation to occur under the guise of recurrence. Old administrative districts become inadequate for some purposes while retaining relevance for others. Administrative action loses its hold on everyday life practices, making social control an improbable aim.

Reconciling administrative districts with the routes traversed by citizens requires painstaking expertise in boundary-making together with an intellectual understanding of boundaries as mundane horizons. Experts in this domain must be sensitive to the intertwining of society and space because thinking of boundaries differently also means thinking of space and society differently. What seems to be crucial in this regard is the ethical and technical talent to discern, in each concrete setting, the emancipative capacity of a boundary from its segregating power. Making disjointed territorialities work together is the difficult task that boundary experts face. In this way, they participate in the production of reasonable boundaries that support civil coexistence without fear of transgression or trespassing.

Hence, responsible experts in boundary-making, advising on administrative partitioning or drawing plans, should balance the advantages provided through the stability of boundaries on the one hand with the advantages provided by adapting old boundaries to new practices. Boundaries are in a constant state of becoming and yet are among the most enduring features of landscapes. Far from being a logical contradiction, this dual quality is entirely consistent with the interpretive framework proposed in this chapter. Boundaries defy pure logic because they belong to the ethical dimension of social life.

19.5 Conclusions

Observing the scene of daily life makes it hard to imagine that we live in a world of diaspora, where people have lost or are losing any spatial bond. A misunderstanding like this almost inevitably occurs if boundaries are taken as a hindrance to social motion, and are confused with material artefacts.

Reframing internal boundaries as everyday life's horizons has a double advantage in theory and practice. Firstly, it keeps the material and social substance of boundaries together, thus paving new ways to understand the relation between society and space. Secondly, it brings the dimension of everyday life to the forefront of border theory, thus making routine motion a major principle of bounding. Reframing boundaries as daily horizons helps to reconsider administrative boundary-making as an attempt to trace the customary territoriality of citizens in order to enforce social control. The mismatching of everyday practices and administrative action inevitably follows in this interpretation from the incremental modification of routines. Social motion is an endless source of unchartered boundaries which set official partitions out of joint.

Experts dealing with unchartered boundaries should be prepared through study and practice in a transdisciplinary environment. Overcoming the compartmentalisation of knowledge typical of modern professionalism is essential to train versatile experts to have the same attention and commitment to social and spatial aspects. Indeed, as subjects of study and action, boundaries do not exclusively belong to any single field of knowledge. Geographic, ethnographic, historical and architectural skills would equally contribute to mastering boundary-making challenges. The field of border studies has developed transdisciplinary approaches for decades. On the other hand, it appears increasingly as a scholarly and often critical enterprise, divorced from boundary-making as technical expertise. Further research on internal boundaries as related to everyday life and local government should help narrow the gap between transdisciplinary scholarship and highly specialised expertise.

Deterritorialisation is a symptom of underlying changes affecting the local patterns of everyday life as well as the global flows of people, culture, information and commodities. However, we are not left only with obsolete boundaries, fortified ramparts (Brown 2010) and hybrid borderlands. Setting apart traditional images of rupture and breakage does not mean that people bounce erratically to and fro like a ball in a flipper game. People ceaselessly produce unchartered boundaries as they go through their daily practices on the move. These nascent boundaries need to be recognised and reconciled with administrative action, while keeping in mind that good citizens make good boundaries and not the other way around, as an old English adage puts it.

References

Andersen DJ, Klatt M, Sandberg M (eds) (2012) The border multiple: the practicing of borders between public policy and everyday life in a re-scaling Europe. Routledge, London

Ashley RK (1987) Foreign policy as political performance. Int Stud Assoc Notes 13:51–54

Balibar E (1998) The borders of Europe. In: Cheah P, Robbins B (eds) Cosmopolitics: thinking and feeling beyond the nation. University of Minnesota Press, Minneapolis, pp 216–233

Balibar E (2009) Europe as borderland. Environ Plan D Soc Space 27:190–215

Bourdieu P (1977) Outline of a theory of practice (trans: Nice R). Cambridge University Press, Cambridge

Brown W (2010) Walled states, waning sovereignty. Zone Books, New York

Clark I (2000) A 'borderless world'? In: Fry G, O'Hagan J (eds) Contending images of world politics. Palgrave, London, pp 79–90
Foucher M (1991) Fronts et frontières. Un tour du monde géopolitique. Fayard, Paris
Foucher M (2016) Le retour des frontières. CNRS Editions, Paris
Giddens A (1984) The constitution of society: outline of the theory of structuration. Polity Press, Cambridge
Gupta A, Ferguson J (1992) Beyond 'culture': space, identity, and the politics of difference. Cult Anthropol 7:6–23
Massey D (2005) For space. Sage, London
McAuliffe M, Triandafyllidou A (eds) (2021) World migration report 2022. International Organization for Migration (IOM), Geneva
Mezzadra S, Neilson B (2013) Border as method, or, the multiplication of labor. Duke University Press, Durham
Miyoshi M (1993) A borderless world? From colonialism to transnationalism and the decline of the nation-state. Crit Inq 19:726–751
Nail T (2016) Theory of the border. Oxford University Press, Oxford
Newman D, Paasi A (1998) Fences and neighbours in the post-modern world: boundary narratives in political geography. Prog Hum Geogr 22:186–207
Ohmae K (1990) The borderless world: power and strategy in the interlinked economy. Harper Business, New York
Paasi A (2009) Bounded spaces in a 'borderless world'? Border studies, power, and the anatomy of the territory. J Power 2:213–234
Paasi A, Prokkola E-K, Saarinen J, Zimmerbauer K (eds) (2019) Borderless worlds for whom. Ethics, moralities and mobilities. Routledge, London
Papastergiadis N (2000) The turbulence of migration: globalization, deterritorialization and hybridity. Polity Press, Cambridge
Prescott JRV (1987) Political frontiers and boundaries. Allen & Unwin, London
Rouse J (2007) Practice theory. https://wesscholar.wesleyan.edu/div1facpubs/43. Accessed 20 Aug 2022
Sack RD (1983) Human territoriality: a theory. Ann Assoc Am Geogr 73:55–74
Schatzki TR, Knorr Cetina K, von Savigny E (2001) The practice turn in contemporary theory. Routledge, London
United Nations. Department of Economic and Social Affairs (2020) International migration report: highlights. https://www.un.org/en/desa/international-migration-2020-highlights. Accessed 20 Feb 2023
Wastl-Walter D (ed) (2011) The Ashgate research companion to border studies. Ashgate, Farnham
Wilson TM, Donnan H (eds) (2012) A companion to border studies. Wiley-Blackwell, Malden

Chapter 20
The Transformational Polarisation as a Factor Stimulating Urban Development in European Cities of Rich Endogenous Values

Zbigniew Władysław Paszkowski

Abstract This chapter examines the transformations occurring in Polish cities today, particularly those with significant endogenous values that are being challenged by modern driving forces advocating for change. The case studies of Krakow, Warsaw and Szczecin are presented alongside examples from Lisbon and Malaga, highlighting the existing phenomena and issues related to the tensions between the desire to preserve the intrinsic values of cities and the ongoing processes of globalisation. The development manifested in spatial transformations alters the scale of urban growth, land use, built structures and the overall form of cities. The challenges posed by the increasing influence of New Urban Tourism on urban structures, as well as issues such as air and water pollution and the post-industrial transition into contemporary urban environments, are key topics of this research discussion. The mixed-method research approach utilised combines quantitative and qualitative data, relying on statistical analyses, urban morphological assessments and observational studies in selected cities. Identifying the polarisation factors that drive urban development can help elucidate the strategic significance of emerging trends and provide a framework for evaluating new policies aimed at enhancing urban environments.

Keywords Endogenous values · Overtourism · Polarisation · Transgression · Urban development · Neoliberalism · Post-socialist cities

---

Z. W. Paszkowski (✉)
Andrzej Frycz Modrzewski Krakow University, Krakow, Poland

G. Cimadomo et al. (eds.), *Urbanism at Borders*, The Urban Book Series,
https://doi.org/10.1007/978-3-032-09594-7_20

20.1 Introduction

20.1.1 Aims

This chapter focuses on contemporary urban developments in selected Polish cities characterised by strong identities and urban forms, shaped by historical development, cultural traditions and specific topographies. These cities are also influenced by global economic trends. The endogenous potential of a city refers to its capacity for internal growth and economic development through its own resources (Bertaud 2004). Endogenous growth theory posits that economic progress is driven by factors such as human capital, knowledge, innovation and entrepreneurship (Black and Henderson 1999).

The study aims to demonstrate that:

- Greater endogenous (cultural, social, topographic) potential leads to increased pressure from exogenous (globalisation) factors on urban transformation.
- Each city has unique characteristics, and uncritical acceptance of globalisation can jeopardise sustainable urban transformation.
- Effective urban development occurs when local endogenous values are balanced with external global factors.
- New Urban Tourism can significantly impact the functionality and structure of cities.

Ultimately, this research seeks to define strategies to address the polarisation between local values and external influences, identify internal transformations and analyse social and economic factors that stimulate local resilience and economic growth (Paszkowski 2011).

20.1.2 Research Methods

A mixed-method approach was utilised, combining quantitative statistical data with qualitative analyses of urban morphology and observations in selected cities. Each city's polarisation factor, which drives urban development, was defined and compared to identify emerging trends and inform urban policy development. The process of designing cities (Cuthbert 2003) is increasingly incorporating a wider range of factors, becoming more open to new trends and challenges. The research method involves analysing the polarisation of transformation processes across spatial, social and economic spheres, identifying areas of rapid and profound transformation that create new structural and economic relationships. This is achieved through observations, strategic policy documents and statistical analyses. The findings are visualised through a polarisation diagram, illustrating the tension between conservative urban stability and the unpredictable acceleration brought by globalisation.

City structures can be analysed across several thematic areas, including historical spatial arrangements, new architectural forms as tourist attractions, public transport accessibility and recreational spaces. While an in-depth analysis of all themes is not feasible in this chapter, it aims to encourage city planners to assess the impact of new trends on urban consistency, social transformation and infrastructure capabilities.

20.1.3 Theoretical Framework

The transformative processes of modern cities—termed the third urban revolution (Asher 2004)—began at the end of the last millennium, altering the scale, form, mobility systems and social relations. Research into the polarisation between local demands and global opportunities for sustainable development is ongoing (Szromek 2012; Krasowska 2019). A key issue is the polarisation observed in many European cities, where historic districts thrive as logistics, commercial and cultural centres, while other areas struggle to maintain their unique human habitats. This polarisation affects spatial, social and economic spheres, heavily influenced by municipal strategies.

Cities are living organisms that cannot remain open-air museums; they must continuously modernise to meet residents' needs. The dynamics of transformation vary by area—stable regions experience gradual, qualitative changes, while post-industrial areas often undergo rapid transformation, significantly impacting the city's development and polarisation between old and new structures. Sustainable urban development amid growing economies in Europe is challenging, particularly in culturally rich urban areas facing strong exogenous pressures for globalisation. These pressures manifest in the establishment of various international enterprises, with increasing interest in culturally attractive locations. The challenge lies in defending endogenous values against globalising influences—whether such defence is necessary or if cultural identity hinders progress remains an open question. Cultural identity is rooted in the history and uniqueness of nations, reflected in linguistic and religious diversity, architectural styles and spatial organisation. Globalisation promotes synergistic cooperation between isolated areas, termed 'islands' (Paszlowski 2019), facilitated by global communication systems and urban management. Modernism and functionalism typified twentieth-century architecture, often disregarding local context. In today's digital age, ideas can be rapidly disseminated, leading to architectural designs that reflect global trends. The 'good city form' (Lynch 1981) is giving place to modern functionality.

European cities with historical legacies must navigate the tension between preserving local values and adapting to globalisation's demands for growth (Paszkowski 2001). The sustainable development is the slogan and the keyword, but not always understood in the sense of urban space continuity. Sustainability means raising the quality of space without changing its character dramatically. Achieving a harmonious compromise between modernity and tradition in a dynamic path of development is an extremely difficult task (Paszkowski 2001). The most striking examples of filling the structure of a historic city with a modern, internationally designed object, is the Museum of Modern Art in Graz, Austria, designed by Peter Cook and Colin Fournier in 2003. A similarly astonishing project is the building of the Szczecin Philharmonic, designed by Barozzi & Vega Architects from Barcelona, selected in the international architectural competition and built in Szczecin in 2011–2014. This facility has won many awards and distinctions, the most important is the Mies van der Rohe Award for European architecture. Both projects are an example of globalisation processes in architectural design and evoke mixed reactions from the public, highlighting the challenges of integrating new developments into existing urban contexts. Similarly, Krakow's ICE Congress Centre and high-rise towers in Warsaw illustrate how global architectural trends that differ from local traditions influence local environments.

Sustainable development should enhance the quality of urban spaces without dramatically altering their character (Kostoff 1992; Llorente 2014). Striking a balance between modernity and tradition is complex (Paszkowski 2001).

These new buildings in Polish cities exemplify international architectural trends, attracting cultural and business tourism, a phenomenon often referred to as the 'Bilbao effect', following the transformation of Bilbao due to the Guggenheim Museum.

However, globalisation has also led to 'overtourism' (Rosa Jiménez and España 2022), which significantly alters city functions and customs. Cultural tourism often prioritises the needs of tourists, leading to the commercialisation of local values and potentially degrading the quality of life for residents (Ba et al. 2022). Residents often react negatively to the concentration of tourist flows, resulting in overcrowding and aggressive commercialisation that diminishes aesthetic experiences and erodes local resources (Postma and Schilder 2007; Postma 2013). This issue is prevalent in many prime tourist destinations across Europe and is increasingly evident in Polish cities such as Krakow, Wroclaw and Zakopane, famous winter capital of Poland surrounded by Tatra mountains.

Addressing the consequences of overtourism requires careful long-term planning (UNWTO 2020). The UNWTO has proposed 11 strategies for mitigating overtourism, focusing on visitor dispersal, local community benefits and improved infrastructures. A case study on overtourism in Wroclaw examined the extent of its symptoms during the 2020 pandemic, revealing both positive and negative effects as perceived by various stakeholders (Fedyk et al. 2020). Although concerns about overtourism exist, the study concluded that the threat is not yet significant, but proactive urban planning measures are essential to prevent future issues. Another study identified five functional causes of social conflicts related to overtourism in Polish cities, particularly Krakow and Poznan, highlighting how value conflicts shape the dynamics of urban tourism development (Zmyślony et al. 2020).

20.2 Case Studies

Poland offers significant tourist appeal, encompassing leisure, cognitive and cultural aspects. Its tourism is notably influenced by the tragic history of Central Europe in the twentieth century. Each city has unique traits affecting mass tourist traffic driven by economic development and globalisation. The impact of global events is evident in the sharp decline in tourist numbers during 2020–2022 due to the coronavirus disease 2019 (COVID-19) pandemic and travel restrictions (Cimadomo et al. 2021). Types of tourism include cognitive, leisure, health, business and religious.

This analysis focuses on Polish cities with varying sizes and cultural potential that rank among the most visited by foreign tourists: Krakow, Warsaw and Szczecin (including the coastal region of Western Pomerania). For comparison, Malaga and Lisbon have been analysed. Tourist traffic in these urban centres is uneven. The summer months attract the highest number of visitors, but the off-season is gradually expanding thanks to year-round tourist infrastructure. More hotels with recreational facilities, such as pools and spas, are being built along the coast, many offering therapeutic services. Increased traffic is also noted during the Christmas and New Year holidays and February winter breaks. Private companies have improved their hospitality and dining offerings, which have enhanced service quality amid growing competition. However, periods outside peak seasons create challenges in maintaining facilities and service standards. Weather conditions significantly impact seasonality, making tourist patterns unpredictable. Cultural and cognitive travel centres, like those in Krakow, are less affected by seasonality, while seaside resorts and recreational goals are more vulnerable. Warsaw, with its emphasis on business, scientific and cultural travel, is the least impacted by seasonal fluctuations.

Polish cities view tourism as a vital opportunity for economic growth, striving to attract visitors with engaging events such as festivals, concerts and sports competitions (Kruczek 2017). Despite the importance placed on tourism in urban development policy, concerns about limiting mass tourist inflow and associated nuisances are occasionally raised in the media, yet increasing action is arising to address these issues.

20.2.1 Case Study 1: Krakow

Krakow is Poland's second-largest city by population and a vibrant hub of science, art and industry, increasingly known for its cognitive, cultural and religious tourism. With a rich history dating back to the tenth century, Krakow was the capital of the Polish Kingdom and boasts numerous historic monuments, the oldest university in the country and beautiful churches. Remarkably, it survived World War II largely intact, which has given it a developmental advantage over other Polish cities like Warsaw and Szczecin.

The city, including the Royal Castle on Wawel Hill, features a well-preserved medieval layout with the impressive Main Market Square. Krakow is the place of life of the Pope John Paul the II and the goal of pilgrimages to sanctuaries located in the city and to the birth place of John Paul II, Wadowice near Krakow. Kazimierz, the former Jewish district of Krakow, houses seven synagogues and the historic Remuh cemetery, with the graves of many important Polish Jews, including Rabbi Moses Isserles which dates back to 1535 (see Fig. 20.1). Cultural, religious and historic features of Krakow are therefore attracting many visitors from all over the world.

Two significant transformation processes have impacted Krakow: industrialisation and a shift towards tourism. During the Soviet era (1945–1989), the establishment of heavy industry, particularly the Nowa Huta steel complex, leads to severe pollution from industrial activities, making Krakow one of the most polluted cities in Europe during the 1970s and 1980s. Recent initiatives to combat air and water pollution have included closing industrial plants, banning solid fuels for heating and encouraging electric vehicles. Since Poland's integration into NATO (1999) and the EU (2004), tourism has surged, with budget airlines opening routes from the UK and Ireland. This influx has led to the touristification of Krakow, urban tourism

Fig. 20.1 Krakow. The Jewish shops reconstruction at the Szeroka Street in Kazimierz District recalling the history of Kazimierz in Krakow the Jewish District. Source: the author

'hypertrophy' (Zmyślony and Kowalczyk-Anioł 2019), particularly in the Old Town and Kazimierz, where urban functions have adapted to cater to tourists. Authorities are modernising public spaces, expanding pedestrian zones and promoting a welcoming city image. The Old Town and Kazimierz have largely become pedestrian areas, with only electric vehicles and horse-drawn carriages allowed. Public transport now relies on electric trams and buses, significantly improving transportation, air quality and noise reduction. Street cafes and restaurants have proliferated, enhancing the city's Mediterranean atmosphere, especially with the Renaissance architecture introduced by Queen of Poland Bona Sforza in the sixteenth century, who invited Italian architects to reshape Krakow's aesthetic.

The Old Town in Krakow and Kazimierz District has become the 'must see' destinations of tourists coming to Krakow, creating huge visitors' congestion in the streets and main squares during peak periods (see Figs. 20.2 and 20.3). Adapting almost all ground floors of tenement houses, outbuildings and passages, as well as courtyards for catering, bars and souvenir shops, distorted the essence of a multifunctional city, living not only from tourism but also having its own 'inner life' of its inhabitants. High rent costs in downtown premises forced the adaptation of their functions to more profitable ones. The historic long time neglected districts Podgórze and Zabłocie were incorporate to the main touristic area with pedestrian public spaces and tourist's services. The urban vehicular traffic has been widely displaced out of the city centre. A full bypass of the city has been built in order to eliminate the transit traffic through the city centre.

Krakow is also a transfer hub to reach world-famous places, such as the Wieliczka Salt Mine, a complex of wooden churches in Podkarpacie listed on the UNESCO World Heritage List and the World War II extermination camp in Oświecim. This place is sacred by the martyr's death of several million inhabitants of Central Europe, mainly Jews and Poles killed by Nazi German occupants during World War II in Poland (1939–1945).

20.2.2 Case Study 2: Warsaw

Warsaw is the capital of Poland and its largest city, a metropolis with the leading functions in Poland of administration, higher education, business, industry, services and trade. But Warsaw is also the important touristic destination. During World War II (1939–1945), Warsaw was almost completely razed to the ground by the retreating Nazi army in retaliation for the Warsaw Uprising (1944). The reconstruction of the medieval city and the Royal Castle has become an important factor in the development of tourism. The history of Warsaw in the twentieth century, like that of Krakow, is connected with the presence of Jews, whose existence in Warsaw ended with the tragic extermination by Nazi Germany in the period 1939–1945.

The number of people visiting Warsaw is largely business trips, as Poland has a very centralised administrative system and most offices and representative offices of

Fig. 20.2 Krakow. The view over the Main Square and the St. Mary Church during the Dragon Parade 2023. Source: the author

companies are located in Warsaw (see Fig. 20.4). A large number of incoming students are made up of students of numerous educational institutions and universities. It is also an intercontinental destination for continuing a visit to Poland in its other regions, due to the best network of Warsaw air connections with distant countries,

Fig. 20.3 Krakow. The Florianska Steet leading from the Central Station to the Main Square, the central public and touristic space in the city. Source: the author

countries of Central Europe and the central location in relation to other destinations in Poland. The Vistula Boulevards have been made available for recreational purposes, creating attractive walking and cycling routes as well as places to relax. The unique attraction of Science Centre on the Vistula embankment became one of the remarkable tourist destination. Opposite to the city centre, the right bank of the Vistula River has been left in its natural state, which is a relaxing contrast to the built-up structure of the city centre.

Warsaw is a much larger city than Krakow, but the rebuilt Old Town is much smaller than historic Krakow. The Old Town in Warsaw has become one of the most crowded places (see Fig. 20.5). Spatial measures were taken to expand the 'tourist capacity' for crowds of tourists flooding Warsaw, visiting for various purposes. The pedestrian zone has been extended along Krakowskie Przedmieście Street and covered the entire Royal Castle Square, as well as part of the New Town. Souvenir stalls were set up along the city walls, which at least partly relieved the narrow streets of the Old Town from tour groups visiting Warsaw. Actions were also taken to make other parts of Warsaw more attractive for tourists. New museums were created, and abandoned industrial plants in the city center were revitalised.

Fig. 20.4 Warsaw. The central business district in the hearth of the Polish Capital. Source: the author

Fig. 20.5 Warsaw. The Royal Castle Square and the Old City view. Source: the author

20.2.3 Case Study 3: Szczecin

Szczecin the capital of Western Pomerania is a seaport and border town adjacent to the Federal Republic of Germany. It boasts excellent motorway connections to Polish seaside resorts, as well as to Berlin and Rostock in Germany. For many years, Western Pomerania has led Polish regions in overnight stays, primarily due to recreational tourism at Baltic Sea beaches and medical tourism. Qualified tourism includes water sports, golf and horse riding. The region's proximity to Germany and Scandinavia means most arrivals are by car, coach, ferry or rail, with the highest influx of foreign tourists coming from these areas. Many visitors to Western Pomerania are from southern Poland, arriving mainly by rail or private car, making airline passenger numbers an inadequate reflection of tourism in the region.

Szczecin does not suffer from an excess of tourists, as most head directly to the Baltic Sea, bypassing the city via the motorway. The port rarely welcomes cruise ships, which are common in many North Sea and Mediterranean ports, making it an 'unattainable dream' for city authorities. The distance of approximately 60 kilometres from the coastline, and the too shallow waterway through the Szczecin Lagoon for large cruise ships, is making entry time-consuming. Consequently, during the summer when Szczecin is at its most beautiful, locals go on holiday while tourists flock to coastal resorts, experiencing overtourism symptoms such as queues at restaurants, high prices and crowded beaches. Both Szczecin and the seaside resorts strive to attract tourists (see Fig. 20.6). A notable example is the Tall Ship Races,

Fig. 20.6 Międzyzdroje. Once a small fisher village at the Baltic sea-side sandy beaches in the West Pomerania Region, transformed into the 'holiday paradise', is under continuous development. Source: the author

recently attracting two million visitors. (see Fig. 20.7). Coastal resort cities in the West Pomerania Region, apart of relaxation on sandy beaches of the Baltic Sea, compete with appealing concerts, promenade constructions, new hotels, aquaparks and other attractions.

Fig. 20.7 Szczecin. The Odra River Boulevard during the Tall Ship Race's final event 2024. Source: the author

Despite Szczecin's lower tourist interest compared to other Polish cities, authorities are working to enhance its image, exemplified by the 'Szczecin - Floating Garden 2050' initiative, aimed at improving public spaces, facades, parks and green avenues. Notably, the transformation of the Oder quay into the Oder Boulevard is becoming a popular meeting and walking spot along the waterfront. Szczecin's spatial structure, historically known as Stettin until the end of World War II (1945), reflects its complex history and strategic location in Europe. The city developed around the medieval castle hill on the left bank of the Odra River at its mouth to the Baltic Sea, experiencing significant growth in the nineteenth and early twentieth centuries due to industrialisation and population increase, resulting in concentric and linear developments along the riverbank, with 'Idealstadtlandschaft' implementation goal in mind (Rodriguez 2015; Reichow 1948).

The city's development was dramatically affected by World War II, which led to its destruction during bombing raids. The transfer of Stettin to Poland resulted from the Potsdam Agreement (1945). It took years for the city to regain functionality, with reconstruction ongoing today. The nineteenth-century districts that survived the war require substantial conservation work to meet modern standards (see Fig. 20.8). The city has lost its original industrial and traditional port character, giving way to new phenomena in tourism, aquatic recreation and urban development. It is to observe the paradigm shift from a production and commercial focus to a more friendly city, relocating burdensome functions to the outskirts. Szczecin aims to develop a new

Fig. 20.8 Szczecin. The nineteenth century tenement houses in the central district of Szczecin preserved from the World War II destruction, now undergoing renovation and adaptive reuse to contemporary needs. Source: the author

image prioritising public spaces that are welcoming to residents, particularly near water, parks and city forests. Attempts to create a large pedestrian zone in the historic centre were stalled by protests from shop owners concerned about reduced income. A potential solution for revitalising shopping and service functions could involve reconstructing main streets, integrating parking options for drivers. The Szczecin Municipality is addressing this through architectural and urban competitions aimed at selecting innovative and functional solutions, incorporating green spaces, pedestrian-friendly pavements and attractive street furniture. Despite the many aesthetic values that Szczecin has and the entire spectrum of attractions, as well as the ongoing revitalization of public spaces to meet the needs of the younger generation of residents and tourists using street bars and restaurants, Szczecin lags far behind Krakow, Lisbon and Malaga in terms of the number of tourists visiting it.

20.2.4 Case Study 4: Lisbon

The degraded historical structure of Lisbon, alongside the preservation of traditions in hospitality, cuisine and the authenticity of its buildings—marked by visible signs of decay—has fostered new values beneficial to urban tourism. These values of

Fig. 20.9 Lisbon. Mercado de la Ribeira. Market interior transformed into a food court with local specialities. Source: the author

authenticity have significantly increased tourist interest in the city. Lisbon has transformed into a hospitality hub, repurposing ground floors for numerous restaurants and cafés, despite some resistance. Tourist interest has altered the city's development policy, necessitating adaptations to accommodate multidimensional tourism. This growth has encouraged local residents to engage in tourist-related activities in transport, art, trade and gastronomy. It has also caused a sharp rise in real estate prices in the city centre, with many properties converted into tourist apartments, generating substantial income for owners. The vibrant city life has attracted young people to study at local universities.

Increasing Lisbon's appeal as a tourist destination and place to live was generated by the 'Madonna effect'—settlement of pop icon Madonna. Simultaneously, Lisbon has become more internationalised, with thousands of foreign visitors and residents from Portuguese-speaking former colonies such as Angola and Mozambique. Urban globalisation is evident in Lisbon's outskirts, where commercial buildings, offices, hotels and housing complexes have emerged. The conservation lobby has prevented significant changes in the historic downtown; however, transformations like the conversion of Mercado de la Ribeira into a large food court have occurred (see Fig. 20.9). Gentrification is gradually taking place in historic areas like Alfama, once the city's most neglected district, now attracting investors due to its charm and lower property prices. Revitalisation efforts in Bairo Alto and

Fig. 20.10 Lisbon. The view over Alfama District, which is undergoing renovation and adaptive reuse to contemporary touristic purposes. Source: the author

Alfama have increased its tourist popularity, blurring lines between local life and tourism (see Fig. 20.10). The influx of tourists is driving displacement of long-term residents, particularly low-income individuals and property owners renting flats for short stays. Many properties are being acquired by foreign investors seeking profitability. Despite that local residents are crucial to Lisbon's vibrant culture, contributing to its atmosphere with their customs and language, they have been pushed to the outskirts, making way for New Urban Tourism.

The city's values, viewed through global cultural tourism, have instigated gentrification in downtown areas. An advantage for short-term tourism is the proximity of Portela Airport; however, noise from air traffic and tourist surges can detract from residents' comfort. These drawbacks are offset by the city's favourable climate, its location at the mouth of the Tagus River and its rich architectural heritage. Urban policies aimed at reducing car traffic in favour of public transport—including trams, buses and new urban bicycle and electric scooter systems—enhance the city's appeal. While New Urban Tourism generates significant profits for cities, it also brings negative consequences, such as rising prices for services and housing, a decline in dining quality and overburdened infrastructure (Pawlicki 2017; Kruczek and Zmyślony 2014). Limitations of tourism space arise from the size of usable areas and their capacity to respond to tourism demands (Szromek 2012). A critical issue is determining acceptable tourist traffic volumes, beyond which it may be excessive.

20.2.5 Case Study 5: Malaga

For decades, Malaga was seen by the international community as little else than an airport on the way to summer holidays' cities Marbella and Puerto Banús. However, in the last 20 years, the city has experienced a metamorphosis, with the rehabilitation of the old town, which is now closed to traffic and the transformation of the old port into a modern marina, with part dedicated to super yachts. The transformation of the centrally located port of Malaga is an example of the city's care for creating a friendly urban environment. An example that Malaga can boast of is the consistent, multi-stage crossing of the borders of the historic area of the city in favour of its enlargement by a seaport (Machuca Santa-Cruz 1987). The first stage of transformation, carried out at the turn of the nineteenth and twentieth centuries, involved the construction of a new port quay by moving the shoreline deep into the sea and creating a city park along the port quay (Andrade et al. 2021). In the next, contemporary transformation stage, the port buildings gave way to a new park and recreation complex with a shaded passage and the opening of a wide view of the port. The transformation of the port quay in Malaga Muelle Uno and Muelle Dos (see Fig. 20.11) is an example of consistent planning strategy, implementing the idea of synergy between the city and the port basin. It contributed to the rise of tourist attractivity

Fig. 20.11 Malaga. The Muelle Dos sun-shaded passage at the waterfront of the former port area, facilitating connection towards the revitalised waterfront area of Muelle Uno and the Malagueta district. Source: the author

and accessibility of the port resulted in the arrival of several cruise ships every day (Chamizo-Nieto et al. 2023). The sudden inflow of great number of tourists reaching six million in 2024 (Muelle Uno 2024) breaks the potential number of tourists that the city can host everyday due to its overall capacity. The current policy of the Málaga municipality appears to focus more on attracting tourists to better distribute visitor flow than on improving the lives of residents (Rosa Jiménez and España 2022). However, it has sparked strong protests from local inhabitants who feel offended by mass tourism, leading to restrictions on short-term tourist rentals.

20.3 Discussion

The case studies of selected Polish cities—Krakow, Warsaw and Szczecin, alongside examples from Lisbon (Portugal) and Malaga (Spain)—reveal various aspects and stages of urban development. These cities grapple with the challenge of preserving their unique values amid the pressures of globalisation, leading to adaptive development that manifests both in historic preservation as spatial transformations, expanded urban areas and changes in building typologies and land use. Renovating historic structures and promoting their beauty has heightened international interest in cultural tourism. However, this focus on short-term tourist profits has led municipalities to prioritise tourist needs over those of local residents, giving rise to 'New Urban Tourism'. This trend creates a disparity between profits and losses, resulting in 'polarisation' (Gravari-Barbas et al. 2019).

Recent developments in cities like Venice (Pawlicki 2017), Barcelona, Brugge and Malaga demonstrate strong resistance to uncontrolled tourist influx, causing overcrowding and conflicts between tourists and residents. Potential social and economic conflicts arise as a direct or indirect result of excessive tourist flow, as illustrated in the accompanying table (see Table 20.1). New Urban Tourism cities are particularly sensitive to seasonal fluctuations, as tourist traffic is unevenly distributed throughout the year, with the main influx occurring in summer. The summer season is extending due to increased year-round tourist infrastructure. However, cities are also vulnerable to sudden drops in tourist inflow, as evidenced during the COVID-19 pandemic, which severely impacted tourist-oriented destinations. Political conflicts, such as the Russian war against the Ukraine, have further affected expected tourism revenues. In Poland, the trend towards New Urban Tourism is emerging more gradually, with the exception of Krakow, a leading destination in this regard. Other Polish cities have not yet experienced the same level of tourist saturation. Metropolitan areas like Warsaw are better equipped to handle fluctuating tourist flows, with a diverse demographic of visitors who come for various reasons beyond tourism. Szczecin is gradually becoming more attractive to tourists, with ongoing discussions about development policies that balance global trends with the protection of local environmental, historical and cultural values. The boundaries of transformational development in cities are somewhat ambiguous. However, there are limits beyond which endogenous values loose relevance, giving way to

Table 20.1 The selected polarisation attributes of urban development strategies of Polish cities

Item	Research	Local policy of the urban conservative stabilisation	Globally driven policy of the urban transformative development	Range of the potential spatial conflicts
1	2	3	4	5
A	Historically shaped city structure	Preservation of historical structures, conservation. Preservation of the historical silhouette of the city	Transformation of urban structures with the replacement of degraded substances	Attitude to historical substance without historical or artistic value
		Reconstruction of demolished objects can be restored (with existing documentation)	Change of building structure in accordance with modern requirements of access to daylight and unobstructed views. Built-in losses implemented according to new architectural concepts	The philosophy of reconstruction vs. new tall buildings Acceptance of new trends in architecture vs. contextual design
B	New architectural structures and forms	New structures adjusted to historical structures (continuation)	New urban structures and facilities designed according to the latest global trends with priority to meet the current and future needs of both city residents and visitors	The ratio of new buildings to the environment Spatial relations of new buildings to historical buildings
		Architectural forms neutral or referring to the environment		
C	Urban public spaces	Preservation, revitalisation of existing forms of public spaces with the characteristic street layout, allowing for minor modernisation changes	Introduction of new forms and solutions in the field of urban landscape architecture—passages, pedestrian routes, exclusive pedestrian zones with the use of underground garages	Improvement of existing recreational areas for better use
		Revitalisation of existing parks with a ban on building new service facilities in their area	Acquiring new spaces in degraded post-industrial areas for the needs of urban recreational spaces Introducing green roofs, water reservoirs and fountains in the urban area	Applying of new landscaping ideas and forms

(continued)

Table 20.1 (continued)

Item	Research	Local policy of the urban conservative stabilisation	Globally driven policy of the urban transformative development	Range of the potential spatial conflicts
1	2	3	4	5
D	Urban service structure	Services traditionally located on the ground floors along street routes or in separate facilities of the historical urban structure or in facilities supplementing the existing structure	Modern forms of commercial and service connected with multi-storey car parks in downtown areas and near peripheral housing estates	Traditional shopping and service centres
		New commercial and service facilities that can be implemented outside the historical structure	A network of service facilities built into existing buildings and as newly designed free-standing facilities	Changes in the pattern of trade services
E	Urban recreation areas	Preservation of the local tradition in the area of historical structures, adaptation of historical, private gardens and parks at palaces and monasteries for public needs	Introduction of linear systems: bicycle paths, walking routes, footbridges over waterways	Transformations of the existing land use for recreation
			Construction of new sports and recreation facilities on the outskirts of the city centre or in their area with the possibility of combining different functions	New locations for modern recreation purposes
F	River and sea waterfronts	Preservation of traditional forms of waterfront management on their coastlines, with admission of technical measures strengthening and improving safety of use	Spatial and functional transformation of waterfront areas with new buildings and boulevards allowing free access to water	Preservation of testimonies of past eras in the field of port services and river or sea trade

commercial interests. This occurs when entire urban districts become temporary venues for vacation rentals, forcing local residents to the periphery and transforming local businesses into tourist-oriented enterprises. This scenario is common in cities with significant New Urban Tourism ratio. Tourists increasingly seek authenticity, venturing into neighbouring districts to discover original local bars and restaurants, signalling potential gentrification.

In summary, the diverse cases and processes of globalisation underline the importance of identifying and monitoring ongoing trends in individual cities, which is essential for municipalities striving for sustainable tourism development. The potential conflicts arising from excessive tourist flow, can be categorised into indirect and direct areas of conflict. Overcrowding in key locations such as city centres, transport hubs and cultural venues leads to significant issues. Inadequate behaviour from tourists, including noise, littering and antisocial activities, further exacerbates tensions. Locals experience a sense of strangeness due to the influx of visitors, often feeling overwhelmed by foreign languages and unfamiliar customs. Privacy concerns arise as tourists encroach on residential areas, leading to feelings of intrusion. Additionally, the general increase in prices across public transport, shopping and leisure activities can strain local residents. The phenomenon of 'tourist gentrification' results in rising rental rates and housing shortages, while competition for resources intensifies due to tourism investments. Finally, the job market is affected, predominantly offering low-skilled positions and often relying on migrant workers (Postma 2013; Zmyślony et al. 2020).

Table 20.1 illustrates the polarisation issues between locally based policy of the urban conservative stabilisation and the globally driven urban transformative development policy in the Polish cities analysed.

20.4 Conclusions

The analyses and case studies highlight the importance of new transgressions in urban areas and the need for an integrated approach aligned with the principles of sustainable development, which is central to European urban planning policy (Krasowska and Paszkowski 2022). The environmental, economic, social and cultural dimensions of urban life are interconnected, and successful urban development can only be achieved through this holistic perspective. Physical urban renewal must be complemented by initiatives in education, economic development, social inclusion and environmental protection, necessitating strong partnerships among local citizens, civil society, industry and government (European Commission n.d.-a).

Furthermore, it is crucial to acknowledge that these are not merely statements; the European Union has prepared financial instruments to support necessary improvements in European cities. For the 2021–2027 programming period, cohesion policy support for sustainable urban development has been strengthened, with member states surpassing the 8% target for allocating ERDF resources, amounting to EUR 24,4 billion aimed at addressing the environmental, social and economic challenges facing urban areas (European Commission n.d.-b). Economic studies from 2000 to 2010 reveal correlations between the growth of tourist activity and various macro-economic indicators in Poland (Szromek 2012). In municipalities that focus on tourism, the potential for increased visitor flow and subsequent income rises. However, while initial growth meets basic city needs, excessive tourism can lead to negative consequences, as demonstrated in the case studies.

It is challenging to generalise conclusions as each analysed city is unique. However, the analysis suggests that external pressures from global phenomena can drive variability and transformation in historically significant Polish cities, with EU Cohesion Funds potentially facilitating this change. Ignoring sustainable development policies may result in unforeseen issues such as increased population, congestion, rising prices and the displacement of residents as properties are converted into temporary tourist rentals. This blurring of boundaries between tourism and daily life can diminish the quality of life (Urry 1995). While becoming new tourist destinations can enhance earning potential for residents, it does not always translate to improved quality of life for all. Studies show that while the wealth of the richest increases, the GDP of average citizens often remains stagnant. Similar to Spain, where many tourist apartments are owned by a few companies, Poland is experiencing a similar trend in urban destinations. The influx of tourists can cause significant disturbances, leading to a situation where, despite financial gains for some, the fabric of the city—rooted in community ties, history and local governance—suffers.

Research indicates that a city's endogenous (cultural, social, topographic) potential influences the pressure from exogenous (globalisation) factors, which can transform cities for business, cultural tourism and asset location. If unregulated, this can lead to over-investment and increased tourist numbers, complicating life for residents. Each city has unique characteristics, structures and limits; succumbing uncritically to globalisation can push cities beyond sustainable development thresholds. The research ultimately aims to outline strategies to address the polarisation between local values and external influences, identifying mechanisms of internal transformation that can help balance these pressures. Individual studies for each city are necessary to correct specific development trends. While cities hold potential as global destinations, there is also a tendency for residents to seek quieter living spaces.

Recommendations for regulating 'overtourism' should be integrated into local planning systems and tailored to each city's scale and character. Advanced analyses of factors driving global changes, which foster both local resilience and economic growth, should be integral to community planning.

References

Andrade MJ, Pedro Costa J, Jiménez-Morales E, Ruiz-Jaramillo J (2021) A city profile of Malaga: the role of the Port-City border throughout historical transformations. Urban Plan 6:3. (2021): Planning for Porosity: Exploring Port City Development through the Lens of Boundaries and Flows. ISSN: 2183-7635. https://www.cogitatiopress.com/urbanplanning/article/view/4189/2200. Accessed 3 June 2025

Asher F (2004) Los nuevos principios del urbanismo. Alianza

Ba C, Frank S, Mueller C, Raschke AL, Wellner K, Zecher A (2022) The power of new urban tourism. Spaces, representations and contestations. Routledge, London - New York

Bertaud A (2004) The spatial organization of cities: deliberate outcome or unforeseen consequence? https://www.researchgate.net/publication/45131759_The_Spatial_Organization_of_Cities_Deliberate_Outcome_or_Unforeseen_Consequence. Accessed 3 May 2025
Black D, Henderson V (1999) A theory of urban growth. J Polit Econ 107(2):252–284. https://doi.org/10.1086/250060
Chamizo-Nieto FJ, Nebot-Gómez de Salazar N, Rosa-Jiménez C, Reyes-Corredera S (2023) Touristification and conflicts of interest in cruise destinations: the case of main cultural tourism cities on the Spanish Mediterranean Coast. Sustainability 15(8):6403. https://doi.org/10.3390/su15086403
Cimadomo G, Jiménez E, Minguet J (2021) Shifting from short term to long term action in post-COVID touristic cities. Drift to exclusion in Seville and Malaga, Spain. In 57th ISOCARP World Congress, Doha, 8–11 November
Cuthbert AR (ed) (2003) Designing cities. Critical readings in urban design. Blackwell Publishing, Hoboken
European Commission (n.d.-a). https://ec.europa.eu/regional_policy/policy/themes/urban-development_en/ Accessed 3 June 2025
European Commission (n.d.-b). https://cohesiondata.ec.europa.eu/stories/s/Sustainable-Urban-development-2021-2027/. Accessed 3 June 2025
Fedyk W, Sołtysik M, Olearnik J, Barwicka K, Mucha A (2020) Overtourism threatens large urban areas: a case study of the City of Wrocław, Poland. Sustainability 12(5):1783. https://doi.org/10.3390/su12051783
Gravari-Barbas M, Jacquot S, Cominelli F (2019) New cultures of urban tourism. Int J Tour Cities 5(3):301–306. https://doi.org/10.1108/IJTC-09-2019-160
Kostoff S (1992) The City assembled. The elements of urban form through history. Thames and Hudson, London
Krasowska K (2019) Rozwój zrównoważony małych miast Pomorza Zachodniego po przystąpieniu Polski do Unii Europejskiej, Szczecin [doctoral thesis]. West Pomeranian University of Technology
Krasowska K, Paszkowski Z (2022) The need to maintain sustainability in the dynamic anthropogenic changes of the natural landscape of Pomeranian Bay in Poland. Sustainability 15(13):1928. https://www.mdpi.com/2071-1050/15/3/1928
Kruczek Z (2017) Frekwencja w polskich atrakcjach turystycznych 2011–2015. Dynamika, trendy, studia przypadku. Proksenia, Kraków
Kruczek Z, Zmyślony P (2014) Regiony Turystyczne. Proksenia, Kraków
Llorente M (ed) (2014) Topologia del espacio urbano. Palabras, imágenes y experiencias que definen la ciudad. Abada Editores, Madrid
Lynch K (1981) Good City form. MIT Press
Machuca Santa-Cruz L (1987) Málaga, Ciudad Abierta. Origen, cambio y permanencia de una estructura urbana. Colegio de Arquitectos de Málaga, Malaga
Muelle Uno (2024) Datos i Cifras. https://www.muelleuno.com/. Accessed 4 May 2025
Paszkowski Z (2001) Tradycja i innowacja w twórczości architektonicznej. Wydawnictwo Uczelniane Politechniki Szczecińskiej, Szczecin
Paszkowski Z (2011) Miasto idealne w perspektywie europejskiej i jego związki z urbanistyką współczesną. universitas, Krakow
Paszlowski Z (2019) Island synergy—a method of creative integration in spatial planning and Urban Management. Przestrzen I Forma 39:9–28. https://doi.org/10.21005/pif.2019.39.A-01
Pawlicki J (2017) Śmierć Wenecji. Polska Newsweek. 9 July. https://www.newsweek.pl/swiat/wenecja-i-barcelona-jak-miasta-bronia-sie-przed-turystami/s1yrtdx
Postma A (2013) When the tourists flew in. In: Critical encounters in the development of tourism., PhD thesis. Faculty of Spatial Sciences, University of Groningen, Groningen
Postma A, Schilder AK (2007) Critical impacts of tourism; the case of Schiermonnikoog. In: Casado-Diaz M, Everett S, Wilson J (eds) Social and cultural change: making space(s) for leisure and tourism. Leisure Studies Association, Bolton, pp 131–150

Reichow HB (1948) Organische Stadtbaukunst von der Grossstadt zur Stadtlandschaft. Georg Westermann Verlag, Braunschweig

Rodriguez JIA (2015) La ciudad de perfiles salvajes. El Idealstadtlandschaft en el urbanismo alemán del siglo XX. Promotora Cultural Malagueña, Malaga

Rosa Jiménez C, España K (eds) (2022) Overtourist City. Estrategias para recuperar la habilidad del centro histórico de Málaga frente a la turistificación. RU Books, Malaga

Szromek AR (2012) Wskaźniki funkcji turystycznej. Koncepcja wskaźnika funkcji turystycznej i uzdrowiskowej. Politechnika Śląska, Gliwice

UNWTO - World Tourism Organization (2020) Overtourism?–Understanding and Managing Urban Tourism Growth beyond Perceptions, Executive Summary. https://www.unwto.org/global/publication/overtourism-understanding-and-managing-urban-tourism-growth-beyond-perceptionsexecutive. Accessed 26 Feb 2020

Urry J (1995) Consuming places. Routledge, London

Zmyślony P, Kowalczyk-Anioł J (2019) Urban tourism hypertrophy: who should deal with it? The case of Krakow (Poland). Int J Tour Cities 5(2):247–269. https://doi.org/10.1108/IJTC-07-2018-0051

Zmyślony P, Kowalczyk-Anioł J, Dembińska M (2020) Deconstructing the overtourism-related social conflicts. Sustainability 12(4):1695. https://doi.org/10.3390/su12041695

Chapter 21
Clashes on the Sunny Side: Social and Territorial Exclusion in the Costa del Sol Touristic Paradise

Jorge Minguet-Medina

Abstract The immediate coast line to the west of Malaga, the (in)famous Costa del Sol, is a 100 km stripe of an international holidays/retirement paradise, where borders do not seem to exist as people of many countries live or holiday in (apparent) harmony. Yet, as foreign habitants have higher incomes, borders reappear around the multinational paradise. Gated communities and huge extensions of—equally gated—golf courses (artificial substitutive of a predated nature) keep visitors isolated in their private spheres, while locals, unable to afford the ever-rising prices of rents and land, now subjected to international competition, are displaced to the outskirts of paradise. And just at the end of this paradise, lies its financial sibling: Gibraltar, where, paradoxically enough, the same economy based on the rejection of borders demands a strict one, as Brexit conflicts stressed. As this chapter tries to underline, all these situations invariably result in a spatialisation of inequality, of the logic of master and servant, as if twentieth century social conquests never happened. Even borders act as servers in the implementation of this logic, which reveals itself as the ultimate target of neoliberalism, the master in its unfettered expansion.

Keywords Gated communities · Social inequality · Tourism urbanism · Southern Spain · Borders

21.1 A Palestinian Introitus

When thinking about borders, the first thing that comes to mind are the big conflictive frontier situations, and there are many current ones to consider (USA/Mexico, China/Nepal etc.). As a researcher on obliteration,[1] I am particularly fond of the

[1]The ensemble of procedures of occultation, erasure and manipulation—and in contemporary terms, power usually means capitalism—naturalises its ideologies, renders them unquestionable

J. Minguet-Medina (✉)
Department of Art and Architecture, Universidad de Málaga, Málaga, Spain
e-mail: jminguet@uma.es

G. Cimadomo et al. (eds.), *Urbanism at Borders*, The Urban Book Series,
https://doi.org/10.1007/978-3-032-09594-7_21

excellent works on the Palestinian West Bank of, mainly, Weizman (Segal and Weizman 2003; Weizman 2007), but also—using imaginative cartographies—Bousac (2011) or Lambert (2009), among others (see Fig. 21.1). At their time, they fighted to unveil a civilian occupation that tonnes of propaganda successfully subtracted for years from the collective imaginary. As we now know, this occupation turned into apartheid and, later, genocide (United Nations 2025: 6), in a terrible evolution that completely surpasses the already daring translations that will be proposed here, based on this previous occupation.

Yet, as Weizman states, 'the source of the term *urbicide*', that he precisely defines as 'the destruction of the condition of plurality that defines a city, did not originate in Belgrade, Mostar Grozny or Gaza' (2004:61) but in situations of peace in supposedly democratic countries. He precisely refers to 'the regenerations and 'hygienic' practices of American urban planning such as those described by Marshall Berman after the aggressive 'cleaning up' of the Bronx' (Weizman 2004:61), but once invented, the same processes have travelled successfully all over the world, just adjusting the level of 'aggressivity' to apply (also regarding the level of obliteration and naturalisation required). The forces that we may see at stake in these extreme quasi-war conflicts are not exclusive of them, not even originated by them, but mostly part of a bigger frame of geopolitical and market-driven agendas, equally naturalised and obliterated.

This chapter will expose how, mutatis mutandis, some of these forces also act in the very close environment of the city of Málaga.

21.2 Introduction to the Costa del Sol

The entire line coast to the west of Málaga till Sotogrande, already in the adjacent province of Cádiz, is called the Costa del Sol (Sun Coast) for evident climatic reasons.[2] The Costa del Sol has been widely known as an international leisure paradise since the decade of the 1950s when renowned international artists chose to spend their holidays here, mostly in Marbella, first and already main spot regarding residential tourism. It has always been promoted as a—at least pretended—sophisticated and gay ambiance, full of parties and joy of life, mostly for the wealthy, where exclusivity has always been, and will keep being, a keyword.

Nevertheless, the wealthy have changed through the ages. From the original jet set of European deposed princes and related nobles (the Hohenlohes, Bismarcks,

---

and even pre-rational truths. My as-yet-unpublished PhD. Thesis "Obliteration in Late Capitalism's Architecture" fully deals with the matter (Minguet Medina 2017).

[2] The denomination has been expanded with tourism to Sotogrande on the West, but also to the whole East stripe (from Malaga to the province of Granada), a more natural and virginal area under the increasing threat of importing the exploitative tendencies of its western sibling. Not to contribute to this expansion, associated with the touristic brand, here the Costa del Sol denomination will only refer to the most touristic West stripe of the province's coastline.

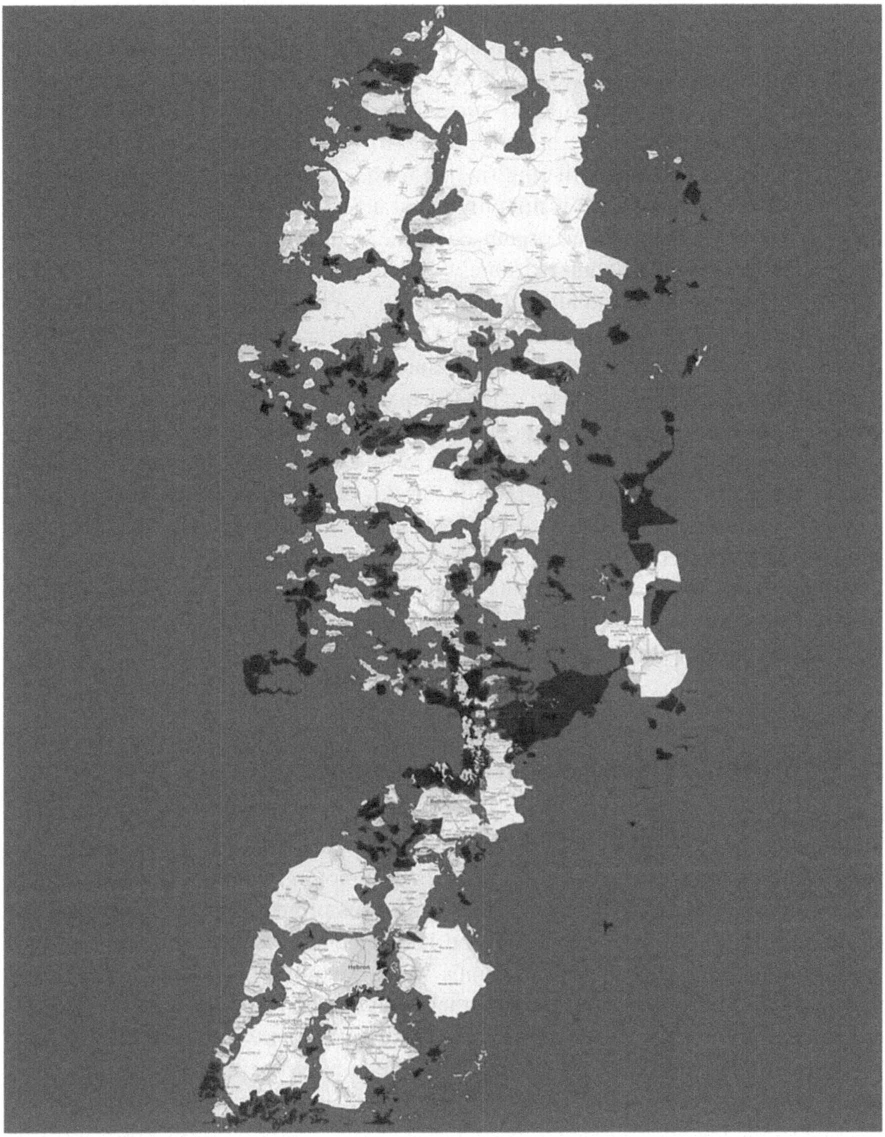

Fig. 21.1 Léopold Lambert, The Palestinian Archipelago (2009), represents the Palestinian West Bank with every Israel colony as a reef and every invaded territory (roads and land under unilateral Israeli control) as sea. Source: Léopold Lambert, Weaponised Architecture: The Impossibility of Innocence, dpr-barcelona, 2012. Creative commons

Soriano y Scholtz von Hermensdorff and de Mora y Aragón) that gave Marbella its primary distinctive impulse, to the Arabian sheiks that joined them later, to the Russian and Ukrainian nouveau riches that seem to proliferate today; the general ambiance has changed several times probably leaving some sense of longing in the air.

There has been an undeniable decay from the gaiety and hippie frivolity of the first 1970s—that happened, not to forget, under the permittivity and protection of an otherwise hard dictatorship—to the extreme ostentation and speculation that came later, and finally to the increasing presence of cartels and mafias that rose crime rate in the area to otherwhere uncompared levels of both frequency and cruelty. As a simple newspaper research will confirm, the Costa del Sol is becoming increasingly related to crime and mafias. It is not only about urbanistic crime, kind of endemic in the area, but also about classic organised crime, like drugs and human trafficking, with increasingly frequent and extremely violent outcomes in the area. As Saviano (2014) pointed out, Mafiosi called the Costa del Sol the *Costa Nostra*, as they used to launder money in the speculative building business. Nowadays international mafias seem to have more to do than just laundry.

As Saviano suggests, urbanism in the Costa del Sol has always been an extremely rentable business, predative and speculative, when not directly illegal, while highly socially accepted. Neoliberal *avant la lettre*, the Costa del Sol, and particularly Marbella at its head, always had faith in 'trickle down economics', that is, the—needless to say, deeply mistaken—belief that the wealth of the richest will somehow bring welfare to everybody. In a land where toponymics are mostly commercial names of private urban developments, when not directly the name of the promoter raised to the status of local hero, it is difficult to say whether prevarication has been more the exception rather than the rule.

21.3 A Preface in the Inevitable Form of Political Corruption

And that has been so, from the very beginning. As early as 1964 the promoter José Banús (one of the previously mentioned hero-promoters) won the consideration of CNIT (Centre of National Touristic Interest) for his project of accommodation and services for as much as 75.000 tourists on 1005 Hectares of land, and with the largest marina of Europe at that time. Banús was a favoured promoter of the Franco regime. He has been involved, among others, in works on the *Valle de los Caídos*—and accused of using prisoner enslaved workforce (Sánchez-Albornoz 2012)—and got the qualification of his CNIT—one of the first ones to be approved and the most ambitious in terms of number of accommodations—less than a year after the promulgation of the law that regulated them. CNIT was an initiative to promote touristic developments via tax reduction and relaxation of urbanistic regulations (Martín and Timón 2006:77). Controversial at the time, it was a very interesting, pre-neoliberal predecessor of later well-known public-private partnerships, as later occurred in BIDs—Business Improvement Districts—and developments as Canary Wharf (Minguet 2014). As the journalist Ana Tudela pointed out:

> It was a bargain. With the Centres of National Tourist Interest, customised town planning plans could be made, soft loans were requested, which in some cases were grants that were thoroughly lost since they were never returned [...] A lot of advantages, associated with the state's commitment to boost luxury tourism to capture foreign currencies. (Calleja 2018)

Fig. 21.2 Nueva Andalucía and Puerto Banús. Source: Ministerio de Medio Ambiente. Guía de playas españolas. https://www.miteco.gob.es/es/costas/servicios/guia-playas/default.aspx

The results were Nueva Andalucía—finished 1978—an urban development that today gives name to one of the five districts of Marbella, and, within it, Puerto Banús—launched in 1970—definitely the perfect symbol for decades of the most luxurious, supposedly elegant but frequently tacky, Marbella's way of life (see Fig. 21.2).

Jumping forward in time, the period from 1991 to 2007 is famous for being the time of G.I.L. (Liberal Independent Group, following the Spanish acronym), a political party with a curious name to be led by a man called Jesús Gil. Mixing and dominating television popularity, football and politics—before Berlusconi did the same in Italy—twice reprieved by Spanish governments of Franco (1971) and PSOE (1994), Gil was an incredible character totally indispensable to understand the Spain of the nineties—its miseries, mainly—and consequently the Spain of today and its endemic malfunctions.[3] The recent and brilliant 4-part HBO's documentary

[3] Reprieve is used here as a translation of *indulto*, the Spanish legal figure for complete pardon, conceded directly by the Government, without the need for further explanations. Conceived to be used only under extremely special circumstances, its abuse from many Spanish governments of different sign—being the double case of Gil of special relevance—gave the legal term shameful connotations.

about him (Bach 2019), significantly called e*l Pionero*, requires a permanent effort of awareness about its non-fictional character, so close it is to its thriller fictional siblings. He was mayor of Marbella during the first 11 years of the period, and later, after being declared ineligible for public office due to his criminal activity, he was succeeded by followers at least as corrupt as himself. From 1995 to 1999 he was also the president of the grouping of municipalities of the Costa de Sol, where his party soon gained relevance and political positions. Having had a provisionally approved PGOU (an urban masterplan for the whole city) since 1986, they never approved it definitively, eluding the required public exposition and sanctioning, and crafting, so, a legal loophole that allowed them to give permits to any urban development just 'by agreement' with each developer. The result is an extremely prolific, completely laissez-faire urbanism that has been elegantly called *a pídola* urbanism (García Vázquez 2008), mimicking a Spanish children game in which one boy jumps over the inclined body of another, and which describes the random, unplanned and unregulated interventions of private developers whereby profit rules over any kind of global results.

The local governments of G.I.L. ended up in one of the biggest urban corruption trials in Spain. The new municipal corporation elected after the debacle proposes the drafting of a new General Urban Development Plan, which first of all focused on detecting the numerous urban infractions that occurred in Marbella during the G.I.L's mandates (see Fig. 21.3). Initially approved in 2007, this Urban Plan reached its definitive approval in 2010, only to be repealed by the Supreme Court in 2015. In their efforts to reverse the scandalous legal situation of the previous responsibles, the planners, the City Council and the regional government had exceeded their attributions, according to the Court. Since then, Marbella has been in a situation of urban planning deadlock, as it has been officially operating with an update of the extremely obsolete 1986 plan, awaiting the drafting of the new plan, the award of which, having faced two appeals, is still ongoing.

Fig. 21.3 Map of illegal interventions in Marbella up to 2007. Green, yellow or red spots indicate the gravity of the infraction. Source: Advance of the general urban development plan of Marbella, 2007

Obviously, the local urbanism of Marbella is not directly translatable to the rest of municipalities in the western Costa del Sol. Nevertheless, it is relevant that the municipality with the longest coastline, and the main touristic and commercial brand of the region has, at this length, functioned over these parameters. The motivations of the rest of municipalities and the pressures suffered by them are similar, when not inherited from those of Marbella, as we will see soon.

21.4 The Costa del Sol Within the Context of Contemporary Global—and Neoliberal—Urban Theories

The urban outcome of so many years of laissez-faire urbanism perfectly fits with much of the contemporary diagnostics about the globalised and neoliberal city. Frequently based on the American Sunbelt Estates conurbations, the terms coined by these theories may significantly apply—as local researchers do—to the Costa del Sol's phenomena. It effectively is a *100 miles city* (Sudjic and Sayer 1993), as local researchers González (2008) and Mediomundo Arquitectos (2006) remark. It is a huge suburbia that does not relate to any centre, but in which 'centrality is virtually ubiquitous, and the solid familiarity of the urban melts into air' as the renowned American geographer, Edward W. Soja (1992) says, in very intentional references to Marx, and even Berman. So, Kling et al. (1995) would call it a 'postsuburbia'*:* an 'environment where what used to be the substance, the urban centre, has become a mere accident; and where the exception, the suburb, has become a rule', as Spanish Carlos García Vazquez recalls (2008).

Local architects Reinoso and Romero (2006) coined the term 'urban territory', in which one wants to see a reference to the classic confrontation between city and country in Lefebvre. Urban territory as something in between urban and rural that is not one nor another, and both at the same time. Richard Skeates simply calls it 'the urban': 'A monster that devours everything and that is swallowing both the city and the countryside and, in doing so, ends the old distinction between the two' (quoted in Graham and Marvin (2001):116–121). A mixing that not even Ginzburg's *disurbanism* could imagine (Mediomundo Arquitectos 2006).

Soja (1992:95) also called it an exopolis, 'the city without, to stress their oxymoronic ambiguity, their city-full non-city-ness', a city that turns itself inside-out and outside-in at the same time. Furthermore, in his later book *Postmetropolis* (2000), Soja compiles a certain state of the art of these kind of characteristics that so many theorists were finding in these new urban sprawls. And there he offers some sharper definitions to describe the peculiarities of our conurbation. We will concentrate on two of them, the two extremes of a pincer that is strangling the territory.

21.4.1 Cannibal City, Golf and Environmental Consumption

'Cannibal city', as described by the Californian sociologist Mike Davis (1998),[4] refers to the consumption of natural as well as human landscapes by this ever-sprawling urban phenomena. He narrates the process of urbanisation of California, years in advance from that of the Costa del Sol but, nevertheless, quite similar, if not its forerunner. The unplanned suburban sprawl squanders huge extensions of landscape by turning them into a non-ending extension of low-density houses, just the starting point of a spiral of resource consumption. Extensive and disorganised as it is, (post-)suburbia's mobility completely depends on the private automobile, requiring wide roads to assimilate an oversized traffic (one man, one car), a myriad of little streets to arrive to every familiar house, and then, also parking: at home, but also to buy, at work, in every service… up to seven parking spaces for each car, as Jane Holtz Kay stresses (García Vázquez 2008), a huge extension of impermeable asphalt. Then, infrastructures are needed: electricity, ICT, gas, water and water treatment to be supplied and maintained for so many and so sprawled; a huge expense for municipalities, as the downsizing cities of the industrialised north of USA have recently proven (Peck 2015). Areas without water supply are even more harmful, as they take the water directly from very limited and fragile aquifers that may, at the same time, be contaminated by their own wasted waters. And then, on the top of that, other apparently unrenounceable requirements: a pool for every single house, foreign gardening species, huge grass extensions, and the king of them all, golf courses.

Golf is a very relevant element in western *Costa del Sol*, also advertised as *Costa del Golf* (see Fig. 21.4). Originated in Scotland as a logical outcome of its landscape, the joyful aspiration to play it in the sun results in an obvious—but insistently repeated—environmental contradiction. As a target destination for British and Nordic visitors, the Costa del Sol, has 48 golf courses, (50% of total Andalusia, and a much higher concentration than in North Spain, where climatic conditions are not so contrary to the needs of golf).

Maintaining such extension of artificial landscape in such a dry country has an obvious environmental cost. It takes 12.5% of the total consumption of water of the area, whose origin in recycling processes is somehow contested. It might not be considered a big figure, just by not keeping in mind that it refers to areas with recurrent annual hydrological deficit (Villar 2010:213–18). Nevertheless, water usage is not the only impact of the spread of golf over the territory. Golf courses substitute big areas of natural landscape with artificial simulations coming from distant latitudes that obviously alter the ecosystem. Furthermore, they substitute open natural spaces with privatised elitist fenced ones, only accessible at high economic rates, becoming barriers against the continuity of the natural landscape and biological environment. Last but not least, and to close the vicious circle, golf courses have

[4]This reference is mentioned by Soja as a compilation of Davis' works. We also cite here other previous Davis' originals as (Davis 1994, 1995).

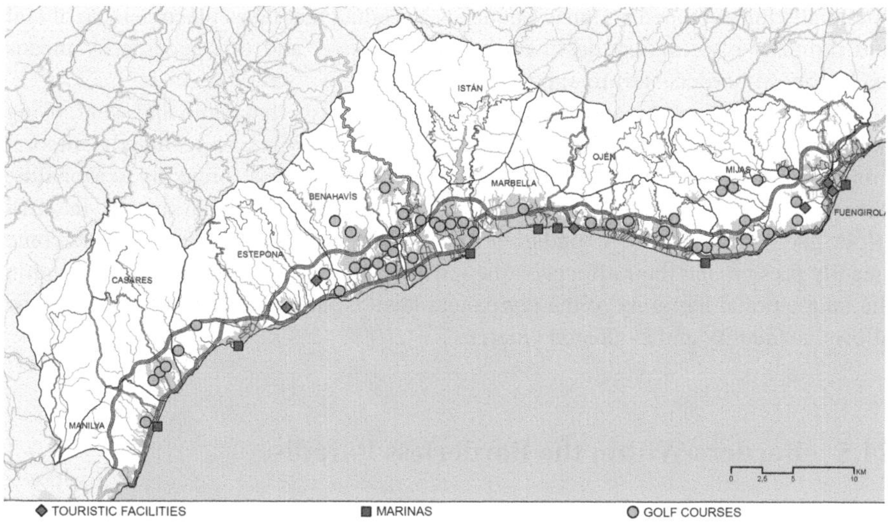

Fig. 21.4 Golf courses in the Costa del Sol based on data from Marbella's Municipal Urban Planning. Source: Territorio y Ciudad, SLP

been extensively used as settlers in the touristic colonisation of the hinterland. In areas too far from the coastline, where not even sea views nor other natural attractions were available, the ubiquitous, artificial nature of golf courses was used as a substitutive stimulus to unceasing urbanisation, now freed from any natural limitations (Villar 2010; Joyanes 2014). As Davis (1998) advises in *Ecologies of Fear*, nature, violently displaced and repressed, may produce violent reactions and increase the frequency of disasters like fires or floods. Particularly, the latter ones have become increasingly regular as an always-increasing uphill surface became urbanised and, consequently, impermeabilised in an area where tempestuous, overwhelming rains suddenly follows drought. Those disasters, as Davis advised, only increase the need for protection, security and the resulting privatisation. And that leads us to the next important contemporary urban concept extracted from Soja's referred compendium (2000).

21.4.2 Privatopia

Coined as a term by McKenzie (1994), the general phenomenon is present all over our culture and has been studied by very reputed theorists. As we started saying, being conceived as an international luxury destination, the keyword in the tourist orientation of the Costa del Sol is exclusivity. It is found in almost all advertisements. The luxury tourist wants the benefits of the location (climate, sea, nature, golf, sports...) but not necessarily mixing up with the local life of a foreign, potentially undeveloped country. Luxury is a product that takes place outside of real and ordinary life. Luxury destinations are exclusive, and this necessarily means exclusion, so borders start to appear.

Initially, there were mainly visiting, seasonal tourists that came and went, using hotels or rented apartments for just a limited period. They were a secluded, temporal minority. Life went on. But after years of tourism success, breaking seasonality, cheaper flights and growth —1.9 million passengers at Malaga airport in 2018 (Rodríguez 2019)—, the boundaries that visitors bring with them have become steadily present and their effects on the territory increasingly unsustainable. Within the international harmony of the permanent leisure paradise that borderless Europe allows, inequality and exclusion emerge.

21.5 Borders Within the Borderless Paradise

On the tourist rental side, the boom of Airbnb and the like has extended the tourist market beyond the more traditional housing one. Every house is potentially a tourist apartment and, as such, it is no longer subject to the local housing market, but to the much more inflationist international tourist market. There is a jump in scale up to, minimum, the European Community market, which is free but not equal. This means that rent prices will automatically rise to be cheap enough for a British or a Danish, but unaffordable for locals. Málaga has become the fourth province in price of housing rentals, with Mallorca—same problem, higher level—being the first, and reaching almost 2.5 times the rental prices of the adjacent and also coastal province of Granada. Following the same report, the financial exertion of rental prices in Marbella is the second of all Spain municipalities, reaching 156.08% of the average income, almost five times the reasonable partake of the total, which must be around 34% (Mitula Group 2018). Estepona, its neighbouring city, is the fifth, with 106.36%, and closing a leading group of touristic destinations. The city has become unaffordable to its original citizens, forcibly uprooted from their own hometown. This has made true what Reinoso and Romero (2006) said some years ago, almost as a *boutade*: 'Capital Málaga—the provincial capital, at the extreme east of Occidental Costa del Sol—really works as a dormitory town of the Metropolitan Zone of the Costa del Sol', and farther and lesser localities will soon add up.[5]

But this will already expand exclusion since, as in its American models, mobility resides—exclusively—in private cars, and not having one means to be excluded to a badly reputed slow and limited bus service. Trains only reach Fuengirola, quite close to Málaga, and are difficult to extend due to the chaotic and massive sprawl of the city. Even cars are classed among those using the free but very dangerous and

[5] In the same work, and together with Joaquín de Salas (2006), they talked about 'freelance city', referring to the economy of workers in the area, and anticipating, already during good times, the precarisation of work to come with the crisis.

collapsed A-7 highway and those using the fast, highly taxed AP-7, both dividing as much as connecting territory.

Residential exclusion is not restricted to the Costa del Sol, nor even of touristic destinations, but to every city submitted to global markets rather than local ones. Local workers cannot afford their rents. Who will clean, who will serve? ‘Riches in turmoil as local workers abandon London’, sarcastically enounces a simulated newspaper of an activist group, with the image of a starving fake Boris Johnson looking for something to eat among the garbage containers, as no one served his meals (Bangla Sanglap Desk 2016). Inequality driven to exclusion is a sign of our times, it is happening in every big city and it is reaching nightmare overtones in locations as Hong Kong, as shown in the fantastic work of the Chinese photographer Benny Lam (Lam 2017a, b).

There are many borders in ‘privatopia’. More than in regular tourism, contemporary Costa del Sol expands over a model of residential tourism, that is: foreign residents that buy houses to live in, at least during part of the year. The housing market, as well as the rental one, becomes therefore international and unaffordable for most locals. The financial exertion to buy a house in Marbella, measured by the time an average family needs to pay the mortgage dedicating all their incomes to it, is the third of Spain with 42.68 years. That means about 125 years, if we limit mortgage expenses to one-third of income, as recommended and verified by banks when granting mortgages. The Costa del Sol building market is reserved for very affluent locals, but mostly for northern immigrants, for which a house in the sun costs less than one in their cold countries. Extreme paradox: the southern sun becomes a patrimony of the northerners.

However, they want the sun, the weather, the sea, but not necessarily the country, not necessarily all that Spain means. They—we all—frequently are ‘mixophobic’, in Bauman terms; they want to live in their own *spheres*, in Sloterdijk’s sense. They create national communities of expats (they have a funny word to call themselves, and naming means excluding), frequently with their own clubs, shops, schools and clinics, reproducing their habits, uses and communities in the sunny South, taken only as an improved background for unchanged routines. Real estate operators, aware of their taste, guide them to buy within their own national community settlement, frequently in gated communities, excluded and secure.

To name this privatopias, García Vázquez (2008) takes a term from Kotkin (2000) that very specifically fits our proposals: *the Valhallas*, Nordic paradises in the sunny south. This has been, as described, the growing pattern for the Costa del Sol. In the coastal municipalities, where growth started long time ago and has spread over time, these existing and numerous enclosed paradises are mixed with hotels, dwellings and other developments, and not so easy to trace,[6] but in recent developments, farther from the coast, they have become monoculture.

[6]The lack of transparency in town halls as Marbella’s one, where information, if existing, is extremely difficult to get, makes the task even harder. On the contrary, Benahavís information—even if not complete—has been easy to find and use.

21.6 Benahavís as a Perfect *Valhalla*, and the Uncanny Case of Its Palestinisation

Urban Planning Professor Pedro Górgolas Martín, whose analyses are sharp and technical, wisely signs Benahavís as the municipality that better exemplifies these tendencies. To the north and over the municipalities of Marbella and Estepona, Benahavís has wonderful sights of the Mediterranean, but no territorial contact with it from its sloppy beautiful landscape. Its registered population growth goes from 2116 inhabitants in 2001, to 7989 in 2018, the 64% of which (5095) were foreigners.[7] A spectacular demographic growth of 377% almost strictly based on immigration, that leads to an even bigger urban sprawl. From 1993 (date of approval of its General Urban Plan) Benahavís has urbanised 69 hectares a year, covering an extension like cities with 15 times its population, as the nearby Algeciras. Furthermore, it has done it with a ratio of 2412 square metres of urban land by registered inhabitant, which means an incredible density of two dwellings by hectare of 'urban' terrain. The respective parameters for a reasonably compact city are 100–150 sq.m./inhabitant and 40–45 dwellings/ha (Górgolas 2019). With 15 luxury housing developments organised around nine golf courses and built under these unparalleled parameters of expansion, Benahavís has become the Valhalla itself. A vast expanse of paradise for the 1%, inaccessible to the rest, has been irreversibly privatised for the benefit of a few, stealing it from both nature and society.

Saving huge distances of suffering, Benahavís has its similarities with Palestinian West Bank as theirs was also a supposedly pacific 'civilian occupation' (Segal and Weizman 2003) in which the land that has been privately 'colonised' will not easily, if ever, return to the community. Assuming this, a map like those of Boussac and Lambert could be traced (see Fig. 21.5); one in which the privatised gated communities and golf courses would turn into water, like Atlantis disappeared from the territory. Rather than an archipelago, Benahavís will look like a marshland, with huge lagoons breaking the continuity of the landscape and through which communications and roads become tortuous.

Looked over in Google Earth, both Israel civilian colonies in the West Bank and the sunny Valhallas of Benahavís look similar: hilly shaped red roofs suburbia which seems to be the shape of power. The top of the hills, Weizman explains, has always been related to geostrategic power, since they control the surrounding territory, as Israel colonies do with the nearby Palestinian cities and camps, or Benahavís mansions do over the fantastic sea views, the contemporary exclusive symbol of economic power. Funnily enough, the way to detect both is the contrary. In the West Bank, the little Street View man only enters Israel territory.[8] Benahavís' Valhallas, oppositely, are inaccessible even to Google and its street documentation viewer.

[7] Data from the Instituto de Estadística y Cartografía de Andalucía.

[8] This is under evolution from 2017, since Google Earth incorporated Ramallah to Street View, but still noticeable.

Fig. 21.5 Benahavís marshlands. Bousac/Lambert's type of plan in which gated communities and Golf courses inside Benahavís municipality turn into water. These water bodies have been named before the names of their original real-estate developments. Please, verify how Puerto Banús could fit inside one of them, to get an idea of its scale and extension. Highlighted in the center, the original town of Benahavís. Data: Benahavís town hall, Google Earth. Image base: Google Earth. Source: Jorge Minguet

21.7 Expanding the Model

Even if Benahavís is the most representative case of this kind of urban sprawl, it is not an isolated one. If we check out the expansivity index of the nearby municipalities in similar conditions (see Table 21.1), we will verify that Benahavís is only an element of a series. The 'urban expansivity index', as defined by Górgolas (2019), is the relation between the urban land (*suelo urbano*) and the land expected to be urbanised—that is, the building land (*suelo urbanizable*)—during the development of the General Urban Plan, and it expresses the will to grow of the urban extension inside the municipality. It is limited by regional planning (regulation 45.4 of the Territorial Ordenation Plan of Andalusia, POTA) to a maximum of a 40% of the existing urban land, and systematically and abusively exceeded by those municipalities of the Costa del Sol that, as Benahavís, have a promising hinterland to explode, or even first line undeveloped lands, on those farther from the relevant original nodes.

Except for Casares—that Górgolas studies as a successful case of wise decreasing model reacting to the 2008 financial crisis situations and new needs—and Ojén

Table 21.1 Expansivity indexes of the municipalities of the Costa del Sol. Data: Municipal Urban Planning. Source: Territorio y Ciudad, SLP

Municipalities	Urban land area	Total area of land for development (zoned, sectorised and non-sectorised)	Area of land for development/urban land
Benahavls	1.720.79 ha	2640,07 ha	153,42%
Casares	777,60 ha	54,8732 ha	7,05%
Estepona	1.404,95 ha	1.734,58 ha	123,46%
Fuengirola	784,054 ha	183,064 ha	23,35%
Istan	36,28 ha	132,50 ha	365,22%
Manilva	598,34 ha	1.903,10 ha	318,06%
Marbella	4.799,69 ha	1.344,53 ha	28,01%
Mijas	2.633,74 ha	2756,75 ha	104,67%
Ojen	258,98 ha	112,26 ha	43,34%
TOTAL	**13.009,44 ha**	**10.861,73 ha**	**83.5%**

at the limit of POTA's recommendations, the cities similar to Benahavís (Estepona, Manilva and Istán) have from very high to extreme indexes. Those reflect the greedy expectation of an unceasing increase in the housing market, based on the permanent arrival of foreign newcomers,[9] in a non-ending loop of land consumption, and a definitive dependence of the local economy on the foreign inversion to come. They reflect the total submission of the municipalities of the Costa del Sol to the neoliberal and unsustainable model based in selling the natural and communal resources to the very rich, following the requirements of the market and not those of their citizens.

Górgolas proposes a new urbanism that tries to 'make urbanism for citizens, and not as a financial asset to be packaged and securitized' (Górgolas 2019: 89) using Casares' turn to degrowth as a model, in an exposition that should be on the desk of most majors and urban planners of the area. On the contrary, as he knows, this is an 'addition of cities for non-citizens' (Mediomundo Arquitectos 2006)—and continues to act mostly like this. As already mentioned, when defining the Costa del Sol, Romero and Reinoso called it 'Urban Territory', which '(...) does not mean that it is a city. That is, a place where there are citizens with rights and duties, because many are not even registered', they said. The model responds mostly to the requirements of those non-citizens who will not take part of the share of communality. So, paradoxically, in its submission to the capital, town halls end up drained of resources, both natural, sold to the rich, and monetary, never returned from them in many cases: unsustainability driven to its highest degree. Due to this situation, but also to the aforementioned historical and recurrent illegal speculation and expensive urban sprawl, the citizens of the Costa del Sol pay a lot of taxes for relatively scarce public services and spaces, not to mention the absolute lack of housing policies, so necessary to compensate the exponential increase of rental prices, also seen before. In the Costa del Sol, as Trump intended to do in Mexico, those on the bad side of the border pay for the wall.

[9] A reference to the film *Welcome Mr. Marshall!* by Luis García Berlanga (1953), which offered an early critique of Spain's naïve yet opportunistic reliance on foreign aid as a means of economic survival.

21.8 Obliteration: Neoliberalism Gets Inside our Minds

Paradoxically, citizens, for their part, faithful believers in the neoliberal dogma of trickle-down economics, attach themselves to their immediate survival—and precarisation—against long-term thinking, and vote for those who push the machinery forward. They probably think *pourvu que cela dure*, as Letizia Bonaparte, or as Madame de Pompadour: *après moi, le déluge*. Marx (1986: vol. 1 part 3, chap. 10, secc. 5), who said that this late phrase was 'the watchword of every capitalist and of every capitalist nation' would be terrified by neoliberalism's ability to permeate the minds of the new order of precariat.[10]

As a meaningful example of this, the obituary of the ex-major of Benahavís in 2014 consisted of just two messages: one, he was the father of the present major—already on charge, what means 24 years of uninterrupted mandate between the two of them— and two, he completely cleaned up the indebted town hall accounts during his term, figures included (Extremera 2014). No further personal references to him nor to his legacy as a major beyond genealogical succession and positive economic balance in a perfect confluence of both the traditional and modern neoliberal right ideologies adequately subsumed and transmitted along the journalistic chain to the collective mind.

Furthermore, and as dreams are the last frontier of neoliberal expansion (Crary 2013), in a strange sort of 'trickledown dreaming'—that instead of its economic sibling seems to really exists—the elitist dream of cosmopolitan leisure paradise for the riches anyhow permeates down to locals. They, apparently assuming real-estate advertising, get infected of the maked up dream of being fortunate for living—or surviving, who cares—in paradise. This phenomenon of masking precarity, has been named as the 'myth of good-living' (García Vázquez 2008: 93. See also Reinoso and Romero 2006; de Salas 2006).

So, somehow, everybody is happy in paradise, even if the apple is more than bitten. How is life outside its borders?

21.9 Other Borders

Just travel to the West end of the Costa del Sol residential paradise, and there lies Gibraltar, its financial sibling. Besides, it is the fulfilled dream of a borderless European Community, a border, an old fashioned geopolitical one, a colony,[11] as those supposed to be ended during mid-twentieth Century. If the lack of borders produced so many borders, may this physical border, somehow, be emancipating?

---

[10] Updating capitalism to neoliberalism, as well as proletariat to precariat.

[11] These areas are referred to as 'territories without self-government, subject to decolonization' by the United Nations, or as overseas territories under EU interpretation—even though no concrete process of decolonisation is currently envisioned. See Miguel and Abellán (2019).

A spot taken from Spain in 1713, extremely little but strategically positioned to control the strait, the difference that Gibraltar introduces in its environment has always been conflictive, and not only by geopolitical reasons. Differences in the taxation system between Gibraltar and Spain (or Gibraltar and the UK, to which it belongs) are huge, with a corporation tax of 10% (25% in Spain, 20% in UK), a top tax of 20% over highest gross incomes (51.9% in Spain, 45% in UK) and no capital gains tax, wealth tax, sales tax or value added tax. Even not fitting the EU and OCDE parameters for fiscal paradises—which significantly also excludes Switzerland—Gibraltar acts locally as one, a fiscal vortex attracting every inversion in the area[12] and beyond, economically draining its surroundings.

At the other side of the border, La Línea de la Concepción is one of the poorest municipalities of Spain with a 34.99%—the first in Spain—in terms of unemployment among cities with more than 40,000 inhabitants (Expansión 2024). Such high shares of unemployment, as generally known, are only possible in the presence of mature underground economies which, in the case of La Línea, is usually imputed to drug traffic across the strait, as the news regularly confirm.

With about 10,000 of the 62,940 La Línea's inhabitants—about half of the remaining active population[13]—working in Gibraltar, its economy is completely dependent on that of the colony, while much poorer. The income per capita in La Línea is about a third of that of Gibraltar (Cañas 2018). So, if borderless Europe spatialised inequalities, asserting submission in a sort of a soft, obliterative manner, this border, reinforced by artificial economic measures, does it in a very sheer way.

Brexit, and especially the fear of a hard one, intensified this submission by making the people of La Línea beg for abiding the very conditions that exploit them, just when negotiations for their betterment were in a sweet spot (Peláez 2019). 'Better slave than dead', should they think. A hard Brexit will further increase the economic and financial gap between both sides of the border, while complicating, if not extinguishing, the labour conditions of cross-border workers—also a life-or-death problem for Gibraltar's economy. Brexit arrangements relating to Gibraltar are, at the time of the latest revisions to this publication, still pending (González 2024).

21.10 Conclusions

Borders have historically reflected balances of power. So, in the era of capitalism as the main ruling power, they just spatialise its logics of inequality and domination that inevitably emerge in every situation. That's how the dream of market and

[12] The number of enterprises located in Gibraltar is also in dispute, from the 13,736 active tax paying enterprises recognised by Gibraltar (InfoGibraltar 2019), to the 55,000 or even 80,000 mentioned by the Spanish press (ABC, 2018; El País, 2019). For 30,000 registered inhabitants, even the lesser figure shows an abnormal concentration.

[13] Extracting the population younger than 20 and older than 65 (following the Regional Institute of Statistics of Andalucía), and subtracting the 30% of unemployed, the total active population of La Línea is about 20,000 people, only a third of the total population.

movement freedom has, after years of blind faith in neoliberalism—even before it was named so—permeated all over society, from the leader to the latest citizen, through trickle-down dreaming, more than economics. It led to a certain 'Palestinization' of the international leisure paradise of the Costa del Sol, territorialising power and exclusion through the emergence of apparently soft, but equally inviolable borders and subtractions, that is: in the classic, old-fashioned way. Paradoxically in terms, but not with the final target, old-fashioned borders as those of Gibraltar equally assert inequalities in the balance of power.

The logics of 'Actually Existing Neoliberalism' (Brenner and Theodore 2002), that conquered the rests of democracy getting inside our minds and dreams through insistent obliteration, adapt themselves to every scale or context, insistently repeating its same mantra under different shapes. Brenner, among others, wisely points out the question of the need for politics of scale—state interventionism not only in the national scale but also in new supra et infra national organisations able to adapt to and solve the new emerging problems, at their appropiate scales. But this requires a high degree of political will and collaboration among administrations that, oppositely, seem to belong more with competitive neoliberalism.

Local communities and municipalities rebel to possibly emancipating higher scale logics in order to keep their short-sighted, submissive, precarious and unsustainable, even if currently functional economies of dependence. *Pourvu que cela dure…*

La Línea, despaired of waiting for the state to favour them, despite its own geopolitical interests, demanded political autonomy (Cañas 2018), to predictably adjust its economic conditions to Gibraltar, what would have probably caused nothing but a slight geographical displacement of the problems from their municipality to the next one. Even if, fortunately, both political and legal powers of the nation denied them this possibility (Website of the General Council of the Judiciary 2023), the mayor persists in his attempts (A. R. 2024). The municipalities of the Costa del Sol, oppositely spoiled by a regional government that cares of its correct—lucrative but sustainable—development, systematically disobey their regulations in favour of the unsustainable, predative frequently illegal and submissive model that we have been extensively depicted here.

Higher scales of the state, equally submitted to the very same global and neoliberal political economy pressures, frequently lack both the authority and the political will to fulfil their tasks in the new politics of scale announced by Brenner (1999) and his likes. The disobedience of the Costa del Sol municipalities has been as ignored by the regional government as its regulations were by them. Spanish negotiations with Gibraltar frequently submit long-term solutions in favour of La Linea's immediate needs, instead of displaying a liberating long-term economic aid plan for them.

Moreover, the annulment of Marbella's General Urban Plan, on the grounds that it had exceeded the powers of the local and regional administrations, was never followed by a State Plan to assume those powers and address the existing problems. Consequently, these problems remained confined to the local level, whose administrations, having been legally stripped of their authority precisely for exceeding their competences, remained unable to provide effective solutions.

Fig. 21.6 The double fence of Melilla after a collective jumping over it. Source: fronterasur Creative Commons

Growing in scale to the highest form of government reached in the area to the moment, the EU just provides a proportional increase in the violence in which borders are expressed. The effort to achieve a paradisiacal Europe without borders—which, as we have seen, is repeatedly cracked by the economical forces of the unified market, and its own internal contradictions, such as Gibraltar and other tax havens—is based on the extreme use of force towards the outside world. The border fences of Ceuta and Melilla—Spanish Gibraltars in África,[14] the only terrestrial borders with the southern continent—are as high and menacing as they can be,[15] and still insufficient to dissuade the desperate immigrants to keep on trying to cross it, risking and sometimes losing their lives (see Fig. 21.6). The difference at both sides of the border is sometimes that from death to life, as they come from zones whose conflicts are not unrelated to Western countries industrial and commercial interests. By sea, things are even worse. 2297 attempting immigrants died in the Mediterranean in 2018, a descending figure from the terrible 5143 dead of 2016. NGO's rescue boats are repeatedly denied docking in EU ports, and legally pursued

[14] They are technically different but geographically similar kind of colonial insertions. Here, there is no need for artificial financial techniques to establish a huge, life or death difference.

[15] The persistent promises of the Spanish government to remove the cruel and unpopular concertina wires are repeatedly delayed and rely on the reinforcement of the measures on the Moroccan side. A superficial makeover to translate the cruelty outside the European border.

for saving lives in the Mediterranean, as if EU economies could not afford to assume this 2297 people among its 508 million (0.00045% of the population, maths does not lie).

'Actually Existing Neoliberalism' seems to control every scale just because, after decades of unceasing obliteration, it reigns in our minds and creates our dreams. We measure everything, from life and death at our frontiers, to the lives of our majors just through sheer, short-sighted economy. If we hope to change these dynamics in favour of more human and sustainable ones, our politics of scale should start at the smallest scale: that of our minds and our dreams. Expelling neoliberalism out of them should be at our reach, though not without effort. Let's take back control.

Acknowledgements This research has been founded by the Research Project PID2023-151195OA-I00 financed by MICIU/AEI /10.13039/501100011033 and by FEDER, UE. The author would like to thank the University of Seville professor and expert urban planner Pedro Górgolas Martín for his help in properly documenting this article with bibliography, data and images, the latter coming from his office Territorio y Ciudad, SLP.

References

ABC (2018) Gibraltar, el "paraíso fiscal" de la discordia. ABC. https://www.abc.es/espana/abci-gibraltar-paraisofiscal-discordia-201811231109_noticia.html. Accessed 8 Mar 2024

A. R. (2024) Franco retomará el proyecto de la ciudad autónoma para La Línea si hay referéndum en Cataluña. Europa Sur. https://www.europasur.es/lalinea/Franco-proyecto-ciudad-autonoma-referendum-cataluna_0_1865514547.html. Accessed 8 Mar 2024

Bach E (2019) El pionero. Documental. Home Box Office (HBO)

Bangla Sanglap Desk (2016) Rich in turmoil as workers desert London. Bangla Sanlap. http://www.banglasanglap.co.uk/?p=3741 Accessed 11 July 2020

Bousac J (2011) L'Etat-Archipel de Palestine. Blog. Obgeographiques. http://obgeographiques.blogspot.com/ Accessed 29 June 2020

Brenner N (1999) Globalisation as reterritorialisation: the re-scaling of urban governance in the European Union. Urban Stud 36:431–451. https://doi.org/10.1080/0042098993466

Brenner N, Theodore N (2002) Cities and the geographies of "actually existing neoliberalism". Antipode 34:349–379. https://doi.org/10.1111/1467-8330.00246

Calleja J (2018) 'Playa Burbuja': 15.000 kilómetros en ruta para explicar el 'boom' del ladrillo en el Mediterráneo. Verne, September 1

Cañas JA (2018) La república independiente (y "harta") de La Línea. EL PAÍS. November 28

Crary J (2013) 24/7: late capitalism and the ends of sleep. Verso, New York

Davis M (1994) Cannibal City: Los Angeles and the destruction of nature. In: Smith EAT, Ferguson R, Museum of Contemporary Art (Los Angeles, Calif.) (eds) Urban revisions: current projects for the public realm. Museum of Contemporary Art and MIT Press, Los Angeles, Cambridge

Davis M (1995) How Eden lost its garden: a political history of the L.a. landscape. Capital Nat Social 6:1–29. https://doi.org/10.1080/10455759509358648

Davis M (1998) Ecology of fear: Los Angeles and the imagination of disaster, 1st edn. Metropolitan Books, New York

de Salas J (2006) Posición afirmativa de la potencia urbana de la Costa del Sol. In: Atributos Urbanos. Centro Andaluz de Arte Contemporáneo, Sevilla

El País (2019) Gibraltar sin paraíso. El País. https://elpais.com/elpais/2019/03/04/opinion/1551720370_448737.html. Accessed 8 Mar 2024
Expansión (2024) Paro por municipios 2024. datosmacro.com. https://datosmacro.expansion.com/paro/espana/municipios. Accessed 8 Mar 2024
Extremera F (2014) Fallece el exalcalde de Benahavís Antonio Mena Gracia. La Opinión de Málaga. March 15
García Berlanga L (1953) Bienvenido Mr. Marshall. Uninci
García Vázquez C (2008) Análisis del Territorio 2. Costa del Sol: Asentamientos. In: Costa del Sol: Arquitectura, Ciudad y Territorio. Consejería de Vivienda y Organización del Territorio. Junta de Andalucía, Sevilla
González P (2008) Propuesta. Costa del Sol—Ap7. Antecedentes, presentes y futuros. In: Costa Del Sol: Arquitectura, Ciudad y Territorio. Consejería de Vivienda y Organización del Territorio. Junta de Andalucía, Sevilla
González JR (2024) Gibraltar se juega evitar un Brexit duro en los próximos seis meses. El Debate. https://www.eldebate.com/espana/andalucia/20240101/gibraltar-juega-evitar-brexit-duro-proximos-seis-meses_163682.html. Accessed: 8 Mar 2024
Górgolas P (2019) Del «urbanismo expansivo» al «urbanismo regenerativo»: directrices y recomendaciones para reconducir la herencia territorial de la década prodigiosa del urbanismo español (1997–2007). Aplicación al caso de estudio del litoral andaluz. Ciudad y territorio. Estudios territoriales Vol. LI: 81–100
Graham S, Marvin S (2001) Splintering urbanism: networked infrastructures, technological mobilities and the urban condition. Routledge, London, New York
InfoGibraltar (2019) Sobre los paraísos fiscales y Gibraltar. Blog. Infogibraltar. https://infogibraltar.com/documentacion/a-fondo/sobre-los-paraisos-fiscales-y-gibraltar/ Accessed 11 July 2020
Joyanes MD (2014) Génesis y Evolución del Golf: de fragmento de paisaje a pieza de ordenación del territorio. Incidencia de los campos de golf en la conformación del espacio turístico Costa del Sol. Málaga (tesis inédita). https://riuma.uma.es/xmlui/handle/10630/8107
Kling R, Olin SC, Mark O (1995) Postsuburban California: the transformation of Orange County since World War II. University of California Press, Berkeley
Kotkin J (2000) The new geography: how the digital revolution is reshaping the American landscape, 1st edn. Random House, New York
Lam B (2017a) Boxed in: life inside the 'coffin cubicles' of Hong Kong—in pictures. The Guardian. http://www.theguardian.com/cities/gallery/2017/jun/07/boxed-life-inside-hong-kong-coffin-cubicles-cage-homes-in-pictures. Accessed 11 July 2020
Lam B (2017b) Benny Lam nos ofrece una nueva perspectiva de las «casas ataúd» de Hong Kong. National Geographic. https://www.nationalgeographic.es/photography/2017/07/benny-lam-nos-ofrece-una-nueva-perspectiva-de-las-casas-ataud-de-hong-kong. Accessed 11 July 2020
Lambert L (2009) The Palestininan archipelago. In: Weaponized architecture: the impossibility of innocence. dpr, Barcelona
Martín LG, Timón DB (2006) Los Centros de Interés Turístico Nacional y el despegue del turismo de masas en España. Investigaciones Geográficas:73–93
Marx K (1986) *El capita*l. Orbis, Barcelona
McKenzie E (1994) Privatopia: homeowner associations and the rise of residential private government. Yale University Press, New Haven
Mediomundo Arquitectos (2006) Atributos Urbanos. Centro Andaluz de Arte Contemporáneo, Sevilla
Miguel RD, Abellán L (2019) La UE definirá a Gibraltar como "colonia" por la presión de España si hay Brexit duro. El País. https://elpais.com/politica/2019/02/01/actualidad/1549024494_425881.html Accessed 13 Mar 2020
Minguet J (2014) Urbanismo participativo o urbanismo democrático. Crisis y crítica. Revista Cidades 11:196–233
Minguet Medina J (2017) Obliteración en la arquitectura del tardocapitalismo. Málaga (tesis inédita). https://riuma.uma.es/xmlui/handle/10630/15485?show=full Accessed 8 Feb 2024

Mitula Group (2018) Informe Trimestral sobre el Precio de la Vivienda en España 2o Trimestre 2018. Madrid
Peck J (2015) Framing Detroit. In: Who saved the City? vol 1. University of Edinburgh: Global Justice Academy
Peláez JM (2019) Gibraltar ya no es un paraíso fiscal. elEconomista.es. https://inspectoresdehacienda.es/wp-content/uploads/filr/1436/20190312_ElEconomista_tribuna_Gibraltar_pelaez_.pdf Accessed 13 Mar 2020
Reinoso R, Romero JM (2006) Territorio ZoMeCS. Atributos Urbanos de la Zona Metropolitana de la Costa del Sol. In: Atributos Urbanos. Centro Andaluz de Arte Contemporáneo, Sevilla
Rodríguez JV (2019) El aeropuerto de Málaga logra un nuevo récord en 2018 con más de 19 millones de pasajeros. La opinión de Málaga. January 14
Sánchez-Albornoz N (2012) Cárceles y exilios. In: Biblioteca de La Memoria 29, 1st edn. Editorial Anagrama, Barcelona
Saviano R (2014) Roberto Saviano: "El dinero que ha hecho explotar la burbuja española provenía del narcotráfico" Interview by Jordi Évole. La Sexta
Segal R, Weizman E (eds) (2003) A civilian occupation: the politics of Israeli architecture. Babel, Tel Aviv
Soja E (1992) Inside exopolis: scenes from Orange County. In: Sorkin M (ed) Variations on a theme park: the new American city and the end of public space. The Noonday Press, New York
Soja E (2000) Postmetropolis: critical studies of cities and regions. Blackwell Publishers, Oxford
Sudjic D, Sayer P (1993) Hundred-Mile City. Flamingo, London
United Nations (2025) Annual Report of Committee on the Exercise of the Inalienable Rights of the Palestinian People to the General Assembly (A/80/35). United Nations. https://www.un.org/unispal/document/ceirpp-2025-annual-report-01sep2025/
Villar A (2010) Territorio, turismo y paisaje: el proceso de urbanización en el litoral de Andalucía. El papel de los campos de golf, PhD. Thesis. Universidad de Sevilla. http://www.juntadeandalucia.es/turismoycomercio/publicaciones/143334969.pdf
Website of the General Council of the Judiciary (2023) El Tribunal Supremo confirma la negativa del Gobierno a autorizar una consulta popular en La Línea de la Concepción (Cádiz) para ser comunidad autónoma. Poder Judicial. https://www.poderjudicial.es/cgpj/es/Poder-Judicial/Tribunal-Supremo/Noticias-Judiciales/El-Tribunal-Supremo-confirma-la-negativa-del-Gobierno-a-autorizar-una-consulta-popular-en-La-Linea-de-la-Concepcion%2D%2DCadiz%2D%2Dpara-ser-comunidad-autonoma
Weizman E (2004) The evil architects do. In: Koolhaas R, McGetrick B, OMA, Nationalgalerie, AMO (eds) Content: triumph of realization. Taschen, Köln, pp 60–63
Weizman E (2007) Hollow land: Israel's architecture of occupation. Verso, London; New York

Chapter 22
The Social Frontiers of the Sun-and-Beach Tourist Space: Duality and Socio-Spatial Fragmentation in Paradise

Carlos Rosa-Jiménez

Abstract The sun-and-beach massive tourist space is socioeconomically a highly aligned space that builds physical urban borders, socio-spatial segregations and labour exploitation. Through a historical review, this chapter analyses this problem and establishes three major urban processes: (1) the dual city, (2) the functional division of space and (3) urban isolation bubbles. The first process is based on the invisibility of the existing city and on the aesthetic construction of an artificial urban scenery for the tourist. In the second process, functionalism created a bipolar urbanism between worker settlements and hotel complexes. Finally, in the third process, the social discontinuity with the territory generated tourist sites disconnected with the cultural reality of the host country. This chapter concludes that the basis of the problem lies in the role of the state as guarantor of social equilibrium, and describes three projects that propose a tourism project open to all social classes and the breakdown of social and economic boundaries of tourism activity: La ciutat de repòs i de vacances in Barcelona (Spain) as an example of tourist city without social classes; the modernisation of the bourgeois spa city in La Grande Motte (Languedoc-Roussillon, France) and finally, the Regional Plan for Agadir (Morocco) that recovery the waterfront as a new meeting place for the tourist and the local population.

Keywords Enclaves · Tourist development · Dual city · Fragmentation

22.1 Introduction

Since the mid-twentieth century, the development of the economic activity of tourism has had an impact on the physical transformation of cities and territories worldwide. In addition, important spatial and social boundaries were generated in the territories where it was settled due to the tourist's own nature. While the traveller

C. Rosa-Jiménez (✉)
Institute for Habitat, Territory and Digitalization, Universidad de Málaga, Malaga, Spain
e-mail: cjrosa@uma.es

G. Cimadomo et al. (eds.), *Urbanism at Borders*, The Urban Book Series,
https://doi.org/10.1007/978-3-032-09594-7_22

moves for various reasons (leisure among them), it is integrated into the cultural reality of the destination; it uses lodgings or residences, and even family homes; the tourist buys an experience through an economic transaction, which has been advertised, based on the tourist resources of a territory, using existing urban structures or creating new ones.

One of the strongest developments of tourism in the twentieth century was characterised by the sun-and-beach tourism on the warm coastlines. This tourist space has two economical and social intrinsic tendencies. First, as an economic activity, it tends to be a physically enclosed space, where it can be adequately exploited; therefore, it creates special conditions of safety, comfort and security for the tourist. An antecedent to the current sun-and-beach resorts is found in medieval European monasteries, where according to Ruiz Esteban (2013: 22): 'the perimeter marks the properties of the religious community [...] and, in a private context, compartmentalize the units that make up the enclosure'. These are precise limits that clearly delineate what is inside and what is the exterior. The physical limit can be a wall, a fence (opaque, or transparent), but they can also be natural elements such as forests, locations where natural conditions make access impossible (such as cliffs in coves), or simply isolated locations from populated settlements.

Second, tourist space tends to be a socially segregated space, as a consequence of its origins within the wealthy and elitist idle class. Tourism was born as an activity of the upper and elitist class, the only ones with sufficient purchasing power to carry out a leisure displacement prior to mass tourism; and this marked elitism continues to have an impact on the type of urban space generated by tourism. The democratisation of tourism from the second half of the twentieth century was parallel to the birth of sun-and-beach mass tourism, and triggered the development of an important economic and spatial transformation of many coasts worldwide. Democratisation meant the approach of the spatial formulas developed by the elite—aristocrats and increasing numbers of the bourgeoisie—to the middle classes, since tourism expands from the wealthy classes to the middle classes, from top to bottom (Urry 1990; Boyer 1999).

This chapter is structured in two parts. In the first, we will analyse the current segregation of the tourism space as a three-phase evolution: (a) social tourism segregation, (b) economic segregation and functional duality and (c) economic segregation and social enclaves. In the second part, three sun-and-beach tourism projects are analysed as examples of alternatives to each type of segregation. Finally, the conclusions are presented.

22.2 The Socioeconomic Segregation of Tourist Space: From the Dual City to the Tourist Encapsulation

22.2.1 Socio-Spatial Segregation: The Coastal Tourist Urbanisation

Privacy and segregation are qualities of elitist tourism that are architecturally manifested in the tourist urbanisation. This is closed par excellence, and the higher the social level, the greater its tendency to isolation, avoiding contacts with the local population. The creation of tourist urbanisation is an evolutionary process in the behaviours of the social elite that we differentiate into four levels. The origin is found in *differentiated enclosed spaces* (type 1), which correspond to specific recreational areas located within palace complexes delimited by walls or defensive systems. A well-known example is the Generalife real estate farm in the Alhambra, where the importance of its gardens and viewpoints stood out. A second type is named *isolated enclosed spaces* (type 2), which are the second residences of the nobility and the bourgeoisie, located in the urban peripheries and composed of a residential palace with a garden, in enclosed and fenced spaces. From here a third type emerges, defined as *the grouped enclosed space* (type 3), formed by the concentration of second residences (type 2) that constitutes the classic tourist urbanisation, and that imitated the construction of pavilions and palaces in the coast carried out by the royal houses. The aristocracy and the upper bourgeois promoted these urbanisations. In Spain, we have examples such as the urbanisation of S'Agaró (1924) promoted by the businessman Josep Ensesa and designed by the architect Rafael Massó. It is an urbanisation of isolated houses or chalets, whose isolation and protection of the well-off class were achieved with the repopulation of pines and the purchase of the entire surrounding environment (Barba and Pié 1988). This model begins to bloom throughout the Mediterranean, where several examples stand out. In Sardinia (Italy), Prince Karim Aga Khan presided over the consortium of companies that developed the urban plan of Costa Smeralda, a luxury complex that includes private access by jets and helicopters. On the Costa del Sol (Spain), Alfonso de Hohenlohe bought the 'Santa Margarita' farm (1953) to found the Marbella Club Hotel, beginning the elite development of urbanisations in Marbella (Málaga).[1]

[1] *NE:* See also Jorge Minguet-Medina Chap. 20 in this volume.

22.2.2 *Urban Social Segregation: The Dual and Bourgeois Coastal spa City*

Many European seaside and coastal cities experienced an important tourist development that had its peak in the nineteenth century, due to the role of royal houses, being the first to build pavilions and palaces on the coast. The implementation of this elitist tourist space on the existing city had an impact on the preceding socio-economic reality. The coexistence with the new social class resulted in the development of a dual city. Since spatial limits cannot be made in public space, segregation occurs on an architectural scale with restricted access to the wealthy class in tourist facilities (hotels, casinos...), and on the urban scale, with the development of *concealment strategies of reality*. The concealment of inadequacy was one of the classist and elitist characteristics of tourist destinations prior to mass tourism (Pié and Rosa-Jiménez 2014). An example of a dual city is found in the German city of Wiesbaden, which evolved from a thermal town from the eighteenth century to a residential city, *ennobled* by the presence of royal houses. The duality supposed a clear social and physical division of the urban space. In 1805, the traditional city became *invisible* by prohibiting traditional activities—such as the passage of cattle—and, by the creation of aesthetic ordinances (1811) where it was forbidden to lay clothes outdoors (Sica 1981). The physical duality was established with the creation of a new residential-tourist space of isolated houses with gardens in the form of mansions. An urban space is formed by closed residential units, in which the garden is the protagonist, being the stage where games, parties and meals are held. As a dividing element between the two urban realities—the historic city and the new tourist urbanisation—a large linear park is built where the casino is located as the main tourist facility.

The shift of this dual model to the coast, in search of cold sea baths, evolved into a new model: a tourist space built on the edge of the sea. An example is the city of Nice, a prototype of a coastal seaside city of the nineteenth century. In the city map of 1891, *La Promenade des Anglais*[2] was the public corridor that separates the urban realm from the beach, and it is the axis that articulates the important growths of villa-with-garden neighbourhoods; but unlike Wiesbaden, the development of leisure facilities (theatres and public establishments) is highly increased as is the number of urban hotels (Nice had 61 urban hotels). A model that is replicated in numerous coastal spa cities and continues to preserve social duality by rendering the existing social reality invisible. For example, in the case of Malaga, the poor- and low-income population was a problem of security (theft), social (begging) and aesthetic (it broke the imaginary of paradise) in its process of transformation into a spa city. As a solution, the Propagandistic Society of Climate and Embellishment of Malaga (1897) that was born with the intention of converting this city into the 'Nice

[2] Plan de la ville de Nice avec indication de tous les noms de rues, établissements publics, monuments, hôtels et banques (1891). Available at https://www.lexilogos.com/nice_plan.htm

of the South' proposed the imprisonment of the poor and beggars in order to favour bourgeois tourism.

22.2.3 Socioeconomic Segregation: Functional Duality in the Sun-and-Beach Tourist Space

World War II was a turning point in the development of mass tourism, supported by the right to paid vacation for workers, the creation of the Welfare State in Europe, along with the consolidation of a growing middle class. Tourism ceases to be an activity reserved for the elite and becomes a massive economic activity called the tourism industry.

In this context, the enclosed tourist hotel has become the main protagonist of coastal developments up to the present day. Unlike the urban grand hotel, the beach hotel develops an important leisure programme in the landscaped areas—which increase considerably in surface—creating closed pieces of large plots, very inspired by the sanatoriums (Jiménez and Vargas 2017), and organised around the pool. The tourist hotel maintains the same dual social relationship of the owner and the service staff of the bourgeois housing. This social duality is transferred to labour relations, since the workers are usually the local inhabitants who do not need much professional qualification, while the owners are the guests of the hotel.

Functionalist architecture strengthens this duality: Louis Kahn's theory of separation of served and server spaces—as a principle of organisation—assumes the existence of two overlapping buildings (Ábalos and Herreros 2000:140). In the same way, in tourist spaces there are two superimposed realities: the front and the back. According to MacCannel (1973) the tourism front area is the place where hosts and guests meet, and the tourism back area is the place where workers retire.

Along these lines, Aldape Pérez (2010) deepens this analogy in relation to the architectural program. If in bourgeois housing, the *front* is the meeting place of the hosts with the guests, while the *back* is the place of rest and relaxation of the service; in a hotel, the *front* is the meeting point of the clients with the service staff (receptions, hallways, lounges, restaurants, playgrounds and leisure areas), while the *back* includes kitchens and laundry rooms.

Tourism expands this model also in the territory. Goffman (1959) has already celebrated a social division between front and back regions. An example of this planning is found in the Territorial Plan of the Costa del Sol designed by Gómez y García de la Buelga (1966). The plan projects a tourist model based on residential urbanisations and tourist facilities. This plan is based on the network of existing urban settlements, giving each of them a tourist function. Thus, it differentiates between: (a) service population centres (server spaces) and (b) tourist poles (spaces served). The tourist poles are located on the coast (front), while the service population centres are located both on the coast and in the interior (back). It should be noted that most of the interior service villages also have the category of typical towns, so, in addition to concentrating the population of services, they are also a tourist destination that completes the sun-and-beach offer—tourists opt for service cities because of their sociocultural value. However, this duality did not necessarily imply a distribution of tourists and residents. On the Costa del Sol of the 1960s and 1970s, the influence of tourism meant a social revolution and a very positive opening in a country subject to a dictatorial regime and a Catholic national ideology (Rosa-Jiménez and García-Moreno 2018).

22.2.4 Socioeconomic Segregation: From Enclave to Coastal Tourist Encapsulation

Tourism enclaving is a process of the spatial and social segregation of tourists from local residents (Edensor 1998; Weaver 2005; Schmid 2008). Edensor (1998) classifies tourist space as either *enclavic* or *heterogeneous*. While heterogeneous spaces have more imprecise limits and a lower degree of planning, *enclavic* spaces have clear and well-defined limits. They are also 'carefully organized and designed, regulated, planned, commercialized and privatized' (Edensor 1998:45). The design of the enclaves encourages segregation between tourists and workers, while allowing the exploitation of local attractions (cultural and natural) and local labour (Manuel-Navarrete 2016), as a continuation of the colonialist model based on relationships master-server (Mowforth and Munt 1998), and with little economic benefits for local communities (Freitag 1994). In the *enclavic* character of organised tourist spaces 'tourists are shielded from potentially offensive sights, sounds and smells' (Edensor 1998:45).

This kind of segregation can be observed mainly in developing countries. A paradigmatic case study is Cancun (Mexico), promoted by the state institution responsible for planning and development of tourism projects: the National Tourism Promotion Fund (FONATUR), which facilitated—since 1950—the entry of international capital as an engine of economic development. The Master Plan of Cancun (1982) combines social with functional duality in the same urban project with a strong socioeconomic specialisation. Cancun is planned in three main sectors: (a) the linear tourist sector along the beach, formed by a mixture of luxury tourist and residential accommodation, where the main hotels are concentrated in closed enclaves with access to private beaches; (b) the residential one for the upper middle class in the form of residential super-zones and (c) the housing for the lower class, in the Puerto Juárez neighbourhood designed to accommodate hotel workers (Aldape Pérez 2010: 56).

Cancun uses the middle-class sector to act as an urban barrier between the tourist corridor (front) and the working-class habitat (back). A spatial isolation accompanied by typological, density, and land-use segregation, where there is no point of interaction between tourists and locals except through the guest–service relationship in hotels. Although the Master Plan reserved 11 public beaches as meeting places between locals and residents, today many of them have been parcelled and sold, and only four of them have services for recreational use (Aldape Pérez 2010:158–9).

Brenner (2005) states that typical problems of similar tourist projects, especially social segregation, burgeoning shantytowns and the formation of economic enclaves, could not be solved. The benefits derived from tourism activity reproduce capital accumulation schemes and the economic and social benefits do not reach the local population (Arroyo Arcos et al. 2015). On the contrary, López-López et al. (2006) show an increase in spatial segregation due to the development of activities in favour of tourists and locals with high purchasing power, reinforced by

> [...] the strategic location of the resorts on uneven terrain, which allows these centres to take over large expanses of beachfront and render them inaccessible to the local population. (López-López et al. 2006:162)

Tourist enclaves in developing countries evolve into a new model that we call tourist encapsulation, which supposes a radicalisation in the spatial and social segregation of the enclaves. The most evident result is the tourist urbanisations' evolution into gated communities. This process coincides with a new generation of hotels with a luxury experiential environment that has three key components: immersion and multisensoriality, theming and hyper reality (Batat 2019).

Minca (2000) has studied in Indonesia the process of re-territorialisation of mature tourist spaces towards extreme forms of spatial segregation, both functional and physical. Unlike the enclaves, here 'micro-universes are recreated; an implosion which attempts to capture and reconstruct a timeless and deterritorialized essence of place' (Minca 2000:390). This process is the consequence of the inability of tourism development to generate a parallel social development in its environment, but rather, quite the opposite, a deepening of poverty and other social impacts. The physical limits of the hotels no longer define an economic activity, but are already defensive limits of an environment that has become hostile to tourists by beggars, street-merchants and souvenir-sellers. This leads to an increase in security, with the development of private security forces, due to the insecurity and social deterioration—drug use, prostitution and loss of identity—generated by tourism activity in the social environment between tourists and local population.

The enclaves become shelters of the problems generated by the tourist activity itself. But it is also accompanied by a functional segregation in the territory that defines:

> [...] a reconfiguration of the surrounding areas as well: assigning peripheral functions to nearby urban centres, generating specialized satellite poles and decentralizing into luxury enclaves privileged segments of the tourist demand; structuring, according to hierarchical principles, a space dominated by tourist territoriality. (Minca 2000:392)

22.3 The Social State as Guarantor against the Breaking of Limits

The state has the ability to influence the configuration of tourist spaces. In developed countries, their capacity is much more intense than in developing countries, where security problems even force the state to promote the segregation of tourists with the local population (Schmid 2008), or attraction of capital has facilitated the development of colonialist models based on the duality of tourist space. Tools available to the state for breaking consolidated boundaries can take the form of legal measures or planning and projection strategies. An example of the legal tools is the Spanish State Coastal Law. This regulation grants ownership and public access to the land domain of the coasts, even in the case of concessions such as marine urbanisations. In this way, Puerto Banús (built in 1970), a world leader enclave for elite tourism, where the most exclusive luxury stores are located, is a space open to tourists of any social level. A space created by and for the elite, where local people and tourists can freely access through its streets and docks. For planning and projection tools, we will analyse three cases as alternatives to the three types of segregation discussed above: the rupture of the dual city, the rupture of the functional city and the rupture of the enclavic city.

An example of the rupture of the dual city was proposed in the functionalist project of the satellite city of *La ciutat de repòs i de vacances* (1931–1928) close to Barcelona (Spain). Prior to World War II, the Catalan architects of the GATCPAC (The Group of Catalan Architects and Technicians for the Progress of Contemporary Architecture) designed a workers' tourism model radically opposed to the bourgeois seaside resort. Based on a space of second open and equal residence, intended for the enjoyment of the beach and sports, the community block prevails over single-family housing. Its objective is to 'satisfy a social need of the middle and working classes, and it does not need casinos and luxury hotels at all' (GATCPAC 1932: 25). The main novelties of the project were (a) the suppression of the seafront as a social scenography of the bourgeois elite, (b) the suppression of the hierarchical urban project and (c) a strong functionalist zoning articulated in four zones. The first one was the bathing area designed to accommodate mass holiday tourism, particularly those engaged in water sports. The others were the weekend areas, consisting of sport areas and small housing units; residential areas, with hotels, modest houses, camping and pine parks and finally, the rest-cure area with sanatorium hotels.

France raised the rupture of the functionalist duality with the Languedoc-Roussillon territorial project, which included the tourist city of *La Grande-Motte* (1960), designed by architect Jean Balladur. La Grande-Motte is a reinterpretation of the bourgeois model of a spa town, characterised by zoning principles (four zones: campsites, villa, marina and the large park), a strong development of infrastructures and a unified design project in collaboration with landscape designer Pierre Pillet.

La Grande Motte shows the social optics of a society that is modernising at a frenetic pace (Prelorenzo and Picon 1999:8). Unlike *La ciutat del repòs i de vacances*, this project involves a reinterpretation of the spa town, but with a breakdown of social segregation in favour of the French and European middle class. On the one hand, the residential project (similar to the stations of the nineteenth century) seeks to offer French middle-class citizens all urban comforts. On the other hand, the grand hotel trilogy, based on casino and bathrooms of the great seaside resorts, is replaced by a congress centre, sports areas, a church and a market. But, like the *La ciutat del repòs i de vacances* project, there is no promenade where the main buildings are located (Grand hotel, casino and bath establishment). However, unlike this one, no real estate was located in a frontal position but perpendicular, all in order to avoid the privilege of sight and in this way democratise visuals to the sea (Prelorenzo and Picon 1999:168–9).

Finally, an example of the rupture of the enclavic encapsulation is the Regional Plan for Agadir (2003) by Flora Pescador and Vicente Miravalle. Both propose the recovery of the waterfront as a new meeting place for the local population. They designed the construction of a new maritime promenade in a mature tourist space based on enclaves, which occupied the entire maritime front of the city. This operation allows citizens to enjoy their beaches Picon hitherto restricted for tourist use and hotels Picon and to recuperate the relationship with the sea.

22.4 Conclusions

The role of the social state is fundamental in breaking the boundaries created by tourism. When the state has exclusively supported economic development, without considering social development, tourism enclaves have been enhanced. These places, although highly planned, are based on a functional and spatial duality, and on the economic specialisation of the tourist territory.

The result is a social division and the inability to generate a public space for social relations. Eliminating the boundaries and limits of tourist spaces is a complex undertaking that requires striking a delicate balance between the economic specialisation of tourist developments and the social coexistence of the public space. While the economic aspect of tourism is undeniably important, the solidity of the social state serves as the foundation for a fairer and more balanced tourist space. When tourist developments are concentrated in specific areas, it can lead to the exclusion or marginalisation of the local population. Local residents may face rising costs of living, displacement or loss of cultural identity as the tourism industry takes precedence. The social fabric of a place can be strained when public spaces become commercialised and primarily serve the interests of tourists, rather than fostering a sense of community and inclusivity.

The problems analysed in this chapter, as well as some examples of tools developed to avoid the emergence of socioeconomic barriers, are summarised in Fig. 22.1. Nonetheless, prioritising the solidity of the social state is essential to ensure a fairer and more balanced tourist space. This means considering the needs and well-being of the local community alongside economic considerations. Public spaces play a

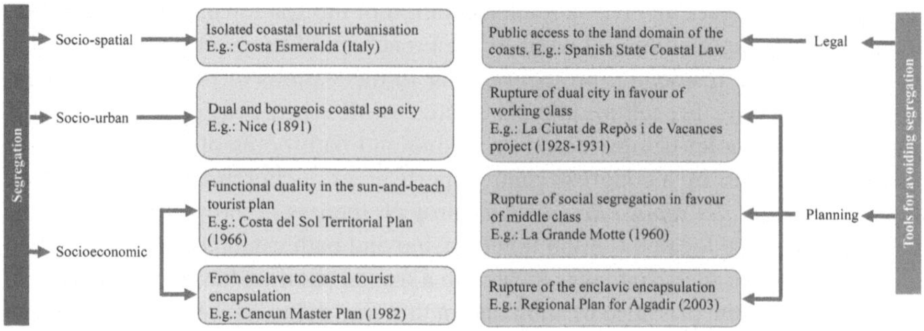

Fig. 22.1 Main segregation process in sun-and-beach tourist space and tools for avoiding these processes

vital role in facilitating social coexistence and fostering a sense of community. They should be protected and preserved as areas for social interaction, cultural exchange and the well-being of all residents and visitors. Balancing the economic interests of tourism with the needs of the local community requires careful management of public spaces.

In conclusion, eliminating the boundaries and limits of tourist space requires striking a delicate balance between economic specialisation and social coexistence. While the economic aspects of tourism are important, it is essential to prioritise the solidity of the social state to ensure a fairer and more balanced tourist space. Inclusive planning processes, protection of public spaces and investment in social infrastructure are key elements in achieving this goal. By taking into account the needs of both tourists and locals, it becomes possible to create a tourism industry that benefits all and contributes to the overall well-being of the community.

References

Ábalos I, Herreros J (2000) Técnica y Arquitectura en la ciudad contemporánea (1950–2000). Editorial Nerea, Hondarribia

Aldape Pérez G (2010) La configuración del espacio turístico en Cancún, Quintana Roo, México. Thesis archive Polytechnic University of Catalonia. https://upcommons.upc.edu/handle/2117/94293?locale-attribute=en. Accessed 26 June 2019

Arroyo Arcos L, López López A, Segrado Pavón R, Serrano Barquín R, Frausto Martínez O (2015) Diferencias socio-territoriales en Tulum. Una ciudad organizada a partir del turismo masivo litoral. CULTUR: Revista de Cultura e Turismo 2:4–28

Barba R, Pié R (1988) Espontaneísmo, recursos y problemas del turismo costero catalán. Urbanismo COAM 4:40–46

Batat W (2019) Experiential setting design. In: Batat W (ed) The new luxury experience. Springer, Cham, pp 91–112

Boyer M (1999) Histoboyerire du tourisme de masse. Presses Universitaires de France, Paris

Brenner L (2005) State-planned tourism destinations: the case of Huatulco, Mexico, tourism. Geographies 7(2):138–164

Edensor T (1998) Tourists at the Taj: performance and meaning at a symbolic site. Routledge, London
Freitag TG (1994) Enclave tourism development: for whom the benefits roll? Ann Tour Res 21:538–554
GATCPAC (1932) La ciudad de reposo que necesita Barcelona. AC Documentos de Actividad Contemporánea 7:24–31
Goffman E (1959) The presentation of self in everyday life. Doubleday, New York. Guston
Gómez y García de la Buelga J (1966) Plan general de ordenación urbanística de la Costa del Sol Occidental. Arquitectura 85:45–52
Jiménez E, Vargas IC (2017) Hotels & Sanatoria. The influence of tuberculosis on mass tourism architecture. Hist Cienc Saude Manguinhos 24(1):243–260
López-López Á, Cukier J, Sánchez-Crispín A (2006) Segregation of tourist space in Los Cabos, Mexico. Tour Geogr 8(4):359–379
MacCannel D (1973) Staged authenticity: arrangements of social space in tourist settings. Am J Sociol 79:589–603
Manuel-Navarrete D (2016) Boundary-work and sustainability in tourism enclaves. J Sustain Tour 24(4):507–526
Minca C (2000) 'The Bali syndrome': the explosion and implosion of 'exotic' tourist spaces. Tour Geogr 2(4):389–403
Mowforth M, Munt I (1998) Tourism and sustainability: new tourism in the third world. Routledge, London
Pié R, Rosa-Jiménez C (2014) La cuestión del paisaje en la reinvención de los destinos turísticos maduros. Málaga y la Costa del Sol. ACE: Archit City Environ 9(25):303–326
Prelorenzo C, Picon A (1999) L'aventure Du Balnéaire. La Grande Motte de Jean Balladur. Parenthèses, Marseille
Rosa-Jiménez C, García-Moreno AE (2018) El impacto de las TICs en el imaginario urbano. Nuevas herramientas digitales en la dialéctica del territorio líquido frente a la ciudad tradicional. Imagonautas 11:82–106
Ruiz Esteban N (2013) En los límites de la Arquitectura. Espacio, sistema y disciplina. PhD Thesis. Polytechnic University of Catalonia. http://hdl.handle.net/2117/94911. Accessed 22 May 2019
Schmid KA (2008) Doing ethnography of tourist enclaves: boundaries, ironies, and in-sights. Tour Stud 8(1):105–121
Sica P (1981) Historia del urbanismo: siglo XIX. Instituto de Estudios de Administración Local, Madrid
Urry J (1990) The tourist gaze: leisure and travel in contemporary societies. Sage Publications, London
Weaver A (2005) Spaces of containment and revenue capture:"super-sized" cruise ships as mobile tourism enclaves. Tour Geogr 7(2):165–184

Part VI
Community Struggles for Local Inequities

Chapter 23
The Regeneration of Place with Community-Based Architecture

Vilmos Katona and Bogdán Funk

Abstract This study focuses on community-based architecture and seeks to show the theoretical and practical aspects of the topic. The authors try to bridge theory and praxis by providing multiple parallel approaches. The outlined agenda for the regeneration of place in this study is deeply rooted in the contemporary understanding of architectural phenomenology and the latter's possible mathematical metaphors. The authors extend the topic of genius loci also to places where it is not expected, and, based on less-known realised projects, they claim that terrain vague or border zones may host creative community activities. These may develop new forms of collaboration, define novel aesthetics and connect to the local cultural heritage less academically. Moreover, they find that combining natural materials and contemporary design skills may facilitate intercultural knowledge exchange. Besides the author's personal experiences as designers and builders, a realistic account of the local community feedback is also an integral part of this chapter.

Keywords Architectural theory · Place-making · Regenerative architecture · Community-based architecture · Architectural education

23.1 Introduction

Organised space can be seen as a product of centralised political power (Lefebvre 1974), but it has its borders. Therefore, the space lacking control is considered weak (de Solà-Morales 1996), and borders are understood as a resistance to uncertainty.

V. Katona (✉)
Research Institute of Art Theory and Methodology, Hungarian Academy of Arts, Budapest, Hungary
e-mail: katona.vilmos@mma-mmki.hu

B. Funk
József Cziráki Doctoral School of Wood Sciences and Technologies, University of Sopron, Sopron, Hungary

G. Cimadomo et al. (eds.), *Urbanism at Borders*, The Urban Book Series,
https://doi.org/10.1007/978-3-032-09594-7_23

Terrains between borderlines are always defined as 'nobody's land'. However, borders also lie within the fabric of society, which implies that the nexus between precisely defined circles of people or activities bridges social weaknesses and uncertainties. This depicts a 'dark tree' of voids with roots from outside and growing into the centre, shooting a branch structure on each scale. The architecture of voids deriving from this is a fractal of the unknown.

Another interpretation of borders implies a different viewport, because nobody's land is a potential chance for occupation due to the absence of order. A clearing in the forest welcomes a new hut to be built (Sharr 2006), an airport lounge allows travellers to regenerate, a state border offers refugees the promise of sanctuary and, finally, demarcation zones often become the stages for ecological revival. If not part of a larger body, life wants to evolve, form communities and extend its spatial structures, thus it encourages the generation of new places and interaction. The act of building reproduces its archetypes from what is already there, nature and culture, given that the latter is capable of migrating. Thus, the structure of voids excluded by sociocultural borders openly invites knowledge, habits and techniques from abroad, yet always remains local due to its limited material resources.

23.2 Borders in Art Philosophy

In his famous essay, *The Origin of the Work of Art*, Martin Heidegger (1971) suggests that *techne* is not a craft of creating art products, but a continuous play between two extremities: being and non-being. That is to say, the work of art unveils itself from obscurity, and, with the same effort, it fades out the material, or banishes it from existence. However, in exchange for this 'unethical' move, we may again focus on the materiality of the work of art, and forget about its more complex entity. Heidegger emphasises this dual play instead of the product and its meaning, suggesting that art is never finished, but has a dynamic existence. It would be easy to say that art, and by extension architecture, has an immanent 'spirit'. That is the case when we read the story told in its language, the story of coming forth into being. The observation of architecture appearing in and reshaping the material world involves both the creator and the recipient. They both witness the exact sacred moment, rather than making or perceiving a product. What is embarrassing about this moment is that buildings, like works of art, catch people before they can perceive them in detail (Böhme 1991). This is why coming into existence falls under the a priori domain. Therefore, the border between experience and non-experience is tied to the one between existence and non-existence. This has many implications, since a priori has always been a problem of phenomenology.

It is not clear how to outline the knowledge before experience. For instance, is the colour (whether the red is red or the blue is blue) experienced, or is it just known? How would we tell since we never experience pure colours independently from physical shapes and mental connotations? Yet, we claim absolute confidence when defining colours or describing atmospheres (Böhme 1991), so this discrepancy needs to be resolved. As von Hildebrand (1976:63–138) suggests, the intelligibility of a priori is not a special ability of a few but is available to everyone. This also implies that the 'secret' in the origin of the work of art is not concealed, indeed, since everyone has direct knowledge about it, even those not able to perceive the work of art through the five senses. Following the same reasoning, one must not cross a border to know it is there. But borders are to be crossed, because this is the fundamental law of phenomenology. It is highly motivating that borders as peripheries become centres whenever crossed (Katona 2018), which describes the spring nature of art's origin. This description also leads to another question: Is crossing borders a kind of art? This may depend on the definition and, again, the new frontiers that the work of art seeks to reach. Crossing cultural conventions, for example, is not necessarily art in a phenomenological sense, because cultures, such as theologies, are parallel systems (Heidegger 1998a), and there is no real ontological difference between them. However, establishing a new *modus vivendi* may exceed the limits of unawareness and begin to conflict with other ways of living. Nor in this case does the act of crossing have to do with the consequences of cultural confrontation, but the phenomenon of appearance. The latter tells its own story without regard to its later accidental oppositions. Cultures are predominantly present in all artworks, weaving and integrating the context, thus forming a place for memories. In this way, 'new' memories may emerge from the unconscious and hidden.

This is also a source of discomfort when trying to preserve information and avoid oblivion. Information carried by messages, symbols and narratives nourishes ways of living. Borders enable them to do so, as the earth provides the foundation (see Fig. 23.1) and stabilises a village hut (Heidegger 1978; Sharr 2006). One could aptly ask whether art is timeless, then? We cannot answer this with certainty. At least it must have an ignition point in time, since it creates its era and chronology. Hence, it does not conquer the totality of being. Still, it is somewhat similar to a half-line or, using a more complex mathematical model, a torus, provided that art's ignition point also serves as a ceasing point. This was best described by Heidegger (1967:36-40) with the antagonism of *Anwesen* and *Abwesen* (emerging and perishing presence) (Hernández 2011).

Architecture may have supposed universal qualities such as beauty, yet it is essentially a poetry of matter-form *(eidos)* (Heidegger 1998b:210), and magical details (Frascari 1996). Even Nietzsche acknowledges the latter's idea, suggesting that beauty is secondary to the stability and power that tectonic details deliver (Harries 2008). Geometry and ratios may capture the 'consumer' with beauty (Salingaros 2018), but their proper function is to create order (McEwen 2018). It is problematic if this order derives from the inner nature of existence or is used externally as a product by the nihilist mind that is eager to rule (Cacciari 1982). Whatever the answer, borders drive art and may differentiate between these approaches, depending on whether the creator is considered a poet or a planner.

Fig. 23.1 Residents at the Women's Centre in Rufisque, Senegal, a symbol of urban palimpsest in marginalised societies. Source: Helena Sandman

23.3 The Mathematical Problem of Place

Logically, we cannot generate an atmosphere out of nothing, for it is already there when we focus on the building as a product. It is not primarily because the house's materials have their subtle frequencies (Zumthor 2006:29–33), but because the building is an intersection of more complex entities. As Christian Norberg-Schulz (1980:17) explains in his book *Genius Loci*, buildings only concretise what is already there in the forms of nature and cultural background, or, let us say, they 'speak out' the implications of the two. Nevertheless, how should we grasp the predominant atmosphere of nature and culture without the work of architects?

Urban landscapes may 'reprint' natural landscapes, such as the Via Sacra in Rome, which reprint the forest and ravines under the rural plateaus, embracing the whole landscape of the *Campagna*. Monuments may represent mountains, like the Egyptian pyramids in Giza, or honour their insurmountable magnificence, like the 18th dynasty mortuary temple of Hatshepsut beneath the cliffs at Deir el-Bahari, near the Valley of the Kings. Most of history's best-known architectural monuments lead back to nature, as the interior 'forests' of Gothic cathedrals may provide plenty of examples. Even today, parametricism has virtually met the growing algorithms of plants, fungi and crystals (Gruber et al. 2011).

Cultures have also ruled nature-inspired plans as artificial landscapes of human society. According to Norberg-Schulz (1980:179), the identity of a place is determined by location, general spatial configuration and characterising articulation. The latter means well-known features (Norberg-Schulz 1976), such as the roof pitch, the window frames and the height of cornices or plinths. Kenneth Frampton (1983) picked up this thread and stated that all these characteristics may be derived from natural and cultural circumstances. Annual rainfall and temperature may determine the inclination and the cladding of the roof, not to mention the average size of the family that occupies the attic. It is well known, for instance, that a chalet in Switzerland has three storeys: the basement traditionally used for storing hay, the elevated ground floor serving as a living room and the bedroom on the upper level. In comparison, the average dwellings of the Pueblos in the Southwestern United States are flat-roofed, densely occupied and attached closely to each other for the different climatic conditions, materials and lifestyle of their people.

The stories imprinted in stone and space exist before builders manifest them. These narratives are inherent in the natural and cultural context of the place, assuming that the place is the hidden conjunction of the two. To make this conjunction understandable, one must construe it through construction (Frascari 1996:500). It means that one either reads or writes (builds) architectural details. However, the question of the need for architecture to express the complexity of a place arises.

Why here-and-now instead of a subtle, or rather metaphysical, omnipresence? This argument can be illustrated with the following mathematical analogy. In the non-professional use of mathematical language, we often mix the concepts of dimension and extent; however, this confusion may camouflage the true nature of elemental geometry. By mistake, a dot is usually called zero-dimensional or dimensionless, since it has no extent in any of the dimensions. A dot can possess as many dimensions as we like, because it can be any of the nodes inside a multidimensional system. It is interesting that we quite generally indicate that point with a cross. It means a point lacks the extent, but we may recognise it as an intersection of two straight lines. We also need to note here that a line has an extent described with one dimension, but straight lines are not necessarily one-dimensional (1D). They may be parts of any multidimensional system, just like dots.

Further, straight lines with a single extent also intersect two planes with two-dimensional extents. This defines a series: every fundamental element with extent n is an intersection of two basic elements with extent $n + 1$, given that n is a natural number. Following this rule, the plane must be an intersection of two different Cartesian spaces, which is not easy to imagine. It gets even harder as we increase the number of extents: space is the intersection of two 4-dimensional spaces, the theoretical realms of polytopes (Coxeter and MacDonald 1973). We live in a three-dimensional space where architecture is composed. We can apply the former mathematical analogy if we consider it an intersection of two higher entities (nature and culture). Supposing that nature and culture are 4D entities, they are incomprehensible to our senses so long as they do not meet. They create a less complex, tangible and three-dimensional intersection when they do. Thus, architecture manifests what is already there (see Fig. 23.2). However, to experience the once imperceptible, a conjunction must occur between the higher entities. This is a sacred moment, when the 'gods', invoked by man, marry to establish a new era on earth (Eliade 1959:32–36). According to our analogy, the role of mathematics, often associated

Fig. 23.2 METI School before completion, Rudrapur, Bangladesh. Source: Kurt Hörbst

with modern science and technology, is not a discipline distant from the matter of place. Instead, it gives us an alternative language to grasp the meaning of *genius loci*.

23.4 The Birth of Nowhere

This sacred act had outlined both the archaic world and the classical antiquity of the Greeks and Romans. *Genius loci* also inspired Medieval Christianity, although it introduced a new type of transcendental sacredness. This made a clear difference between the celestial and the earthly natures of humans, places and arts, and thus created a new set of things considered unworthy to call sacred. The shift gave up the view that connected the totality of the known world with one of cosmic origin, for a view that selected 'good' out of the whole and left the rest out of consideration. Theology sought only the absolute Good, and relegated the 'meagre' to vanish as profane. Places of worship integrated social life, market and politics under the vaults of enthralling basilicas (Földváry 2002), yet their auras did not cover the whole city with its candlelit taverns, filthy alleys and dormant houses.

The *polis* lost meaning and no longer integrated nature, countryside and urban environment, drawing borders between centres and peripheries. Cathedrals were interpreted as exclusive sources of light (Timmermann 2000); however, the ideals of the sacred city prevailed in the visions about Heavenly Jerusalem. We may read about these ideals and the strictness of transcendental borders in the works of Dante

Alighieri, best known for his *Divine Comedy*. In his description of the spiritual world, the profane have no salvation, yet the chosen ones rejoice in eternal peace. Such a universe lacks balance and transparency, and there are only a few distinguished adventurers, like Dante himself, who may visit all its subrealms. Led by Virgil, Dante meets the once magnificent persons trapped in the fallen world, and presumes to bring their fame back to light to compensate for the terrible losses of knowledge. During the evolution of medieval Christianity, the spirit of place was replaced by memorial spaces. While the depth of poetry quarantined the concept of immanence, the cult of relics moved masses of pilgrims to meet Christian ancestors. On the contrary, the Eucharist was omnipresent in every church and was renewed by every holy mass. There were no longer differences between places regarding the one and unquestionable essence, transcendence. Thus, the 'universal place' was born, and churches appeared to formulate an almost universal liturgical space as its physical frame. *On the other hand, Genius loci* sank into the collective subconscious of the Middle Ages—Dante's *Inferno* (Brockman 2017)—with all its human nature and uniqueness.

Interestingly enough, after the great century of Historicism, modernity returned to the formal concept of the ancient Roman city without a real cultural cohesion (Norberg-Schulz 1971:96–109; 1980:189–195). High-rise architecture of the West, then the East, occupied orthogonal urban grids, while the Gaussian skylines of modern cities literally portrayed real estate price diagrams. In opposition to contemporary downtowns, the generic city of the suburbs (Koolhaas 1995) became horizontal fields of economic fluctuation. Thus, modern megalopolises grew liberally, but denaturalised from any local character (Ricoeur 1965). The glass 'cathedrals' of the modern age (Droste 2002:18) were similar to the Christian ones in one aspect: they were both symbols of omnipresent order, different from the organically innate. Moreover, profane order sprawled as a shadow effect of the former ecclesial and cultural conquest as soon as the latter lost momentum. Consequently, the architecture of roughly the last millennium has set ideals somewhat against the traditionally heterogeneous *genius loci* —we cannot agree with Frédéric Debuyst's (1997) primaeval *génie chrétien du lieu*—yet it is nothing to compare with the effects of the last century's globalisation. Despite their universalism, Christian churches could aptly imply 'otherness'. Symbols of the fallen world were selected, 'baptised' and reused, yet globalisation—moreover, 'mondialisation'—could not interpret anything beyond the economic system of objects (Baudrillard 1996). Far from the nobility of creation, the realm of objects has psychologically affected consumers. Collection and meaning by empty connotations replaced integrity. Such an amnesiac conglomeration takes no place. It is ephemeral, flexible and nowhere.

Since place and existence are interconnected (Heidegger 1978), it is no wonder that the new horizon of objects shifts from the physical to the virtual. A digital cloud has no topography, but it has a structure. Files are absent yet always available (Heidegger 1993). Similarly to these tendencies, mass industrialism, once producing uniform (Benjamin 1969) yet durable elements from 'real' materials, began using 'fake' and disposable ones. According to Jean Baudrillard (1996:38), the fast spreading of plastic is a profane reappearance of the Baroque belief in the universal material. Plastic consequently set up the best metaphor for consumerism, and advanced to its universal acceptance. Decades later, we notice the adverse effects of easy-to-dispose unnatural materials, since they fill the oceans, pollute the shores and make endangered animals suffer. Life in the post-industrial civilisation is unaware and disregards the ends of its consumer chain. Litter goes nowhere, people think, as nature is not a part of their mind map, narrowed to understand only the positive side of well-being and technical progress. As a benefactor of otherness and place, nature is expelled from the idealistic metropolitan 'heaven' to the subconscious. However, natural and cultural sustainability needs awareness of the whole, not just parts of it. We need to ask if we can regenerate memories, thus places, by renouncing our ideals of a careless and, in the end, nihilist civilisation (Cacciari 1982). Practical knowledge may be the solution.

23.5 Places under Regeneration

Global interests avoid the peripheral fields' quasi-space because it is not a potential market for producing uniform products. A regenerative strategy, therefore, will be kept intact by globalism until it starts to sell talent, as a trademark (Čeferin 2008:51). As soon as local technology promotes itself as 'the alternative', it exchanges its natural character for an aesthetic one. This is how it loses flexibility and freezes evolution, thus becoming unaccommodating to the challenges of the alternating local conditions (Katona 2018). In his essay about weak architecture, de Solà-Morales (1996) illustrates that there is always a chance for rebuilding, even if there are no examples of how it should be done. Solà-Morales speaks about the hidden reserve of vacuity in wastelands, for example, 'drosscapes' (Berger 2006) or fields hit by disasters. Until these territories are reutilised (erased/replanted) for the global economic interest, they appear infertile, marginal and in-between. But still, this lack of interest preserves the peripheries from being utilised generically. The drive behind any alternatives could be community, meaning the real, not virtual, social connection between people living in close neighbourhoods. However, according to specific experiences, communities do not usually form spontaneously, but because of proactivism from outside the region.

Regeneration of these peripheries brings together foreign architects, voluntary students and social workers who work within the local communities. In most cases, workers are related to each other and the users of the edifices they plan and build together. The workshops of Martin Rauch operating in Schlins and several other villages in the Vorarlberg region of Austria are examples of this (Kapfinger 2003). Rauch made plans for the rammed earth walls of a funeral chapel in Batschuns 2002, designed by Stefan and Bernhard Marte. The project revitalised the building community by inviting people to participate in the construction. Rauch taught villagers to build from local soil, getting them involved in erecting walls layer by layer, so he appeared as a mastermind among the villagers.

The same applies to many building projects outside the 'developed world'. Examples appear in some parts of Africa where Hollmén Reuter Sandman Architects designed social complexes for African women and children in need. The Finnish team studied the lives of local people and involved them in the building process. Their credits include the TunaHAKI Centre orphanage project in Moshi, Tanzania and the Rufisque Women's Centre, 2001 (see Fig. 23.1), which integrate both the local brevity and simplicity of design and the traditional village planning of the place. The fenced, square-type ground plan inspired the conventional structure of the local Ukumbis (Hollmén 2005), which is an apparent attempt to revitalise cultural memory, or the spirit of place.

In Africa, Diébédo Francis Kéré, a native of Burkina Faso, organises regenerative projects from Berlin, where he runs an international office. After graduating from the German capital, he completed his studies and built the first primary school in Gando as his diploma project in 2004. The school complex with its extensions, for example, library and teachers' accommodation, serves a common purpose: to give his folk the opportunity to develop living standards using local cultural resources and the knowledge he gained from more technologically advanced societies. The latter does not manifest as the installation of Western industry—he would hardly find a market for that, nonetheless—but instead, the translation of culture into technology, and technology to culture (Moore 2001). This transparency may use the most immediate devices, such as a stick Kéré (2014) uses to explain plans sketched in the dust, and, of course, the regional mud building technique he combines with brick and wood to replace the environmentally unfriendly concrete.

Socially involved and environmentally conscious architecture and urban planning probably peaked at the 2016 Venice Biennale, curated by Chilean architect Alejandro Aravena. Still, it is worth remembering former encounters with the same idea. Samuel Mockbee with his Rural Studio, for example, has been working with architectural students and volunteers from around the world to develop housing conditions, social life and infrastructure in places like Auburn, Greensboro, Akron and Newbern (Alabama, USA), among others. His public buildings and the outcome of the 20 K housing project are examples of applying simple but effective architectural knowledge. The latter, on one hand, allowed accommodation for residents who had a steady, though low or modest, income and were unable to obtain adequate housing through conventional financing in the United States. On the other hand, this progress has reinterpreted Alabama's architectural character using handmade wood structures and recycled materials. Whether avoiding economic collusion or mitigating humanitarian and natural disasters, regenerative architecture may differ in its purposes, outcomes and intensity. However, the architect's role and relation to the place are theoretically more essential in deciding. By definition, building communities are always local, yet planners may be foreigners. Local architects may have a deeper connection to the people and the enhanced region, but 'imported' knowledge is invaluable for peripheral societies. Among our examples, we may consider Martin Rauch, Samuel Mockbee and Diébédo Francis Kéré as locals. The same applies to Kengo Kuma's cultural revitalisation in Japan after the Fukushima disaster 2011.

Kuma's office developed temporary objects of use, such as paper tents that were easy to fold and unfold anywhere, but Kuma's most realistic intervention was methodological. His East Japan Project (2019) was launched to root out traditional artisans near the damaged zone. In Sendai and other parts of the Tohoku Region, they found the last sewing, texturing and basketry artisans, and encouraged them to make products for the Fukushima reconstruction. With a team of architects and designers, they started to work together and exchanged traditional craft knowledge with contemporary experiences, thus saving a dying local culture. Such methodological approaches may emerge from deep insight and learning about the place.

In Vietnam, efforts like Vo Trong Nghia Architects' experimental 'S-House' projects incorporate the same design strategy (Nghia et al. 2016). The architect uses mangrove palm leaves to resist the severe weather conditions in the Mekong Delta, and designs open ground plans to adapt to the humid climate. Therefore, community-built houses are close to nature, yet accommodating and available for low-income people. Hollmén Reuter Sandman Architects are among the 'foreign' Europeans, but they are not alone in their effort. Their Finnish colleagues, Sami Rintala and Hans Skotte, and architectural students from Norway were invited by TYIN tegnestue Architects (2012) to participate in re-planning a 'Safe Haven' orphanage in Myanmar, 2009. Sixty Norwegian companies and individuals supported the project, and, despite the tiny scale of edifices built with mixed techniques of traditional materials and modern structural elements, empowered the life of the community of children.

The efforts of Anna Heringer and Eike Roswag from Germany also manifested deep knowledge about local human and material resources in Rudrapur, Bangladesh, in 2007. The concept behind their handmade METI School (see Fig. 23.2) was to communicate and develop knowledge and skills within the local population so that they can make the best possible use of their available resources. Our next two projects represent similar efforts from our personal experiences.

23.6 Passive Solar Schools in Zangla and Tanpo, Ladakh Region, India

In Zanskar valley, in the remote region of Ladakh (India), winters last from September until May, and snow covers the only road connecting the area to the closest town. Winter temperatures are often below −40 degrees Celsius, and due to being above the treeline, there is no firewood available. During these months, local families use only one room of their houses, which they can heat with dried yak faeces, which they collect during the summer and dry on their rooftops. The same material is used for cooking, and there is none left to heat the school building, which is closed during the coldest days. As a result of an initiative by the local school staff, the school's open patio was covered with a transparent plastic sheet to make it warmer with solar energy. Csoma's Room Foundation (2019) decided to build a new school for the community (see Fig. 23.3), which can be occupied also during the winter months without using any energy for heating.

They received permission and a suitable building plot from the king of Zangla, and soon the new school started to take shape using the minimal selection of local materials. Traditional building methods were combined with simple contemporary design and fundamental physics. Firstly, the volunteers, who were mostly students from around the world, had to learn the traditional technologies from the local people, such as making adobe blocks (see Fig. 23.4). They also built the roofs together out of wooden twigs laid in between the beams, and covered the structure with a mixture of dried straw, grass and clay, applying the last layer of soil wet to increase its solidness.

The school is a simple, single-storey building with two entrances for the four classrooms and 240 square metres. The south-facing walls are fully glazed to maximise the surface area for absorbing solar energy for the passive heating of the rooms. The division walls are made of adobe blocks, while the wall facing the hillside is made of stones because of the danger of avalanches. The building was completed during three summers between 2013 and 2015 with the cooperation of international volunteers and local artisans.

With this passive solar technology, the classroom temperature during the day does not sink below −5 degrees Celsius anymore, which is already comfortable for children conditioned to low temperatures. The following year, another, smaller school was built in the nearby village of Tanpo, with a single 64 square metres classroom (see Fig. 23.5). Csoma's Room Foundation coordinated both projects in Ladakh.

Fig. 23.3 Passive Solar School in Zangla, India. Source: Balázs Mohai

Fig. 23.4 Drying adobe blocks in the Himalaya. Source: Szilvia Odry

Fig. 23.5 Building team at the Solar School of Tanpo, India. Source: Szilvia Odry

The cooperation was beneficial for both parties, and it went far beyond completing school. The volunteers have learned some local building technologies and how to live in close symbiosis with nature, even in primitive conditions. They knew that some things they took for granted—such as 24/7 electricity and drinking water from the tap—are indeed luxuries not available to most of the earth's population. On the other hand, volunteers have shown the local artisans how to build more precisely using simple measuring tools and water levels. Furthermore, as an indirect effect on the local architecture, passive solar technology is trending as huge glazed surfaces on the south facades have started to replace the traditionally narrow openings.

There was an initial plan to involve the builders and the whole community by inviting each family to prepare an amount of adobe blocks for the construction. Still, it did not succeed as only a few families have made their share. This was an important lesson for the project. In this level of society, disadvantaged people fight for their and their families' everyday survival and have no capacity left to do unpaid jobs, even if it is for their community. They did not even grasp the concept of volunteering and found it hard to understand why Europeans would travel a long way to work pro bono for people they had never seen before.

23.7 Zero Energy School in Koforidua, Ghana

The development team of Newill Academy (2019) faced very different challenges. The school is located in the agglomeration of Koforidua, a relatively large and easily accessible city in the Eastern Region of Ghana, where a wide variety of construction materials are available. The climate is tropical, with a long, hot and dry season and a shorter, humid and rainy season. The school is a non-profit private school founded in 2010 as an initiative by a local teacher in an empty storage building next to his mother's house. He aimed to provide quality education for families who suffer social disadvantages in the area. Over the years, the school became overcrowded, and some classes were held in corridors or temporary shelters. The extension of the school facilities began with a local draughtsman's plans for an unsustainable building with statically under-dimensioned structures. At this point, a group of students volunteered to join the development team and revised the plans concerning sustainability and climatic conditions in close cooperation with the local management (see Fig. 23.6).

The small building plot and the partially finished base slab were the main limitations during the design process. The three-storey building with the L-shaped floor plan was kept out of the original proposals. Still, the openings and the corridor system were optimised, thus creating community spaces and developing the connection with the courtyard. The main design issues in a tropical climate are the provision of shade and ventilation, which were realised by applying deep-set louvre windows, large roof overhangs, perforated classroom walls and an interleaved bamboo facade in the corridor. Due to the number of storeys, traditional materials were limited to the facade and the interior. Reinforced concrete and handmade concrete blocks were used for the load-bearing structures.

Unfortunately, an architect from the more developed part of the world still sets the example, or at least acts as a catalyst. People of previously colonised countries are unconsciously used to identifying development with 'western solutions', so they try to copy what foreigners bring, often without fully understanding all

Fig. 23.6 Construction at Zero Energy School, Koforidua, Ghana. Source: Bogdán Funk

implications. The bookshelf of the Ghanaian Draughtsman—usually the designer of smaller buildings—is packed with magazines showing North American family houses to impress the clients and gain some inspiration. The building solutions developed over the centuries have been forgotten as new trends emerge. Local materials such as clay and bamboo are seen as outdated, the material of low-income people. Modern trends eclipse traditional methods. The volunteer team realised their huge responsibility in this situation. They tried cooperating closely with the local community to understand architecture and feel the building as their own. The team hoped that local materials and sustainable solutions would become trendy again, also under a different composition.

To provide natural ventilation, the roof is elevated from the top slab and covered by a metal sheet. It is the cheapest available roofing material, but it warms up quickly under the sun. This disadvantage was turned into an advantage as it can work as a solar chimney by drawing hot air from the classrooms connected to the elevated roof with ventilation shafts. The small holes on the walls are for continuous fresh air intake. The large overhanging eaves evoke traditional roofing. On top, a single, wide roof plane connects the lower and higher building parts, making rainwater harvesting easier.

This school is also being built with volunteers and local professionals. The school building has twelve classrooms on three floors and a total area of 650 square metres. Construction was started in 2016, and the building was inaugurated in January 2022. Since the school has to operate continuously, the work has to be done little by little during the summer holidays, which are relatively short in Ghana. The semi-finished classrooms were already in use before the official opening (see

Fig. 23.7 Classroom in Zero Energy School, Koforidua, Ghana. Source: Mensah Göllnitz

Fig. 23.7), and, despite the raw walls and lack of carpentry, they provided a better environment for studying than the initial structure offered. There was also an unsuccessful effort to involve the parents and the community in the construction work. However, volunteering in Ghana is also seen as a hobby for bored, young Western people.

23.8 Conclusions

Place derives from both nature and culture and can be manifested by architecture. This act of coming into being crosses the border between existence and non-existence, moves into the world of art and engenders the state of 'otherness'. Mathematically, this otherness is analogous to the case of two 4D polytopes that become three-dimensionally construable by an intersection. The spirit of place had been cultivated as immanently sacred until Christianity introduced its new, other-worldly idea of transcendence. Consequently, this led to the generalisation of place, then the disappearance of the concept of *genius loci* in the modern era. In our globalised civilisation, borders between urban and rural worlds are phenomenological rather than topographical. The vacuity of the rural or post-industrial zones reveals hidden cultural and environmental reserves for the regeneration of place. These forgotten regions benefit from a new kind of community-based architecture as opposed to the commercial mechanisms of the global building industry. From this point of view, borders are synergistic factors of architecture.

Our case studies have shown that community-based architecture is key in defining new places or regenerating old ones. This activity takes advantage of neglected zones, borders, local human resources and available building materials. Our experiences show that communities do not form or act randomly. They require the initiative of coordinators, who can interact with the local people. Consequently, the regeneration of a place requires the crossing, yet not necessarily the elimination, of cultural boundaries.

The exchange between regional building traditions and advanced technical tools may preserve dying crafts and encourage sustainable design. Regenerative architecture also attracts students and volunteers who seek inspiration from the local communities and their natural and built environments. They experience that the language of architecture facilitates understanding even when spoken language may not. Borders inspire place-making, but when a new *genius loci* exists, it no longer supports or maintains cultural distance.

References

Baudrillard J (1996) The system of objects. Verso, London. (1968)

Benjamin W (1969) The work of art in the age of mechanical reproduction. In: Arendt H (ed) Illuminations. Schocken Books, New York, pp 217–252. (1935)

Berger A (2006) Drosscape. In: Waldheim C (ed) The landscape urbanism reader. Princeton Architectural Press, New York, pp 197–217

Böhme G (1991) Über Synästhesien. Daidalos 41:26–36

Brockman DD (2017) A psychoanalytic exploration of Dante's the divine comedy. Routledge, London

Cacciari M (1982) Nihilismo e progetto. Casabella 46(483):50–51

Čeferin P (2008) Far from home: contemporary Slovenian architecture in the making. In: Herrle P, Wegerhoff E (eds) Architecture and identity. LIT Verlag, Berlin, pp 47–56

Coxeter H, MacDonald S (1973) Regular polytopes. Dover Publications, New York

Csoma's Room Foundation (2019). https://csomasroom.org. Accessed 30 Aug 2019

de Solà-Morales I (1996) Weak architecture. In: Whiting S (ed) Differences: topographies of contemporary architecture. MIT Press, Cambridge, pp 57–72. (1987)

Debuyst F (1997) Le génie chrétien du lieu. Les Éditions du Cerf, Paris

Droste M (2002) Bauhaus 1919–1933. Taschen, Köln

East Japan Project (2019). www.e-j-p.org. Accessed 30 Aug 2019

Eliade M (1959) The sacred and the profane: the nature of religion. Harcourt, Brace & World, New York

Földváry MI (2002) "Istenünk tornácaiban": Az osztott templomtérje lentőségéről ("in the porches of our Lord": on the significance of divided church space). Magyar Egyházzene 9(2):171–182

Frampton K (1983) Towards a critical regionalism: six points for an architecture of resistance. In: Foster H (ed) The anti-aesthetic: essays on postmodern culture. Bay Press, Port Townsend, pp 16–30

Frascari M (1996) The tell-the-tale detail. In: Nesbitt K (ed) Theorizing a new agenda for architecture: an anthology of architectural theory 1965–1995. Princeton Architectural Press, New York, pp 500–514. (1984)

Gruber P, Bruckner D, Hellmich C, Schmiedmayer H-B, Stachelberger H, Gebeshuber IC (eds) (2011) Biomimetics—materials, structures and processes: examples, ideas and case studies. Springer, Berlin

Harries K (2008) The need for architecture. In: Solomon NB (ed) Architecture: celebrating the past, designing the future. Visual Reference Publications, New York, pp 19–30
Heidegger M (1967) Being and time, trans. John Macquarrie and Edward Robinson. Blackwell, Oxford. (1927)
Heidegger M (1971) The origin of the work of art. In: Poetry, language, thought. Harper & Row, New York, pp 143–162. (1935)
Heidegger M (1978) Building dwelling thinking. In: Krell DF (ed) Heidegger, Martin basic writings: philosophy—collected work. Routledge & Kegan Paul, London, pp 319–340. (1951)
Heidegger M (1993) The question concerning technology. In: Krell DF (ed) Basic writings. Harper Collins, New York, pp 307–342. (1954)
Heidegger M (1998a) Phenomenology and theology. In: McNeil W (ed) Pathmarks. Cambridge University Press, Cambridge, pp 39–62. (1927)
Heidegger M (1998b) On the essence and concept of physis in Aristotle's physics B, I. In: McNeill W (ed) Pathmarks. Cambridge University Press, Cambridge, pp 183–230. (1939)
Hernández JP (2011) How presencing (Anwesen) became Heidegger's concept of being. Universitas Philosophica 28(57):213–240
Hollmén S (2005) The strength of a woman. Architecture Plus 51(10):26–29
Kapfinger O (2003) Living in earth. Architecture d'aujourd'hui 346:102–109
Katona V (2018) A search for genuine regionalism: a regenerative agenda for the peripheries. In: Zaman QM, Troiani I (eds) Transdisciplinary urbanism and culture: from pedagogy to praxis. Springer, Cham, pp 213–226. https://doi.org/10.1007/978-3-319-55855-4_1
Kéré DF (2014) Architecture is a wake-up call. Louisiana Channel, Louisiana Museum of Modern Art. https://channel.louisiana.dk/video/diebedo-francis-kere architecture-wake-call-0. Accessed 30 Aug 2019
Koolhaas R (1995) Generic city. In: Sigler J (ed) Small, medium, large, extra-large: Office for Metropolitan Architecture. 010 Publishers, Rotterdam, pp 1246–1264
Lefebvre H (1974) La production de l'espace. Anthropos, Paris
McEwen IK (2018) Symmetry takes command: the case of the Roman capitol. Symmetry Cult Sci 29(3):365–388
Moore SA (2001) Technology, place and the nonmodern thesis. J Archit Educ 54(3):130–139
Newill Academy (2019). http://www.newillacademy.org. Accessed 30 Aug 2019
Nghia VT et al (2016) S-House. A&U 550:92–101
Norberg-Schulz C (1971) Existence, space and architecture. Praeger Publishers, New York
Norberg-Schulz C (1976) The phenomenon of place. Arch Assoc Qtly 8(4):3–10
Norberg-Schulz C (1980) Genius loci: towards a phenomenology of architecture. Rizzoli, New York
Ricoeur P (1965) Universal civilization and national cultures. In: History and truth. Northwestern University Press, Evanston, IL, pp 276–277. (1961)
Salingaros NA (2018) Applications of the golden mean to architecture. Symmetry Cult Sci 29(3):329–351
Sharr A (2006) Heidegger's Hut. MIT Press, Cambridge
Timmermann A (2000) Architectural vision in Albrecht Von Scharfenberg's Jüngerer Titurel. In: Clarke G, Crossley P (eds) Architecture and language: constructing identity in European architecture c. 1000—c. 1650. Cambridge University Press, Cambridge, pp 58–71
TYIN tegnestue Architects (2012) Safe haven library and bathhouse, Ban Tha Song Yang, Thailand. A&U 499:40–43
von Hildebrand D (1976) Was ist Philosophie? Verlag W. Kohlhammer, Stuttgart
Zumthor P (2006) Atmospheres: architectural environments—surrounding objects. Birkhäuser, Basel

Chapter 24
Urban Evolution of the Informal Settlements in Ibagué, Colombia: Strategies and Policies

Ingrid Carolina Vargas-Díaz and Eduardo Jiménez-Morales

Abstract Since the middle of the twentieth century, the influx of human migration in Latin America has contributed to the proliferation of large-scale informal settlements. In many cases, increasing inequality results in socioeconomic residential segregation, generating gaps among the different groups of a community. This phenomenon is expanding in Latin America. In cities of considerable size, experiences from some integration and careful interventions with the existing urban structure have been enriching. The *Favela Bairro*–Rio de Janeiro programme, the PUI–Medellin, the MIB–Bogota or *Chile Barrio* have had enough budget and financing possibilities. However, there is less experience in the strategies to be used in intermediate, which has been increasing during the last decades. This chapter analyses the development, evolution and causes of the informal settlements in the intermediate city of Ibagué, Colombia. It tries to find some applicable strategies for studying national and local policies. The research establishes an approach to the aspects contributing to the formation of informal settlements and its possible links to socioeconomic residential segregation. The research concludes by proposing a neighbourhood improvement plan to integrate those slum areas. As part of the strategies involving the community in constructing and planning their environment, the municipal administration—city council—and inter-university cooperation are evident.

Keywords Informal settlements · Integration strategies · Intermediate cities · Participation · Colombia

I. C. Vargas-Díaz (✉)
Independent Researcher, Malaga, Spain

E. Jiménez-Morales
Department of Art and Architecture, Universidad de Málaga, Malaga, Spain
e-mail: eduardo_jm@uma.es

G. Cimadomo et al. (eds.), *Urbanism at Borders*, The Urban Book Series,
https://doi.org/10.1007/978-3-032-09594-7_24

24.1 Rapid Urban Growth and Socioeconomic Residential Segregation

Since the middle of the twentieth century, the significant migration processes in Latin America have contributed to rapid urban growth and huge informal settlements. In 1950, the urban population of Argentina, Brazil, Chile, Colombia, Mexico, Peru, Ecuador and Venezuela was 47%, and the rural population was 53%. However, in 1980, the proportion of urban inhabitants in these areas increased to 70%, rapidly exceeding the rural population (United Nations 1997, 2008a, b). Nowadays, the urban population in Latin America reaches 82% of about 650 billion inhabitants (Siclari 2022).

The harmful effects of World War II, the debt crisis of the 1970s, economic changes and the Washington Consensus policy implemented by the World Bank with the International Monetary Fund contributed to the import of goods and an unequal modernisation of agriculture (Jaramillo 1993; Stiglitz 2007; Portes and Roberts 2008). The peasants and small-scale farmers became poorer because they could not be competitive in the global market. Their only alternative for survival was to create substantial market links between the countryside and the city (De Oliveira and Roberts 1996). This dependence on the city added to the panorama of increasing opportunities in urban centres due to industrialisation and the improvement in quality of life, which had the effect of attracting new permanent inhabitants and expanding and increasing the central and peripheral slum areas of the city.

One of the reasons for voluntary rural–urban migration is the lack of opportunities in rural areas, such as employment, services, infrastructure and urban equipment. Service data analysis between rural and urban territories indicates the absence of investment and inequality. The following figure shows a comparison of public services between rural and urban areas of nine Latin American countries (see Fig. 24.1). The coverage data in rural areas was 51.2% for water, 29.5% for sewage, and 73.6% for electricity; while in cities was 92.2% for water, 77.1% for sewage, and 97.1% of electricity (United Nations 2008a, Cepal 2007). This unequal distribution dooms an agrarian population that sustains the financing of the development and well-being of the cities in Latin America. Although the benefits of the construction of hospitals, industries, housing and schools directly pertain to the urban population, the payments in external credit do not come from large cities of industry but primarily from significant agrarian exports of the peasantry (Nyerere 1988).

Civil wars and repression have also contributed to the growth of informal settlements in Latin America (Davis 2006). This situation is demonstrated sharply in Colombia, where the armed conflict between the guerrillas (ELN and FARC) and militia has generated forced displacement of the peasant, afro-Colombian and native population towards informal settlements in primary and medium-sized cities. In some of these settlements, there is a lack of infrastructure and public services and a high risk of landslides and flooding. Furthermore, a high socioeconomic residential segregation is generated partly because urban planning and municipal administration rarely consider these areas and their inhabitants. According to Katzman (2003), this concentration of urban population in spaces of a socially homogeneous

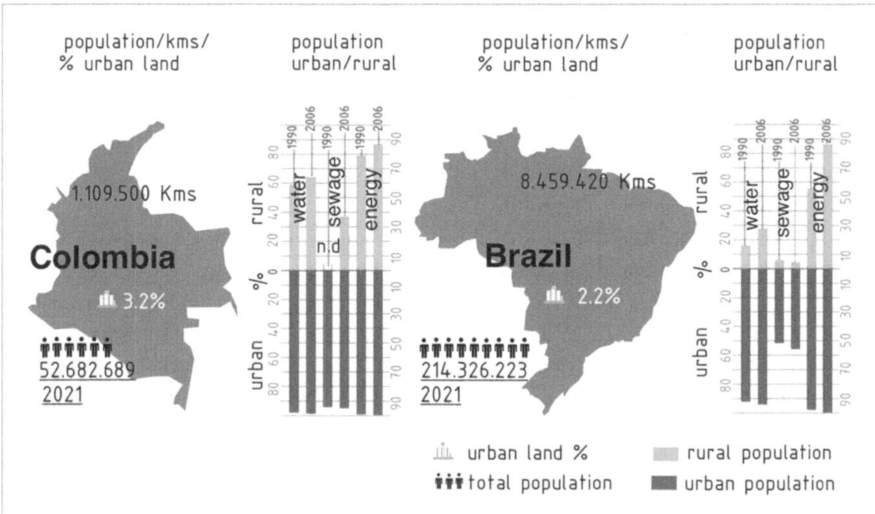

Fig. 24.1 Coverage of public services between rural and urban areas (water, sewage and electricity) between 1990 and 2006. Source: Ingrid Vargas

composition produces socioeconomic residential segregation. The population is in distinct urban areas, disconnected from other socioeconomic urban classes. This isolation has substantial negative consequences in society and results in significant educational segregation and fragmentation in public services. In addition, the segregation produces a fracture in employment opportunities that makes access to the formal labour market very difficult and, according to Arriagada and Rodríguez (2003), it also exerts a considerable influence on the risk of being a victim of violence and on the increase in the murder rate. Many of the inhabitants of these settlements tend to accept and approve illegal attitudes in their neighbourhoods and come to include them in their behaviour patterns. However, these violent attitudes, as opposed to the norms accepted by other neighbourhoods in the city, contribute to the isolation of these sectors and cause recoil on the part of the formal neighbourhoods. The rise of criminality generates increased fear, manifesting in the progressively protected gated communities of the middle and upper classes. The isolation of the urban population in gated communities is not limited to the upper classes; the middle classes in Latin America are more frequently acceding to these forms of housing (Janoschka 2002). In other words, the various manifestations of violence, which result in urban segregation through a cyclic effect, continuously contribute to the increase in isolation.

Instead of appreciating the societal value and aesthetics of singular processes in irregular settlements, society usually has a contemptuous attitude. However, according to Berenstein (2001), some of these settlements are an important part of the cultural and artistic heritage of the cities. They are built following peculiar survival forms, such as solidarity networks, and organised through collective and individual commitment. These forms of survival allow them for long-term sustainability.

The experience in some Latin American cities is enriching. The analysis of some programmes shows that the most comprehensive and respectful interventions that integrate these communities with the urban structure are the most successful. One of the most successful schemes is the neighbourhood improvement programme. Besides providing infrastructures and services, these programmes are usually focused on the social field. Some of them promote the formation of associations and labour training, supporting community empowerment and job creation.

The Inter-American Bank of Development usually finances these programmes and have been established in several Latin-American countries (see Fig. 24.2). Among them Colombia with Urban Integral Projects (PUI) in Medellin, which started in 2004 as a part of the strategy for neighbourhood improvement in the city, and *Proyecto Sur con Bogotá* (South Project with Bogota), which began in 1999 and was developed within the cooperation framework between the governments of Germany and Colombia until 2006 (Alcaldía de Medellín 2008). Venezuela also implemented the *Programa de Habilitación de Barrios* (Programme of Neighbourhood Physical Rehabilitation) and *Programa Tricolor* (Tricolored Programme). The development of the former was implemented between 1999 and 2005 and succeeded in approaching improvement through a more open vision involving the community, the government and academia (Baldó 2007). Yet, the programme that has often been considered, focusing on urban and social improvement, as a paradigm of neighbourhood improvement in Latin America, is *Favela Bairro* (Slum Neighbourhood) in Rio de Janeiro-Brazil (Conde and Magalhaes 2004). This programme represented a clear shift from implementing ad hoc interventions to establishing more coordinated metropolitan actions, which considered the existing urban infrastructure, community participation and the future maintenance of improvements.

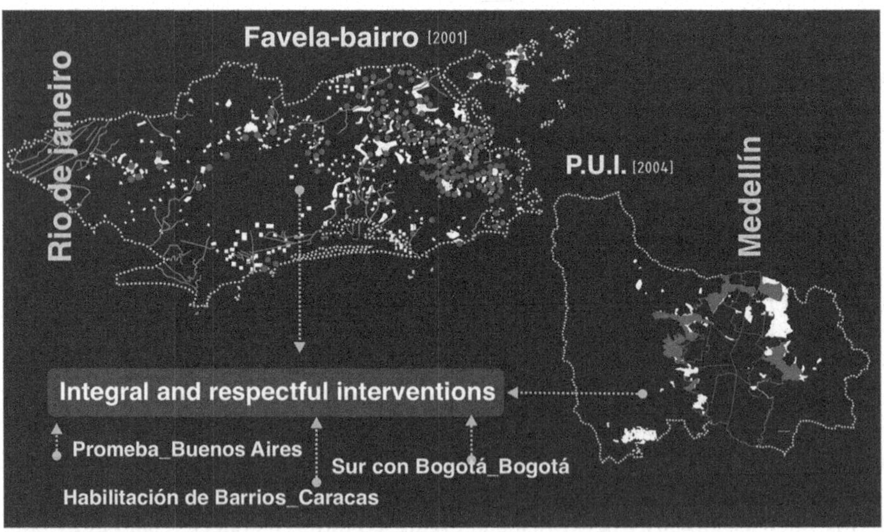

Fig. 24.2 Neighbourhood improvement programmes in Rio de Janeiro and Medellín. Source: Ingrid Vargas

Although some of these plans have also had negative impacts, such as the migration of inhabitants to other areas of the city or abandonment of urban works carried out after the improvements, the general achievements have been very positive. According to Brakarz (2002), the most significant impact of Neighbourhood Improvement Programmes is integrating the informal settlements into the formal city from different fields—physical, urban, social and political. Many of these programmes have set the basis for design integration strategies. Some of these lessons are the importance of regularisation, which becomes a civic duty after urban interventions; extending and relaxing the period between interventions to obtain the strengthening of improvement processes; and promoting community empowerment to manage their long-term development (Universidad Nacional de Colombia 2009, Torres 2009). Furthermore, it is necessary to interweave the physical–territorial issues with the historical-socio-cultural aspects in implementing future programmes (Martín 2008). Although these processes of urban integration have usually been established in large cities, the same problems exist in small towns and even more so in medium-sized Latin American cities. The latter, characterised by a population of between 100,000 and one million, have also become important poles of attraction, representing 32% of the total Latin-American population. Having these developing cities low financing possibilities and limited budgets, the question arises of how to implement integration alternative policies with these constraints (Banco de Desarrollo de América Latina 2019).

In order to have a better understanding of informal processes in medium-sized cities and the possibilities of urban integration of these areas and their population into the formal tissue, this research focuses on the study of this phenomenon in the city of Ibagué-Colombia.

24.2 Ibagué: A City Built from Informality

Ibagué, Colombia, as the capital of the Tolima state, has approximately 543,94 inhabitants (ORMET Tolima 2022). Ibagué is a medium-sized city bounded on the south by the *Combeima River*, on the north by the *Tutelares Hills* and on the west by an agricultural plain. Between 1940 and 1950, the city showed the first informal growth in the peripheries caused by the migratory processes from rural violence. Civil wars and repression have also facilitated urban informality in Latin America. This situation is even worse in Colombia, where armed conflict—between guerrilla and paramilitary groups—has generated a forced displacement of peasants and indigenous people from rural areas towards the main cities, increasing the size and population of slums in urban peripheral areas.

According to Neira (1995), between 1949 and 1958, the violence in Tolima was so overwhelming that it obtained first place in mortality nationwide. After 9 years, these migrations increased the city's population by 66%. Insurmountable problems in rural regions confronted the peasant population and began to invade urban areas, supported by politicians, labour unions and some associations (Gutierrez 2003).

Between 1960 and 1970, the city government attempted to solve the problem by offering urban land subdivided into small plots in the peripheral areas, but without any public services—electricity, sewage and water supply—or infrastructure. These settlements experienced a long and critical abandonment phase, becoming ghettos before they were consolidated as neighbourhoods. Therefore, this policy solved the initial housing emergency but, at the same time, led to the uncontrolled growth of the peripheral areas. Some of these settlements were located on the banks of the *Combeima River*, the Chipalo River and the *Tutelares Hills*.

Although informal settlements decreased between 1970 and 2000, they increased again around 2008. Data from 2000 indicated 36 informal settlements; in 2008, there were 65 (Alcaldía Municipal de Ibagué and Departamento Administrativo de Planeación Municipal 2008a). Since 1990, the state has played a significant role in stimulating the demand for housing, leaving the supply of dwellings in the hands of private companies, planning few social housing projects and promoting an unregulated market. In addition, the upsurge of violence in rural areas has generated new displacements of population towards the city. Therefore, the demand for housing solutions has been growing at the same pace as informal settlements in peripheral areas. By December 2015, Ibagué had received approximately 64,500 victims of forced displacement (RUPD 2018). Although the government reached a peace agreement in 2016, by 2018, the city had received 5200 more victims. Although the city government has attempted to find solutions to the growth of informal settlements, they have not solved the situation completely. The quantitative housing deficit was estimated at 27,000 housing units in 2008 (Alcaldía Municipal de Ibagué and Departamento Administrativo de Planeación Municipal, 2008b), reaching 60,000 dwellings in 2018 (MC Noticias 2018).

The situation was made even worse, as the unemployment rate in the city was the highest in the country for a decade. This rate reached 17.6% in the city in 2010, far above the national average of 11.8% (DANE 2010). As of 2022, the unemployment rate was 16.5%, and Ibagué was the second city with the highest rate in the country (DANE 2023). This unemployment has promoted underemployment and the expansion of informal markets as a legal alternative for survival in the city. Furthermore, the development of gated communities has increased in recent years as a residential alternative (see Fig. 24.3). Private companies offer these solutions. At the same time, public housing is usually located in remote urban expansion areas with an average of 22.5 square metres in parcels of 50 square metres (Vargas et al. 2010), which is inadequate for the average of 3.6 persons per dwelling. This undersized programme becomes pressing when the low-income population usually has large families and develops its work at home.

Currently, the *Plan de Ordenamiento Territorial de Ibagué* (Land Use Planning of Ibague) (Alcaldía de Ibagué and Departamento Administrativo de Planeación Municipal 2017) has implemented an integral improvement treatment for housing legalisation. However, it has not originated any particular neighbourhood improvement programme for the city. However, by 2009, the Ministry of Housing claimed that a pilot improvement scheme had been pre-approved for two neighbourhoods in Ibagué–Modelia and Las Delicias (CONPES 2009).

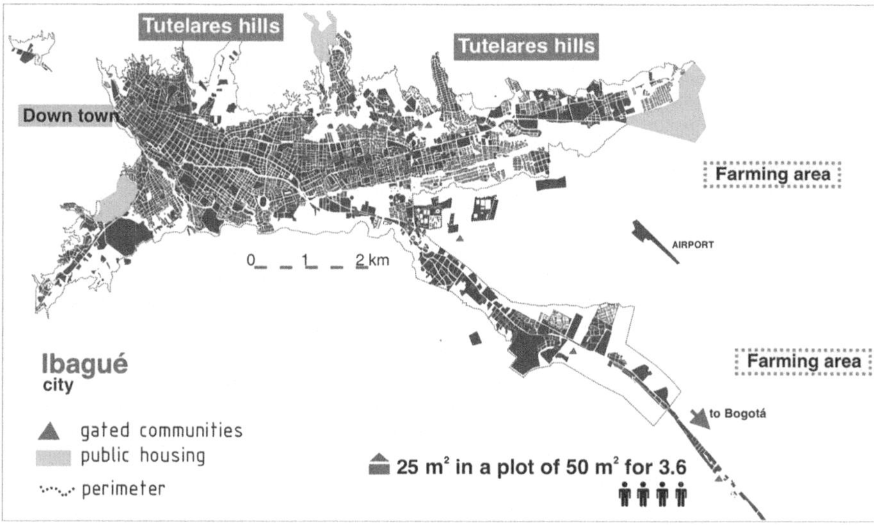

Fig. 24.3 Location map of public housing and gated communities in Ibagué Source: Ingrid Vargas

Modelia, as a part of the seventh district and Las Delicias of the sixth district, had and still have significant problems related to urban violence, lack of services and segregation. Although the government had planned an investment of 980,000 Euros, the city lost this opportunity due to problems with the tender in 2016. However, the implementation of the neighbourhood improvement programme is still necessary.

24.3 Las Delicias as a Field of Action

Las Delicias neighbourhood is a peripheral and isolated area in Ibagué, whose population is, according to *Instituto Interamericano de Derechos Humanos* (2002), characterised by many inhabitants displaced by rural violence. It was the destination of the displaced population from rural areas in the middle of the twentieth century, and years later, at the beginning of the twenty-first century, it received more population due to the armed conflict.

The neighbourhood, located on *Tutelares Hills* in the northern part of the city (see Fig. 24.4), was created in 1945 and belongs to the sixth district. This district has a stark contrast in the quality of life of its inhabitants. Five thousand of its 46,900 inhabitants live in high-risk areas, 20,000 drink non-potable water and 48.9% of its neighbourhoods have an informal origin and lack infrastructure (Secretaría de Planeación Municipal 2011). However, 15.4% of its territory belongs to the wealthiest population of the city (Gobernación del Tolima and Fundación Social 2004). Las Delicias has two parts: a formal sector and an informal sector. In 2012, the neighbourhood had a population between 4000 to 4500 and an average of 5 inhabitants

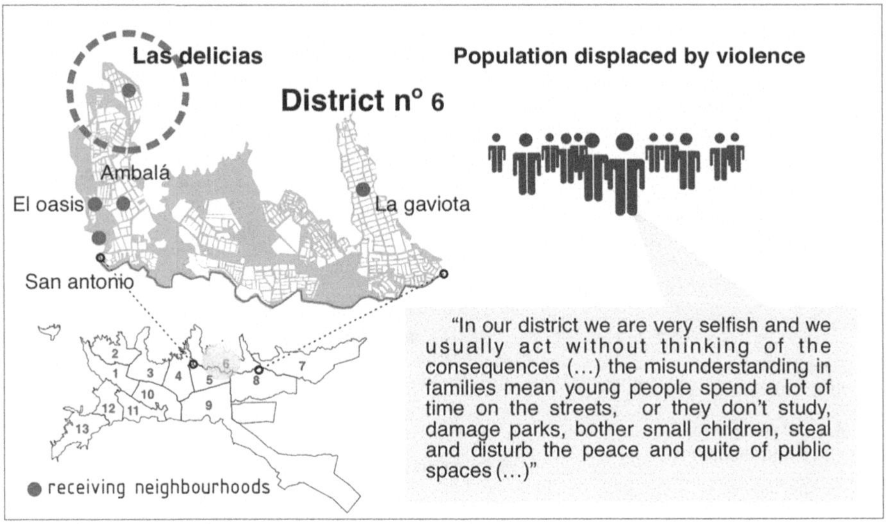

Fig. 24.4 Location of Las Delicias, rivers, streams and receptor areas of displaced population. Source: Eduardo Jiménez and Ingrid Vargas

per dwelling (Vargas 2014). It presents serious socioeconomic deficiencies and significant urban and environmental degradation. The pollution of La Balsa Stream, the construction of dwellings in high-risks areas and the use of public spaces for drug addiction and other illicit activities cause these problems. In addition, neighbouring communities usually stigmatise their populations.

However, most of the problems are situated in the social field. The complex socioeconomic situation, added to the intense internal armed conflict in the country, has promoted the implementation of criminal activities and the presence of armed groups, such as *AUC—Autodefensas Unidas de Colombia* (United Self-Defense Forces of Colombia). Although this organisation was demobilised in 2009, it has had a significant impact on the increase of violence and the creation of gangs.

A 22-year-old resident explains this situation:

> When we arrived in the neighbourhood, dismembered bodies appeared, half buried in the courtyards, dead on the streets […]. Two years ago I started committing crime by stealing drugstores, gambling establishments, supermarkets, etc. Of ten children, three become drug addicts starting with cocaine base, glue and finally they steal for vice. Of ten girls, one becomes a prostitute. (Pedreros 2011)

In 2004, Fundación Social, a Colombian nongovernmental organisation, carried out a symbolic act against violence in the district. There, a community leader expressed the problematic situation:

> In our district we are very selfish and we usually act without thinking of the consequences […] the misunderstanding in families means young people spend a lot of time on the streets, or they don't study, damage parks, bother small children, steal and disturb the peace and quiet of public spaces […]. (Gobernación del Tolima and Fundación Social 2004:102)

The arrival of the forced displacement population has produced a fracture in the social fabric and disturbed the community sense. The lack of guidelines for their integration usually increases socioeconomic segregation. The current associations cannot achieve their goals due to structural problems like high rates of unemployment, low school attendance rates and a high proportion of single-parent families headed by mothers. This reality involves territorial conflict and inequality in the distribution of wealth. The councillor of the sixth district remarks the absence of territorial belonging:

> We are running out of natural resources, such as water and forest, because there is an environmental ignorance and an absence of belonging in the neighbourhood. There are bad people who ignite the forest only to create problems for neighbours. There are people like that. This is the reason because we decided to build the gate and situate the guard in the forest edge […]. (Gobernación del Tolima and Fundación Social 2004:101)

The solution is not to take measures that stop the enjoyment of public spaces nor isolate the population that has been displaced by violence. Considering this idea, in 2004, the community, the University of Ibagué and la Fundación Social collaborated on a Participative Plan for the Sixth District (PP6) and made some general community planning assessments. In PP6, a broad diagnostic of the population's quality of life was made, including the areas of education, sanitation, social and economic, amongst others. However, the data are generic, and the community's neighbourhoods are of different socioeconomic strata.

Architecture students participated in the needs assessment and the diagnosis of some urban areas. The following were some of the reasons given by residents of Las Delicias that explained the neighbourhood degradation in 2004: lack of identity, lack of knowledge of the urban environment, minimal participation of the community in neighbourhood planning and the precarious integration with the urban setting. In 2012, to find integration alternatives for urban improvement in Las Delicias, some meetings were held with neighbourhood associations, some universities in Colombia and Spain and the municipal administration of Ibagué.

24.4 University Cooperation Project as a Proposal

To resolve this problem, a University Cooperation Project was suggested between the University of Granada and its International Centre of Cooperation for Development (CICODE), the University of Ibagué and the National University of Colombia. The authors presented the project 'IX Call for submitting university development cooperation projects' in 2009. The universities and research groups offered technical support and coordination. The will of participation, at that moment, of the Municipal Administration, added political support and possibilities of long-term sustainability. The main objective of this proposal was to contribute towards urban and environmental improvement in Las Delicias with community participation. It also had four specific objectives:

1. To consolidate links between the neighbourhood, the university, neighbouring communities and the municipal administration.
2. To improve social integration in the neighbourhood.
3. Urban and environmental improvement in public spaces.
4. Establish an organisation mediating between the community and the municipal administration.

The project's long-term outcomes aim to reach the following results:

- Establishing and consolidating an Urban Observatory of Informal Settlements in the University of Ibagué.
- The improvement of public spaces with high levels of conflict and stream pollution.
- The implementation and the achievement of mechanisms for participation and urban environmental awareness.
- The improvement of inter-institutional coordination between organisations and the community.

The project took advantage of the location of the University of Ibagué—near Las Delicias—and the good experiences gained with some local risk management courses in the neighbourhood. The goal was to improve the urban and environmental conditions by strengthening their inhabitants, the interaction between academy-community and the participation of municipal planning. The project suggested a series of Las Delicias activities resulting from a neighbourhood's needs assessment. In addition, the central relationship between problems identified by the community and the causes of such problems were considered. The following decision tree (see Fig. 24.5) is a model that shows the causes and effects of a given problem and establishes the basis to design a specific activity to address the problem (Camacho et al. 2001). The objective was to design and implement small interventions to generate improvement processes over time. The interventions were meant to be low-cost actions that should be financed initially by CICODE and continued by Ibague's Municipal Planning.

To determine and validate the relevance of currently addressed problems, the project proposed:

1. A technical participative diagnosis with the community.
2. An awareness campaign among the community aimed to recognise the urban environment.
3. A series of meetings will be held between representatives of both the community and participating universities to define the interventions.

These actions would initially be concentrated on the bank of *La Balsa Stream* and other public spaces. To monitor the project and achieve sustainability, the idea was to design a Neighbourhood Improvement Plan in conjunction with the community. A systematic and independent examination should evaluate the project to determine its impact's efficiency and its objectives' relevance. Finally, all the

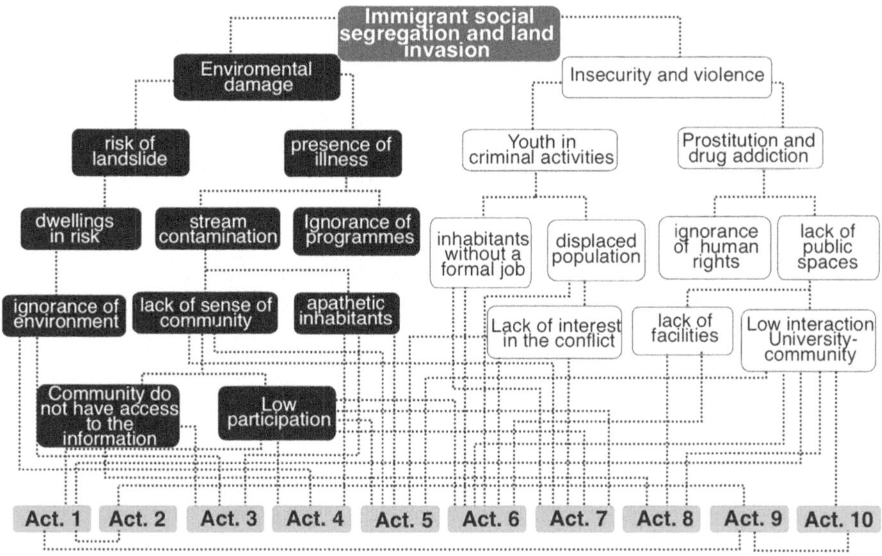

Fig. 24.5 The decision tree model of the sixth district. Logical Framework Approach. Source: Ingrid Vargas

interventions would involve its inhabitants—paid and volunteer work—and their wishes to advance.

There was no plan for significant interventions in advance because the main objective was to implement changes gradually in cooperation with the inhabitants. However, this did not imply a lack of coordination. The project was to be developed according to the logical framework approach (Camacho et al. 2001). This method is a way of structuring the main elements in a project, highlighting logical links between intended inputs, planned activities and expected results. All the activities were programmed and inserted into a planning matrix. The indicators, means of verification and assumptions for each item were analysed.

The activities were integrated into a schedule, and the cooperative project budget was programmed (see Fig. 24.6). This budget had a breakdown of expenditures and a justification of every item: equipment, inventoried material, staff, consumable goods, journeys and management cost. The project's duration, with international support, was approximately 2 years, and CICODE's funding provided 20,000 Euros per year. Feasibility factors were essential elements in the design of the cooperation project. Eight factors related to the UN Sustainable Development Goals (SDGs)e described below:

1. Political support (SDG 16. Peace, justice and strong institutions).
2. Institutional aspects (SDG 16. Peace, justice and strong institutions).
3. Sociocultural aspects (SDG 11. Sustainable cities and communities. Make cities and urban settlements inclusive, safe, resilient and sustainable).
4. Gender focus (SDG 5. Gender equality. Empowering all women and girls).

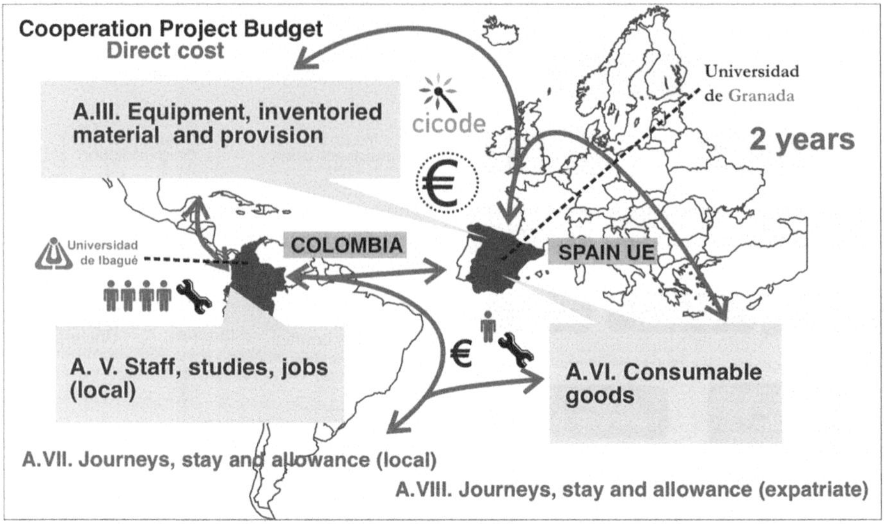

Fig. 24.6 The cooperation project included direct and indirect costs. Commitments between the University of Granada and the University of Ibagué (counterparty). Source: Eduardo Jiménez and Ingrid Vargas

5. Technological factors (SDG 16. Peace, justice and strong institutions).
6. Environmental factors (SDG 13. Climate action. Take urgent action to combat climate change and its impacts).
7. Economic and financial factors (SDG 10. Reduce inequality).
8. Inhabitant suitability (SDG 11. Sustainable cities and communities and SDG 10. Reduce inequality).

Furthermore, the interventions would be distributed throughout the territory like a kind of 'urban acupuncture', as sustainable activities necessary to revitalise, cure problems, make improvements and create new positive actions in the neighbourhood (Lerner 2005). However, how could we attain sustainable interventions in an isolated and passive community? The solution envisioned was to establish a permanent organisation: an 'Urban Observatory of Informal Settlements in the University of Ibague'. It would administrate, evaluate and provide institutional support. It would be an urban platform that would educate the population about their environment, stimulate their participation in the planning and executing strategies, and help them attain sustainability. The observatory would be an independent body considering improvement alternatives proposed by the universities, the community and municipal planning.

As a designing centre of participative strategies, the observatory would promote a multicultural meeting, and architecture students would be able to provide a more expansive vision of society. This was necessary to improve a city, which needed similar interventions in 65 informal settlements (Alcaldía Municipal de Ibagué and Universidad de Ibagué 2008). These actions are still needed because in 2022, the

Municipal Government reported 14 invasions in different districts of the city (El Nuevo Día 2022). Located at the University of Ibagué, the observatory would become a mediator agency that would be able to design sustainable solutions. Also, it would observe the urban processes of informal settlements in the city. This action was necessary to learn about their history, population, customs or culture and to obtain a classification of a complete historical-socio-cultural and physical–urban mapping.

Based on a University Cooperation Project, this proposal took advantage of the possible 'share of knowledge' of the Urbanism Laboratory—University of Granada, UGR Spain—and the Habitat Research Group of the National University–UN Colombia. It also involved the human resources of the University of Ibagué and the support of the Municipal Planning Office. Each of these entities was responsible for the development of the project. The University of Ibagué, with the Laboratory of Urban Studies of the University of Granada, should have consolidated the observatory, coordinated its activities, incorporated students in the work and kept in continuous contact with the community. Rather than being passive, the community would have the opportunity to express their opinions, clarify the real environmental needs and participate in every activity. Similarly, municipal planning would participate in and financially support the activities and interventions implementing the Neighbourhood Improvement Plan. The focus of this proposal was to create a propitious climate to generate new dynamics for the city. This experience had the potential to be replicated in other middle-sized cities. This would be possible by establishing a mediator organisation such as the Urban Observatory, which can coordinate the available institutional skills and talents and create links between municipal administration and the community.

Finally, this project was about breaking down barriers fragmenting urban neighbourhoods. The project strived to recognise that the responsibility for production in the city belongs to the collectives—the government and society. Only through acceptance and encounter with 'the other' can a more democratic, peaceful and sustainable city be built.

References

Alcaldía de Ibagué and Departamento Administrativo de Planeación Municipal. (2017) Plan de Ordenamiento Territorial de Ibagué. Documento Técnico de Soporte. https://cimpp.ibague.gov.co/wp-content/uploads/2017/10/Documento-T%c3%a9cnico-de-Soporte.pdf. Accessed: 14 Sept 2019

Alcaldía de Medellín (2008) Modelo de buen gobierno y desarrollo social integral. Empresa de Desarrollo Urbano de Medellín. http://www.edu.gov.co/index.php?option=com_content&view=article&id=106&Itemid=73. Accessed 20 Oct 2017

Alcaldía Municipal de Ibagué and Departamento Administrativo de Planeación Municipal (2008a) Revisión y Ajuste de Plan de Ordenamiento Territorial. Alcaldía de Ibagué, Ibagué

Alcaldía Municipal de Ibagué and Departamento Administrativo de Planeación Municipal (2008b) Inventario de tenencia y accesibilidad de la vivienda en la ciudad de Ibagué. Alcaldía de Ibagué, Ibagué

Alcaldía Municipal de Ibagué and Universidad de Ibagué (2008) Revisión y ajuste del Plan de Ordenamiento Territorial Cartografía de asentamientos precarios, componente urbano: U16. Alcaldía Municipal de Ibagué-Universidad de Ibagué, Ibagué
Arriagada C, Rodríguez J (2003) Segregación residencial en áreas metropolitanas de América Latina: magnitud, características, evolución e implicaciones de política. Serie Población y Desarrollo N° 47. CEPAL/Naciones Unidas, Santiago de Chile
Baldó J (2007) El programa de habilitación de barrios en Venezuela. Ejemplo del proceso de construcción y de administración de los recursos por parte de comunidades organizadas. Tecnología y construcción 23:9–16
Banco de Desarrollo de América Latina (2019) Ciudades intermedias y desarrollo en América Latina. https://www.caf.com/es/actualidad/noticias/2019/02/ciudades-intermedias-y-desarrollo-en-america-latina/. Accessed 13 Mar 2023
Berenstein P (2001) Estética das Favelas. Arquitextos, 13. Vitruvius article. https://www.vitruvius.com.br/revistas/read/arquitextos/02.013/883. Accessed 11 Sept 2019
Brakarz J (2002) Ciudades para todos. La experiencia reciente en programas de mejoramiento de barrios. Banco Interamericano de Desarrollo, Washington D.C.
Camacho H, Cámara L, Cascante R, Sainz H (2001) El enfoque del Marco Lógico: 10 casos prácticos. Cuaderno para la identificación y diseño de proyectos de desarrollo. Fundación CIDEAL, Madrid
Cepal (2007) Anuario estadístico de América Latina y el Caribe. Naciones Unidas. http://www.eclac.cl/cgibin/getprod.asp?xml=/publicaciones/xml/8/32598/P32598.xml&xsl=/deype/tpl/p9f.xsl&base=/deype/tpl/top-bottom.xsl. Accessed 28 Oct 2017
Conde L, Magalhaes S (2004) Favela Bairro: uma outra história da cidade do Rio de Janeiro. Vivercidades, Rio de Janeiro
CONPES - Consejo Nacional de Política Económica y Social (2009) Documento Conpes 3604: Lineamientos para la consolidación de la Política de Mejoramiento Integral de Barrios-MIB. Bogotá: Ministerio de Ambiente, Vivienda y Desarrollo Territorial
DANE (2010) Data on indicators of the labour market 2001–2008. Departamento Administrativo Nacional de Estadística. http://www.dane.gov.co/index.php?option=com_content&task=section&id=19&Itemid=42. Accessed 23 Mar 2013
DANE (2023) Data on indicators of the labour market 2022. Departamento Administrativo Nacional de Estadística. https://www.dane.gov.co/. Accessed 20 Mar 2023
Davis M (2006) Planet of slums. Verso, London
De Oliveira O, Roberts B (1996) Urban development and social inequality in *Latin America*. In: Gugler J (ed) The urban transformation of the developing world. Oxford University Press, Midsomer, pp 253–314
El Nuevo Día (2022) El complejo panorama de las invasiones urbanas en Ibagué. https://www.elnuevodia.com.co/nuevodia/especiales/483192-el-complejo-panorama-de-las-invasiones-urbanas-en-ibague. Accessed 7 Apr 2023
Gobernación del Tolima and Fundación Social (2004) Plan Participativo de desarrollo comuna 6. inédito, Ibagué
Gutierrez A (2003) La gran migración. In: Academia de Historia del Tolima (ed) Compendio de Historia de Ibagué. P&G Editores, Ibagué, pp 31–137
Instituto Interamericano de Derechos Humanos (2002) Resolución Defensorial Regional n° 005. www.iidh.ed.cr/.../Resolucin%20Regional%20005%20%20Desplazados%20Tolima.htm. Accessed 13 Sept 2011
Janoschka M (2002) El nuevo modelo de la ciudad latinoamericana: fragmentación y privatización. Revista Eure. https://doi.org/10.4067/S0250-71612002008500002
Jaramillo S (1993) El desenvolvimiento de la discusión sobre la urbanización latinoamericana: hacia un nuevo paradigma de interpretación. In: Jaramillo S, Cuervo L (eds) Urbanización Latinoamericana. Nuevas Perspectivas. Editorial Escala, Bogotá, pp 9–44
Katzman R (2003) La dimensión espacial en las políticas de superación de la pobreza urbana, Serie medio ambiente y desarrollo N° 59. CEPAL/Naciones Unidas, Santiago de Chile

Lerner J (2005) Acupuntura urbana. Instituto de Arquitectura Avanzada de Cataluña, Barcelona
Martín Y (2008) Hacia la práctica de una arquitectura dialógica en la transformación de los barrios populares. Revista Argos 45:21–43
MC Noticias (2018) Ibagué tiene un déficit de 60 mil viviendas de interés social: Humberto Quintero. Youtube. https://www.youtube.com/watch?v=BB805OZSVaw. Accessed: 17 Aug 2018
Neira H (1995) Tolima en la encrucijada de la agricultura. Pijao Editores, Ibagué
Nyerere J (1988) Let us pay heed to the peasant. In: Gugler J (ed) The urbanization of the third world. Oxford University Press, New York, pp 38–39
ORMET Tolima (2022) Data on total population in Ibagué 2022. Observatorio Regional de Mercado de Trabajo. https://ormettolima.org/ibague/. Accessed 24 Mar 2022
Pedreros D (2011) Mito de los derechos humanos en el barrio Las Delicias de Ibagué. Youtube. https://www.youtube.com/watch?v=LrFLfHDw0gs. Accessed 12 Sept 2019
Portes A, Roberts B (2008) La Ciudad Bajo el Libre Mercado: La Urbanización en América Latina Durante los Años del Experimento Neoliberal. In: Portes A, Roberts B, Grimson A (eds) Ciudades latinoamericanas: un análisis comparativo en el umbral del nuevo siglo. Miguel Ángel Porrúa, México, pp 19–74
RUPD - Registro Único de Población Desplazada. (2018) Data of the national registry of displaced population. Población desplazada en Ibagué. https://www.unidadvictimas.gov.co/es/registro-unico-de-victimas-ruv/37394. Accessed 7 Sept 2018
Secretaría de Planeación Municipal (2011) Indicadores del Desarrollo Territorial para Ibagué. Alcaldía de Ibagué, Ibagué
Siclari P (2022) Nueva Agenda Urbana en América Latina y El Caribe: implementación 2018–2022. https://www.urbanagendaplatform.org/sites/default/files/2022-02/Latin%20America%20 Regional%20Report.pdf
Stiglitz J (2007) Making globalisation work. Penguin, London
Torres C (2009) Hábitat y vivienda: Pobreza Urbana y Mejoramiento Integral de Barrios en Bogotá. Universidad Nacional de Colombia, Bogotá
United Nations (1997) Population by sex, residence and intercensal rates of increase for total population, each census: 1948–1997. Demographic Yearbook. Historical supplement. http://unstats.un.org/unsd/demographic/sconcerns/densurb/urban.aspx. Accessed 28 Feb 2009
United Nations (2008a) Urban population, development and the environment 2007. UN document. http://www.unpopulation.org. Accessed 28 Mar 2010
United Nations (2008b) Global report of human settlements 2007. Enhancing urban safety and security. http://www.unhabitat.org/downloads/docs/GRHS.2007.8.pdf. Accessed 14 June 2010
Universidad Nacional de Colombia (2009) Pronunciamiento Público sobre MIB y ciudad. In: Torres C (ed) Foro público por Arte de Barrio Foro público por Arte de Barrio. Memorias. Mejoramiento Integral de Barrios en Bogotá. Universidad Nacional de Colombia, Bogotá, pp 99–108
Vargas I (2014) Activación urbana en asentamientos informales. La ciudad de Ibagué como laboratorio de experimentación. PhD. Thesis. University of Granada, Spain
Vargas I, Jiménez E, Grindlay A, Torres C (2010) Procesos de mejoramiento barrial participativo en asentamientos informales: Propuestas de integración en la ciudad de Ibagué [Colombia]. Revista INVI. https://doi.org/10.4067/S0718-83582010000100003

Chapter 25
Artivists of the World, Unite! The Story of Stuttgart 21

Iraj Esmaeilpour Ghoochani

Abstract Stuttgart is a city where its past fights with its future. Many citizens are against the Hausmannian dream of replacing their historic terminus train station with a tunnel known as Stuttgart 21 (S21). S21 has many other subsections in its master plan, including a new city library and a large shopping centre (Milaneo). The behaviour of the mass (*das Volk*) in response to S21 could be divided and studied into two major categories: those who are opponents of the S21 and those who are already the end-users of the entertaining facilities provided by the first completed phase of the project (Milaneo shopping centre, city library etc.). Both of these behaviours are changing the political and urban scene of the city through very complex dynamics. This chapter is aimed at a better understanding of this process. The anthropological materialism framework of Walter Benjamin's Arcades Project has been taken as a methodological reference, as well as his urban theory. His conceptualisation of 'phantasmagoria' plays a central role in studying mass culture as a set of ever-changing images like those seen in a dream projected upon the urban scene. Interpreting these ever-changing and entertaining dream objects that have mesmerised the young generation globally is crucial for everyone in the pedagogical or prevention work field. The argument of this chapter is based on a strong criticism of the left side of the political spectrum, including the greens. Because of its extra focus on work and economic capital, the idea is that the Left has a significant theoretical and empirical lack of appropriation (*Aneignung*) of urban spaces in the twenty-first century, keeping its own bordered ideology. This chapter conclusions recognise the failure of an alternative phantasmagoria or 'dream' needed to fight the neo-capitalistic leisure industry.

Keywords Activism · Public space · Protest · Phantasmagoria · Entertaining culture

I. E. Ghoochani (✉)
InsideOut Advisory Service, STJG (Stuttgarter Jugendhaus gGmbH), Stuttgart, Germany

G. Cimadomo et al. (eds.), *Urbanism at Borders*, The Urban Book Series,
https://doi.org/10.1007/978-3-032-09594-7_25

25.1 The Story of Stuttgart 21

Once upon a time:

> *Wolf*gang Schusster, the Lord Mayor of Stuttgart, sat together with the other politicians and investors, who thought with each other: "How can we make even a better trade?" Then, they came up with the idea: they build a new train-station, because then the railway tracks of the old station, will be demolished. And under the part where the tracks were, is the fillet of the whole city and it would be free and they can build houses on it. Oh, [...] no houses for the people to live inside, but office-buildings, some houses for the rich people, shopping centers, etc. [...] and then, they started to make their big big business. (Anonymous interview 2018)

This is the story of Stuttgart 21 (S21) when heard from its opponents. They mostly believe that the city already has a very well-functioning train station, and the only reason this station should be displaced by a station with much fewer tracks is the greed of the ruling class that represents a few and one that is separated as a bordered social entity away from the popular mass. One of the interviews held on the occasion of this research highlighted these thoughts:

> The old train station was very well [...] technically extremely well made and organised and many trains can drive in at the same time to Stuttgart [...] now the station should be placed under the ground... under the ground is for any human being uncomfortable [...] We do not want to go under the ground [...]. (Anonymous interview 2018)

Stuttgart is one of the last cities where people could still say 'We'. They are willing to share the same future, dismissing S21 as an imposed form of future initially decided by the ruling class and real estate advocacy groups and avoiding popular consultations. However, what can be read in this chapter is also a self-critic of all the activists who have lost the opportunity to investigate better what is happening to their values, for instance, democracy: participation of the masses in what the central station, as the city's most important urban space, is supposed to be in future. Stuttgart 21 is now an exemplary lesson for the importance of the citizen's participation in urban development projects:

> The decade-long German railway and urban-development project Stuttgart 21 provides a lesson in how vital citizens' participation is to success in major infrastructure projects: citizens and advocacy groups that had not been involved at the beginning of the project spoke out against the mounting expenses and environmental impact associated with Stuttgart 21, and the project turned into a political lightning rod. It was a key factor in the 2011 state elections. (Laartz and Lülf 2014)

Stuttgart citizens have protested every Monday at 18:00 in *Schlossgarten* for more than a decade against this imposed future (see Fig. 25.1). They held their protests on Mondays intentionally to follow the Monday demonstrations of East Germany, especially that of Leipzig in 1989. They have celebrated more than 450 demonstrations. In one of these demonstrations, there was an intense clash between the demonstrators and the police. The Stuttgart 21 demonstrations reflect the citizenry of a healthy city. It is also the disappointed reaction of citizens who still believe in politics. However, it also talks about the intrinsic dilemmas of an anti-capitalistic utopia

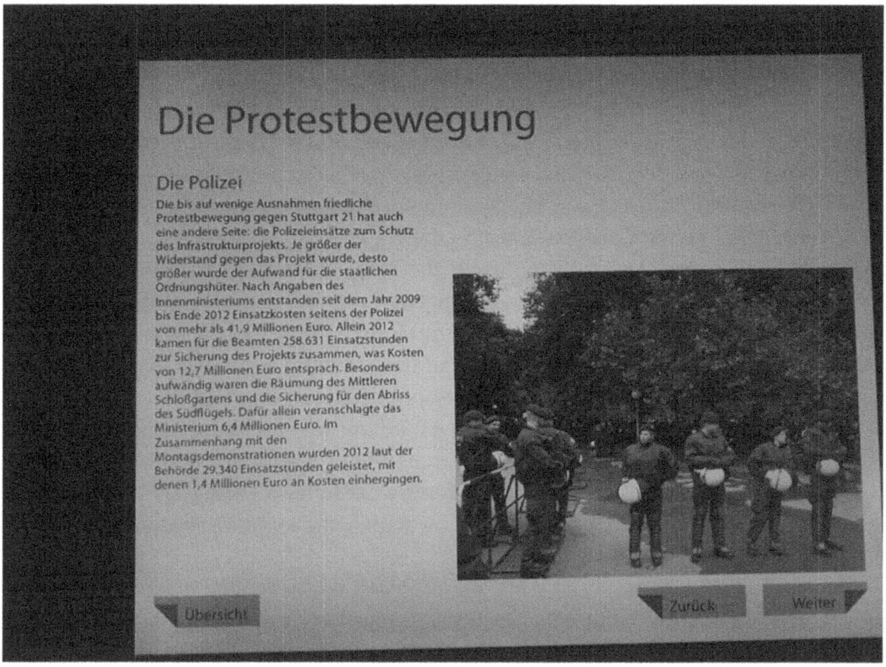

Fig. 25.1 The tower of the central station is appropriated for a free entrance museum for S21 (above: a page from these digital demos). The establishment packs and appropriates all of the protests paradoxically inside the propaganda for the project with an utterly fatherly patronising tone that sounds very provocative. (Source: the author)

that is on the brink. Stuttgart is revealing one of the longest resistance cultures in the era in which *All that is solid melts into air*[1] and this makes the position of Stuttgart as a metropolis and S21 as a metro-political project very special. Hannelore Schaffer, a professor in literature in one of Monday's demonstrations, proclaimed:

> Stuttgart has the worst reputation of all metropoles [...] It is considered suburban, compulsively neat, boring [...] But Stuttgart has the best theater audience in Germany. This bourgeois activity now turns into a city-political interest; the citizens are not to be tricked with deception. And station 21 is a delusion—like an absolutist castle (Kümmel 2010).

This very short piece of Prof. Schlaffer's speech contains three revealing signifiers: theatre audience, delusion and absolutist castle.

Hannelore Schlaffer encourages theatrocracy by praising the demonstrators as the best theatre audience. Theatrocracy, as suggested by J. Ranciere's, is a situation in which the relation between the players (politicians) and the audience (mass) is reversed in favour of a better staging equality (cf. Hallward 2009): 'Now it is our turn to play'. The Stuttgarters are considered the most enthusiastic theatre audience.

[1] This is the title of a book by Marshall Berman, paraphrasing a quote taken from the *Communist Manifesto* by Karl Marx and Friedrich Engels See: Berman (2010).

Now, they see themselves on stage. They are doing their best to celebrate themselves despite this state of emergency. They are aware that the rest of the country is watching them. However, 'The revolutionary theatre in its entirety [is] founded on an ignorance of the profound historical realities that rendered it possible'. (Ranciere 2006:15) In our case, this *historical reality* is the power of dreams and delusions: the fundamental need of human beings to be mesmerised or even fooled by a fantasy. 'S21 is a delusion'; it is a Hausmanian dream, and as such, its proposers are possessed by the myth of progress, seeing themselves as visionaries who are in good pace with the future. The fight between the opponents of S21 and its designers, investors, immobile lobbyists etc. is a fight between two distinct realms of being: a fight between a wakeful people who want to take back their (histori) city and a dream that has already marked the future of the city with a number that stands for a whole century—creating a bordered and opposed line of ideology.

Without a proper and collective theory of dream, we would not be able to see the essence of S21 as a dream as well as the dynamic of resistance against S21, which is, as we will see, essentially a war between wakefulness (being an audience of a real catastrophe) and dream (as a scene on which the symbols become more real than reality).

The local Green party—the only major party opposing the scheme—was the most voted in the 2009 municipal elections. It marked the first time the Greens won the most votes in a German city with more than 500,000 inhabitants (Novy and Peters 2012:136). The resistance against S21 could be grossly considered a green movement. However, 77% of demonstrators arrange themselves to the left of the political spectrum (Rucht et al. 2010). They sing: We are the people, we are the money (*Wir sind das Volk, wir sind das Geld*). Peter Kümmel has seen the paradox: 'it is the choir song of a new bourgeois self-confidence'. (Kümmel 2010)

The paradox is that they are too obsessed with money as the most materialistic form of capital. At the same time, other forms are also available, namely symbolic, social and cultural capitals (Bourdieu 1997). Wolf Reuter is correct in saying that the fundamental structure of the conflicts of S21 is centred around economic issues (Reuter 2001:29). They are still fetish worshippers of metal money (cf. Marx 1959:53). As such, they are not even a developed money nation and could not fight a dream that will cost them more than 8000 million euros.

Similia Similibus Curantur: like is cured by like. Following this old wisdom, the American dream of globalisation, which is now blooming in the middle of the city, could not be defeated with another dream. Any life form which is empty of dreamy elements is boring, and the mere focus on money is dangerous because 'the money is in a preventive way can put itself in the centre of all life interests, on the other hand, this is just the barrier against which almost all human relationship fails'. (Benjamin 1972–1989, Band IV: 96).

25.2 'Left is dead! Long live the Left!'

Die Kunst ist tot! Es lebe die Kunst! (Art is dead! Long live art!). This motto of the Dadaists brings us directly to the essential conflict between a particular art language as a form of 'life' and the potentially infinite forms that life can ever take, and finally, *life* will supersede every arid, putrid, fetid form because 'life', as such, should either create forms or [at least be able to] move in [side given] forms (cf. Simmel 1921:45). Isn't this vital motto of the Dadaists that once successfully saved Western art from its dead-end applicable to the predicament that the Left, as a form of life and culture, got stuck into it? 'Left is dead! Long live the Left!'

The Left had once the wild aura of intellectualism. This aura is now exhausted, and the right-wing has appropriated it in various forms of cultural capitalism, for instance, by reproducing the design logic of a shopping centre in the architectural design of the city library. The leftists today are mostly a mass of insular people awakened amid the global capitalistic dreams: their life is like a wake-walker (in comparison to dream-walkers) whose values are in direct opposition to the neo-capitalists' dream as a myth. However, in their unprepared combat, they have neither a 'collective dream' nor a consolidated class identity like the proletarians of the twentieth century to bring them together with the bond of the 'collective work'.

To see this tragic situation, it would be enough to compare the K21 project—K21 is a term used to refer to the alternative proposal to the Stuttgart 21 railway project advocating for the renovation of the existing railway station in Stuttgart (*Stuttgart Hauptbahnhof*) instead of building a new underground station—proposed by the opponents of S21 with the resourceful design of S21.[2] The alternative proposed project K21, which has marked its name in the logo of the protest against S21 (see Fig. 25.2), is impoverished of new visions (dreams). For example, compare the glassed roof used in both projects as the primary design element (see Fig. 25.3).

In contrast to K21, which traditionally uses glasses—in the same way that the city of Paris had once used it at the beginning of the twentieth century to bring the sky on the earth as a practical realisation of the paradise (Buck-Morss

Fig. 25.2 Symbol of the protest against Stuttgart 21. The symbol shown here includes the website for Terminus 21 or Kopfbahnhof 21, which is still accessible (last accessed on 29.03.2023), and serves as a useful supplementary source for examining their approach as opponents of Stuttgart 21. (Source: https://en.wikipedia.org/wiki/Stuttgart_21, Public domain)

[2] K21 advocates for renovating the existing railway station in Stuttgart (Stuttgart Hauptbahnhof) instead of building a new underground station. The 'K' in K21 stands for 'Kopfbahnhof', which means 'terminus station' in German, and '21' refers to the year 2021, which was the originally planned completion year of the Stuttgart 21 project. See one of the latest digital brochures of the K21 project at: https://fuer-k21.de/

Fig. 25.3 Above: Photomontage of possible platform roofing under K21. Below: Visualisation of the train station, considered by the specialist judge for the architectural design competition a 'non place'. (Source: https://de.wikipedia.org/wiki/Kopfbahnhof_21 and Plan b, Stuttgart, Public domain)

1989:107)—S21 is promising to bring the sky onto the under-earth. K21 appears traditional and conservative in comparison to S21. Nevertheless, the Left was not this traditional and empty of new visions at its launch. Vladimir Tatlin's tower, commissioned to create a monument to the Bolshevik Revolution, is solid proof of the time that an oneiric fantasy once possessed communism.

Tatlin's tower was originally designed to make a dwarf out of the Eiffel Tower in Paris: a towering symbol for the sublimation of history, historical determinism, fulfilment of the world spirit (*Weltgeist*) etc. (see Fig. 25.4). The phantasmagoria behind Tatlin's tower is the myth of the progress of history onto its predetermined endpoint. To see the great potential of this myth, it would be enough to look into the artistic movements like the Russian Avant-garde, Russian Futurism and so on, all directly or indirectly influenced and flourished in the early times of the Russian communist empire. Even today, to think of a commune is a beautiful, inspiring dream: this naïve utopian instance that visualises a compromise between mutually exclusive pairs of egotism and collectivism. Furthermore, commune is not the only dreamy term used in communism. Marx has proved that he is a master in understanding the importance and function of symbols in human life by introducing the fetishism of commodities as the 'enigmatic character [to] the product of labour'. He invites the oneiric idea of fetishism into the political economy as an inseparable part

Fig. 25.4 Tatlin and Eiffel: Two iconic structures. The Tatlin Tower was designed by Russian artist Vladimir Tatlin in 1919 as a monument to the Third International and was meant to symbolise the new Soviet era. In contrast, the Eiffel Tower was built in 1889 as a centrepiece for the Universal Exhibition in Paris and was primarily intended as a tourist attraction. (Source: Punin, Nikolai. 1921. *Tatlin (Protiv kubizma)* [Татлин (Против кубизма)]. Petrograd: Gosizdat, 1. Public domain. Eiffel Tower: Public domain)

of the—production of—commodities. In fetishism, the commodities take the instances of a dream because the dream is a space where the symbols become more real than they ever symbolise: 'A space in which the Word takes Flesh' (Esmaeilpour Ghoochani 2017:38). To live consciously with the 'free intellect' as a working alternative for the old rotten class ideologies, for example, to praise laziness and leisure as the authentic social attachment instead of work, it is crucial first to overcome the mesmerising power of the symbols fuelled by the myth of progress.

25.3 The Right of Laziness and the Dreaming Myth of Progress

The Left has once focused on a sort of deterministic ideology that takes history as something natural. However, there is nothing natural in history; instead, this is nature that paces forward historically. It was the genius of Walter Benjamin, the modern philosopher of the twentieth century, who put commodity at the centre of his focus and abstracted phantasmagoria out of ideology as an ideological process that mesmerises the masses in a dream-like manner. Benjamin takes the debris of mass culture seriously as a vital source of philosophical truth. When Walter Benjamin was focused on Phantasmagoria and the figure of *flaneur*, the central part of the left political spectrum was excessively focused on work and labour, neglecting the importance of the dream, leisure and laziness in human life. However, the Left itself is not the only alternative to the Right. *The Right to Be Lazy*, the work of Paul Lafargue, Marx's son-in-law, is a good example that takes issue against the 'myth of progress' instead of 'work and money' as the essence of capitalism. In his Arcades project, Walter Benjamin inspects the traces of this 'progress myth on the city's body'.

Now it is time to visit this ghost of progress as the new working ideological process or phantasma(goria) hidden behind the number 21. Perhaps the designers of S21 see themselves as visionaries who are directly in touch with the God of progress. The supporters of K21 have much less 'appeal to dream' to argue against these Hausmanian visioners of S21. Drawing from Bourdieu's classification of class orders, the *ruling class* demonstrated higher levels of initiative, ingenuity, imagination, and operational capabilities through the S21 project compared to the K21 project, which was intended for the 'masses'. The oppositional group advocates renovating the terminus train station as a viable alternative to the construction of S21, estimated to cost 3739 million Euros in September 2010. Even though K21 was 349 million Euros cheaper at that time, this argument appears to be significantly less compelling, as it overlooks the crucial role played by 'dream' as a construction material.

Topology is not necessarily a science about the properties of space; when drawn on its Greek roots, topology becomes a place (*topos*) that speaks (*logos*): 'Saying of a place' (Malpas 2006:33). 'Saying of a place' is also the shortest definition of a dream. On the other hand, Stuttgart is like a guitar; the railways are its strings, and S21 is a big hole creating resonance. Then, we can also imagine it as a 'Saying of a place': a vocal cord that the 'Truth' is now talking through its throat. The central station has become a site that speaks volumes about the collective psychological states of the citizens, providing insight into their collective psyche. Moreover, new eclectic urban places are produced where their 'dialectic of images' is homologous to a *rebus*. According to Freud's theory of dreams, the dream is also a rebus (Freud 2015:235). Walter Benjamin's method in the Arcades project was to interpret these urbanistic dreams by dismantling the ideological conductions behind phantasmagorical appearances. He handles the city like the objective unconsciousness of the mass.

'Unconscious speaks' or to be more precise: 'It speaks' (*ca parle*) (Lacan 2007:682, 838). Analogous to this quote of Lacan, Benjamin has developed many innovative concepts to make the voice of the space hearable: *Leibraum* (body-space), *Bildraum* (image-space), *Phantasmagoria* etc., The people share the urban space just like language, so the city is, in a significant sense, the objective unconsciousness of its citizens. Benjamin brings his adventurous study of commodities in connection to dreams as the situation of a human on its way to becoming a new *(zeit)geist* or humanity: one epoch dreams of the next. Benjamin puts phantasmagoria in place of ideology and brings the modern urban space into speech through this topological turn. Phantasmagoria is a kind of 'magic lantern' that creates its (spatial) reality by reproducing dream-like images. The film, as a sort of art that is produced—reality is based directly on reproducing images—is a living metaphor for phantasmagoria.

25.4 Stuttgart 21 Is a Dream

There are many voices against S21.[3] The voice of reason that strives to remain both modern and independent is particularly significant. In many interviews with those intellectuals who are against the project, one can detect a similar pattern: 'Everybody who has a little wit is against this project'. 'Look at the Weissenhofsiedlung! Stuttgart is already one of modernity's most monumental icons, originally designed by Bauhaus. More modernity is not necessary'. 'Central Station is a modern architectural symbol'. 'They have collapsed the north wing of the station. We have shown patience but let's see what will happen if they touch a tree in our parks…', etc. Stuttgart is a site of modernity, where the mythological future promised by S21 is challenging its historical present. So, history becomes spatial, in the same way that Paris was, and for Benjamin, it is a vitrine for the history of Europe in the twentieth century. It is possible to ask oneself if there is something not settled in the twentieth

[3] The following link contains a series of cartoons that depict the chronological development of the tensions arising from S21: https://www.koufogiorgos.de/anti-stuttgart-21/

century that we must remember in the twenty-first century to understand S21. The answer may be found in the Arcades project. The Arcades project of Walter Benjamin is a materialistic philosophy of mass culture which is constructed from a collage of historical material: every remainder of the buildings, technical equipment and commodities that are phenomenologically handled like a set of dream objects that cry for interpretation. Benjamin's goal is not to reproduce the dream but its expulsion: the dream images will be drawn into a waking state through dialectic images. Here, the word 'awake' is synonymous with historical knowledge:

> In the dialectical image the past of a certain epoch [...] is only before a very definite epoch: namely, in which humanity, rubbing its eyes, recognises precisely this dream image as such. It is at this moment that the historian takes over the task of dream interpretation. (Benjamin 1972–1989:580)

S21 is a dream because of a symbol acting autonomously and unrelated to the mass's participation. The city's historical background is labelled here with histori(city) for more emphasis on their intrinsic relation. The myth of progress is the hidden engine of S21 as a Hausmanian dream. S21 could not work without the digit 21 as the symbol of progress. Parallel to this, the myth of progress cannot manifest itself anywhere better than German Railways: a ready-made icon of speed with ideal customers who can change their train in three minutes. In contrast to this icon of progress, the Left has another icon: *Angelus Novus* (New Angel), a monoprint by artist Paul Klee (see Fig. 25.5). Otto Karl notes that Benjamin's reading of this image (Benjamin 1972–1989: 697 f.) is at the base of its success in becoming an icon of the Left (Werkmeister 1997:9).

S21 has divided the city and the masses in a very essential way. It is not just about opponents and supporters of the project. It is about the conductions of a psyche that, notwithstanding its issue against S21 is a consumer of the urban space provided by S21, including the new Milaneo shopping centre, the fast-food cathedral inside the city library etc. Following the track of Walter Benjamin in his Arcades project, it is possible to differentiate all these into two significant realms of dream and wakefulness to see how the whole 'thing' is correlated to a critical but less appreciated issue of 'leisure'. To see the dynamic of this correlation, it is recommended that the commoditised time spent for work, sport, leisure and play be mapped onto the commoditised place occupied by S21. Before starting with the analysis of S21 under the light of a commodity theory, some descriptive materials are necessary.

25.5 Entertaining Culture (*Bespaßen-Kultur*)

'Man is just a tireless seeker of pleasure' (Freud 2015:91). However, the challenge of contemporary culture and education is that 'fun' and pleasure come first after hard work. Work in the broadest sense of the word: homework or a work of art that takes a long or tedious time. Boring is an important part of learning.

Fig. 25.5 *Angelus Novus*, 1920, Paul Klee. (Source: Public domain)

Bespaßen-Kultur, as the new dilemma of our time, has already given its contagious aura to our educational system: the children should be entertained. For a better understanding of the hidden ideology behind this culture of entertainment, it is important to engage seriously with the concept of leisure and understand the way Walter Benjamin sees the issue (see Fig. 25.6).

If we flip this diagram horizontally, it could be mapped (as the commoditised time) upon the general diagram for the commodities suggested by Prof. Susan Buck-Morss (see Fig. 25.7) as the hidden structure behind the whole Arcades project. There are a lot of monumental or historical instances inside the district, and their contrast with the plan of S21 produces a dialectic *of* 'seeing'. To describe the nature of this dialectics, the following diagram is borrowed from Prof. Susan Buck-Morss's seminal work, which reveals the hidden structure of Walter Benjamin's Arcades project.

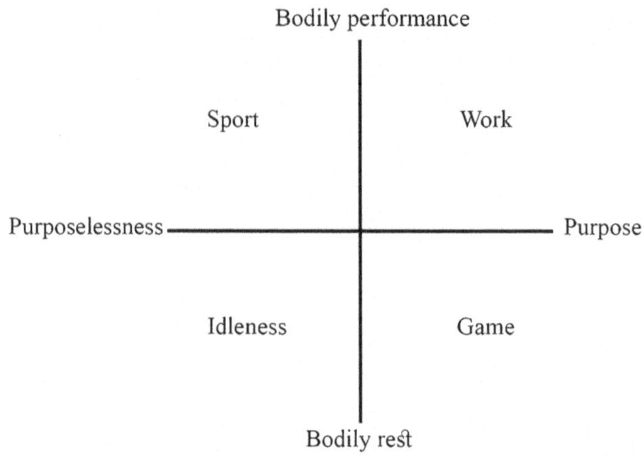

Fig. 25.6 The original German version of this graph can be found in Benjamin's collected works in Volume I, originally published in Frankfurt am Main by Schurkamp in 1972–1989 (1991), on page 1177. (Source: the author)

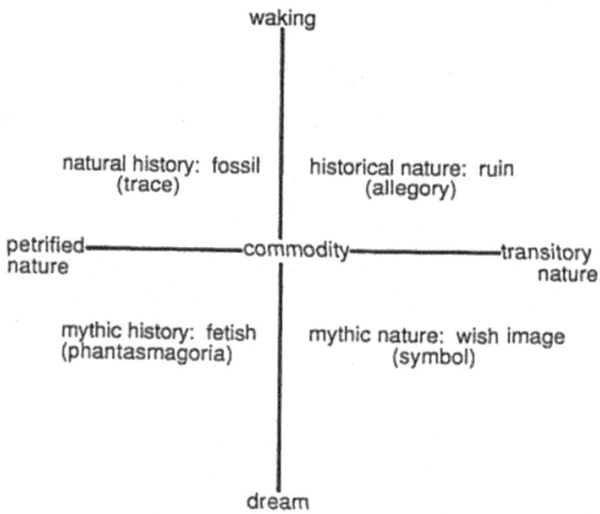

Fig. 25.7 The hidden structure of the Arcades project. (Source: Buck-Morss 1989:211)

As stated before, in Benjamin's philosophy, the commodity is that enigmatic 'thing' (*das Ding*) that adjusts the position of everything around itself in four major parts:

1. *Natural history; fossil (trace)*: The upper left side of the diagram is about those parts of the city that are worthy of being retained and survived in the new version of the city: one engraved column here, one historical gate there etc. These pieces of the past are analogous to fossils reflecting the natural pace of the city in its

Fig. 25.8 'This is not a nike'. Nike (the logo) is a dream object: a symbol more real than [the use value of] what it symbolises. (Source: the author)

trace of evolution. In S21, all terminus platforms should vanish unless platform number 16. It is supposed to remain untouched and take the role of a trace.

2. *Historical nature; ruin (allegory)*: The north wing of the central station and all 15 platforms are examples of ruin: those parts that humans, as the cultural extension of nature, give back to nature.
3. *Mythic history; fetish (phantasmagoria)*: The city library and the Milaneo shopping centre take on the role of mythic history. They are two different phenomenal attributes of the same thing: They carry the phantasmagoria of the myth of progress and give the illusion of living inside a cube of knowledge (city library) or a real pleasure island (Milano) of a global village.
4. *Mythtic nature; wish image (symbol)*: The *wish image* is the product of the *mode industry*, for instance, whatever is served and represented inside the phantasmagorical buildings of the Europaviertel under the ocean of reified signifiers (see Fig. 25.8), like the brand names written above the shops: Nike, H&M, Apple etc.

The upper side of the diagram belongs to the wakeful people, having *Angelus Novus* as their modern totem. The wakeful people are the antithesis of the S21 city designers and its supporters, who are labelled as dreamers at the lower part of the diagram. The word 'dream' is related to the concept of *flanerie*: the dream walkers of the urban space. The young generation, who spend considerable leisure time strolling through shopping malls with their clique, are also dreamers of the same order. 'Dream' is also used here in the same sense that the myth of progress once possessed Baron Hausmann.

In our Benjaminian analysis of dream and wakefulness webbed around S21 as a project that is selling itself as the dream of a century, we should keep in mind the danger of taking front: Benjamin's concept of 'the structure of awakening' is also the structure of drowning, drowning into self-image, losing the self-awareness that a *flaneur* may enjoy. Before long, it was an essay he had in mind: 'Paris Arcades: A

Dialectical Fairyland'. In the last episode of this essay, the dialectical relationship between the city library and the Fairyland of Milaneo is visited.

25.6 Last Station: City Library (*Stadtbibliothek*)

God's not dead; it is our unconsciousness. Even inside a secular culture like German culture, the unconscious remains divine, and as we will see, the new city library framed in the S21 project is solid proof of this. The same could be said about the displacement of divinities and religious devotions and worship onto commodities: fetishism. The city library (see Fig. 25.9), humorously labelled with funny names such as *Bücherknast* (the prison of the book) or *Stammheim Zwo* (Stammheim prison Nr. 2), is located in the middle of *Europaviertel*. It was the first finished part of the S21 plan. The architecture of this city library, a perfect cube, is neither inviting to stay nor to leave. It lies somewhere between being a home and a station. It is both like a Buddhist temple and a shopping mall.

The city library is an excellent location for a stroll and, much like a shopping centre, provides an opportunity to observe other individuals strolling through the area, thus making the observer's body an extension of the surrounding space. It is built in complete contrast to the traditional library desks, intentionally designed to deprive the eye of watching things other than the book. The unspoken intention behind the design appears to be an effort to blend the phantasmagorical atmosphere of the *Europaviertel* with the ambience of a library, given that both 'reading' and 'shopping' are popular leisure activities. However, reading remained separate from popular culture during the past century, unlike cinema, which shares the same essence of phantasmagoria as a magic lantern. This embodiment and union of the city library as the first finished sub-project of S21 for the leisure industry of its neighbourhood is what makes the S21 very special as a project that uses both intellectualism and Zen-Buddhism for its legitimation:

> The spatial expectation of the entering visitors to see and find books is disoriented. The visitors find themselves enveloped by a four-story 14-meter cubed void [...] It's a perfectly detailed and executed pure geometrical white cube. Relocated in some Zen-moment, the readers find themselves standing on an empty terrazzo floor looking at a tiny square pool with a small fountain. This opening scene invites the entering readers to clear their minds, to turn their backs on the urban sounds and impressions. (Bleeckere and Gerards 2017:166-7)

Slavoj Zizek, renowned philosopher and irrepressible cultural critic, considers the ethic of Buddhism as the spirit of neo-capitalism in the same way that the protestant ethic was the spirit of capitalism: 'In capitalism, Buddhism, the antimaterialist doctrine par excellence, finds itself in support of the most materialist of economic ideologies' (Pound 2008:131). Zizek also argues that the only possible choice is between neoliberalism and Asian values (Literaninovinym 2011). We are witnessing the collapse of these two into one doctrine in the city library.

Fig. 25.9 Above: The City Library. Below: The Milaneo Shopping Centre. (Source: Flickr and Wikimedia Commons, Public domain)

25.7 Conclusions

Shopping centres resemble pleasure islands globally, spreading the neo-liberal order and values. This makes them also a perfect target for terrorism. To understand this fact, it would be enough to look at the number of rampages and crazed actions in shopping centres on a global scale (LaTourrette et al. 2006:63ff.). What is missing today is the experience of an intensive insight into the timeliness of an authentic Left theory (cf. Buck-Morss 1989:25). It is through this authentic Left that we may hope for authentic prevention work because hate is an inevitable product of the non-participatory attitude of the ruling class. As dreamers, it sees no strong opponent on their way to *progress*: the Left today is not the opposite of the Right. Most of the leftists today are no longer those intellectual modernist dreamers who take issue against conservatism. They have taken the position of tradition instead, and in doing so, they are no longer in a battle with the establishment in the sense of a class conflict.

Looking at two left architectural expositions (K21 and Tatlin's Tower), the declining route of an ideology that still perceives its (histori)city in terms of a Hegelian tradition of positive dialectics has been shown. However, there is no necessity for things to turn in a certain way. This domain of contingency that opposes every necessity of a wakeful life was assigned here with the realm of dreams visited in leisure time. Without an authentic Left theory, most leisure time is already managed by liberalistic leisure industries. However, in this void resides a big chance for artivists (activist+artist) to enter the scene if equipped with a proper theory of anthropological materialism. Using the dialectic of images as a Benjaminian method, we can promptly follow the upheaval (*Umwältzung*) forms of the mass and the establishment.

The global revolution in mass intelligence calls for a revolution in our understanding of social class and class conflicts. Here, the 'free intellect' should work consciously or unconsciously on behalf of a class (Benjamin 1972–1989, Band VI:781). An artivist possesses such free intellect. This is why the motto of the communist manifesto (1848) 'Workers of the World, Unite!' is suggested to be variated onto 'Artivists of the World, Unite!'

References

Benjamin W (1972–1989) Gesammelte Schriften. Schurkamp, Frankfurt am Main

Berman M (2010) All that is solid melts into air: the experience of modernity. Verso, London/New York, p 1989

Bleeckere S, Gerards S (2017) Narrative architecture: a designer's story. Routledge, New York

Bourdieu P (1997) Die verborgenen Mechanismen der Macht: Schriften zu Politik und Kultur 1. VSA-Verlag Hamburg, Hamburg

Buck-Morss S (1989) The dialectics of seeing: Walter Benjamin and the Arcades project. MIT Press, Cambridge

Esmaeilpour Ghoochani I (2017) Bābā Āb Dād: The phenomenology of sainthood in the culture of dreams in kurdistan with an emphasis on sufis of qāderie brotherhood. PhD. Thesis. LMU München: Faculty of Philosophy, Philosophy of Science and the Study of Religion, Munich

Freud S (2015) Interpretation of dreams, translated by AA. Brill. Dover Publications, Inc, Newburyport. 1900

Hallward P (2009) Jacques Rancière: history, politics, aesthetics. Duke University Press, Durham. 1905

Kümmel P (2010) Der Gesicht Sagt Nein! Zeit Online. 9 September 2010. https://www.zeit.de/2010/37/Stuttgart-Bahnhof. Accessed: 28 Aug 2019

Laartz J, Lülf S (2014) Partnering to build smart cities. Mckinsey and Company. https://www.mckinsey.com/~/media/mckinsey/dotcom/client_service/Public%20Sector/GDNT/GDNT_SmartCities_v5.ashx. Accessed: 28 Aug 2019

Lacan J (2007) Écrits. The First Complete Edition in English, W. W. Norton & Company, 1st edn. http://users.clas.ufl.edu/burt/Lacan.pdf. Accessed: 17 Jan 2016. (1966)

LaTourrette T, Howell DR, Masher DE, MacDonald J (2006) Reducing terrorism risk at shopping centers: an analysis of potential security options. Rand, Santa Mónica

Literaninovinym (2011) Slavoj Žižek—China, "Asian Capitalism" and our lack of ideology. https://www.youtube.com/watch?v=4ADfH9Rt6pc. Accessed: 29 Aug 2019

Malpas J (2006) Heidegger's topology: being, place, world. Massachusetts Institute of Technology, Cambridge

Marx, K. (1959). Economic and philosophic manuscripts of 1844 (M. Milligan, Trans.; D. J. Struik, Rev.). Progress Publishers. (Original work written 1844; first published 1932); URL: https://www.marxists.org/archive/marx/works/download/pdf/Economic-Philosophic-Manuscripts-1844.pdf

Novy J, Peters D (2012) Railway station mega-projects as public controversies: the case of Stuttgart 21. Built Environ. https://doi.org/10.2148/benv.38.1.128

Pound M (2008) Zizek: a (very) critical introduction. William B. Eerdmans Publishing Company, Cambridge

Ranciere J (2006) Hatred of democracy. Verso, New York

Reuter W (2001) Öffentlich-privates Partner-schaftsprojekt 'Stuttgart 21'. Konflikte, Krisen, Machtkalküle 145:29–40

Rucht D, Baumgarten B, Teune S (2010) Pressekonferenz 27. Oktober 2010, Wissenschaftszentrum Berlin für Sozialforschung: Befragung von Demonstranten gegen Stuttgart 21 am 18.10.2010. Wissenschaftszentrum Berlin für Sozialforschung, 27 Oct 2010. http://www.fjnsb.org/sites/default/files/downloads/handout_stuttgart21_rucht_2010.pdf. Accessed: 28 August 2019

Simmel G (1921) Der Konflikt der modernen Kultur (Ein Vortrag). Duncker & Humblot, Munich

Werkmeister OK (1997) Icons of the Left: Benjamin and Einstein, Picasso and Kafka after the fall. University of Chicago Press, Chicago

Chapter 26
Exploring the Significance of Intangible Cultural Heritage Within the Confines of the Neighbourhood: An Examination of Participatory Techniques Used in Fontanalla, Malaga (Spain)

Francisco José Chamizo-Nieto, Nuria Nebot-Gómez de Salazar, Marta Cornax-Martín, and Carlos Rosa-Jiménez

Abstract The involvement of local communities in the management of intangible cultural heritage (ICH) has been claimed, not only by the United Nations Educational, Scientific and Cultural Organization (UNESCO), but also by experts and researchers in the last decade. If there is awareness of this necessity of locals' involvement, there is no consensus about common criteria and methods on how to carry it out, especially in the case of small urban areas and vicinities. Aiming to fill this gap, this research explores some participatory actions with local communities in the neighbourhoods to identify and inventory their ICH. A participatory process has been carried out in Fontanalla area within the city of Malaga, including experts' opinions and local knowledge. This innovative experience consists of different events where citizens and neighbours play an active role in order to identify their cultural heritage and showcase it to the rest of the city. This case study aims to establish some of the bases that can be extrapolated to other neighbourhoods in order to showcase and enhance those areas in terms of their cultural identity and also to identify their undervalued heritage.

Keywords Intangible cultural heritage · Citizen participation · Identity · Cultural mapping · Neighbourhood regeneration

F. J. Chamizo-Nieto (✉)
Department of Urban and Regional Planning, Universidad de Sevilla, Sevilla, Spain
e-mail: fchamizo1@us.es

N. Nebot-Gómez de Salazar
Department of Art and Architecture, Universidad de Málaga, Malaga, Spain
e-mail: nurianebot@uma.es

M. Cornax-Martín · C. Rosa-Jiménez
Institute of Habitat Territory and Digitalisation, Universidad de Málaga, Malaga, Spain
e-mail: martacornax@uma.es; cjrosa@uma.es

G. Cimadomo et al. (eds.), *Urbanism at Borders*, The Urban Book Series,
https://doi.org/10.1007/978-3-032-09594-7_26

26.1 Introduction

26.1.1 Enhancing Intangible Cultural Heritage as an Opportunity to Trigger Participatory Processes

Intangible Cultural Heritage (ICH) is defined in the *Convention for the Safeguarding of the Intangible Cultural Heritage* as:

> The practices, representations, expressions, knowledge, skills – as well as the instruments, objects, artefacts and cultural spaces associated therewith – that communities, groups and, in some cases, individuals recognize as part of their cultural heritage. (UNESCO 2003:2)

This is clear desire in that framework to showcase and enhance the need to preserve, document, research, promote, foster, transmit, revitalise and increase awareness of ICH (Frieri 2014) for its 'safeguarding':

> Measures aimed at ensuring the viability of the intangible cultural heritage, including the identification, documentation, research, preservation, protection, enhancement, transmission, particularly though formal and non-formal education, as well as the revitalization of the various aspects of such heritage. (UNESCO 2003:3)

Thus, safeguarding of ICH is a tool to construct the necessary conditions for it to continue to exist, along with encouraging communities to disseminate and recreate the expressions, representations and practices that comprise their living cultural heritage (van der Hammen et al. 2015).

The protection of this heritage requires in-depth research due to the lack of previous experiences fostering the safeguarding with guarantees. Therefore, it is complicated to classify and manage ICH, due, fundamentally, to the problems as regards its conceptualisation and its great diversity of underlying expressions of the community involved (Mariano et al. 2014). This is, partly, down to several reasons: (1) there is a direct interdependency between the intangible cultural heritage and the natural and cultural heritage, a fact that has led to a distinction being made after the *Convention concerning the Protection of the World Cultural and Natural Heritage* (UNESCO 1972); (2) even though there is a common roadmap for all the State Parties defined in the 2003 Convention, a specific methodology has not been defined regarding how to put that safeguarding into practice—aside from any possible variations according to the nature of what is likely to be ICH, as that task is the national competence of each of the State Parties (UNESCO 2003); and (3) the drawing up of inventories—the means of cataloguing the ICH defined in the convention (UNESCO 2003)—is a challenge given the open nature of this document that is free to the interpretation of the states (UNESCOCAT 2011), a fact that makes safeguarding something experimental and testing on a trial-and-error basis while clear and specific criteria have not been defined to pinpoint the ICH.

At a national level, regarding the duty of the Spanish Public Authorities pursuant to Sect. III of the *Convention for the Safeguarding of the Intangible Cultural Heritage* (UNESCO 2003)—endorsed in 2006—the *Spanish National Plan for the Safeguarding of the Intangible Cultural Heritage* (Ministerio de Educación, Cultura y Deporte—Gobierno de España 2011) was drawn up in order to establish a common framework for the whole national territory in order to thus adjust the provisions of the convention to the situation in Spain. Therefore, the Safeguarding of the Intangible Cultural Heritage Act 10/2015, of 26 May, was passed, where ICH was defined as in the 2003 Convention, such as the uses, representations, expressions, knowledge and techniques that local communities, groups or even individuals who give them cultural heritage value on their own. In this regard, the following are known as intangible assets: (1) oral traditions and expressions, (2) performing arts, (3) social practices, rituals and festive events, (4) knowledge and practices concerning nature and the universe, (5) traditional craftsmanship, (6) cuisine, culinary dishes and food, (7) specific use of the natural landscape, (8) collective ways of socialising and organisations and (9) sound expressions, music and traditional dance (BOE 2015).

This national framework likewise defines the following criteria to value ICH: (1) undeniable prominence of the community, (2) imminent danger of disappearing, (3) specificity, (4) continuity, (5) forms of transmission, (6) own traditional organisation, (7) involvement of participants, (8) diversity of multisensorial expressions, (9) own spatial frameworks, (10) temporal integrity and internal rhythm, (11) relevance of the objects and (12) autonomy (Carrión 2015). This list reveals the importance of the local population being able to establish which one will be allocated the ICH category:

> Intangible Heritage is never more reproducible than by its own protagonists, who are its incumbents, and they must decide about the change or the tenure of its expressions. Therefore, any cultural policy linked to the safeguarding of Intangible Cultural Heritage must be discussed with and accepted by the creators and bearers of its expressions. (Carrión 2015:28) [own translation]

Not only the community—'those who embrace an expression as their own and as part of their cultural references, of their identity and collective memory' (Frieri 2014:35) [own translation]—endorse and nurture the traditions, though it is also the *raison d'être* of its persistence and its maintaining and continuity in time. In order for a certain expression to be categorised as ICH, it must be first recognised by the individuals or communities that create, maintain and transmit that heritage. Nobody else can determine that heritage without the recognition of those groups. Therefore, in order to maintain or safeguard the ICH, its continuity needs to be guaranteed by means of a skill and knowledge transmission process of the cultural representations that must be steered and critically taken into account (Bakar et al. 2011).

Taking, therefore, the safeguarding of the ICH as an opportunity to trigger participatory processes opens up the way to participation spaces where people can be identified and affirmed through initiatives such as the 'route for the participatory construction of safeguarding projects': 'this tool offers the possibility to think about each of the aspects of an initiative to safeguard the ICH and develop them in a participatory manner' (Frieri 2014:113) [own translation]. As reflected in Article 15, *Participation of communities, groups and individuals*, of the Convention:

> Each State Party shall endeavour to ensure the widest possible participation of communities, groups and, where appropriate, individuals that create, maintain and transmit such heritage, and to involve them actively in its management. (UNESCO 2003:7)

26.1.2 Artistic Craft Activity

The expressions of ICH are linked to the different fields where the knowledge, know-how and practices are developed, namely: (1) languages and oral tradition, (2) social organisation, (3) traditional knowledge about nature and the universe, (4) traditional medicine, (5) traditional production, (6) techniques and traditions associated to manufacturing craft items, (7) popular arts, (8) recreational and festive events, (9) collective traditional religious events, (10) traditional techniques and know-how associated to the habitat, (11) culinary culture and (12) ICH associated to cultural spaces (Frieri 2014). Special interest is placed within that classification on artistic craft activity: 'set of community and family traditions associated with the production of fabrics, ceramics, basketry, decorations and, in general, craft utilitarian objects' (Frieri 2014:37) [own translation].

Popular arts and craft traditions consist of a collective practice of artists and artisans whose collective and creative work is aimed at new artistic expression where to continue reasserting themselves and, thus, keep their trade alive (van der Hammen et al. 2015). Focusing on the case of Fontanalla neighbourhood in Malaga, this type of manual activity has seen how industrial production has eroded it away: a fact that can be seen from the very few people who continue to live from this practice—only some families have passed their knowledge on from parent to child—which calls into question the continuity of this craft activity (Asociación de Artesanos Malagueños 2019). Therein lies the importance of working on this specific line of artisan trades. Andalusia is one of the Spanish Autonomous Regions with the greatest artisan tradition, accounting for 19% of the national total, and employing 20,000 people. Therefore, 10 Areas of Artisanal Interest are located there, together with 201 workshops and 49 recognised masters (Junta de Andalucía 2019). In addition, there are a total of 142 trades—though currently 159 are registered (Junta de Andalucía 2019)—included in the List of Trades of Andalusia (Consejería de Turismo, Comercio y Deporte—Junta de Andalucía. 2008): pottery, jewellery, carpentry, ceramics, metalworking, basketmaking, glass decoration, joinery etc.

Within this context, the need to address methodologies to safeguard these popular arts is clear. Therefore, they need to be based on their three main pillars: (1) the creative community, (2) the popular artists and (3) the objects themselves. However, even though those three stakeholders can be identified in any field of action where a methodology to safeguard artistic artisan activity can be tested, there is a historical depth of tradition that requires this methodology to be adapted to each context, working directly with the local population. Therefore, inviting the citizenry of this participatory process to enhance their artisan activity allows the construction of a context methodology and the creation of a space for dialogue between the different stakeholders in the collective process to determine, document and foster the craft and artistic expressions of ICH (van der Hammen et al. 2015).

Within the methodological process developed in the case studies of La Chamba, Carmen de Viboral and Ráquira, ceramic-based artisan communities in Colombia (van der Hammen et al. 2015), a series of spheres of action have been developed that would be taken into account in the methodology developed in the Fontanalla neighbourhood (see Table 26.1).

This methodological route starts with the local community, beginning with a sociogram to show which stakeholders are intervening and are interested in the theme to, thus, demonstrate that each artisan-individual is part of creative communities that stretch beyond the exclusive production of craft pieces. In a second phase, the aim is to reconstruct what is remembered with the help of photographs to identify traditional trades and artisans: illustrious figures that are part of the collective memory, evolution of the technique in the production process. Therefore, holding workshops that shift away from marketing and increasing production—an approach that originated from the 'value of change' where the craft product is rather seen as another good to be consumed by tourism (Lefebvre 2017)—facilitates greater citizen implication as they see that there is a 'healthy' interest in their art from the 'value of use' (Lefebvre 2017), in other words, from its symbolic value and as part of their identity. Finally, in a third phase, the focus is on highlighting popular art and taking it from the workshop context to, thus, open it up to the neighbourhood and to the city, seeking greater attachment by the citizens and creating synergies between the different local artists and artisans to construct communities.

The contribution of this research is focused on designing and testing a participatory process to identify the ICH at neighbourhood scale, within the consolidated city. In this process, and in contrast to the previously described methodology, a local community not only of artisans but also of local residents, institutions and neighbourhood associations is involved from the start. On the other hand, a very open basic methodology is defined so that it can be agreed and adapted to the needs and

Table 26.1 Methodology phases developed in the Fontanalla neighbourhood

Phase	Subphase
(1) Characterisation of the popular art and craft tradition	(1.1) Review of sources
	(1.2) Mapping stakeholders involved
(2) Establishing local research	(2.1) Identification of trade and artisans
	(2.2) Identification of recognised pieces and objects
	(2.3) Retrieving what people remember in talking shops and workshops with the help of archive and own photographs
	(2.4) Preparing documentary communicative pieces
(3) Local management to showcase popular art and the craft tradition	(3.1) Retrieving collective spaces for artisans
	(3.2) Interventions in the public space
	(3.3) Establishing demonstration workshops

Source: Authors

reality of the neighbourhood. The testing of the participatory process has led to some conclusions, achievements and shortcoming, which may be useful for future participatory processes in other neighbourhoods. This real experiment conducted in Fontanalla sought to provide information on the problems and difficulties found, and aspects to be taken into account when identifying and protecting the ICH in neighbourhoods.

26.1.3 Case Study: Arrabal de Fontanalla Neighbourhood, Malaga

The Arrabal of Fontanalla in Malaga is an old neighbourhood outside the walls of the ancient Moorish city (Machuca 1997). It was an industrial area in which the main activity was the craft of pottery, which lasted from the Arab period until the eighteenth century. This craft activity justifies the discovery of a multitude of medieval archaeological ovens in the different excavations carried out. Proof of this is the discovery of many medieval archaeological kilns (Gerencia Municipal de Urbanismo de Málaga 2009). The artisan culture continues to be deeply rooted and there are local residents who still work as artisans to earn their living (PTV Málaga 2019). From an urban point of view, the neighbourhood lacks a quality public space, as there are hardly any well-kept and pleasant places to stay. There is also a need to improve the many abandoned and empty plots, many of which require previous archaeological campaigns. This promotion of the artisan practice is supported by numerous local initiatives, such as the recent *Arq & Crafts Conference* organised by the Malaga Artisan Association (Asociación de Artesanos Malagueños 2019): talks given by master artisans and the opening of the trade exhibition at the Malaga Crystal and Glass Museum.

26.2 Objectives

The main objective of this research is to develop a methodology to identify the ICH of neighbourhoods taking the Arrabal de Fontanalla neighbourhood in Malaga as a case study, in order to design a methodology that can be extrapolated to other case studies. The preliminary research methodology will be developed by the technicians or team of experts as a proposal: the objective is to determine a preliminary methodology or central methodology by a team of specialist technicians as a baseline to identify the ICH. The aim is also to redefine this core methodology with the local community through meetings, gatherings and workshops to engage local residents: that residents and the local community take ownership of this process to identify ICH and be part of the investigation process, thus defining a context methodology close to its reality.

26.3 Methodology

26.3.1 Phase 0: Past Experiences

Malaga School of Architecture lecturers and students have already conducted several studies involving the Fontanalla neighbourhood. The first led on from an urban planning workshop run by Juana Sánchez, Nuria Nebot and Susana García. The project proposed, as part of the subject *Urbanism III (course 2012–2013)*, focused on teaching and practical sessions on urban recycling. Thus, during a quarter, the Arrabal de Fontanalla neighbourhood became the centre for reflections and experimentation regarding cities and way of living there. The student projects were exhibited, under the title *We have something to say: the Funtanalla neighbourhood deserves a special plan*, in June 2013 at the Malaga Crystal and Glass Museum: a meeting place where local residents could discover development opportunities for the neighbourhood from the exhibited projects (Sánchez et al. 2013).

Subsequently, during the 2013–2014 academic year, the urban planning department at Malaga School of Architecture, set, as a theme for the end-of-degree project, the study and proposal in this area of the city entitled: *The revitalisation of Malaga Old Town: The Fontanalla artisan neighbourhood*. Several years later, in June 2016, ‘The Mystery of Fontanalla’ exhibition, where students could exhibit their projects, was held to mark the eighth anniversary of the Crystal and Glass Museum. The exhibition sought to highlight the problems of this historical neighbourhood and, above all, to show possible solutions getting the residents involved. As part of these sessions, a roundtable was held and the local residents were invited to participate. Solutions were there discussed to improve this urban space, with the possibility of turning it into an artisan neighbourhood, or its inclusion in the immediate tourist setting, given its proximity to the city centre (Universidad de Málaga 2016).

26.3.2 Phase 1: MethodologyDesign

A phase when specialists and experts on intangible heritage in neighbourhoods prepared a preliminary methodology or core methodology is performed prior to the pooling with the community. In the case study of Fontanalla, the group of technical specialists is made up by professionals from different fields, whose research and proposal work is as follows:

1. Architects: Analysing and assessing the neighbourhood urban evolution, run-down points requiring improvements and spaces of opportunity within it.
2. *Sociologists:* Study of the community and of different urban groups. Analyse the different social levels and the cohesion between them.

3. *Artists:* Expertise, promotion and dissemination of tradition and local trades.
4. *Other informants:* Historians, archaeologists, teaching staff, citizens etc., whose work will contribute to the analysis and assessment of the neighbourhood in their fields.

The ICH of the neighbourhood was identified by means of research and information gathering thanks to different archive document sources. This phase is carried out at the following work points:

1. Bibliographic research on ICH: A bibliographic search was carried out on the neighbourhood, its history and cultural heritage. Different sources were used: books and guides on the history of the city and the area, historical maps, local websites, strategic plans, etc.
2. Informal meetings with mixed working group: A series of informal meetings have been held with the agents of the mixed working group. In these meetings, the aim of the project was explained: to recognise the intangible heritage potential of the neighbourhood, and to make it known to the local community in order to preserve it. One of the objectives of these meetings is to obtain the vision of the participating agents on the values with heritage potential in the neighbourhood.

26.3.3 Phase 2: Implementation

A series of activities were considered and organised after analysing and defining the arts and traditions inherent to the artistic artisan activity. The different actions emerged from the proposals in the previous development phase (1) from the technicians and local communities. They are as follows:

1. *Group walks*: Urban strolls where citizens are invited to discover the neighbourhood guided by informants explaining the history of each of the points or milestones on the set route. This experience or urban action is similar to the Jane's Walks (http://www.janejacobswalk.org), a global initiative that invites people to think about the city and to be part of a reflective look at the urban environment, whose objective is to create a local community of critical residents (Jiménez, 2016). This critical action will allow, as a field work, a shared account of the life in the neighbourhood to be built.
2. *Building collective 'memory'*: Activities in order to recover the memory of the neighbourhood:

    - Photographic collection: Historical review by looking at photographs to identify typical characteristics of the neighbourhood, accounts, stories, anecdotes and folk legends. This compilation occurs through meetings, workshops and talk shops to get together and share experiences, and through virtual platforms where photos—along with information in the form of blog entries—are uploaded to construct a local archive or repository.

- Workshop or Discussion group: The photos collected are used to trigger the memory that will allow us to produce a chronological timeline to organise the history of the neighbourhood and identify the different tangible and intangible heritage elements.
- Public exhibition: After selecting and identifying those photos endorsed by the local community, there will be a public exhibition of the compiled and screened material. Therefore, a public exhibition open to the residents of the city and neighbourhood can be held to thus open up the debate.

3. *Final workshop*: In order to determine those heritage elements, a final meeting will be held in order to open up the debate and reach conclusions from all the activities organised. The final goal of this workshop will be to identify the intangible and tangible heritage to conduct an inventory of the intangible heritage. This workshop will end the methodological route of the ICH of the neighbourhood and will act as a catharsis and culmination of the whole participatory process led by the local community. Other data collection tasks will be implemented:

   - Preparation of inventory sheets for an ICH catalogue: Based on the information gathered in the previous phases, its analysis and classification, the authors will prepare a series of datasheets for each element with cultural heritage potential, in accordance with the criteria and aspects to be considered determined by United Nations Educational, Scientific and Cultural Organization (UNESCO).
   - Dissemination of ICH in digital inventories: The different elements with heritage potential will be registered and mapped in the Information and Management of Historical Heritage platform of the municipality of Malaga (http://malakanet.adabyron.uma.es).

26.4 Discussion

This research project is currently in development up to including the Group walks activity. Even though the aforementioned methodology route has not been completed, its development has been critically assessed to weigh up the level of suitability of the participatory process in question to safeguard the ICH. The participatory actions carried out do not constitute a closed process, but they rather aim to promote new spaces for participation that may have continuity over time, even if some of the agents involved do not continue, such as external agents (university and technical advisors). Precisely, the proposals of the second phase should be developed by the local community without the presence of external actors.

26.4.1 Adapting of a Core Methodology by the Local Community

Counting on the local population—residents, institutions, artisan guilds, traders etc.—is fundamental to build a context methodology. Preparing the participants sociogram, in the case of the Fontanalla neighbourhood, was a simple task thanks to the involvement of the neighbourhood associations, institutions and local businesses: including the Arrabal de Fontanalla Neighbourhood Association, Malaga Artisan Association, the Arrabal–Development and Action Research Association, Malaga Crystal and Glass Museum, Vicente Espinel Secondary School—popularly known as Gaona High School—and the *VIARCA: Malaga Artistic Stained Glass* team of artists. They are all a sample of the social fabric of the neighbourhood and make up a representative part of the local community.

Before proceeding to start any activity with the local community for the safeguarding of the ICH, it is necessary to clarify what ICH is and what its purpose and importance are. Possible ways include workshops or discussion groups by means of an assembly methodology, where doubts can be raised or suggestions regarding the concept made, always avoiding an exclusively theoretical approach to be able to reach all citizens. To do so, several meetings and encounters with the different local groups took place in the Fontanalla neighbourhood. It was positively surprising that a large majority of the participants were familiar with the concept, partly because the local artist and artisan guild were used to dealing with the concept. In any event, irrespective of there being no specific ICH immersion workshop, graphic material, in the form of a *flyer*, was prepared, providing information about the ICH, what the objective was and how to participate (see Fig. 26.1).

Right from the presentation of the project, the members of the different representative neighbourhood groups expressed great interest in the project and always embraced it: taking decisions about the result of the activities to be implemented, showing great involvement due to their great attachment to the neighbourhood and the trust endorsed by prior experiences set out in Phase 0.

26.4.2 Activity (a): Group Walk

A series of preparatory meetings were held with the different groups involved in order to arrange the guided group walk. During those sessions, the local community—represented by the participating institutions and associations—established the route and agreed on the content to be explained at each of the milestones and which figures—historians, archaeologists, teachers and residents, among other informants—would tell the anecdotes, events and stories associated to each of the proposed sites. Those preparatory meetings were very interesting given all the contributions from the debates and discussions, as it would be very hard to recover and document that information if it were not thanks to local citizen participation. All this

Fig. 26.1 Back of the *flyer* prepared for the *Arrabal de Fontanalla Intangible Heritage* session on 12 January 2019. (Source: Authors)

documentation is the basis of a list of elements likely to be classified as ICH, as an inventory of the intangible heritage of the neighbourhood. In the specific case of the Fontanalla neighbourhood, this list was prepared, but without established clear criteria regarding what is and what is not an expression of ICH. This is because, even though the local community had been told about the concept and everything involved, there was no desire to check the debate and it was preferred that the discussion flows without filters.

As regards the latter, the role that the Gaona High School plays is as a benchmark in and outside the neighbourhood: an eighteenth century building that a priori is taken to be tangible heritage with special mention of its architectural value and its cloister with murals. Such is the attachment that the neighbourhood residents have for this centre, that it was established as a milestone in the group walk meetings held. Despite the doubts that this point may raise as it is something tangible and therefore tangible heritage, the many generations' life histories mean it is a container of an important period of the life of most of the local residents. This case raised doubts as to whether or not this milestone should be included given the blur between tangible and intangible heritage. Finally, at the request of the groups involved in the preparatory workshops, it was set as one more stop within the established route. However, even though it is complex to dissociate one heritage from another as one is a container and content of the other on many occasions, this fact invites us to reflect on which their clear differences are with regard to future activities.

As regards to the development of the route, one important aspect to highlight, apart from the major points to visitors, is its planning: which content to be shown, who should speak about it, how long each stop should last, which sequence to follow etc. Not only were the milestones chosen, but they also specified which part of the area around the milestone should be visited; Cloister of Gaona High School, crypt of the Church of San Felipe Neri etc. Furthermore, the group of narrators for each of the milestones was decisive: historians, archaeologists, restaurateurs, teachers and ‘lifelong’ residents of the neighbourhood. The involvement of this heterogeneous group and its experience and knowledge of the place was crucial to specify and define the route to be followed. An information flyer was prepared to facilitate this step and it was distributed to the visitors so that they could discover which milestones they were going to visit and in which order (see Fig. 26.1). After the walk, the participants were offered refreshments prepared by the neighbourhood residents at the VIARCA workshop: an ideal opportunity to spend a pleasant moment while looking at the work of the artisan group, while at the same time creating community.

The visit itself was very enriching: the participants themselves were creating personal experiences to be added to the tales about life in the neighbourhood. All these contributions were documented in photographs and videos as evidence for the activity. This initiative was underpinned by the call to take part in a photography contest to illustrate what each person on the walk considered to be ICH.

Overall, a total of 120 people took part in the urban walk and it was considered a success from that aspect. The initiative was very well received by the citizens and it attracted different figures, politicians, local authority technicians, architects, teachers, archaeologists, artists, artisans etc., a fact that meant the neighbourhood could be showcased to the city. Furthermore, the social media dissemination was very active, Malaga University’s press room allocated a section to the event (Universidad de Málaga 2019) and local stations and channels reported the activity by showing interest in the initiative (PTV Málaga 2019).

Critically speaking, the activity proved to be successful, even though there was a lack of planning regarding the time spent at each stop: a fact that delayed the arrival to other points set on the route, causing attendance to be hindered on the last stages of the route. A survey would also have been a valuable feedback to the activity: impressions, suggestions, ideas for improvement etc. They would have provided much information on the outcome of the walk, as they would have been an opportunity to learn of the successes and errors and foster the change for the better. The aim of the guided group walk through the neighbourhood was not only to address the ICH expressions related to artisan culture, but also to openly embrace different group expressions and spheres of the community (see Fig. 26.2).

Fig. 26.2 Post of the Arrabal de Fontanalla Intangible Heritage Day on 12 January 2019. (Source: the authors based on a photograph by M. Mayorga)

26.5 Conclusions

26.5.1 Participatory Dynamics: Towards the Construction of a Context Methodology

The participatory dynamics occurring through different workshops and collective meetings outlined a methodological route close to the specific reality of the Fontanalla neighbourhood. Those discussions revealed the lack of criteria regarding the intangible cultural heritage concept: a concept that seems understandable to the population, when we are actually not capable of defining it or specifying criteria for its classification. Therefore, the involvement of a multidisciplinary group with experience in managing the heritage in all the phases of the inventory—identifying, recording and cataloguing—is required, as the Fontanalla experience requires a screening of those 'ICH assets'—without fixed criteria—contributed by the participants. In this regard, the '0 workshop' proposal would have been correct and it would have focused on how to involve the local community in detecting those ICH expressions. This workshop would also facilitate the managing, steering and defining of the other activities defined in Phase 3 and, thus, condition the research from the start.

26.5.2 Sense of Attachment by the Local Community

The presentation of a core methodology to the Fontanalla local community—with possibilities of adapting to the specific reality of the neighbourhood—for the safeguarding of the ICH has facilitated the involvement of local groups and levels of participation that have greatly exceeded any type of expectations. The ownership of the project by the residents and their key role guaranteed the autonomy of the research and the freedom of action by the local population. Furthermore, the preparatory workshops managed to create community in the neighbourhood, along with showcasing the different situations and initiatives of the neighbours. In particular, special mention should be made of the artisan activity of VIARCA, the Arrabal de Fontanalla Association and the cultural work of the Malaga Crystal and Glass Museum.

26.5.3 Safeguarding the ICH as a Trigger of Participatory Processes

Safeguarding ICH is an opportunity to generate attachment for the place to which a community belongs. It does not only seek to identify those inherent expressions of the intangible heritage per se, but also, by involving the local population, links and ties of identity are built, along with attachment and the sense of belonging, both to the community and to the place, the traditions and the customs. The experience in Fontanalla has enabled the neighbourhood to engage their citizens and to reassess its distinctive features and inherent values. In this regard, the ICH becomes a pretext to bring together the local community, nurture its relationship with the place and foster social cohesion, all of which is thanks due to the inherent processes and dynamics of citizen participation culture.

Acknowledgements The Chair of Emerging Technologies for the Citizenry, University of Malaga and Malaga City Council have funded this research. We thank Eva Morales Soler for her contribution in the methodology definition phase, the Fontanalla residents who took part in the activities, and the institutions and entities involved: the Arrabal Fontanalla Resident and Trader Association, the Malaga Crystal and Glass Museum, VIARCA Stained Glass Workshop and Vicente Espinel Secondary School. We would also particularly thank Víctor Heredia, Carmen Peral, Joaquín Gallego and Paz Unguetti for their collaboration.

This research was funded by University of Malaga under Grant B3. First Research Program (B3-2019_15) and FEDER funds (UMA20-FEDERJA-131).

References

Asociación de Artesanos Malagueños (2019) Jornada Arq & Crafts: El valor cotidiano de la artesanía. http://coamalaga.es/agenda/actividad-cultural/arqcrafts-el-valor-cotidiano-de-la-artesania/May. Accessed: 23 Mar 2023

Bakar AA, Osman MM, Bachok S (2011) Intangible Cultural Heritage (ICH): understanding an expression. In: International Conference on Universal Design in Built Environment (ICUDBE2011), Kuala Lumpur, Malaysia

BOE—Boletín Oficial del Estado (2015) Salvaguardia del Patrimonio Cultural Inmaterial. [Ley 10/2015]. DO: Jefatura del Estado—BOE, n° 126. https://www.boe.es/buscar/act.php?id=BOE-A-2015-5794. 27 May. Accessed: 25 Mar 2023

Carrión A (2015) Plan Nacional para la Salvaguardia del Patrimonio Cultural Inmaterial. Ministerio de Educación, Cultura y Deporte—Gobierno de España, España. https://doi.org/10.4438/030-16-425-3

Consejería de Turismo, Comercio y Deporte—Junta de Andalucía (2008) Repertorio de Oficios Artesanos de Andalucía. Junta de Andalucía, Sevilla. http://www.juntadeandalucia.es/turismoycomercio/publicaciones/32557.pdf. Accessed: 25 Mar 2023

Frieri S (2014) Manual de herramientas participativas para la identificación, documentación y gestión de las manifestaciones del Patrimonio Cultural Inmaterial. Ministerio de Cultura & Tropenbos Internacional Colombia, Bogotá. https://pcisa.files.wordpress.com/2014/02/manual_de_herramientas_pci_ilandocs.pdf. Accessed: 25 de March 2023

Gerencia Municipal de Urbanismo de Málaga (2009) Catálogo de protección arqueológica. Ayuntamiento de Málaga, Málaga. https://urbanismo.malaga.eu/modificacionpgou2008/catalogos/CATALOGO%20ARQUEOLOGICO%20COMPLETO%20(MAYO%202009).pdf. Accessed: 25 March 2023

Jiménez S (2016) Cómo hacer un paseo de Jane. Ministerio de Educación, Cultura y Deporte—Gobierno de España, Madrid. http://laaventuradeaprender.intef.es/guias/como-hacer-un-paseo-jane. Accessed: 25 Mar 2023

Junta de Andalucía (2019) Tradiciones populares. https://www.juntadeandalucia.es/andalucia/cultura/tradiciones.html. Accessed: 13 Aug 2019

Lefebvre H (2017) El derecho a la ciudad. Capitán Swing, Madrid

Machuca L (1997) Málaga, ciudad abierta. Origen, cambio y permanencia de una estructura urbana. Colegio de Arquitectos, Málaga

Mariano M, Endere ML, Mariano CI (2014) Herramientas metodológicas para la gestión del patrimonio intangible. El caso del municipio de Olvarría. Rca Rev Colomb Antropol 50(2):243–269

Ministerio de Educación, Cultura y Deporte—Gobierno de España (2011) National plan for the safeguarding of intangible cultural heritage. http://www.culturaydeporte.gob.es/planes-nacionales/eu/dam/jcr:73de51a5-978f-47c4-a67a-42b20e09fe41/08-inmaterial-eng.pdf. October. Accessed: 25 Mar 2023

PTV Málaga. 2019. «Tu barrio»: Arrabal de Fontanalla. https://www.youtube.com/watch?v=r48_yewL3RY. 25 January. Accessed: 25 March 2023

Sánchez J, Nebot N, García S (2013) Ensayos de reciclaje urbano como experiencia docente realizados por estudiantes de Arquitectura. Aprendiendo del barrio artesanal de La Funtanalla. In: Greencities & Sostenibilidad: Inteligencia aplicada a la sostenibilidad urbana de la Convocatoria de Comunicaciones y 1ª Bienal de Edificación y Urbanismo Sostenible. Malaga. http://aula-greencities.coamalaga.es/wp-content/uploads/2014/06/29.-Greencities2013-Comunicaciones_Ensayo-de-reciclaje-urbano-como-experiencia-docente.pdf. Accessed: 25 March 2023

UNESCO—United Nations Educational, Scientific and Cultural Organization (1972) Convention concerning the Protection of the World Cultural and Natural Heritage. [WHC/72/WS/1]. DO: Conventions and Agreements of a Standard-Setting Nature adopted by UNESCO. https://unesdoc.unesco.org/ark:/48223/pf0000002091?posInSet=33&queryId=e4c3ee4f-1ce0-434a-afeb-fac73e7d7598. 16 November. Accessed: 25 Mar 2023

UNESCO—United Nations Educational, Scientific and Cultural Organization (2003) Convention for the Safeguarding of the Intangible Cultural Heritage. [MISC/2003/CLT/CH/14]. DO: Conventions and Agreements of a Standard-Setting Nature adopted by UNESCO. https://unesdoc.unesco.org/ark:/48223/pf0000132540_spa. 17 October. Accessed: 25 Mar 2023

UNESCOCAT (2011) Metodología para el inventario del Patrimonio Cultural Inmaterial en Las Reservas de la Biosfera. Centro UNESCO de Cataluña, Barcelona. https://ich.unesco.org/es/BSP/metodologa-para-realizar-inventarios-del-patrimonio-cultural-inmaterial-en-reservas-de-biosfera-la-experiencia-del-montseny-00648. Accessed: 25 March 2023

Universidad de Málaga (2016) La ETS de Arquitectura colabora en el VII Aniversario del Museo del Vidrio y Crisal de Málaga. Sala de presna de la Universidad de Málaga. https://www.uma.es/sala-de-prensa/noticias/la-ets-de-arquitectura-colabora-en-el-vii-aniversario-del-museo-del-vidrio-y-cristal-de-malaga/. 27 May. Accessed: 25 March 2023

Universidad de Málaga (2019) Los vecinos de la Asociación Arrabal de Fontanalla ponen en valor el patrimonio cultural inmaterial. https://www.uma.es/sala-de-prensa/noticias/los-vecinos-de-la-asociacion-arrabal-de-fontanalla-ponen-en-valor-el-patrimonio-cultural-inmaterial/. 15 January. Accessed: 25 Mar 2023

Van der Hammen MC, Frieri S, Sánchez LF, Peña A (2015) Opciones y acciones para la salvaguardia del PCI: una compilación de experiencias. Ministerio de Cultura & Tropenbos Internacional Colombia, Bogotá. https://openlibrary.org/works/OL20088537W/Opciones_y_acciones_para_la_salvaguardia_del_PCI_una_compilación_de_experiencias_-_1._edición. Accessed: 25 Mar 2023

Part VII
Change of Scale: Focus on Dwellings

Chapter 27
Privacy and Gender: Segregation in Bedouin Settlements in Dahab, Egypt

Eman Sherif

Abstract Dahab is a small coastal city located in South Sinai, Egypt. Dahab has witnessed urban developments led by the state's plan that was initiated after the reclamation of Sinai in 1979 and is now part of new development plans that are targeting the southern part of the Sinai Peninsula. The native inhabitants of Dahab, the Bedouins, were and still are impacted by these development plans. The implemented state's approach to development was based on global standards of functions, planning and architecture. The Bedouins, on the other hand, lead lives based on their norms and traditions. Privacy and gender segregation are two of the most critical aspects of their culture. The research highlights the representation of privacy and gender segregation in contemporary Bedouin dwellings outside and inside Dahab. To do so, the study focuses on exploring and documenting samples of homes inhabited by the Bedouins, explicitly focusing on the manifestation of privacy and gender segregation. The research relies on unstructured interviews to gather the data about sociocultural norms and customs. Official documents and literature are used for demographics, plans and historical studies. Observation research through first-hand observations, photographs, illustrations and sketches from site visits is used to map the layout and design of Bedouin dwellings and highlight the implementation of privacy and gender segregation in different scenarios. The outcome of this research will inform policy and decision makers about the local context. Hence, the information could help consider the Bedouin community and its needs in the new development plans.

Keywords Housing · Gendered space · Bedouin · Privacy · Segregation

E. Sherif (✉)
Arab Academy for Science, Technology & Maritime Transport, Cairo, Egypt

G. Cimadomo et al. (eds.), *Urbanism at Borders*, The Urban Book Series,
https://doi.org/10.1007/978-3-032-09594-7_27

27.1 Introduction

27.1.1 Dahab (Egypt): Geographic Location and Demographics

The Sinai Peninsula, situated in the northeastern region of Egypt, is a geographic bridge connecting Africa to Asia and spans an expanse of 61,000 square kilometres. South Sinai is divided into eight sectors: Nuweiba, Dahab, Sharm el-Sheikh, El Tur, Ras Sudr, Abu Zenima, Abu Rudeis and Saint Catherinne. The focus of this study, Dahab, is a quaint coastal city nestled along the shores of the Red Sea, encompassing an area of 1130 square kilometres in South Sinai. Positioned between the cities of Nuweiba to the north and Sharm el-Sheikh to the south, Dahab offers a picturesque setting for exploration and analysis (see Fig. 27.1).

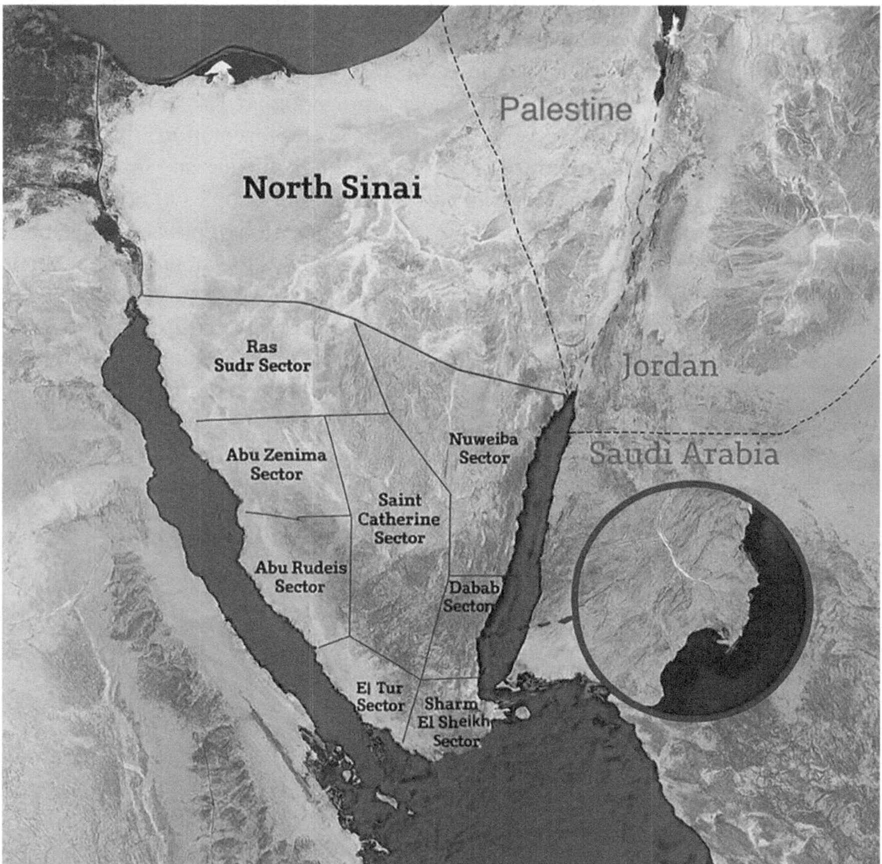

Fig. 27.1 The Sinai Peninsula map with the highlighted South Sinai sectors and a zoomed image of Dahab. (Source: the author based on Google Earth 2024)

The Bedouin community is an ethnic community that roamed around North Africa and the Middle East. The Bedouins are nomads of three generations; the first generation was born around 1914, the second generation by the 1940s and the third generation by the 1960s. The third generation became coastal dwellers by 1972 (Ragab 2011). The Bedouin community follows a tribal structure, is self-governed and lives based on norms and culturally oriented laws and regulations (al-qawanin al-'urfiyya) (Karkabi 2013). To understand Dahab and its people well, it is important to comprehend and digest the place's historical background and the events it has witnessed over the years. Accordingly, the following section explores Dahab's historical background and its impact on development.

27.1.2 History and Background

As a border area, the Sinai Peninsula witnessed a history of battles and occupation. After the 1979 Egypt-Israel Peace Treaty, the state initiated a development plan aiming to increase the population and settlement in South Sinai as a means of protecting the eastern borders from reoccupation (Hemdan 1980; MOHPUC 1997). As part of South Sinai, Dahab was affected by the state development plan. The following explains the timeline of political events and development in Sinai based on (Lavie and Young 1984).

1956–1967. Before 1956, the Bedouins roamed around Sinai. Between 1956 and 1967, the government constructed a road that connected Suez to Sharm el-Sheikh for security reasons (Lavie and Young 1984).

1967–1979. Israeli forces occupied Sinai between 1967 and 1979, during which time 21 settlements were built for political, economic and security reasons. These reasons included disconnecting the Peninsula from the rest of Egypt by encouraging the Bedouins' self-governance (Al-Khashab 2008). During this phase, the Israelis introduced the tourism industry in South Sinai by building tourist-oriented settlements (Ali 1998). This phase was divided into two parts; between 1967 and 1972, the area witnessed low investment and development. The Israelis developed a road between Eilat and Sharm el-Sheikh, moving the economy to the east coast. Later, between 1972 and 1979, the Bedouins were exposed to the westernisation introduced by the tourism industry. This exposure caused a conflict between cultural survival and economic well-being (Lavie and Young 1984). Realising the importance of tourism, the Bedouins started settling along the east coast by building straw huts and shacks (Ali 1998). During that period, the Egyptian government had concerns regarding the Bedouins and decided to try to integrate them into the plan of developing Sinai (Lavie and Young 1984).

1979–1988. During this phase, the Bedouins started to settle (Lavie and Young 1984). Between 1979 and around 1990, the Egyptian government began to have an apparent role in developing Sinai (Ali 1998). The political side of the development plan was to stabilise the eastern borders. Accordingly, the government started to encourage Nile Valley migration to increase Sinai's population. The tourism industry was the tool of attracting people to move and settle in Sinai (Hemdan 1980; Karkabi 2013).

1988–2011. Between 1988 and 2005, the tourism industry flourished enormously (Ali 1998), and by the 1990s, foreign migration increased (Karkabi 2013). The 2011 Egyptian Revolution led to major political and economic shifts. The country's instability affected the economy, hence development. The number of tourists from 2010 to 2011 dropped by 24.5% (SSG 2012).

2011–2019. Due to economic instability, statistics show that the rate of tourists and inhabitants decreased after the revolution. Tourism decline resulted in a high migration rate, leaving South Sinai. The political situation stabilised in Egypt, but it is taking time to recover after the significant drop in the number of inhabitants; in which the number of residents increased from 3649 in 1996 to 7494 in 2006, but the number later in 2017 declined to 2934 (MOHPUC 1997; CAMPAS 2010, 2017; SSG 2011). The state announced its interest in developing the Peninsula in 2014 (SIS 2020).

2020–2024. According to the South Sinai Governorate website, the number of residents of the southern part of the Peninsula reached 108,761 in 2024 (SSG 2024). Further data exploration for this study stopped in 2017, but according to history, all events impact development. Hence, the situation during the coronavirus disease 2019 (COVID-19) phase is worth exploring and understanding.

The tourism industry was and still is the main force for developing Dahab. The next section explores the tourism industry as the main engine for development and the progression of the industry in Dahab. The section also presents the aim behind development and a brief glimpse of the plan previously set for Dahab.

27.2 Tourism Development Plans

The tourism industry targets economic growth and has grown enormously (UNWTO 2013). Tourism is a fast-growing industry that sustains economic development (Saha and Yap 2015). Tourism is also defined as an economic activity that leads to social and economic changes that rely on the environment, economy and society (Zaei and Zaei 2013). Sinai was undeveloped due to the political unrest it had gone through, being the battlefield in three wars (1956, 1967 and 1973). Sinai finally settled on the Camp David Accords in 1979 and became free of Israeli occupation in 1982. The development approach was not new to South Sinai; the tourism industry was first introduced in South Sinai in 1970 by the Israeli occupation in the form of huts and services. The government's plan to develop South Sinai into a major international tourism destination was to create the area, attract foreign investment and avoid reoccupation. This idea was in progress through the 1980s, but the significant step of developing the Gulf of Aqaba was in 1994 (Karkabi 2013).

The state introduced a development plan that was presented as a concept in the early 1980s, studied and designed between 1994 and 1997, implemented in 1997 and was to be achieved by 2017. The plan encouraged the movement and settlement of people in Dahab by introducing many job opportunities (MOHPUC 1997). There were several reasons for that plan, regarding Sinai in general and Dahab in particular. The state was generally trying to connect Sinai to the rest of Egypt socially and economically, and attract the population from all over Egypt, hence reducing the load on the delta. In addition to that, the plan targeted the encouragement of investment in Sinai. Regarding Dahab, the plan aimed to allocate land for the tourism industry to sustain economic development, to use Dahab's natural assets as an attraction point and to protect unplanned areas from informal growth. Accordingly, the plan was designed to transform Dahab from a small town into a major tourist destination (MOHPUC 1997). The development plan aimed for 20% of the surface area of Egypt to be occupied instead of the 4% by the Nile Valley. South Sinai was seen as a hope for fulfilling such a plan (Ragab 2011). Accordingly, with the increase of migration, a diversity of cultures occurred, which impacted each other and the built environment. Three stakeholders contribute to the development of the tourism industry in South Sinai: the government and decision makers, the locals and the tourists. The government's interest in Sinai's development is political and economic. The role of the tourists is to spend time at the destination, influencing economic growth. The locals also participate in developing the host area's tourism industry (Ibrahim 2009). Tourism development plans are considered guidelines that orient ideas and resources of development, but not a decision-making process (UNWTO 2013). According to this, several political events halted the plans, and these plans were not followed up; hence, they were not adjusted or refined to adapt to the current situation and needs. As a result, the existing state of Dahab is different from the envisioned plan.

The approach of the state's plan was based on the global standards of functions, planning and architecture. In contrast, the emerging urban reality is based on the local context, sociocultural norms and customs. The differences between the state's approach and the urban reality are apparent. For example, three scenarios will be examined in urban planning: the planned division, the existing state division and the actual division known to people. The state plan divided Dahab into sectors, neighbourhoods and districts (see Fig. 27.2). The government studied Dahab for development by classifying it into zones. Dahab was split into the Bedouin housing, Assalah, including Mashrabah and Masbat, Magles El Madeena, Lagoon and Wadi Qunai (see Fig. 27.3) (MOHPUC 1997). The existing scenario, which grew from the previous classifications, is somewhat different; existing divisions are not into sectors and neighbourhoods but are more classified into the main areas' growth. The Bedouin zone, which is considered a part of Assalah, continued to grow despite the events that halted the development plans. This shows Bedouins' interest in settling and urbanising. North of Assalah by the bay is the Eel Garden area, which is named after the dive site. South of Assalah are Lighthouse, Masbat and Mashrabah, which are treated as different zones, not as part of Assalah (see Fig. 27.3).

According to the previous discussion, the plan being developed currently should approach Dahab with great sensitivity and understanding, not just of the people and

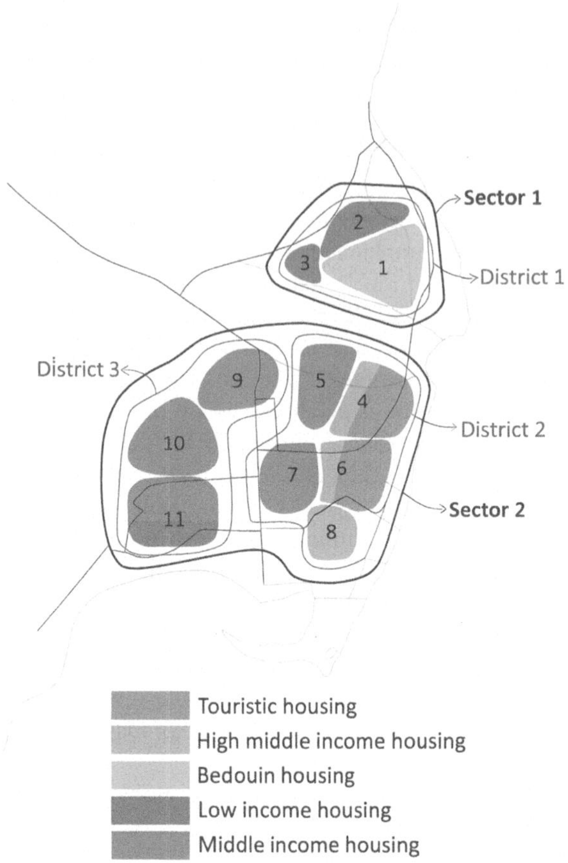

Fig. 27.2 The zoning designed by the state for Dahab in the 2017 development plan. (Source: (Sherif et al. 2016)

their culture but also of their decisions regarding the development of their built environment.

The following section explores and documents the transformation of Bedouin dwellings in Dahab to digest the existing context and understand the impact of people's needs on the built environment, specifically the Bedouin community. The documentation highlights the important aspects that are constantly present in the Bedouin dwellings, such as gender segregation and privacy.

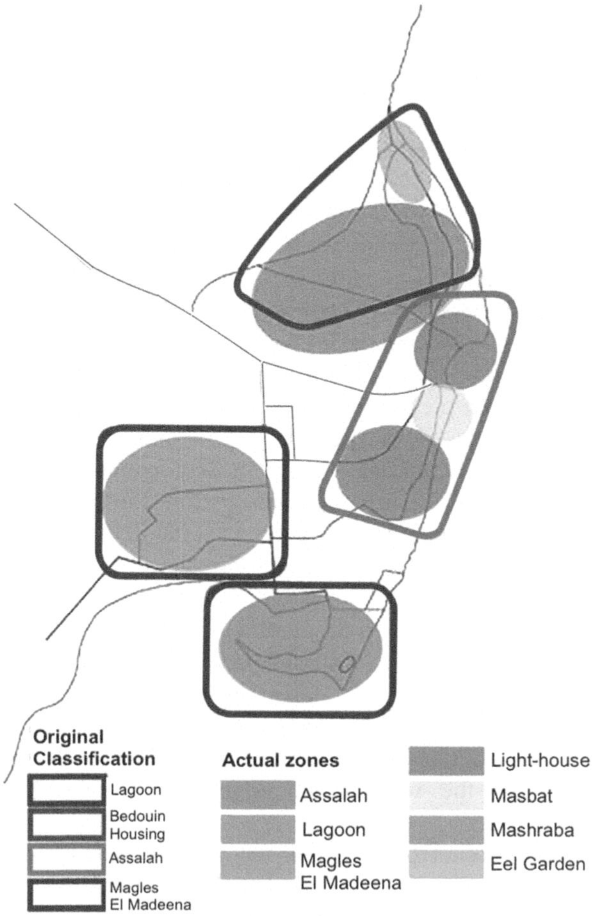

Fig. 27.3 The original classification by the state (MOHPUC 1997) and the actual developed zoning after a survey in Dahab by 2017. (Source: the author)

27.3 Dwellings Documentation

The following part of this study illustrates the transformation of Bedouin dwellings and examples of contemporary Bedouin dwellings inside and outside Dahab. The constraints of this section are represented in the lack of documentation and the minimum data available on the Bedouins of Sinai and their dwellings. Accordingly, this section is based on unstructured interviews with Bedouins of various age groups, written literature describing parts of the Bedouin lifestyle and dwellings and observations from site visits. The author illustrates that most sketches are based on oral and written descriptions and documentation of data gathered from literature and during site visits.

27.3.1 Early Bedouin Dwellings' Transformations

The following section shows the three phases of the transformation of Bedouin dwellings. This part of the study is mainly based on unstructured interviews and narratives by the elder Bedouins in Dahab passed on by many generations. Accordingly, the exact phases of dwelling transformation are not dated, but the sequence is according to the passed-down narratives. The first phase is before settlement, pre-1973, when the Bedouins were still nomads (nomadism). The second phase is between the early 1970s and the late 1980s, covering the end of nomadism and the beginning of sedentariness. The third phase represents the increasing population concentration in Dahab and the growth of urban areas as people transitioned from a rural or non-settled lifestyle to living in a more urbanised context, which occurred after 1988 till the present (urbanisation).

There are several models of transformation of Bedouin dwellings, from bayt al-sha'r (Bedouin tent) to the urbanised version of contemporary dwellings. The Bedouin tent is considered the first Bedouin dwelling; it is a light structure that they carried around roaming the lands, made of animal hair. The tents were placed in a straight line beside one another facing the same orientation with the core household—usually where the senior relatives and their families dwell—is located in the centre (Abu-Lughod 1999) as the central aspect of organisation of tents was privacy (see Fig. 27.4) (Tarawneh 2000). As extracted from interviews, a large tent called *maq'ad* was placed far from the other tents to welcome strangers. The variation of distances between tents is dependent on the power of the tribe; the stronger the power, the greater the distance (Tarawneh 2000). The number and size of the tent were relevant to the wealth and status of the owner (Tawfiq 1994; Tarawneh 2000).

Bayt al-sha'r is different from a regular tent as it is made of black goats' hair, camels' hair, or wool carried on structural elements. As illustrated in Fig. 27.5, the basic tent design was divided into two parts defined by structural elements. Literature presented the same description of the Bedouin tent, but the structural

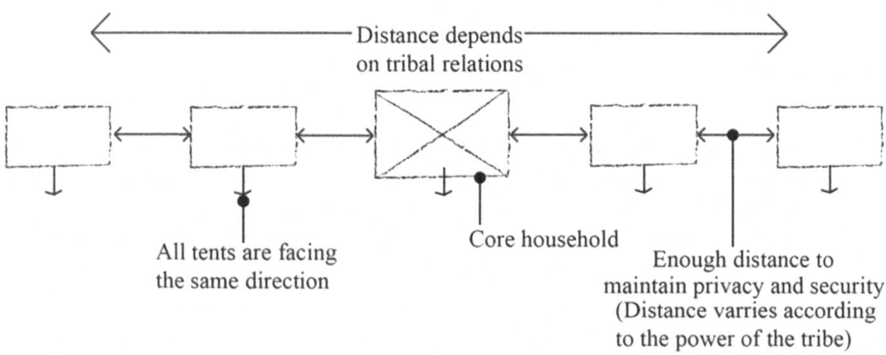

Fig. 27.4 Tents arrangement diagram. (Source: Sherif et al. (2017) based on description by Lila Abu-Lughod (1999))

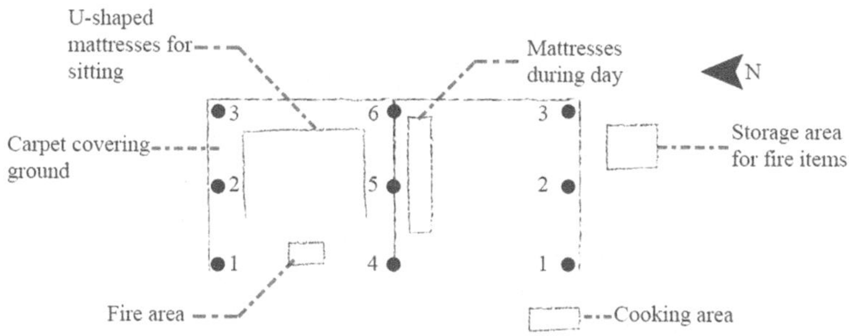

Fig. 27.5 Bayt al-sha'r, a Bedouin tent diagram. 1-2-3: Inclined structural elements, 4-5-6: Main structural elements. (Source: Sherif et al. (2017)) based on the descriptions of Tarawneh (2000); Shuqaïr (1991); Hess (2012))

elements were named differently. The given diagram is an adaptation from both descriptions of Johann Jakob Hess and Na'ūm Shuqaïr (Shuqaïr 1991; Hess 2012). Structural elements 1, 2 and 3 are inclined structures located on both sides of the tent and structural elements 4, 5 and 6 are the main structural elements located in the centre of the tent. There are various types of tents based on the number of spaces and the space definition of the unit (Shuqaïr 1991: Hess 2012). The following description is cited from (Sherif et al. 2017) based on literature by Na'ūm Shuqaïr (1991), Johann Jakob Hess (2012) and Musa Tarawneh (2000). The design of a basic tent, which consists of two parts, includes the left section for men and the right for women and family, separated by a partition located between structures 4, 5 and 6 and a removable fabric is placed on the front side of the tent for privacy and protection from foul weather. The men's side was for male guests, while the women's was used for cooking, as a living area during the day, and as a sleeping area for the whole family. The men's area was furnished with a carpet, a U-shaped seating area and a fire area in front of the men's zone. On the other side, the women's area was furnished by mattresses, a cooking area was located in front of the women's section and a storage area for fire items was situated on the side of the women's area (Sherif et al. 2017).

The following documentation of dwellings is based on unstructured interviews, stories and memories from Bedouins in Dahab, which were conducted between 2013 and 2016. Some of the descriptions are supported by photos similar to the original forms. With the beginning of settlement, straw huts started to occur. When tourism was introduced, lightweight structures made of straws provided basic shelter for Bedouins by the coast. The Bedouins built those huts by the coast to offer services to the tourists. As per the informal interviews, the straw huts are considered the first evidence of the transformation of Bedouin dwellings and settlement from movable tents to fixed structures. Straw-covered structures are still used today as seating areas for tourists by the coast.

The model that was developed next was *Bayt al-tïn,* which is a dwelling made of mud brick. The dwelling was usually divided into three spaces in addition to a detached cooking area; the spaces were divided between men and women (see

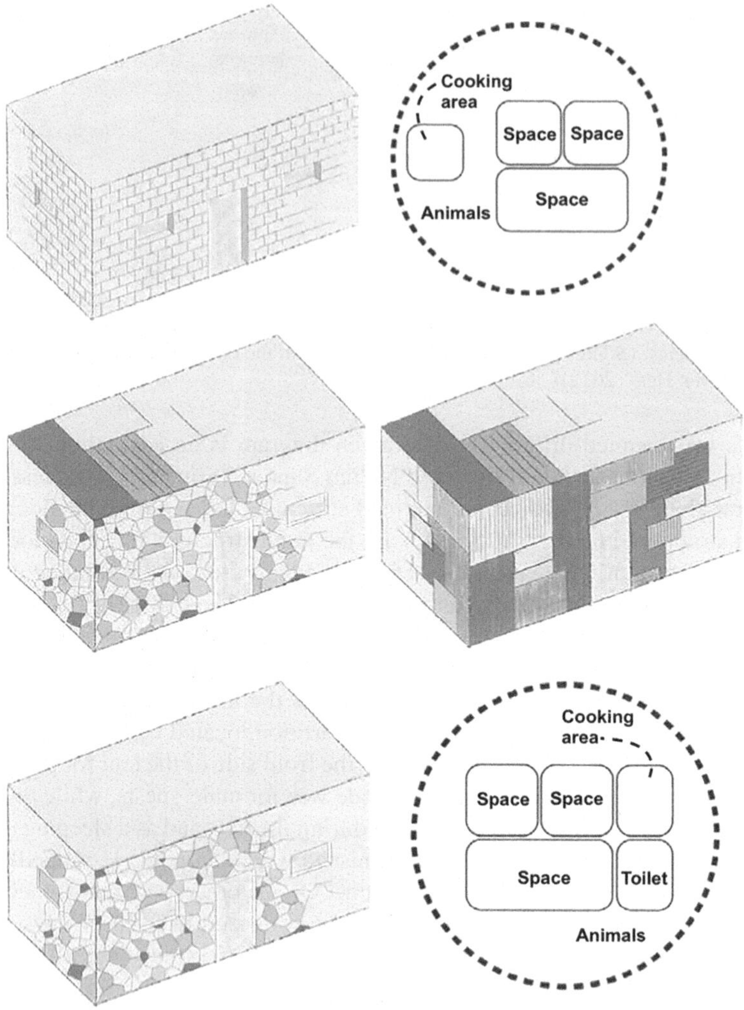

Fig. 27.6 Reconstruction sketches of four types of Bedouin dwellings: Bayt al-tīn, Bayt al-ḥajar wa al-ṣfīḥ, Bayt al-ḥajar, and Bayt al-ṣfīḥ. (Source: the author)

Fig. 27.6). With the growth of settlement, the Bedouins started to explore new materials to use in building their dwellings to create stronger and more stable units. *Bayt al-ḥajar wa al-ṣfīḥ* (see Fig. 27.7); a dwelling made of stone walls and tin ceiling, *bayt al-ṣfīḥ*, which was made of only tin and bayt al-ḥajar, which was made of stones, were developed with the same division system and zoning. Each model consisted of three spaces, an attached cooking area and a toilet (see Fig. 27.6).

Fig. 27.7 Images of an old bayt al-ḥajar wa al-ṣfīḥ located in Ras Abou Galloum, North Dahab. (Source: the author, 2016)

27.3.2 Contemporary Bedouin Dwellings' Transformations

The documentation of contemporary Bedouin dwellings includes examples from both outside and inside the city (see Fig. 27.8). The houses located outside the city, away from the dense urbanised context, are built by the Bedouins themselves as a residential area occupied only by the same tribe or the same family members. The featured examples inside the city are categorised into two types: the first is built by the state and the second type is built by the Bedouins themselves. The first two featured models are from outside the city; the last three are from inside the city.

The featured dwellings outside the city are part of a family and tribe settlement, and all the members living in the settlement are related. The family lives detached from the dense urbanised city, with no services located close to the area. The following section will illustrate two neighbouring dwellings in the same settlement (see Fig. 27.9).

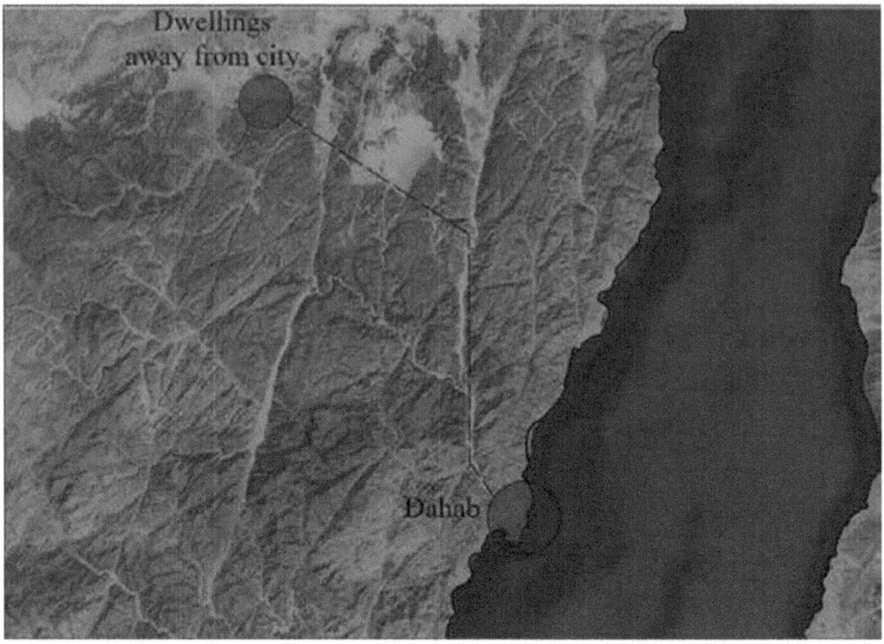

Fig. 27.8 Map showing the location of Bedouin dwellings inside and outside the city. (Source: the author based on Google Earth 2017)

Fig. 27.9 Map showing the location of the studied Bedouin dwellings in the settlement outside the city. (Source: the author based on Google Earth 2017)

27.3.2.1 Outside the City: Dwellings Built by the Bedouins—Dwelling 1

The dwelling has an L-shaped plan consisting of a living area, three multi-functional rooms, a kitchen and a toilet on one side. The other side comprises a winter *maq'ad* (social space) and a baking area. A summer *maq'ad* for women is located in front of the dwelling. In contrast, the men's *maq'ad* is situated on the side of the dwelling (see Fig. 27.10). The images in Fig. 27.11 show simplicity of execution, openness and basic furnishing. Multifunctional rooms are furnished with rugs or storage for goods. The kitchen is clear due to the presence of a stove and a sink, and the toilet is recently renovated and has an embedded floor unit.

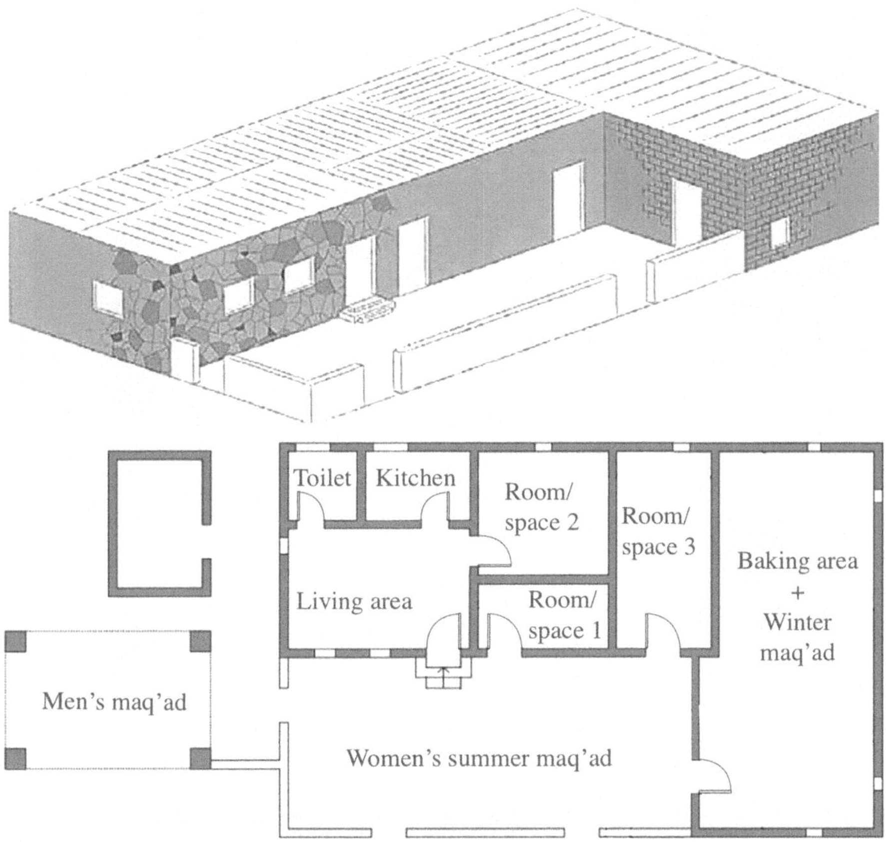

Fig. 27.10 Sketch of dwelling 1, a dwelling built by the Bedouins outside the city. (Source: the author)

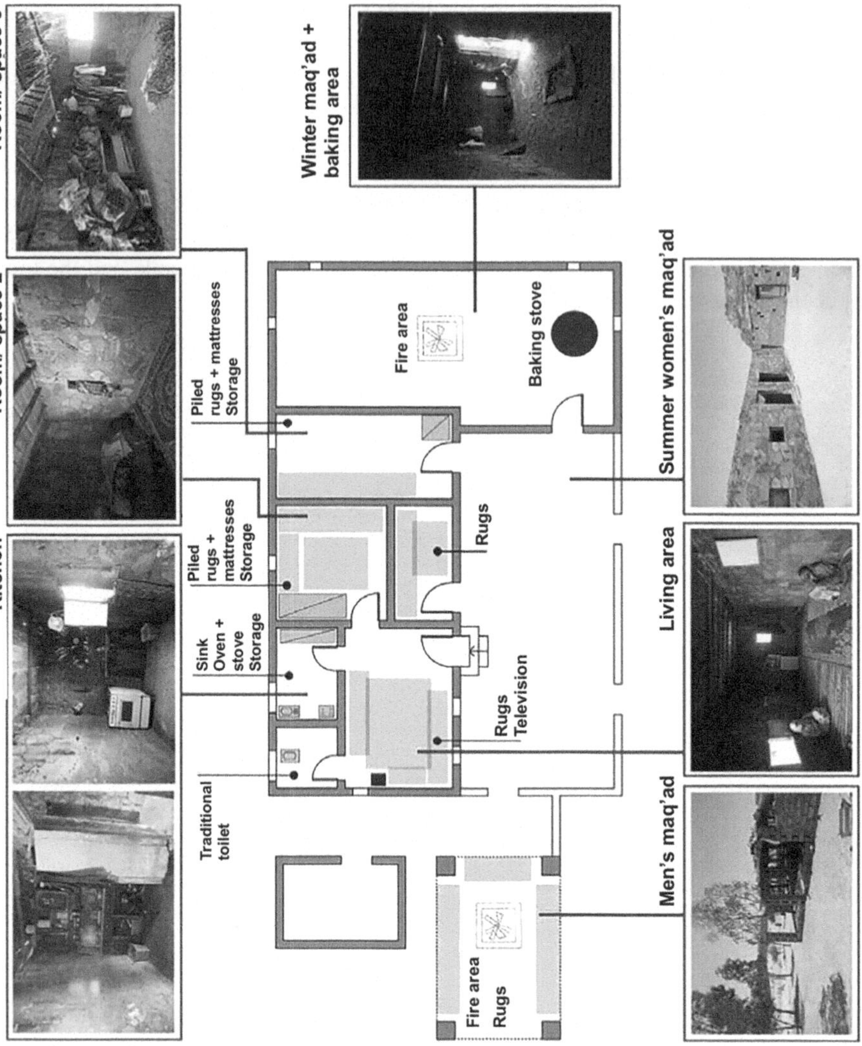

Fig. 27.11 Furnished plan with images of dwelling 1, a dwelling built by the Bedouins outside the city

27.3.2.2 Outside the City: Dwellings Built by the Bedouins—Dwelling 2

The dwelling has a rectangular plan and consists of an entrance lobby, three multi-functional spaces/rooms, a toilet, a kitchen and a common space between the rooms. The outdoor area is used as a *maq'ad* for women and a space for animals (see Fig. 27.12). Furnishing the dwelling is basic with rugs, storage units and mattresses. No specific functions are clear in spaces except for the kitchen and toilet (see Fig. 27.13).

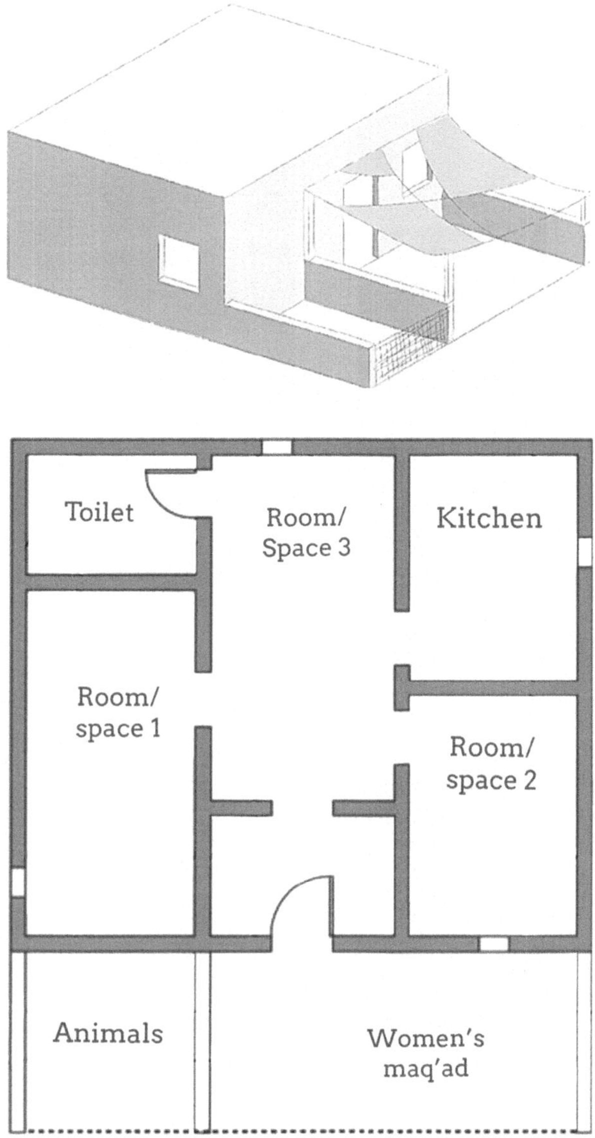

Fig. 27.12 Sketch of dwelling 2, a dwelling built by the Bedouins outside the city. (Source: the author)

The following section presents three examples of Bedouin dwellings located inside the city. The first two dwellings were built by the Bedouins themselves and located in Assalah. The dwellings built by the Bedouins illustrate their needs and culture and the manifestation of important aspects expressed in the built environment (see Figs. 27.14 and 27.15).

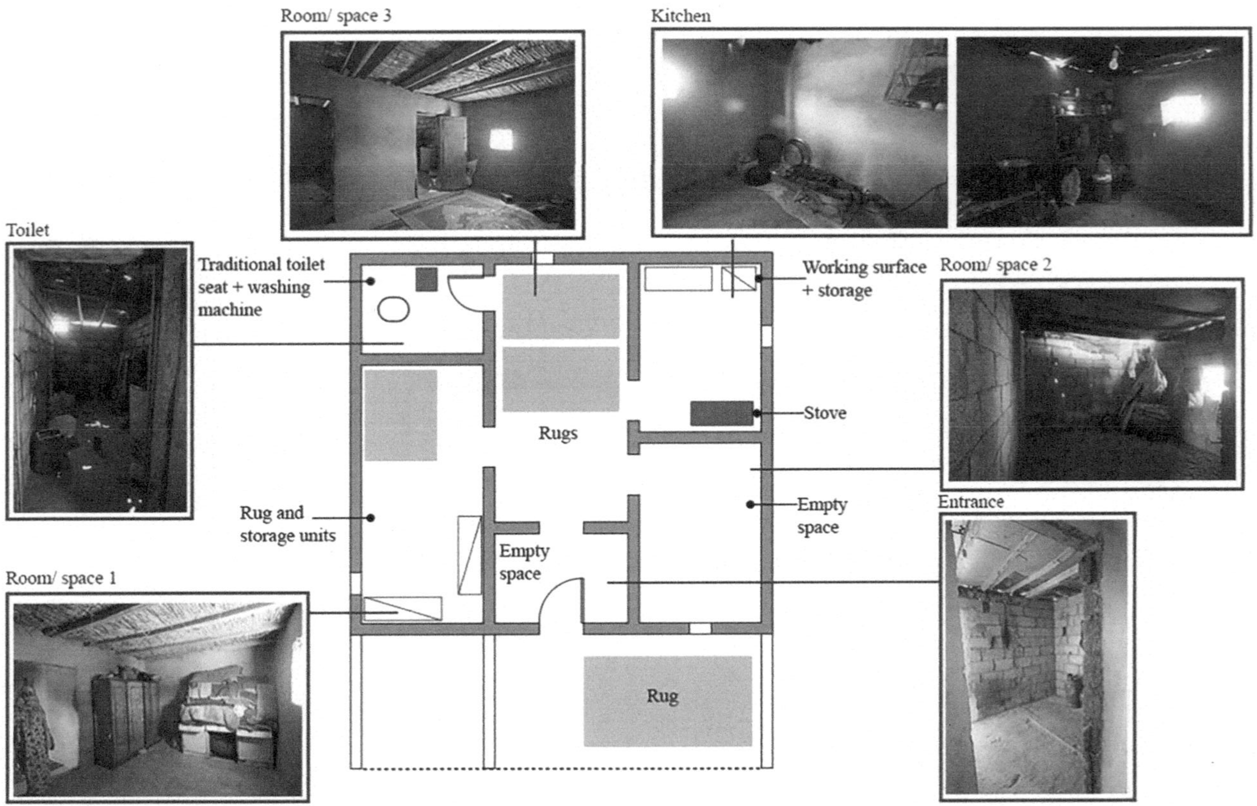

Fig. 27.13 Furnished plan with images of dwelling 2, a dwelling built by the Bedouins outside the city. (Source: the author)

Fig. 27.14 Map showing the location of the featured Bedouin dwellings inside the city. (Source: the author based on Google Earth 2017)

Fig. 27.15 Map showing the location of the featured Bedouin dwellings inside the city. (Source: the author based on Google Earth 2017)

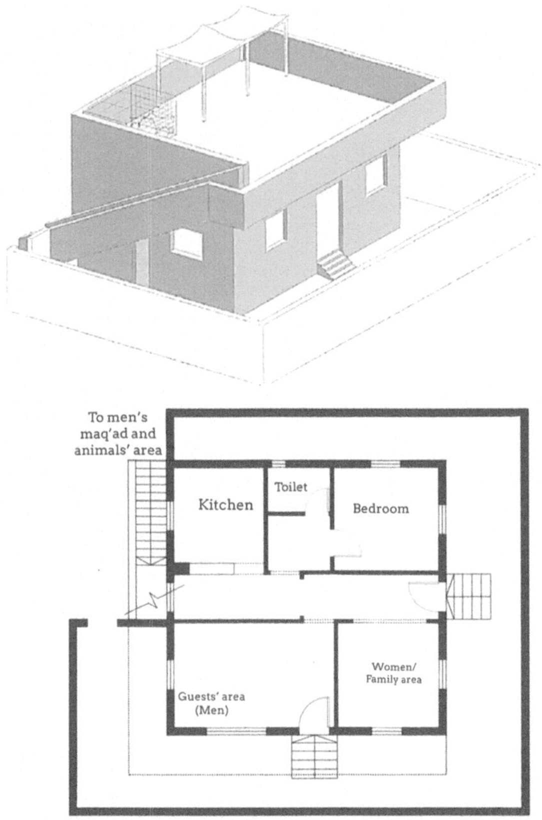

Fig. 27.16 Sketch of dwelling 3, a dwelling built by the Bedouins inside the city. (Source: the author)

27.3.2.3 Inside the City: Dwellings Built by the Bedouins—Dwelling 3

The dwelling has a rectangular plan with an access to property and another access to the rooftop from the outside. The dwelling itself has two entrances and consists of a guests' area for men, a women and family area, a kitchen, a toilet, a bedroom. The maq'ad and the animals' area are located on the rooftop with a separate entrance (see Fig. 27.16). The dwelling is furnished with basic cushions, mattresses and rugs in the guests' and living areas. The kitchen, toilet and bedroom are influenced with Western habits (see Fig. 27.17).

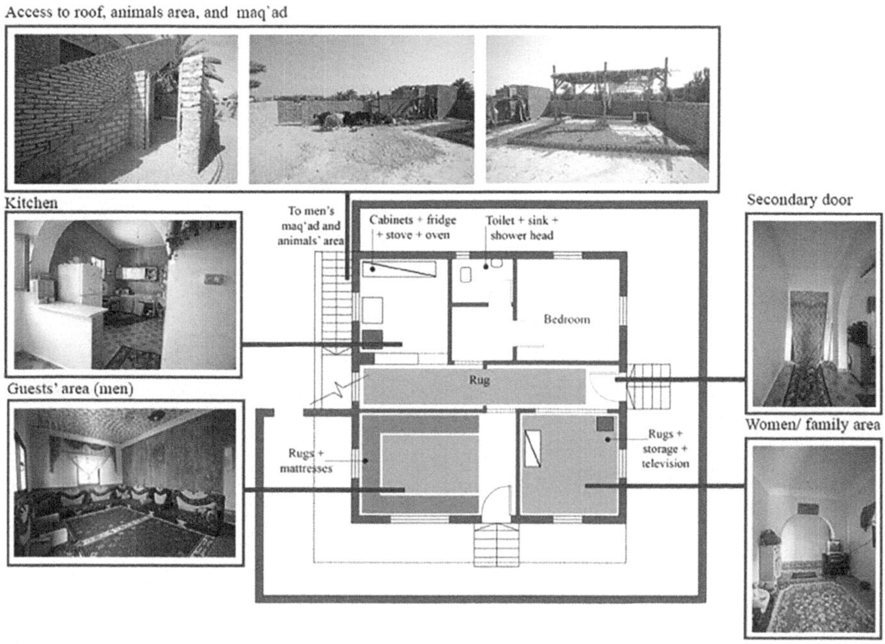

Fig. 27.17 Furnished plan with images of dwelling 3, a dwelling built by the Bedouins inside the city. (Source: the author)

27.3.2.4 Inside the City: Dwellings Built by the Bedouins—Dwelling 4)

The family house built in the late 1980s was extended by the son in 2004. The old structure was renovated and spaces were added to the existing structure. Instead of two spaces, the dwelling transformed into five spaces. The original dwelling consisted of two multifunctional spaces/rooms used for sleeping, eating and gathering. The new addition had more definition to the spaces; a toilet and kitchen were introduced. The other three spaces were specified as a bedroom, a socialising space for women and a socialising space for men (see Fig. 27.18). The guests' and living areas are furnished with rugs, cushions and mattresses. On the other hand, the toilet, kitchen and bedroom are also westernised (see Fig. 27.19).

27.3.2.5 Inside the City: Dwellings Built by the State—Dwelling 5)

This example is different from previous ones; it is an apartment in a building built by the state, not a detached unit inside the city (see Figs. 27.20 and 27.21). The identical group of 10 buildings each consists of three floors and each floor has two apartments. The original dwelling consists of a guests' area for men, a living area, a bedroom, a toilet and a kitchen. The design of the apartment is made by the state but adjustments were made by the inhabitant to fit their lifestyle. The kitchen was open

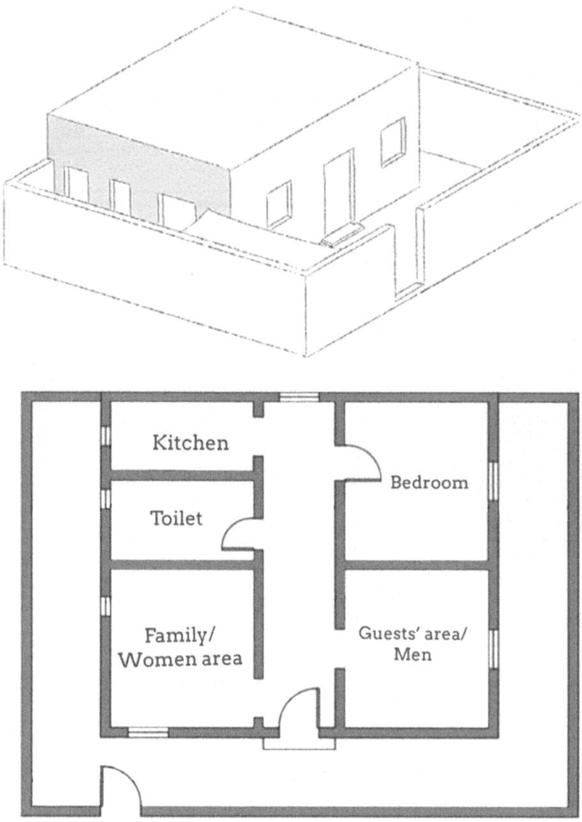

Fig. 27.18 Sketch of dwelling 4, a dwelling built by the Bedouins inside the city. (Source: the author)

to the entrance corridor and the guests' area had access to a terrace attached to it (see Fig. 27.22).

The inhabitants made some changes to accommodate their needs. The kitchen entrance was moved to being open to the corridor leading to the bedroom. A light partition was added to define the guests' area and the terrace was enclosed to the indoor space (see Fig. 27.22). The plan was furnished and finished in a modern way. The bedroom, kitchen and toilet are westernised. The guests' area is finished with rugs, carpets, cushions, mattresses and a fire pot that is traditionally used to make tea when men are around, similar to a *maq'ad* (see Fig. 27.23).

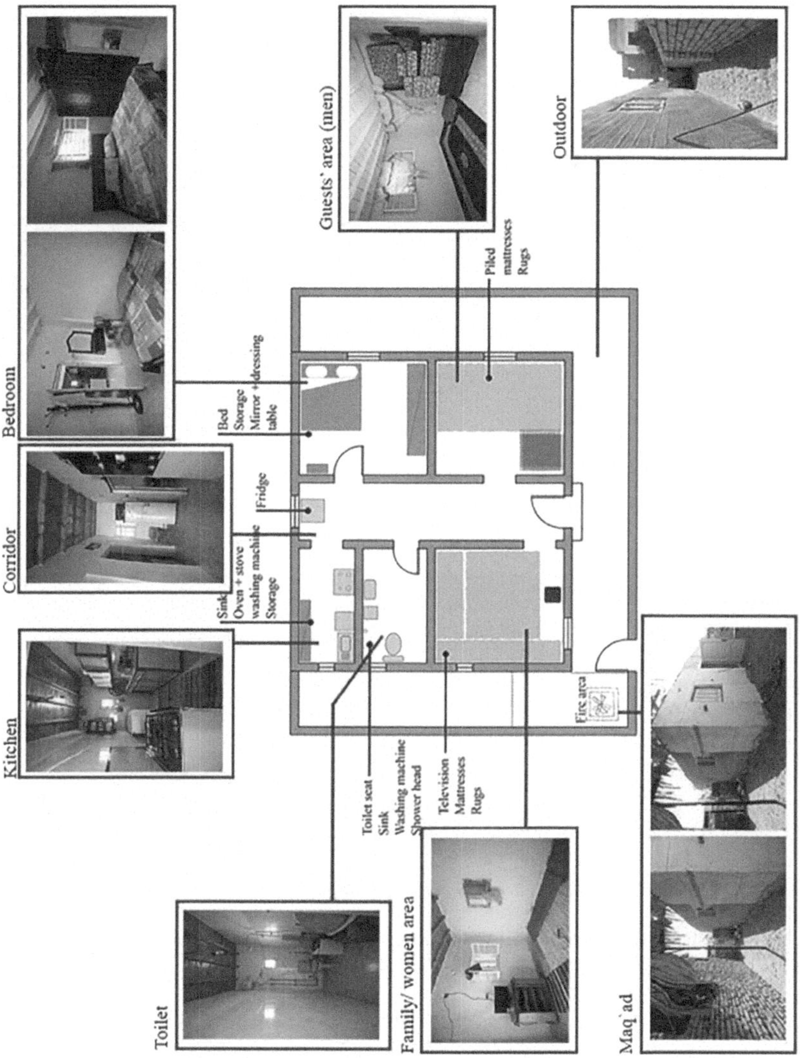

Fig. 27.19 Furnished plan with images of dwelling 4, a dwelling built by the Bedouins inside the city. (Source: the author)

Fig. 27.20 Map showing the location of the featured dwelling inside the city (dwelling 5) in Eskan Mubarak. (Source: the author based on Google Earth 2017)

Fig. 27.21 Map showing the equal units including (dwelling 5) in Eskan Mubarak. (Source: Sherif et al. (2016))

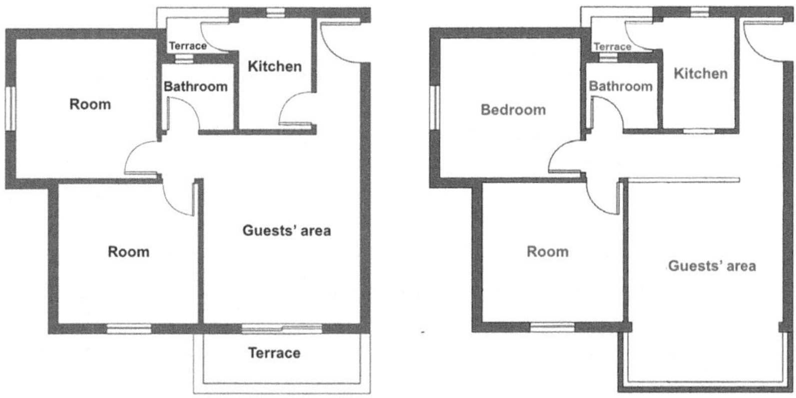

Fig. 27.22 Left: Dwelling 5 plan before adjustments. Right: plan after adjustments. (Source: the author)

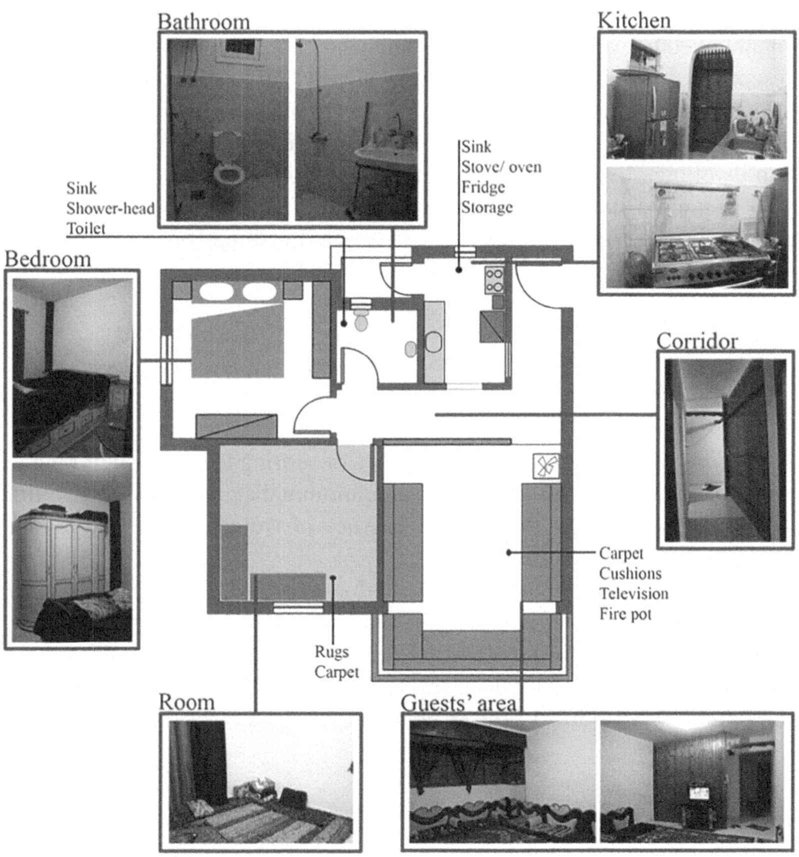

Fig. 27.23 Furnished plan with images of dwelling 5, a dwelling built by the state and inhabited by Bedouins inside the city. (Source: the author)

27.4 Privacy and Gender Segregation: Analysis of Contemporary Buildings

People, culture and the environment are strongly interconnected (Altman and Chemers 1989), with the physical environment serving as a reflection of various lifestyles and behaviours across different cultures. Cultural concepts such as privacy influence the way individuals interact with their surroundings (Rapoport 1983; Altman and Chemers 1989). For the Bedouins, privacy and gender segregation have long held significance, evident in their dwellings even prior to the transition from nomadism to settlement and urbanisation.

This section delves into two key aspects of Bedouin culture: privacy and gender segregation, as reflected in their dwellings' spatial organisation. It examines the way these aspects are represented in Bedouin architecture, emphasising their physical manifestations in various examples. In the models presented in the progression of Bedouin dwellings, privacy and gender segregation were constant aspects occurring in all the models. The two aspects were clear in the various old models, starting with the choice of location and orientation of the dwellings, in addition to the distance maintained between them. Privacy and gender segregation were also shown in the presence of a physical separation between men and women in the form of clear space definition with a fabric partition in case of tents and in the form of rooms in the other models. The following part of the study is an analysis of contemporary Bedouin dwellings highlighting the manifestation of gender segregation and privacy. The featured examples are analysed into four main categories: distance between dwellings, physical boundaries between dwellings, openings and accessibility and gender segregation in spaces.

27.4.1 Distance Between Dwellings

Outside the city, the Bedouins have the luxury of settling based on their own terms. According to Bedouin norms and laws, the minimum distance between dwellings is 50 metres creating a buffer that ensures privacy of women inside and around the dwelling (see Fig. 27.24).

On the other hand, inside the city, the Bedouins are surrounded and exposed by foreign neighbours. To maintain privacy, the Bedouins tend to cluster their dwellings close to each other to maintain the privacy of the dwellings and the women moving between houses. The close proximity between dwellings creates narrow pathways that protect women from being exposed by strangers (see Fig. 27.25).

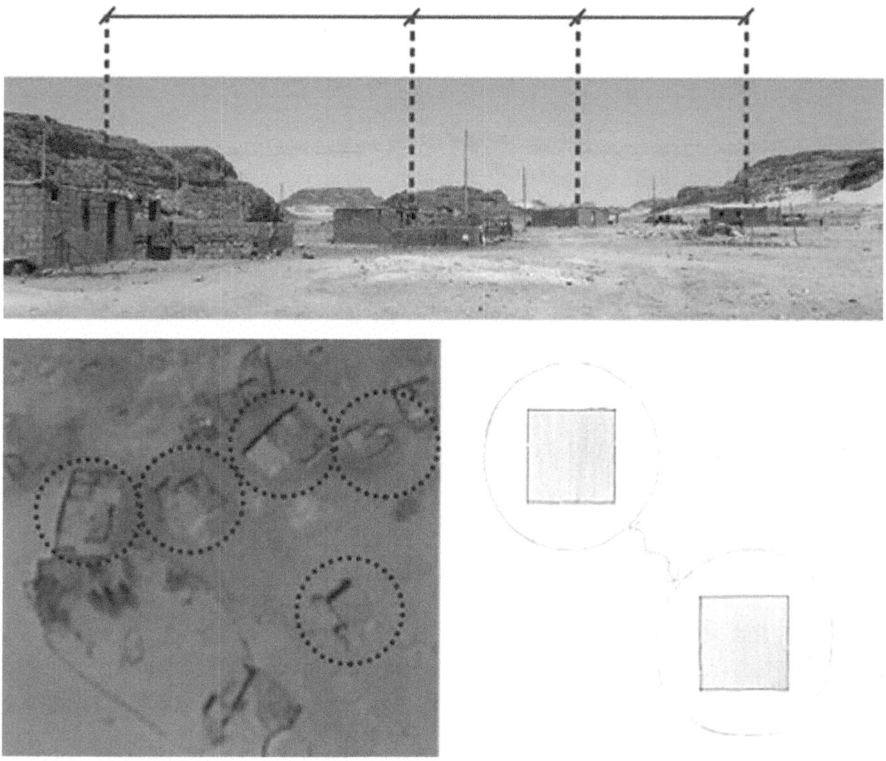

Fig. 27.24 Bedouin dwellings outside the city: large proximity between dwellings creating privacy. (Source: the author)

27.4.2 Physical Boundaries Between Dwellings

Outside the city, the distance between dwellings provides the needed privacy for the people inside the house. Accordingly, fences or physical boundaries around the dwellings are low or not needed (see Fig. 27.26). On the other hand, inside the city, the dwellings are constructed close to one another and sometimes even attached. Other dwellings are separated by narrow roads or corridors. In both cases, fences around dwellings are high to ensure the privacy of the dwelling (see Fig. 27.27).

27.4.3 Openings and Accessibility

The window heights in the featured examples vary. Within the city, windows are strategically positioned at higher levels to minimise the visibility of interior spaces. A height of 1.20 metres is not much higher than the usual 0.9 metres, but it is high in relevance to the Bedouin furnishing style; the use of mattresses and cushions for

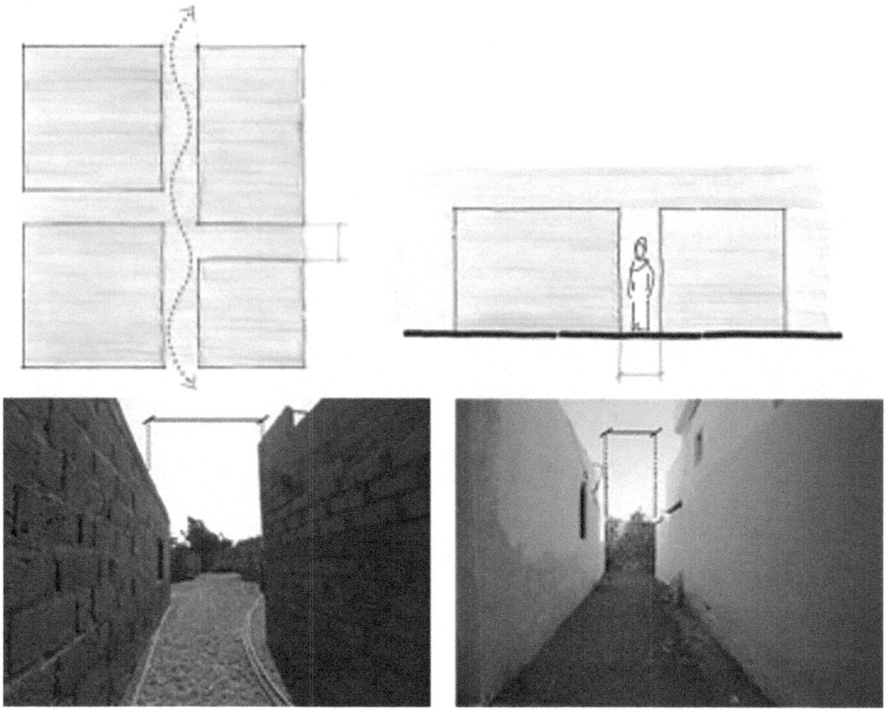

Fig. 27.25 Bedouin dwellings inside the city: close proximity and narrow pathways. (Source: the author)

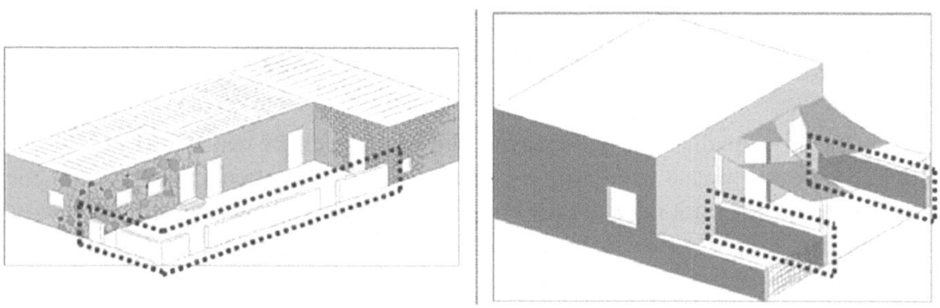

Fig. 27.26 Low fences around dwellings outside the city, dwellings 1 and 2. (Source: the author)

seating (see Fig. 27.28). In dwelling 4, there is an old opening on a very low level where men used to sit on mattresses and communicate with any passerby before urbanisation and building fences (see Fig. 27.29). Similarly outside the city, the openings are not very high due to the large distance between dwellings that ensures privacy (see Fig. 27.30).

The original design of dwelling 5 has a terrace open to overlook the main street but after interviewing the dwellers, the wife explained that there is no use of a

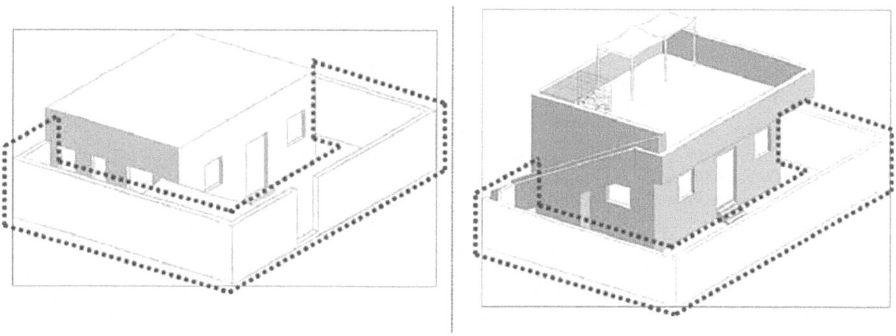

Fig. 27.27 High fences around dwellings inside the city, dwellings 3 and 4. (Source: the author)

Fig. 27.28 Windows on high level in dwellings inside the city to ensure privacy, dwellings 3 and 4. (Source: the author)

terrace due to the importance of privacy. Based on that, the couple decided to enclose the terrace and add it as an extra space serving the guests' area. The users accommodated the space according to their needs and priorities. This alteration was common between the Bedouins living in these units.

Accessibility to dwellings in some cases show consideration of privacy. In dwelling 3, women are provided an access different from the men's to ease their movement outside the dwelling (see Fig. 27.31). In addition to that, the open men's

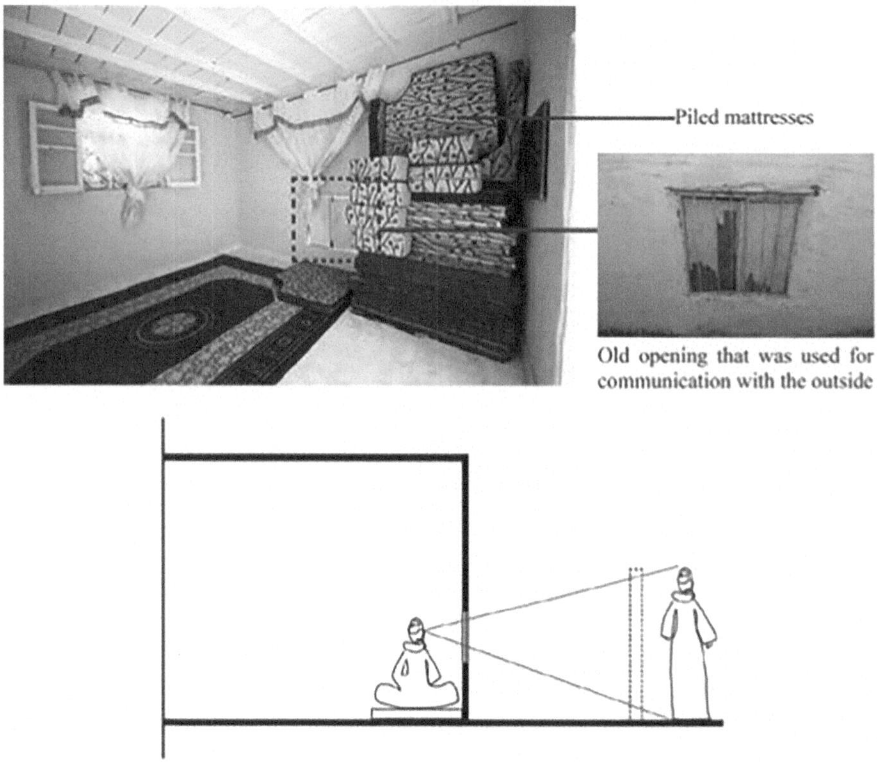

Fig. 27.29 Old opening that was used for communication before urbanisation, dwelling 4. (Source: the author)

Fig. 27.30 Windows on low level in dwellings outside the city, dwelling 1. (Source: the author)

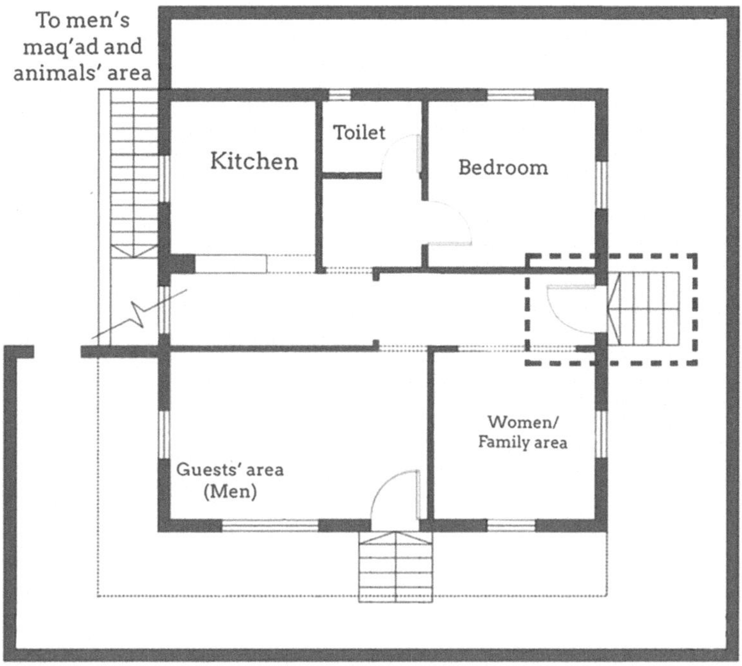

Fig. 27.31 Interior privacy measures in dwelling 3, separate access for women. (Source: the author)

maq'ad in most cases is located outside by the main entrance and specifically in dwelling 3 the space is provided with a separate entrance accessed from outside the dwelling on the roof to ensure the privacy and segregation between men and women (see Fig. 27.32).

27.4.4 Interior Boundaries

Privacy is important inside the dwellings as it is around them. In some cases, the social space for men is located inside the house. Accordingly, privacy measures are taken inside the space as well through partitions, curtains, or shifted entrances. In dwelling 3, the access overlooking the corridor from the guests' area (2) is shifted away from the access to the private area. In addition to that, access to the private zone (1) is covered by a curtain as an extra measure of privacy (see Fig. 27.33).

Dwelling 5 is different from the rest of the examples, in which the unit is built by the state not by the Bedouins. Accordingly, the distribution and design of inner spaces are different from the dwellings built by the Bedouin themselves. For example, the couple living in dwelling 5 added a partition enclosing the guests' area to avoid exposing the kitchen and the private area. The couple also changed the

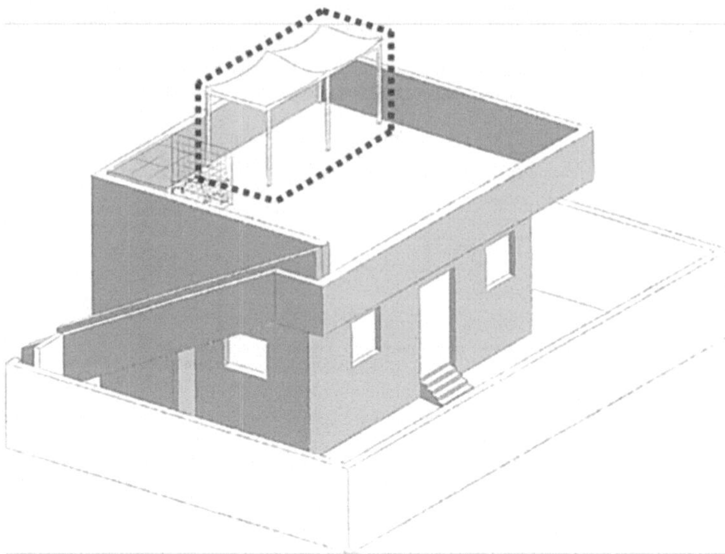

Fig. 27.32 Private access to maq'ad on roof in dwelling 3. (Source: the author)

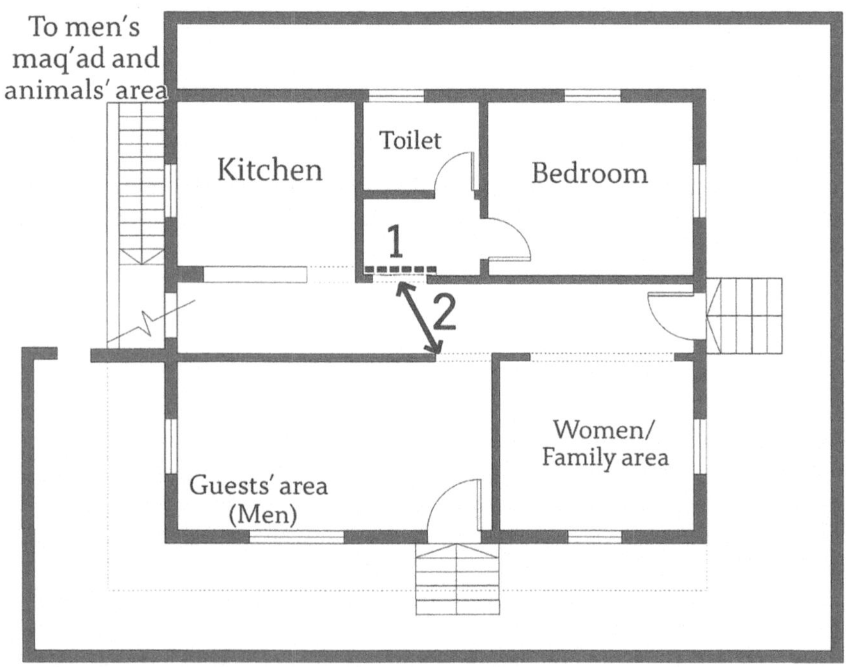

Fig. 27.33 Interior privacy measures in dwelling 3. (Source: the author)

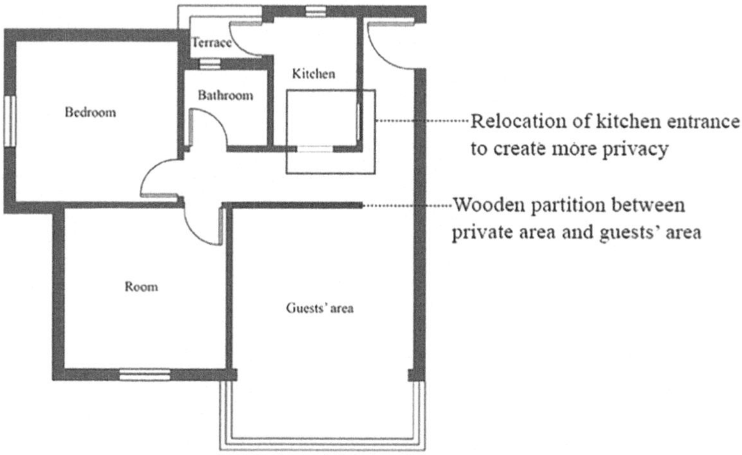

Fig. 27.34 Adjustments made to interior spaces and openings of dwelling 5 to ensure privacy. (Source: the author)

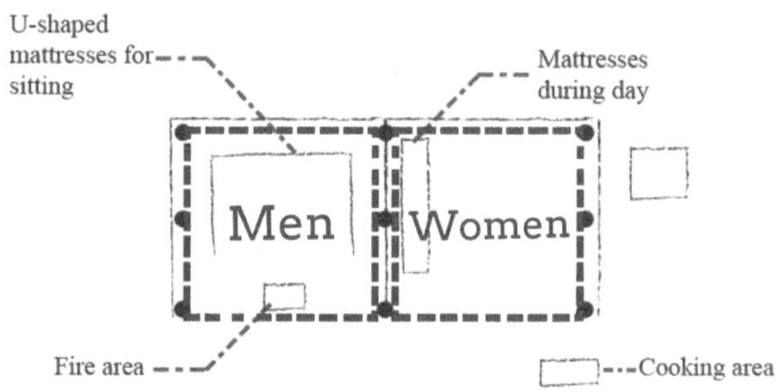

Fig. 27.35 Gender segregation implemented in bayt al-sha'r. (Source: the author)

entrance of the kitchen to overlook the private area corridor instead of facing the front of the main entrance (see Fig. 27.34).

27.4.5 Gender Segregation in Spaces

Gender segregation in spaces is clear in several examples even before settlement such as in bayt al-sha'r (see Fig. 27.35). The dwelling indoor spaces are usually divided into a space for male guests and another for female guests. The outdoors gathering spaces are also divided by gender.

In dwelling 1, the women gathering area is inside the dwelling and their outdoor maq'ad is outside in front of the dwelling. On the other hand, the men's outdoor maq'ad is detached and next to the dwelling and the indoor maq'ad is the attached space to the dwelling. In dwelling 2, men use the maq'ad located away from the

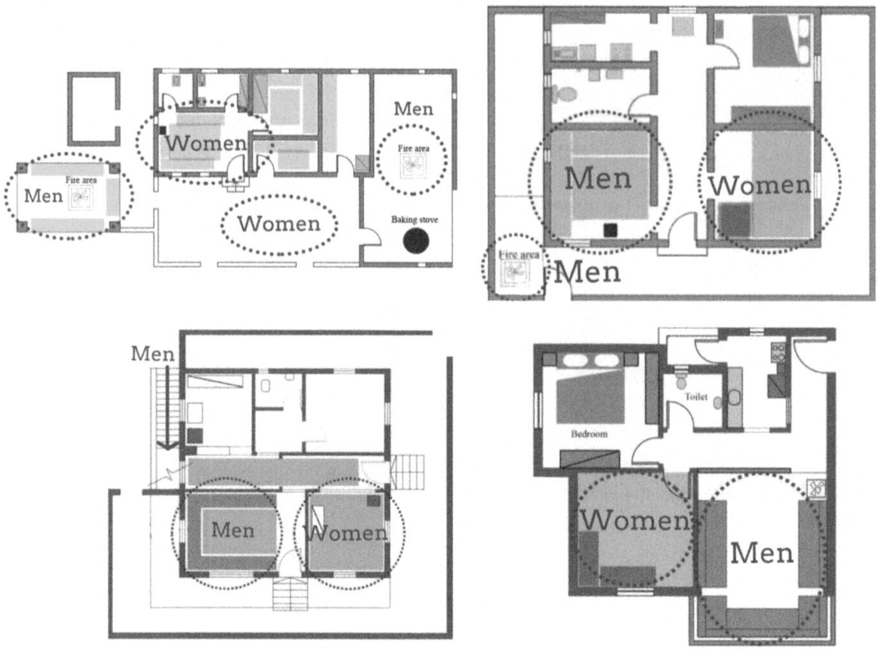

Fig. 27.36 Gender segregation in dwelling 1, 3, 4 and 5. (Source: the author)

dwelling. In dwellings 3, 4 and 5, the men and women gathering spaces are separated (see Fig. 27.36).

Also, the previously mentioned dwellings (1,3,4, and 5) are similar to the old Bedouin tent; the men's area is furnished in a U-shaped or L-shaped setting to insure the ease of communication between men in their gatherings. The women/family areas in the four examples are furnished with rugs or/and piled mattresses to give women the flexibility to use the space as needed; either for spending time with other women, doing errands or spending time with their children (see Fig. 27.36).

27.5 Conclusion

The political history and development timeline of Sinai give a clear background of the relations between events and decisions, plans, social and economic aspects and the physical environment (see Fig. 27.37). In elaboration of the relations explained on the timeline, it is clear that events and political decisions have an impact on the built-environment. For example, with the introduction of the tourism industry during the Israeli occupation, settlement with light weight structures started to occur. Later, during the peak phase of development between the 1990s and the early 2000s, settlement increased and progressed following the emerging tourism industry.

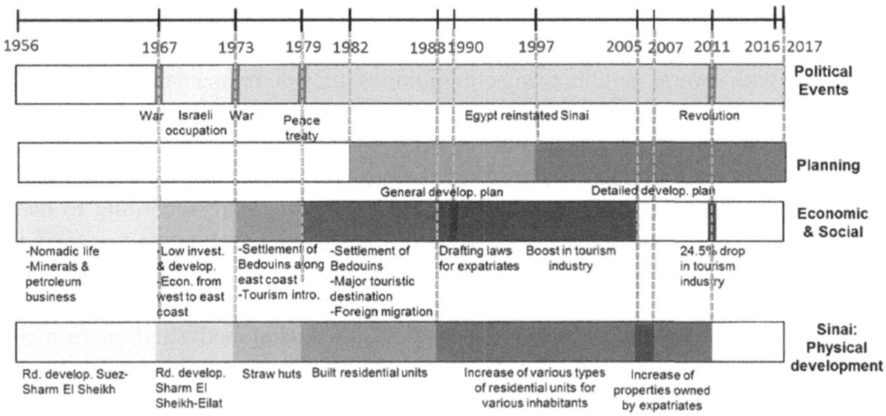

Fig. 27.37 Timeline showing the political history and development in Sinai Peninsula between 1956 and 2017. (Source: Sherif et al. (2017))

Tourism industry has impacted the development of South Sinai economically and physically. Dahab as part of South Sinai has witnessed many changes including the transformation of Bedouin dwellings from tents during the nomadic period till urbanisation. In 2014, Sinai was retargeted for development and the state started its journey for development. Accordingly, this and further research might help in approaching the context and the community in a more sensitive manner.

This research highlighted the impact of the tourism industry and development on the built-environment specifically on a sensitive context like Dahab and the Bedouin community. Tracking the progression of Bedouin dwellings along the years proved that the introduction of development has a role in the formation of their dwellings but does not impact the aspects related to the core of their culture. Understanding the constant aspects in the lives of the Bedouin community helps in constructing a list of the attributes that should be present in the design and planning of Bedouin dwellings. Accordingly, this study could be considered a base for understanding and approaching the development of Bedouin dwellings.

Focusing on the documentation and study of Bedouin dwellings in Dahab, it was noted that privacy and gender segregation are two of the constant aspects that were implemented in all the dwellings. The examples located outside the city are constructed based on Bedouin laws and regulations with respect to privacy and gender segregation. Accordingly, the manifestation of both aspects is clear and simple. On the other hand, the dwellings located inside the city are constrained by land size and neighbouring plots. The neighbours surrounding the Bedouins inside the city are considered strangers as they are not part of the family or tribe. Hence, the Bedouins take more measures in implementing privacy. Literature and interviews also confirm the importance of the two aspects in the Bedouin culture.

The Bedouins deal with the available resources to build a life connected to their norms and culture. In all the presented models of the dwellings, various functions were added and different materials were used along the years. Gender segregation and privacy either in the featured examples, in literature, or even in the narratives were highlighted as key aspects. Even when the Bedouins decided to settle in a dense area with diverse neighbouring communities, they maintained the two aspects; using narrow corridors and high fences for privacy and different rooms and sometimes different entrances to separate men from women. In addition to that, the Bedouins accepted living in a multi-storey building instead of their usual one storey units but accommodated the already built models by the state according to their needs implementing the important adjustments that ensure prolonging privacy and gender segregation, hence fit their lifestyle, for example, enclosing the terrace with the indoor space as it is too exposed for their use. Also, westernisation influenced their lives through the presence of phones, television, social media and the furnishing of private spaces like bedrooms but it did not impact their idea of segregation, privacy and the perception of gathering spaces.

In conclusion, privacy and gender segregation are two of the most important aspects to the Bedouin community. In a context similar to Dahab, prototype buildings are used but not optimum to fit Bedouins' habits. Privacy and its regulations should be considered during the design and planning phase. While designing the interior spaces, gender segregation and privacy inside the dwelling should be put into account to ensure the presence of the quality of life needed by the inhabitants. Accordingly, thorough contextual research is necessary in an area like Sinai with a complicated political background, natural assets and a diverse community that includes the Bedouins. Such research will guarantee the quality of life that serves the physical environment, the inhabitants and the state

References

Abu-Lughod L (1999) Veiled sentiments: honor and poetry in a Bedouin Society. University of California Press, Berkeley

Ali DF (1998) Case study of development of the peripheral coastal area of South Sinai in relation to its Bedouin Community. Master's Thesis. Virginia Polytechnic Institute and State University, Virginia, USA

Al-Khashab OA (2008) Tārīkh taṭaūr hdūd Miṣr al-shrqiyyah, Wa t'thīruh 'alā al-'mn al-qawmy al-miṣrī 1892-1988. (The history of the development of Egypt's eastern borders and its impact on the Egyptian National Security 1892–1988). Dar Al-Shorouk, Cairo

Altman I, Chemers M (1989) Culture and environment. Cambridge University Press, Cambridge

CAMPAS (2010) Egyptian population by age. South Sinai Governorate, Egypt

CAMPAS (2017) A study of the fundamentals and development resources and investment opportunities: South Sinai

Hemdan G (1980) Shkhṣīyat Miṣr, drāsah fī ʿbqarīyt al-makān. In: Egypt's personality, a study of genius loci, vol 4. ʿālam al-kutub, Cairo

Hess JJ (2012) Badu Wast Al-jazira. Adat—tqalid—hkayat wa aghan (Von den Beduinen des Innern Arabiens). Alwarrak Publishing Ltd, Beirut

Ibrahim Z (2009) Tourism development and the environment on the Egyptian Red Sea coast. Master's Thesis. University of Waterloo, Ontario

Karkabi N (2013) Lifestyle migration in South Sinai, Egypt: nationalisation, privileged citizenship and indigenous rights. International Review of Social Research 3(1):49–66. https://doi.org/10.1515/irsr-2013-0004
Lavie S, Young WC (1984) Bedouin in limbo: Egyptian and Israeli development policies in southern Sinai. Antipode 16(2):33–44. https://doi.org/10.1111/j.1467-8330.1984.tb00661.x
MOHPUC (1997) Al-mukhaṭaṭ al-'ām lmadīnat Dahab 2017, al-juz' al-'wal: al-drāsāt al-tskhṭiṭīyah. (The general plan for the City of Dahab 2017, Part One: Planning Studies)
Ragab N (2011) Friction: the interplay of a Bedouin community, the Egyptian Government, local and global NGOs, South Sinai. Master's Thesis. Lund University, Lund, Sweden
Rapoport A (1983) Development, culture change and supportive design. Habitat Intl 7(5–6):249–268
Saha S, Yap G (2015) How does political instability influence the effect of tourism demand on economic growth? NZAE annual conference, Wellington, New Zealand
Sherif E, Sherif L, Abou El-Ela M (2016) Political decisions, social change and planning: the case of Dahab, Egypt. CPUD'16, Istanbul, Turkey
Sherif E, Sherif L, Abou El-Ela M (2017) The impact of tourism on the built environment in Dahab: the case of Bedouin dwellings. Master's Thesis. Arab Academy for Science, Technology, and Maritime Transport, Cairo, Egypt
Shuqaïr N ū (1991) Tārīkh Sīnā Al-qadīm wa Al-ḥadīth wa Jghrāfītahā (The ancient and new history of Sinai and its geography). Dār Al- jīl, Beirut
SIS (2020) ECSS Releases Study on Sinai development since 2014. https://sis.gov.eg/Story/145783/ECSS-releases-study-on-Sinai-development-since-2014?lang=en-us. Accessed: 4 Mar 2024
SSG—South Sinai Governorate (2011) Main services 2011. South Sinai Governorate: Unpublished document
SSG—South Sinai Governorate (2012) Tourism development in Dahab between 2010 and 2011. South Sinai Governorate: Unpublished document
SSG—South Sinai Governorate (2024) Tʻaraf 'alā janūb Sīnā. (Get to know South Sinai). https://www.southsinai.gov.eg/StaticContent/TabContent?ParentId=5. Accessed: 4 March 2024
Tarawneh M (2000) House form and cultural identity—the case of Bedouin housing in southern Jordan. Master's Thesis. McGill University, Montreal, Canada
Tawfīq MN (1994) Al-khdma Al-'jtmā'ya fī Majāl Al-mjtam'āt Al-rīfīa wa Al-bdūīa wa Al-musthḍtha m' Al- trkīz 'lā Al-'tjah Al-tnmūī. (Social Services in the Field of customary, Bedouin and new societies with focus on the development approach). Mktbat Al-'njlū Al-miṣrya
UNWTO—United Nations World Tourism Organization (2013) Sustainable tourism governance and management in coastal areas of Africa. United Nations World Tourism Organization, Madrid
Zaei ME, Zaei ME (2013) The impact of tourism industry on host community. Eur J Tour Hosp Res 1(2):12–21

Chapter 28
Towards Models of Non-fragmented Cities: Continuity and Symbiosis in Brazilian Architectural Practice

Héctor Navarro Martínez

Abstract The connection between a building's interior and exterior spaces is often realised through façade interventions that create a synergy between these two realms. However, this chapter explores alternative strategies that operate from within the building, fostering a state of symbiosis between architecture and urban space. These strategies aim to avoid the fragmentation of the city's landscape and architecture, instead creating a cohesive unit that significantly impacts citizens and the established system of relationships. This chapter illustrates these concepts through a selection of projects: *Theatre-Office* by Lina Bo Bardi, where a street is replicated to create a new kind of theatre; *House in Butantã* by Paulo Mendes da Rocha, where permeable interior façades redefine the boundaries of private spaces to include social ethics and *Havaianas Shop* by Isay Weinfeld, which reimagines retail architecture by transforming its use and meaning. The strategies analysed through these built examples reshape architecture's conception and, more profoundly, incorporate urban and collective dimensions into interior spaces. These operations, understood as forms of resistance due to their unconventional nature, propose an alternative city model, demonstrating that urban quality is shaped not only by urban planning or politics but also by the work of architects, who can integrate the public sphere even in private developments.

Keywords Fragmentation · Domestic space · Dwelling · Façades

H. Navarro Martínez (✉)
Department of Architectural Composition, Universidad Politécnica de Madrid, Madrid, Spain
e-mail: hector.navarro@upm.es

G. Cimadomo et al. (eds.), *Urbanism at Borders*, The Urban Book Series,
https://doi.org/10.1007/978-3-032-09594-7_28

28.1 Introduction

28.1.1 The Role of Façades and In-between Spaces

The boundaries between architecture and urban space define the relationship between indoor and outdoor environments, shaping how users interact with their surroundings. Traditionally, these boundaries have been conceptualised as façades, two-dimensional elements that mediate between private and public realms. Scholars such as Colin Rowe and Fred Koetter (1978), Robert Venturi (1966) and Aldo Rossi (1966) have extensively studied façades, focusing on their function as separators that establish distinct spaces. However, this chapter aims to explore a more nuanced understanding of façades, seeing them not merely as dividing elements but as integral parts of creating 'in-between spaces.' These spaces, which blur the lines between interior and exterior, contribute to a more dynamic and interconnected urban fabric.

The concept of façades as more than barriers aligns with contemporary architectural discourse, emphasising built environments' continuity and permeability. In particular, this chapter introduces the idea of architectural symbiosis, a term discussed by authors like Alejandro Zaera-Polo and Farshid Moussavi (2003) to describe the harmonious integration of architecture with its surrounding environment, where both entities, building and city, mutually benefit from their interaction. This idea extends beyond simple coexistence to suggest that architectural elements, such as façades, actively contribute to and are shaped by their urban context. By treating façades as flexible and hybrid interfaces, the text explores how these elements can facilitate fluid transitions between the private and public domains, promoting a more cohesive and socially responsive urban environment. The notion of hybridisation is particularly relevant here, as it underscores the blending of traditionally separate spatial functions—such as interior and exterior, or public and private—into a unified, integrated experience. Moreover, the selected case studies go beyond merely addressing architectural limits. Through strategic interventions in the interior spaces—what can be termed 'interiorised exteriors'—these cases challenge the traditional dichotomies of interior–exterior and public–private. The primary objective of this text is to demonstrate how architectural elements, specifically façades and thresholds, can evolve from mere physical boundaries into active participants in a symbiotic relationship with the city. However, in addition to this primary goal, there is also a secondary objective: to explore how other types of operations carried out within the architectural piece can model symbiotic relationships strictly from the interior. This dual perspective enhances the architectural experience and redefines the interaction between architecture and urban life, offering new possibilities for how we conceive and inhabit these spaces.

28.1.2 Justification for the Selection of Brazilian Architects

The selection of Brazilian architects—Lina Bo Bardi, Paulo Mendes da Rocha, and Isay Weinfeld—is deeply rooted in Brazilian architecture's broader historical and cultural context, particularly during the mid-twentieth century. This period was marked by the flourishing of the Paulista and Carioca schools, which significantly shaped a national architectural identity. These schools were not just architectural movements; they represented a critical reflection on the role of architecture in society and the city.

The Paulista School, led by figures such as João Batista Vilanova Artigas and reinforced by Paulo Mendes da Rocha, was characterised by a brutalist approach emphasising architecture's social function. The school advocated for ethical and socially responsible architecture deeply intertwined with the needs of the urban population. This approach led to creating buildings that were not just functional but also fostered community interaction and engagement (Giedion 1967; Wisnik 2002).

On the other hand, the Carioca School, associated with architects like Oscar Niemeyer and Lucio Costa, emphasised a more expressive and plastic form of modernism, integrating architecture with the natural landscape of Rio de Janeiro. This school contributed significantly to the global modernist movement while striving to create a uniquely Brazilian architectural language that resonated with the country's cultural and environmental context (Mumford 2000; Deckker 2001).

Despite their differences, these two schools shared a common goal: to define what Brazilian architecture should be and how it should contribute to the creation of cities that reflect the nation's identity and values. The works of Lina Bo Bardi, Paulo Mendes da Rocha and Isay Weinfeld are direct products of this rich architectural heritage. They each represent a critical approach to architecture that challenges conventional norms and seeks to create socially and culturally responsive spaces. For instance, Lina Bo Bardi's work reflects the Paulista School's emphasis on social function, while incorporating her own unique vision that blends Brazilian vernacular traditions with modernist principles (Lima 2013). Paulo Mendes da Rocha's brutalist approach exemplifies the Paulista School's commitment to creating spaces that serve the community, often blurring the lines between public and private realms (Wisnik 2002). While operating in a later period, Isay Weinfeld continues this legacy by integrating simplicity and elegance with a deep sensitivity to the urban context, reflecting the ongoing evolution of Brazilian architectural thought.

The decision to focus on these architects is, therefore, not incidental. Their work is emblematic of a critical moment in Brazilian architecture, when the debate about national identity and the role of architecture in shaping cities was at its peak. Their contributions are significant within Brazil and have influenced global architectural practices. This chapter highlights the broader implications of Brazilian architectural practices in creating non-fragmented cities by focusing on Bo Bardi, Mendes da Rocha and Weinfeld.

28.2 Methodology

28.2.1 Multi-layered Approach

This chapter adopts a multi-layered methodological approach integrating theoretical analysis with empirical case studies. The research begins with a comprehensive literature review, drawing on historical texts and contemporary studies in architecture and urbanism. This review establishes the theoretical foundation to understand how façades and in-between spaces function as critical elements in urban design, particularly in creating non-fragmented cities. The theoretical framework is informed by key concepts from scholars such as Jane Jacobs and Juhani Pallasmaa, whose works provide insights into the role of façades as interfaces that foster urban continuity and social interaction.

28.2.2 Comparative Case Analysis

Following the theoretical review, this chapter conducts a comparative analysis of selected case studies, chosen for their relevance in demonstrating the principles discussed in the theoretical framework. Each case study is examined through a combination of spatial analysis, which explores the physical characteristics and design strategies employed, and sociocultural analysis, which considers how these spaces influence and are influenced by their urban context. The comparative approach allows for identifying common themes and design strategies that contribute to integrating public and private spaces within the urban fabric.

28.2.3 Temporal Dimension Analysis

The methodology also critically examines the temporal dimension, considering how the design and use of façades and in-between spaces have evolved over time in response to changing social, cultural and environmental conditions. This diachronic analysis enables a deeper understanding of the dynamic nature of these spaces, highlighting how historical and contemporary approaches can be synthesised to create more integrated and sustainable urban environments. By analysing the temporal evolution of these architectural elements, this chapter aims to uncover patterns and strategies that have proven effective in fostering urban continuity across different contexts and periods.

28.3 Broader Theoretical Framework

28.3.1 Reinterpreting Façades as Interfaces

Traditionally, façades have been perceived as rigid barriers that separate the interior of a building from its external environment. This view, while functional, limits the potential of façades to contribute to a more integrated urban fabric. Contemporary architectural discourse, however, increasingly advocates for the reinterpretation of façades as flexible interfaces that can foster a more dynamic interaction between private and public spaces.

Jane Jacobs (1961), in her seminal work on urban planning, emphasised the importance of 'eyes on the street', the idea that the interface between buildings and the public realm plays a crucial role in creating vibrant, safe and engaging urban environments. Jacobs argued that active façades, which encourage interaction and visibility, are key to fostering community engagement and urban vitality. This concept challenges the traditional notion of façades as mere enclosures and positions them as active participants in the city's life. Furthermore, the phenomenological approach of architects like Juhani Pallasmaa (2005) adds another layer to this discussion. Pallasmaa highlights the sensory experiences of architecture and the importance of transitional spaces in shaping human perception and interaction with the built environment. In this view, façades are not just physical boundaries but are experienced through touch, sight and movement, contributing to the overall sensory experience of the space.

By reinterpreting façades as interfaces rather than barriers, architects and urban planners can design spaces that promote fluid transitions between the interior and exterior. This approach enhances the functional aspect of buildings and enriches the social and cultural fabric of urban spaces. The theoretical concepts surrounding façades and in-between spaces come to life through their application in diverse historical and contemporary architectural contexts. By examining specific examples, we can better understand how these strategies have been implemented to address urban fragmentation challenges and promote more integrated environments. This section presents a selection of case studies that illustrate these principles in action, ranging from traditional elements in Indian architecture to contemporary projects in Japan and Brazil. These examples bridge the gap between theory and practice, demonstrating the tangible impact of design choices on creating cohesive and socially engaged urban spaces.

28.3.2 Traditional Examples: The Otta *in Indian Architecture*

The *otta* (see Fig. 28.1) is a traditional element in Indian architecture that exemplifies the creation of in-between spaces within urban contexts. As Jain (2002) documented, the *otta* serves as a transitional space that delineates urban areas, providing

Fig. 28.1 Merchants along a street in Bombay. (Source: Edwin Lord Weeks. Public domain)

protection from environmental elements such as sun and rain. More importantly, the design of the *ottoman* promotes social interaction by creating functional areas that encourage use and engagement. This traditional solution reflects a nuanced understanding of environmental adaptation and social needs, demonstrating how architectural elements can contribute to defining the cityscape and fostering intensive urban activity.

28.3.3 Contemporary Examples: Kengo Kuma's Glass/ Water House

Kengo Kuma's Glass/Water House in Japan is a contemporary example that reinterprets traditional concepts of façades and boundaries to create transitional spaces that blend seamlessly with the natural landscape. In this project, water is used as a dynamic boundary element that merges the built environment with its surroundings. This approach enhances the spatial experience and challenges conventional notions of interior and exterior spaces. By integrating natural elements and employing permeable materials, Kuma's design fosters a fluid transition between inside and outside, aligning with the theoretical framework that redefines façades as flexible interfaces.

28.4 Blurring the Lines: Lina Bo Bardi's MASP

The São Paulo Museum of Art (MASP), designed by Lina Bo Bardi, exemplifies an innovative approach to urban space by creating a significant void beneath the building (see Fig. 28.2). This elevated structure, supported by two large concrete pillars, allows the plot that the building would fully occupy to remain open and accessible to the public. The resulting void is transformed into a covered plaza that not only prevents the complete privatisation of the site but also serves as a dynamic public space that connects Avenida Paulista with the area behind the museum, offering a viewpoint over the city.

This design strategy introduces a new typology of urban space that enriches the city's complexity and enhances the urban experience. By maintaining this open, accessible space, the MASP adds a layer of social and spatial interaction that blurs the traditional distinctions between public and private realms. The covered plaza under the museum becomes an active part of the city's public infrastructure, providing a place for gathering, circulation and reflection. Through such interventions, the architecture contributes to creating a more integrated and vibrant urban fabric, reinforcing the role of public spaces in the city's life.

Fig. 28.2 Museo de Arte de São Paulo by Lina Bo Bardi. São Paulo, Brazil. (Source: Nelson Kon)

28.4.1 Synthesis of Traditional and Contemporary Approaches

These traditional and contemporary examples illustrate how façades and in-between spaces can be conceptualised and implemented to enhance urban environments. Traditional elements like the *otta* demonstrate how architectural design can promote social interaction and environmental adaptation. Contemporary projects, such as Kengo Kuma's *Glass/Water House* and Lina Bo Bardi's *MASP*, showcase how modern design strategies can reinterpret and expand these traditional concepts to create more integrated and responsive urban spaces. These cases highlight the various approaches to establishing relationships between public and private spheres, using architectural and natural elements. Kuma and Bo Bardi's projects provide insight into implementing these strategies, offering a wide range of possibilities for enhancing the interaction between built and natural environments. This chapter examines these cases to highlight the importance of considering both historical and contemporary approaches to façade design and transitional spaces. These examples provide valuable insights into how architectural strategies can foster non-fragmented cities, promoting continuity, permeability and social engagement within the urban fabric.

In the following chapter, we will delve into three carefully selected case studies that exemplify the concept of symbiosis between interior and exterior spaces. Unlike conventional approaches that focus on the architectural envelope or boundaries to achieve this integration, these projects demonstrate how symbiosis can be achieved by operating primarily within the interior spaces. This approach challenges and transcends the strict dichotomies of interior and exterior, public and private, by allowing the interior to flow seamlessly into its external environment without necessarily altering the building's façade. The concept of architectural symbiosis, explored by theorists such as Ken Yeang (2006) and Rafael Moneo (2004), is reinterpreted in these case studies as an interior-driven strategy that fosters a harmonious relationship between the built environment and its surrounding context. By focusing on the interior rather than the external limits, these projects create what can be described as 'exteriorised interiors', spaces that dissolve the traditional boundaries and redefine the interaction between architecture and its environment.

This chapter will thoroughly examine these case studies, identifying the design strategies and themes that underpin their ability to blur the lines between interior and exterior spaces. The insights gained from this analysis will contribute to a deeper understanding of how interior spaces can create a more integrated and symbiotic relationship with the external environment, ultimately offering a new perspective on architectural design.

28.5 Case Studies Analysis in Exteriorised Interiors: Strategies for Architectural Symbiosis Beyond Boundaries

One of the most common approaches in contemporary architecture involves integrating elements from the external environment into interior spaces. This strategy, seen in literal and abstract forms, creates interiors that blur the line between the inside and the outside, fostering a dialogue between architecture and nature. Examples of this approach are numerous, ranging from including natural materials such as earth, rocks and vegetation, to more innovative incorporations of water features within built environments.

This practice can be observed in various projects, each demonstrating a unique method of merging interior and exterior spaces. For instance, in the Tanikawa House, designed by Kazuo Shinohara in 1972, an earthen floor is incorporated into the interior, bringing a piece of the natural environment directly into the living space. Shinohara further explored this concept in the second house commissioned by the poet Shuntaro Tanikawa in 1974, where the inclined ground floor introduces a sense of imbalance, intentionally challenging the conventional perception of interior space (Shinohara 2011).

Another notable example is Oscar Niemeyer's Casa das Canoas in Rio de Janeiro, where a large stone is incorporated into the house, intersecting with the glass façade. This solution not only blends the interior with the natural surroundings but also respects the place's history and context through a seamless integration of new architectural actions.

Lina Bo Bardi's SESC Pompeia cultural centre in São Paulo offers another perspective. Bardi transcends the traditional indoor–outdoor dichotomy by including a pond where parents and children can playfully interact with the space, encouraging a new kind of social interaction within the architectural setting. The examples above illustrate the diversity of approaches in creating 'exteriorised interiors,' where the boundaries between inside and outside are intentionally blurred. These strategies open new possibilities for human habitats, challenging traditional binary classifications. The following three case studies delve deeper into this concept, offering distinct methodologies and insights into how architectural design can redefine the relationship between interior and exterior spaces.

28.5.1 House in Butantã (Brazil) by Paulo Mendes da Rocha

Through the work of Brazilian architect Paulo Mendes da Rocha (1928–2021), it is possible to identify the themes developed by the Paulista School, many of them promoted by its leader João Batista Vilanova Artigas (1915–1985): the topographic deformation of the ground level; the unitary volumetry posed an impervious position that contrasted with interiors of great richness and spatial complexity; a structural language that defined internal and external form and relationship and the technique applied to constructive detail.

From the perspective of the Paulista school, architecture should include social work, and the architect's role should be oriented to ensure the interests of the oppressed citizens. Among the many contributions that can be distinguished in Mendes da Rocha's work, the following lines are intended to give value to the space operations he developed in his own home in Butantã, planned between 1963 and 1964. The proposal is defined by a floating concrete volume, where the floor plane is released, generating a shadow space. The project seems to collect Le Corbusier's lessons, who promoted an architecture separated from the ground by pilotis, facilitating the flow of natural light. However, a home is private property. The architect proposed a solution for the plot's perimeter that involves topographic manipulation of the boundary, making it visually evident while formalising it in a way that allows the ground plane to be understood as a continuous element rather than a strict vertical boundary (see Fig. 28.3). This avoids vertical elements that involve more obvious and abrupt interruptions. These actions rework the canonical limit and propose a city model where the ground plane flows and is not interrupted (García del Monte 2011).

Many of the architects of the Paulista School sought to accurately critique the bourgeois housing model in their residential projects and provide solutions for the working class. They tried to undo the rigid partitioning of spaces, move away from the spatial structure of the leftovers or even create homes without a service wing.

Fig. 28.3 House in Butantã by Paulo Mendes da Rocha. São Paulo, Brazil. (Source: https://www.archdaily.cl)

The *House in Butantã* was organised in three areas (see Fig. 28.4). Common spaces are placed facing west; an ample and generous space, well lit and protected by the cantilever of the rooftop. On the façade, mechanisms that facilitate ventilation are included, allowing design and technique to define their style. In the opposite parallel band, a free space is developed, which functions as an access corridor to the bedroom units and a study area. The dimensions, as well as the light and ventilation conditions, allow this duplicity of use. Five bedrooms are between both bands, except the main bedroom, without direct contact with the façade.

The operation highlighted in this case study is the treatment of the limits of these bedrooms. Both the partitions that separate the bedrooms and the walls that become independent from the common side areas do not reach the ceiling (see Fig. 28.5). This detail has many consequences on the functioning of the house. Mendes da Rocha prints in it something that goes beyond the spatial: it raises a house to host a family model whose relations among different members imply a social background. The spatial relationship between the various bedroom units, where the adjoining one can be felt, is similar to what happens in the *favelas*, where constructive definition has a much more communal feeling. Somehow, domestic life is being urbanised, significantly reducing personal intimacy and blurring the classical boundaries of the home concept. Flávio Motta refers to this house as a work that alludes to 'living in the middle of a rationalised favela' (Masao 2004). The spatial rigidity of the bourgeois house, based on individual isolation, provided by the bedroom of abrupt limits, drifts towards a collective and public perspective. And the same principles that govern the proper functioning of a society are transferred to the house: each member's freedom depends on the other's freedom. The norms associated with social ethics print an order in human groups and result in totally exteriorised but spatially introverted houses.

The architect managed to manipulate the boundaries within the house to establish new relationships within the family structure. While criticising the individualism that governs the bourgeois family, it strengthens family ties. A simple action

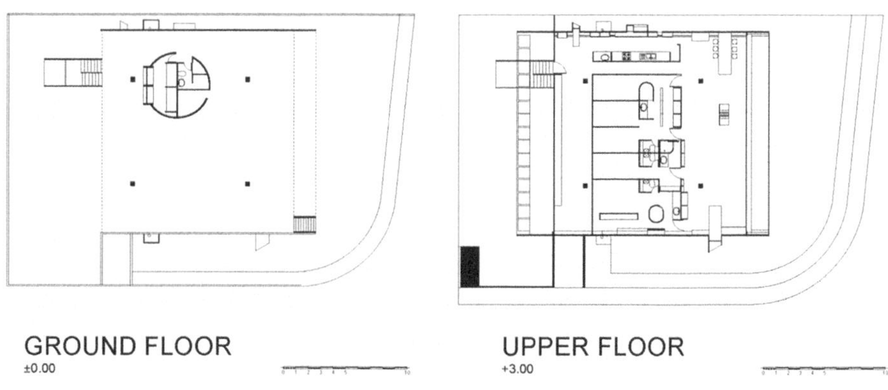

Fig. 28.4 House in Butantã. Plan by Paulo Mendes da Rocha. São Paulo, Brazil. (Source: the author)

Fig. 28.5 House in Butantã by Paulo Mendes da Rocha. São Paulo, Brazil. (Source: https://www.archdaily.cl)

brings up a whole debate about family. It does not matter if it is a criticism of the bourgeoisie or a reference to a traditional construction such as favelas. The important thing is that it permeates the domestic space of values that have to do with social ethics, which makes a community develop through the correct functioning and sensitisation of each one of the beings as individual entities.

28.5.2 Theatre-Office by Lina Bo Bardi

Lina Bo Bardi's work reflects a professional practice constantly seeking the community's good. In her first stage in Italy, she worked as an editor for *Domus* magazine, which Gio Ponti then directed.

From the beginning of her career, the figure of Lina Bo Bardi is that of a woman with a revolutionary posture for her time (Sánchez Llorens 2015). When she worked as director of the women's magazine *Grazia*, in 1941, she carried out a precious 'emergency cradle' project for people without resources, which she described as her first social project. She explained how to build a crib with a fruit box and some other items in the publication. Throughout her professional career, she faced the intellectual problems of the architectural debate of the second half of the twentieth century. For her, architecture, art, material and social culture sought a common ground capable of responding to all the needs of the time (Bardi et al. 2014).

The *Theatre-Office* in São Paulo is the case study presented here. This is a restoration project of an old theatre, as in the case of the cultural centre *SESC Pompeia,* which was previously mentioned, and in many other projects developed in Salvador de Bahia. Lina Bo Bardi decided to maintain the existing building and all that helped preserve the footprint of the place. It is paradoxical how the work of an architect with this attitude towards preservation was the victim of the opposite, as many of the buildings designed by Bo Bardi in Salvador de Bahia are currently in a poor state of conservation. Be that as it may, the *Theatre-Office* is the crystallisation of multiple experiences linked to the theatrical world. She had previously designed sets and costumes and had contact with various professionals involved in experimental projects. The filmmaker Glauber Rocha introduced Lina Bo and José Celso Martinez Corrêa, director and founder of the Theatre-Company, in 1967 (Bardi and Oliveira 2002). The theatre had suffered a fire a year earlier. The spatial concept of the new theatre was raised as a public sidewalk that crossed it from its entrance, in Jaceguai Street, to Japurá Street, located in the background. Later, with Lina in charge, and in collaboration with Edson Elito, it was decided to keep only the brick walls seen from the old theatre, a box 9 metres wide and 50 metres long. The stage was conceived as a street-walkway ramp covering the entire theatre length. In parallel to the longitudinal walls, a detachable blue metal structure generates galleries that occupy the full height of the building. The public, actors or technicians can occupy them, and function as a stage, space for the public or even dressing rooms (see Fig. 28.6).

Fig. 28.6 *Theatre-Office* by Lina Bo Bardi. São Paulo, Brazil. (Source: Nelson Kon)

Lina Bo Bardi designed a theatre where the seating area was transformed into a public place that reproduced urban life. The typical armchair space was not considered in the proposal. That is to say, Bardi was creating a space without spectators: actors, technicians and the public were part of the theatrical performance together with the architecture itself and the objects that were part of the scenery. The theatre model imported from Europe is avoided at all costs, where the audience usually acquires a passive role as spectators, with no interaction with the theatrical function. Bardi designed a space that fostered a dynamic theatre, where all stakeholders participated in the same experience. Access from Jaceguai Street is done through an inclined plane that is part of the scenery (see Fig. 28.7). A wooden floor is used in

Fig. 28.7 Theatre-Office by Lina Bo Bardi. São Paulo, Brazil. (Source: Nelson Kon)

the central axis of the inclined plane, incorporated as a sound resource which functions as a sounding board. The public crosses this space and occupies the sides, as if it were a street. The floor, as is the scaffolding that makes up the galleries, is removable. The pre-existing and the new coexist, as well as the architecture and the city, the actors and the public. It is all part of the theatrical proposal that moves away from mythical stories and impossible dreams to represent content close to the lives of ordinary citizens of São Paulo.

According to José Celso Martinez Corrèa, the theatre was conceived as a *terreiro*, where Camdomblé (an Afro-Brazilian religious tradition) was once clandestinely officiated. The *terreiro* is also an earthen space near the house where many domestic work activities happen. It is where parties and neighbourhood meetings are held (Bardi and Oliveira 2002). The *Theatre-Office* is formed as an enclosure full of exterior references. It is a street stage with the presence of the outside environment: the glazed surfaces that allow the sun to enter, the large vegetation planted on the ground and the sliding roof that gives way to the breeze and rain. Even the scaffolding is an element of the urban landscape. The theatre merges with the urban plot, generating an interface between the city and its inhabitants.

In a text written in 1989 about her works in Salvador de Bahia, the architect includes an appointment with a reference to Le Corbusier that seems to become the guide of her work:

> Theatre that goes out to the squares, to the streets, that invades the city. Chairs and furniture that leave the houses, and people, men, women, children, a whole town that inspired, in 1963, Le Corbusier, when he visited Brazil, in a famous letter to Minister Gustavo Capanema: "Mr Minister, Do not order to build theatres with stages and seats, leave the Plazas, the Streets, the Green, free. Only send "wooden planks" of wood, open to the People, and the Brazilian people will occupy them, "improvising", with their natural elegance and intelligence". (Bardi and Oliveira 2002:184)

28.5.3 Havaianas Shop by Isay Weinfeld

The third and final case study is based on a project by Isay Weinfeld, one of Brazil's most outstanding contemporary architects. Of Polish origin and in contact with the world of cinema, his architecture is influenced by other arts that are part of his global vision: 'For me, architecture is a form of expression, but it is not the only one'. Through its architecture you can see how each project looks for new variables and forms of expression: 'I try not to accept works that relate to the last thing I did... I want to be surprised, to take risks. To err trying something new rather than repeating a formula that works' (Weinfeld et al. 2014:16).

Unlike other Brazilian architects, his architecture does not imply social criticism, does not flaunt poverty and discards structural heroism. His architecture is contemporary, elegant and sober. His housing plans show an architectural practice where spaces are activated by furniture—often designed by himself—or dispense with it depending on the purpose to be enhanced. It concentrates and expands the programme for human habitation, using the exterior, which is included in the inner experience. There is total control, from the most general to the smallest detail, as if it were an orchestra conductor. Weinfeld is also a music composer and director. The aesthetic sobriety is accompanied by an exhaustive definition that continues the wake of São Paulo architects and other figures of Brazilian architecture of the twentieth century, such as Joaquim Guedes or Eduardo de Almeida, among many others.

The project selected here corresponds to a shop of the Havaian flip-flop sandals brand, which is located in one of the most expensive streets in the city of São Paulo, Rua Oscar Freire. The store (2009) is designed as a double-height space excavated in the ground (see Fig. 28.8). There is no façade or shopping window when the store is open. Still, it becomes an extension of the urban area (see Fig. 28.9). The roof is formed by a grid of lattices and skylights that flood the interior space with natural light, and on the sides, two narrow bands of land include vegetation that grows to the ceiling. All project decisions seek to undo the canonical commercial space and reproduce an urban scenario occupied by the brand. With this, the architect aims to generate a point of sale that can transmit the Brazilian environment and not dissociate this footwear from the natural: 'Our biggest challenge was to catch in the architecture the climate that the brand inspires: freshness, informality, comfort, ease, Brazil, the Brazilian' (Weinfeld 2009).

The floors are defined by stone and natural wood flooring, typical scenarios where Havaianas are used. The brand demanded that the store be iconic within the city, for which the architect replied with the most primitive point of sale of the famous footwear: urban markets. This direction includes other independent

Fig. 28.8 *Havaianas Shop* by Isay Weinfeld. São Paulo, Brazil. (Source: https://www.archdaily.cl/)

Fig. 28.9 *Havaianas Shop* by Isay Weinfeld. São Paulo, Brazil. (Source: https://www.archdaily.cl/)

elements that implement the shopping experience: a cube as an exhibitor collects the brand's history and a container displays imported models. A transparent cylinder presents the new collection. In this way, it is possible to activate the floor level and incorporate the brand philosophy into the space dedicated to the store, establishing a new type of relationship with the customer and including a new spatial experience associated with the purchase. Using the difference of dimensions and the suppression of the façade, generating a space enclosure in continuous contact with the outside is possible. No clear limit defines the 'enter' action; instead, it becomes a gradual phenomenon that discovers a new urban space more connected to the city.

28.6 Conclusions

The three case studies presented in this article explore different perspectives on transforming interior spaces to foster specific uses and enhance the relationship with the surrounding urban environment. These interventions demonstrate that public space can be effectively valued and integrated through actions conducted within the architecture itself, challenging the notion that urban dimensions are separate from the city's buildings. The approaches discussed illustrate works where the public and private realms cease to exist in parallel, instead establishing a symbiotic relationship that benefits both.

From this perspective, redefining the use concept becomes crucial, offering new scenarios that encourage dynamic experiences centred on communal interactions. For instance, Mendes da Rocha's redefinition of bedroom boundaries models relationships that diverge from individualism, fostering a sense of community. His design infuses the domestic space with values related to social ethics, promoting communal development through proper functionality. This approach reincorporates an intangible dimension of the urban and collective into the housing space, blurring the lines between the public and the private. A similar transformation occurs in the other two analysed cases. Weinfeld's retail design and Bardi's theatre space are deeply connected to their spatial proposals. In Bardi's theatre, the space establishes a physical connection with the external environment. It creates a domesticated, versatile street capable of adapting to community celebrations that use the theatre to express local experiences and concerns. The theatre thus transcends its role as a self-contained building, merging with the city to create a collective and inclusive experience. Similarly, Weinfeld redefines the shopping experience associated with his store by evoking the concept of the urban market, distancing the space from the privatised environment that often conveys exclusivity. The Havaianas store, representing affordable and accessible footwear, eliminates traditional barriers, proposing a friendlier alternative to the canonical façade, where access is more evident and less exclusive.

In the ongoing evolution of cities, reclaiming urban space as collective property is not just a crucial milestone but a necessary step towards fostering social equity, enhancing communal well-being and promoting sustainable development. By reclaiming urban spaces as shared assets, society can work towards creating more inclusive and resilient cities, where public spaces serve as platforms for social interaction, cultural exchange and community building. This transformation seeks to counteract the privatisation and fragmentation of urban areas, ensuring that the benefits of metropolitan living are distributed more equitably among all inhabitants.

The most abstract limits—those that associate codes and symbols with certain statuses that exclude and marginalise—are dismantled by recreating urban landscapes within interior spaces. These three remarkable projects counter conventional architectural practices, offering resistance that could be the seed of future trends. It is important to note that all the case studies attempt to recover models from the past, where public space and community relations held greater significance within human groups. Western culture, focusing on privatisation and economic exploitation, has often formalised spatial arrangements that reinforce the boundary as an element separating interior spaces from the city. The Western family model promotes individualism and has similarly influenced relationships between individuals and society. By manipulating space and its boundaries, it is possible to create new architectural and urban environments based on a symbiotic relationship that fosters community life, avoiding the dissociation of two spatial entities that, when unified, offer a better alternative for developing communal living.

References

Bardi LB, de Oliveira O (2002) Lina Bo Bardi: obra construída = built work. Gustavo Gili, Barcelona
Bardi LB, Rubino S, Grinover M (2014) Lina Bo Bardi por escrito: tex- tos escogidos 1943–1991. Alias, Mexico
Deckker Z (2001) Brazil built: the architecture of the modern movement in Brazil. Spon Press, Londres
García del Monte JM (2011) Paulo Mendes Da Rocha: conciencia arquitectónica del pretensado. Nobuko, Buenos Aires
Giedion S (1967) Space, time and architecture: the growth of a new tradition. Geoffrey Cumberlege, London
Jacobs J (1961) The death and life of great American cities. Random House, New York
Jain K (2002) Thematic space in Indian architecture. AADi Centre, New Delhi
Lima ZRM d A (2013) Lina Bo Bardi. Yale University Press, New Haven
Masao J (2004) The modern Brasil House. In: Andreoli E, Forty A (eds) Brazil's modern architecture. Phaidon, London, pp 142–170
Moneo R (2004) Theoretical anxiety and design strategies in the work of eight contemporary architects. MIT Press, Cambridge
Mumford E (2000) The CIAM discourse on urbanism, 1928–1960. MIT Press, Cambridge, MA
Pallasmaa J (2005) The eyes of the skin: architecture and the senses. Wiley, Chichester
Rossi A (1966) The archi1tecture of the City. MIT Press, Cambridge, MA
Rowe C, Koetter F (1978) Collage City. MIT Press, Cambridge, MA
Sánchez Llorens M (2015) Lina Bo Bardi. Objetos y acciones colectivas. Diseño Editorial, Buenos Aires
Shinohara K (2011) Kazuo Shinohara: casas = houses. Gustavo Gili, Barcelona
Venturi R (1966) Complexity and contradiction in architecture. The Museum of Modern Art, New York
Weinfeld I (2009) Tienda Havaianas. http://www.plataformaarquitectura.cl/cl/02-26007/tienda-havaianas-isay-weinfeld. Accessed: 2 Aug 2019
Weinfeld I, Blanco R, Fuentes O (2014) Isay Weinfeld, Buenos Aires
Wisnik G (2002) Paulo Mendes da Rocha: Obra Completa. Ediciones Gustavo Gili, Barcelona
Yeang K (2006) EcoDesign: a manual for ecological design. Wiley-Academy, London
Zaera-Polo A, Moussavi F (2003) The function of ornament. Actar, Barcelona

Chapter 29
(Inter)Acting Knowledges: Methodological Research Approach to Bairro da Torre Self-produced Neighbourhood in Greater Lisbon

Rosa Arma

Abstract The chapter explores the methodology employed in researching self-produced neighbourhoods within the context of the author's doctoral research in architecture, which examines the socio-spatial impact of an interactive approach to intervention in these neighbourhoods. Bairro da Torre, a self-produced neighbourhood in Greater Lisbon, served as case study. The account on the author's research methodology in Bairro da Torre—including its theoretical foundations, the techniques applied and the process—offers a critical reflection on the interactive approach and its impact on the research itself. Framed as a dialogue—envisioned as encounter and exchange—the research created a platform for the (inter)action of lay and scientific/technical knowledge. This facilitated the co-production and transformation of both theoretical and empirical knowledge on self-produced neighbourhoods, fostering the imagination of more emancipatory approaches to intervention while also contributing to the author's own transformation.

Keywords Greater Lisbon · Self-produced neighbourhoods · Interactive research methodology

29.1 Introduction

Due to global population growth, the large-scale displacement of (im)migrants and refugees and their influx into cities, alongside inadequate housing responses and processes of racialisation, expulsion, peripheralisation and segregation of vulnerable groups within the context of capitalist production of urban space, there is a global phenomenon that Davis (2006) calls 'urbanisation of world poverty'. This phenomenon leads to the expansion of cities' self-produced borders and increased

R. Arma (✉)
Research Centre for Architecture, Urbanism and Design (CIAUD), Lisbon School of Architecture, Universidade de Lisboa, Lisbon, Portugal

G. Cimadomo et al. (eds.), *Urbanism at Borders*, The Urban Book Series,
https://doi.org/10.1007/978-3-032-09594-7_29

socio-spatial injustice. Pushed to the 'end' of the cities, self-produced neighbourhoods arise on occupied land, often in risky and/ or non-buildable marginal areas, and present significant deficiencies in infrastructure, public space, access to services and facilities, and housing. Situated at the intersection of struggles for urban space, self-produced neighbourhoods are a shelter solution for people with fewer resources, often from vulnerable groups, who are 'on the move' in search of a better life in the cities. The notion of 'self-production' has been used by researchers at GESTUAL—Group of Socio-Territorial, Urban and Local Action Studies (*Grupo de Estudos Socio-Territoriais, Urbanos e de Ação Local*) at the Lisbon School of Architecture. It refers to Henry Lefebvre's notion of production of space (Lefebvre 1974)—not merely to self-construction—and aims to draw attention to the energy invested by the producers of these neighbourhoods, rather than emphasising, as the dominant concepts of informality or illegality do, what they lack, in dialectical opposition to the consolidated city (Raposo 2012; Raposo et al. 2012).

In Portugal, the National Survey on Rehousing Needs (*Levantamento Nacional das Necessidades de Realojamento Habitacional*), the result of a survey conducted by the IHRU-Housing and Urban Rehabilitation Institute (*Instituto da Habitação e da Reabilitação Urbana*) with city councils and published in 2018, concluded that there is a high concentration of housing needs in Greater Lisbon and Greater Porto (IHRU 2018). The survey identified more than 25,000 families across Portugal (13,828 in Greater Lisbon alone) as living in conditions of severe housing need ('families to be rehoused'). The same survey identified 11,999 families (6,280 in Greater Lisbon) living in 'shacks and precarious constructions'. More recent studies (Allegra and Di Giovanni 2024; CNA-PRR 2025), based on the city councils' Local Housing Strategies (*Estratégias Locais de Habitação*), revealed that the number of families living in indecent housing condition[1] in Portugal is significantly higher, reaching a total of 143,283, of whom more than 50,000 live in Greater Lisbon (AML 2022).

Globally, self-produced neighbourhoods are still not recognised as part of the city itself and are often regarded as 'non-places' (Augé 1992), where a 'surplus humanity' dwells—the unwanted by-product of the system (Davis 2006). This is a perspective frequently reinforced by architecture and urban planning, which help to (re)produce the dominant discourse. This discourse aims to homogenise, disqualify and stigmatise self-produced neighbourhoods, and is paternalistic, justifying interventions that range from *tabula rasa* approach to rehousing in suburban and inadequate large-scale social housing estates, as well as to welfare-based survival initiatives. These responses are often grounded in hygienist and technocratic logics. They are driven by a top-down approach that polices and constrains local complexity through the enforcement of norms, in order to legitimise public or private initiatives typically dominated by the logic of capital, with benefits accruing to those groups with greatest resources (Harvey 1980). Although such interventions

[1] Defined by four parameters—Precariousness, Unhealthiness and Insecurity, Overcrowding and Inadequacy—identified by the Decree Law n.° 37/2018.

may, in some cases, provide more dignified housing conditions, they generally fail to guarantee the Right to the City (Lefebvre 1968).

Self-produced neighbourhoods are here understood as insurgent (Holston 1998) and their inhabitants emerge as potential catalysts for change through practices that respond not only to lack of resources and opportunities, but also to autonomy in decision-making (Nascimento 2011), and that are driven by other logics and approaches to socio-spatial production. These include the subversive tactics of everyday life discussed by De Certeau (1984), the creation of support and solidarity networks, the affirmation of values such as care and sharing of knowledge, self-management of resources, implementation of alternative microeconomies and the very act of self-construction, which is insurgent in itself, as it takes place without 'formal' experts (Nascimento 2011). Therefore, self-produced neighbourhoods call for alternative ways of intervening, starting with the recognition of the wide range of socio-spatial production practices embedded in their everyday life and the promotion of inhabitants' participation as co-authors in the transformation of their own space. Participatory approaches to intervention in self-produced neighbourhoods are increasingly evoked. However, they have also been appropriated by top-down and conservative agendas, sometimes becoming mere expedients for the co-optation of consensus. Even projects implemented through interactive and emancipatory approaches are not completely free of conflict, as actors with different interests, logics and approaches are confronted as in a 'local political arena' (Olivier de Sardan 1995:162). The outcome of any participatory process depends on the balance of power among the actors involved (Raposo 2017) and, as Guerra (2010) argues, conflict is always inherent and requires resilience from all participants, ultimately contributing to their empowerment.

This chapter addresses methodological questions that arose during the author's doctoral research in architecture. The research focuses on the socio-spatial impact of an interactive approach to intervention within self-produced neighbourhoods.[2] Bairro da Torre (BDT), a self-produced neighbourhood in Greater Lisbon has been the object of action research conducted by GESTUAL (of which the author is a member), under the project 'Action-Research in Bairro da Torre, Loures. Academic Extension and Methodological and Design experimentation' (*'Ação-investigação no Bairro da Torre, Loures. Extensão Académica e Experimentação Metodológica e Projetual'*) initiated in 2014 and coordinated by architect-urbanist and former professor Isabel Raposo. Since 2016, the author, as a member of GESTUAL, has played an active role in the team carrying out the action research in that neighbourhood. BDT was chosen as case study for the author's research because it has been the object of different participatory essays (varying in scale, approach and actors involved). The account of the field research methodology applied in BDT aims to outline the theories supporting it, critically reflect on the interactive techniques employed, the research process and its actors, and on its impact on the

[2] Provisional title of the PhD thesis: "Democra-city. (*Transform*)*Ações na Cidade Autoproduzida*". The thesis is supervised by Isabel Raposo (FAUL, Portugal) and Juliana Demartini (UNEMAT, Brasil).

transformation of the knowledge about self-produced neighbourhoods, which contributes to new approaches to invervention.

The field research methodology was built on a notion of knowledge produced through interactions—as articulated in Paulo Freire's dialogical approach—namely on the encounters and exchanges between lay and scientific/technical knowledge (Freire 2005), based on the understanding of knowledge as inter-knowledge, arising from the dialogue between different spheres. In line with several authors (including Freire), within GESTUAL's action research and the scope of the author's research, the inhabitants of Bairro da Torre were recognised as real experts, being the (self) producers of their own space, and considered reflective subjects and active agents of the research itself. The construction of trust-based bonds and an 'empathetic chain' (Ferrarotti 2007:27) over time was an essential step for the establishment of a 'meaningful relationship, an authentic interaction' (idem:17) with inhabitants. The proposal to read self-produced neighbourhoods as 'places' (Augé 1992), and to explore alternative ways of intervening through a more interactive and transformative approach based on dialogue (Freire 2005), conceived as a 'space' of encounters and exchanges, stems from the understanding of inhabitants' everyday life and forms of subsistance, resistance, self-determination and agency, revealed through the interactive field research methodology implemented and outlined below.

The chapter starts with a presentation of the case study, followed by a discussion of the field research process, techniques applied and theoretical framework underpinning it, and concludes with a reflection on the capacity of an interactive methodology to inspire the transformation of knowledge and practices towards self-produced neighbourhoods, and on the author's own transformation in the research process.

29.2 The Case Study: Bairro da Torre

Bairro da Torre (BDT) was a self-produced neighbourhood in Greater Lisbon, in the municipality of Loures. It resulted from the occupation beginning in the late 1960s of a plot bordering Lisbon Airport, following an internal exodus of population from the impoverished countryside towards the main urban areas in search of better life conditions. The area was composed of one large plot, owned by the Portuguese government and transferred for use to the airport, and of two other contiguous plots that are privately owned. The neighbourhood hosted a small group of families mostly from São Tomé, who were pushed into the interstices of the urban fabric, along with other immigrants from the former colonies excluded from both the housing market and the eligibility for the public housing system (Ascensão and Leal 2019). They self-produced the neighbourhood from the mid-seventies onwards, in the aftermath of the 1974 Revolution, which marked the end for the fascist dictatorship, and of the decolonisation process, alongside another small group of poor Portuguese families of Roma ethnicity who settled there following family and community ties (see Fig. 29.1).

Fig. 29.1 Bairro da Torre neighbourhood at the borders of Lisbon Airport. (Source: Google Earth 2016)

BDT experienced a series of socio-spatial vulnerabilities throughout its history, like other similar self-produced neighbourhoods that have been neglected (and/or repressed) by public authorities in Portugal. It faced significant deficiencies in terms of infrastructure (access to piped water, sanitation—shortages of sanitary facilities, lack of or limited access to the sanitation network and inadequate sanitation—and electricity supplies), public space (degraded and littered with waste and lacking facilities), roads (unpaved), transport, access to services and facilities in the surroundings, and housing conditions, which were extremely precarious. Most inhabitants had low levels of education and high unemployment rates, particularly among families of Roma ethnicity, relying on social benefits and other forms of assistance. As Raposo (2012) points out, in the Portuguese context, when land tenure is secure, as in Urban Areas of Illegal Genesis (AUGIs), Law 91/95 and its subsequent revisions allow for legal and urban reconversion. In neighbourhoods resulting from the occupation of public or private land (which we refer to as self-produced neighbourhoods, such as BDT), socio-spatial vulnerabilities tend to worsen because there is no provision for legal and urban reconversion. The public intervention approach in self-produced neighbourhoods, consistent with the PER-Rehousing Special Programme (*Programa Especial de Realojamento*) of 1993, has remained largely technocratic and top-down, focusing on their 'eradication'[3] and the dispersed rehousing of inhabitants in distant public housing estates. In the 1990s, hundreds of families lived in BDT, however in 2007 and 2011, Loures City Council rehoused part of the population under the PER, with the immediate demolition of their homes. Some families who were not included in the programme also had their homes demolished and were left without any housing alternative. They managed to rebuild

[3] Degrading term used by the PER itself.

their dwellings within the neighbourhood, albeit in a more precarious manner. Owing to the organisation and resistance of the families and the support of different collectives, residents continued living in the area, neverthless their housing and living conditions remained precarious. According to surveys carried out by GESTUAL as part of the action research project mentioned above, 70 families (around 240 people) were surveyed in 2016. By the end of that year, the neighbourhood experienced an electricity supply outage. In 2018, a major fire broke out due to a short circuit in the precarious electrical connections, leaving 14 families (35 people) homeless. Following this, the Loures City Council, with the support of the IHRU, resumed and intensified the rehousing process, dispersing families across different public housing estates, including some located in other municipalities, although still within Greater Lisbon. The rehousing process concluded at the beginning of 2023, when the last family was rehoused, and BDT was dismantled (see Fig. 29.2).

The 'eradication' of BDT was a top-down process, and the dispersion of its inhabitants across Greater Lisbon disregarded their right to place and, at the same time, their Right to the City (Arma 2023). After rehousing, families' housing conditions improved, however in many cases, their socio-economic situation deteriorated. This solution deprived them of the local solidarity and civil society networks that had previously supported them. Some lost their sources of income, experienced isolation, struggled to adapt to their new neighbourhoods and faced additional costs such as rent, bills and transport (Ibid).

In 2012, families organised themselves into the *Torre Amiga-Moradores do Bairro da Torre* Residents' Association, currently chaired by Ricardina Cuthbert (of African origin) and Maria Cardoso (of Roma ethnicity) as vice-president. The creation of the association resulted from a process of organisation among

Fig. 29.2 '*Para onde vou*?' ('Where do I go?'). Mural painting found in a house in Bairro da Torre expresses the family's precarious housing and living condition. (Source: Rosa Arma)

Fig. 29.3 The *Festa de Nossa Senhora de Madre Deus* in Bairro da Torre. (Source: Rosa Arma)

BDT inhabitants that began in 2011 to mediate the rehousing process and was later formalised to guarantee access to piped water. The association survived the dismanteling of the neighbourhood and, thanks to funding from the Healthy Neighbourhoods Programme[4] (*Programa Bairros Saudáveis*) in 2021, the community house 'Holding Hands' (*Dar as Mãos*) was (re)established inside Ricardina's new home, where the association's activities continue, responding to the needs of an expanded community including new neighbours. Although marked by severe socio-spatial vulnerabilities and injustice, the inhabitants of BDT revealed diverse and rich experiences of appropriating and (self)producing space, personal and collective strategies of resistance in everyday life, forms of community (self)organisation and practices of solidarity and mutual aid, and cultural manifestations that supported both individual subsistence and the existence and resistance of the group as a whole. The *Festa de Nossa Senhora de Madre Deus* is one of such distinctive cultural manifestations (see Fig. 29.3). Since 2014, this feast, imported from São Tomé, has brought together families of African origin and Roma ethnicity from the neighbourhood, as well as hundreds of immigrants from São Tomé who used to travel to BDT from across Portugal and even from France or England to attend the religious celebration, musical performances and to enjoy the traditional food and drink.

That is why BDT attracted the attention of civil society associations such as Habita, of academia (GESTUAL) and the Church (the Comboni Missionaries), which, together with the inhabitants and both central and local public

[4]The *Bairros Saudáveis* Programme aimed at fostering partnerships and local interventions to promote health and quality of life in vulnerable communities, through the funding of collaborative projects, presented by associations, collectives, non-governmental organisations, civic movements, and residents' organisations, supported by local authorities and health authorities.

authorities, sought to improve the living conditions in the neighbourhood and develop a dignified housing solution.

In 2014, following dialogue with the Habita Association, which had supported families resisting the evictions of 2011, GESTUAL launched the action research project mentioned above. That same year, a GESTUAL team, coordinated by Isabel Raposo and composed of professors, researchers and students, carried out a socio-residential survey of the families. This was followed by a territorial, urban, housing and social diagnosis, as well as the proposal of a short and medium/long-term action plan, which began by supporting the Torre Amiga Association in improving sanitation and by seeking funding and low-tech, low-cost solutions for in situ rehousing. Focus groups with families of African origin and Roma ethnicity were also conducted to study dwelling practices in order to support the proposal for in situ rehousing. Collaborative mapping of self-produced houses and small participatory interventions aiming to improve living conditions were likewise implemented. During an initial workshop with students enrolled in the elective course 'Upgrading of Housing Suburbs' (*Qualificação dos Súburbios Habitacionais*), held by Isabel Raposo in the academic year 2014–2015, banana trees were planted to serve as biofilters and a space within the neighbourhood was cleaned to build a football pitch. As a result of a second workshop in the academic year 2015–2016, a Waste Cleaning Campaign by the Residents' Group to Combat Waste, together with GESTUAL and with the support of the City Council, Parish Council, neighbouring companies and the Scouts, was launched as a response to the accumulation of waste issue in the neighbourhood and to mitigate fire and health risks. From 2017 onwards, each house was assessed in terms of structure, roof, walls, floor, comfort and protection from water, wind and animals. Roofs were waterproofed using track canvas and, thanks to a donation, structural elements and internal protection were improved. Collective bathrooms and community washing facilities were designed and, following the electricity outage in 2016, a proposal for legal electricity supply was developed under the Decree-Law 36/2018[5]. A participatory photography process with children was carried out in 2018 to uncover their concerns and aspirations regarding the place they inhabited and to disseminate their (photo)voices (Arma 2022). Several documentary films were produced about the history and living conditions of the neighbourhood, as well as about the interventions it underwent. Hugo Jesus, a member of GESTUAL, developed a proposal for the in situ rehousing of families as part of his Master's Degree in Architecture, based on residents' aspirations and considering both urban criteria and local living practices. Finally, with the support of GESTUAL, Torre Amiga's project 'Holding Hands', mentioned above, was implemented, after receiving funding from the Healthy Neighbourhoods Programme.

All these actions were carried out through close and sustained interaction between professors, researchers and students involved in GESTUAL's action research project, the Torre Amiga Residents'

---

[5] DL 36/2018 approved an extraordinary regime regarding the temporary electricity supply to homes within areas of precarious housing.

Fig. 29.4 Playful moment of interaction between GESTUAL, students and inhabitants in Bairro da Torre. (Source: Rosa Arma)

Association and inhabitants (see Fig. 29.4), as well as other actors such as the City Council and the Comboni Missionaries. The aim was to support inhabitants' struggle for the right to exist and for self-determination—that is, the right to actively and consciously shape one's own future—representing an effort to realise their Right to the City (Lefebvre 1968).

29.3 The Field Research Methodology: Notions, Techniques, Process and Impact

The reading of the case study follows an interactionist approach in which urban space is understood in close relation to its producers (Lefebvre 1974), their logics and everyday practices. According to Raposo (1999), this approach requires qualitative methods of empirical data collection. To this end, the fieldwork was based on the technique of participant observation, which was considered the most appropriate for gathering qualitative information about the case study, specifically the everyday lives of the inhabitants of BDT, its socio-spatial (self)production and GESTUAL's interventions. Similar to BDT, the four cases—Terras da Costa and 2° Torrão in Almada, the Alto da Cova da Moura in Amadora and the co-produced neighbourhood of Cruz Vermelha in Lisbon—selected as references to critically understand the socio-spatial impact of interactive intervention in the self-produced neighbourhoods of Greater Lisbon, served as sites for the author's 'affectation of experience' (Tsallis et al. 2022), which functioned as a methodological device for data production. From a structuralist perspective, the author's methodological approach also considers space as corresponding to the constraints of global structures to which everyday practices are subject. Accordingly, a micro and macro-scale observation of the current socio-spatial production of self-produced neighbourhood in Greater Lisbon was adopted, focusing on the logics, actions and interactions of two specific groups: the inhabitants and the technicians. The tactics of everyday life (De Certeau 1984), the inhabitant's creative resistance, at both individual and collective levels, will be revealed with particular emphasis on 'places for the encounters and exchanges' (Lefebvre 1968:140). Lefebvre's formulation of the Right to the City in 1968 continues to encourage the (re)discovery of the temporalities and spaces of a renewed urban life, defined by use value rather than exchange value, which underpins the logic of capital, and by everyday moments of interaction among human beings, as well as by the *fête*. Although seemingly a playful event, the *fête* becomes a privileged instance of the Right to the City and an opportunity for its fulfilment (Carolino and Biagi 2022). The action of technicians will be analysed in its interaction with inhabitants' everyday practices, examining whether technicians acted against a technocracy that legitimises processes serving the interests of hegemonic actors, or wheter they contributed to the practice of a 'rebel architecture', fostering the emergence of 'spaces of hope' (Harvey 2004), in which dominant modes of socio-spatial production are subverted.

Considering the theoretical lines of thought outlined, the empirical research process in BDT required the collection of existing information (bibliographic, documentary and iconographic, including knowledge already produced by GESTUAL), as well as the production of data in the field. Data collection for the case study extended throughout the entire duration of the research conducted in BDT. While quantitative techniques were also incorporated, the study predominantly followed a qualitative approach, focusing on understanding how people perceive and produce space in the neighbourhood, as well as the interactive interventions it underwent and their impacts.

The author engaged in a dialogue with the inhabitants through the following stages:

- Observation of inhabitants' everyday life (see Figs. 29.5 and 29.6), with records produced through photographs, short videos, audio recordings and field diaries, aiming to adopt a sensitive approach to the case. Observations focused in particular on collective processes of (self)organisation and struggles for access to essential goods and adequate housing, solidarity and mutual aid practices, 'places for the encounters and exchanges' (Lefebvre 1968:140), feasts, 'being a child' in the neighbourhood, micro economies and the self-production of the neighbourhhod and housing. These observations informed a reading of the neighbourhood 'beyond the shacks', undestood as a meaningful place built through interactions. To cater to the distinct concerns, needs and desires of individuals and of the group, and to gain multiple perspectives on the territory, observations were car-

Fig. 29.5 Ricardina Cuthbert's anniversary party in Bairro da Torre. (Source: Rosa Arma)

Fig. 29.6 Around the campfire with a Roma family in Bairro da Torre. (Source: Rosa Arma)

ried across numerous visits and diverse situations since the end of 2016. These included the initial field reconnaissance trips to BDT from Lisbon also by using public transport, the selection of different entry points and neighbourhoods walkthroughs, (participant) observation of Torre Amiga Association meetings (internal or involving other actors), attendance at City Council assemblies together with residents, as well as at feasts and social gatherings, protests, solidarity projects, fairs and mutual aid actions organised by Torre Amiga Association and other actors' initiatives.

- As a member of GESTUAL, the author participated in and monitored the group's action research described above, from late 2016 to 2023, within the framework of both the collective project and her doctoral research.
- As a member of GESTUAL, the author also participated in and monitored different collective initiatives of which the Torre Amiga Association was part, including: the Residents' Assembly (*Assembleia de Moradores*), which brought together residents from four self-produced and rehousing neighbourhoods in Greater Lisbon and its initiative the Caravan for the Right to Housing (*Caravana pelo Direito à Habitação*), which travelled across Portugal to advocate for the Right to Housing, as well as, more recently, the Loures Housing Platform (*Plataforma de Habitação de Loures*); the production of the documentary *Artigo 65*, by Marian van der Zwaan and Kevin Raposo, during which, over a period of approximately six months starting in June 2019, the author accompanied Ricardina Cuthbert, visiting professor Juliana Demartini and the two artists in conversations with 26 inhabitants, representatives of the families still living in the neighbourhood at the time, and with external key actors (see Fig. 29.7); the drafting and management of the 'Holding Hands' project, mentioned above.
- Numerous unrecorded informal conversations took place throughout the field research and were documented in field diaries.

Fig. 29.7 Interview with a Roma family in their house's *alpendre* (the porch) in Bairro da Torre, in the context of the film *Artigo 65* production. (Source: Rosa Arma)

- In-depth semi-structured interviews were conducted across multiple sessions with the current president and vice-president of the Torre Amiga Association, as well as with other inhabitants, considered 'privileged informants'. Audio/ photo/ video recordings were made and complemented by written notes and full transcription. The interviews were carried out in situations of everyday interaction, as a strategy for accessing the universe of local actors within the framework of socio-anthropological research (Olivier de Sardan 2008).
- In-depth semi-structured interviews were also conducted with other actors belonging to the supralocal sphere of the neighbourhood, namely, those contributing to the socio-spatial production and transformation of BDT. Together with the inhabitants, these actors form a strategic group, enabling the collection of a representative range of perspectives for discussion (Olivier de Sardan 1995)

The model chosen for the interviews was an arrative one, in which the interviewer introduces topics to stimulate discussion, allowing interviewees to freely construct their accounts around these topics, sometimes leading to new ones, while occasionally seeking clarification or redirecting the conversation. A flexible script was used, based on a timeline that begins with the interviewees' past experiences, moves through their present and extends to their envisioned futures, in order to elicit their *récits de vie* (Bertaux 1997). The interviews therefore focused on the interviewees' life trajectories, interactions—encounters and exchanges—and the personal and collective transformation generated by these encounters and exchanges, particularly those between internal actors (the inhabitants) and external ones (such as the researchers from GESTUAL). In the case of the inhabitants, the aim was also to understand their everyday life in BDT and, where applicable, in the neighbourhoods in which they were rehoused. Particular attention was paid to their relationship with and perception of space, in terms of interactions with the rest of the city, public space, housing, economic activities, leisure, sociability and participation (or lack thereof) in collective action. The content of the interviews was analysed, cross-referencing statements from the different interviewees as fragments of a collective discourse shaped by socio-spatial relations (Bertaux 1997).

A second participatory photography process—photovoice and photo-elicitation—was conducted in 2021 and involved children who had been rehoused in the meantime. This enabled a comparison between children's everyday life in self-produced neighbourhoods and that of rehoused children in public housing estates, as well as a discussion of changes in their relationship with space and perception of their new neighbourhoods, including play spaces and activities, before and after rehousing. Participatory photography, as a research methodology, fostered dialogue between the researchers and children, thereby supporting the co-production of knowledge (Arma and Andrade dos Santos 2025).

According to Olivier de Sardan (2008), fieldwork involves permanent conceptualisation, (self)monitoring and discussion, which were supported by field diaries filled in with 'thick descriptions' (Geertz 1973), consisting of notes dense with meaning and sketches that recorded the 'affectation of experience' (Tsallis et al.

2022), and by moments of group sharing, which took place mainly within the GESTUAL team, as well as through discussion in broader scientific forums.

Data analysis was based on a continuos process of cross-referencing sources (participant observation, action research, interviews, participatory photography processes, bibliographic, documentary and iconographic materials) with theoretical and critical reflection. As such, the empirical research constituted a reflective practice and a foundational stage of the study enabling an iterative process between the interactions with the field and the theories underpinning the research. This process aimed, as Latour (2008:49) argues, 'to put theory at risk', that is, to allow for the (re)structuring of the research problem and refinement of the object of study, the (re)construction and maturation of the theoretical framework, research questions and proposed hypothesis, as well as the assessment and adjustment of the methodological model, creating opportunities for new lines of inquiry (Olivier de Sardan 2008). Accordingly, the research was established as 'theoretically oriented empirical research', as defined by Costa (1999:7), situated at the intersection of guiding theoretical lines and empirical data collected in the field. It therefore remained open to contextual demands and changes, as well as to intuitions, advances, setbacks and reformulations in the progressive construction—through successive approximations—of the dimensions of observation and analysis, and in the adaptation of data collection tools in response to the analytical requirements of the object of study. The author's initial empirical research plan also underwent several adaptations in response to field contingencies, including the dispersion of rehoused families throughout Greater Lisbon, demolition of BDT, Covid-19 pandemic and changes in the composition and dynamics of the GESTUAL team. Some tasks have not yet been completed, notably the structured return of the research results to the inhabitants of BDT through moments of collective debate. As advocated by Raposo (1999), beyond enabling the verification of results and the collection of new data, returning the knowledge produced aims to strengthen local reflection and to embed the research within local dynamics and collective action. In this process, the inhabitants are actively engaged in knowledge construction and problem-solving, thereby enhancing their capacity to participate in decision-making, particularly in contexts in which they typically lack a voice.

29.4 (Inter)Acting Knowledge: Concluding Notes

The account of the interactive methodology and techniques used to research self-produced neighbourhoods in Greater Lisbon, namely BDT, forms part of the author's broader reflection on processes and projects with greater transformative potential to contribute to a more dialogical, empowered and emancipatory, inclusive and cohesive, just, qualified and sustainable city—a 'democra-city' (as defined within the framework of the author's doctoral research). The hypothesis advanced is that processes and projects grounded in dialogue conceived as (inter)action constitute a necessary condition for intervention in self-produced neighbourhoods, enabling them to function as local laboratories of democracy and to contribute to the democratic transformation of both society and the city, in line with a notion of the Right to the City, which links the transformation of the city to our own (Harvey 2008). The research emerged from the author's search for an understanding of a reality—the self-produced neighbourhoods—that is often forgotten, homogenised, stigmatised, silenced and erased by hegemonic views, with the aim of returning it in all its complexity and richness. It is therefore urgent to focus on everyday life and forms of subsistance, resistance, self-determination and agency among self-producers, in order to cultivate bolder visions that transform the very nature of intervention and reframe self-produced neighbourhoods as opportunities to rethink both society and the city. The self-produced city cannot be approached merely as a technical issue, detached from the temporalities and spaces of its everyday lived experience, but rather within the framework of the imagination of a better society and city, starting from its borders. The research presented here aims to contribute to the production of scientific knowledge that challenges stereotypical views of self-produced neighbourhoods, supported by the proposed field research methodology, in order to help 'rethink' them through an understanding of the quality of their socio-spatial (self)production. While avoiding the temptation of voyeurism or romanticisation, an interactive field research methodology based on interaction can transform both theoretical and technical knowledge about self-produced neighbourhoods. 'Rethinking' them enables alternative approaches to intervention that are more closely alignedwith local practices and aspirations, approaches that can dignify and enhance both the neighbourhoods and the everyday lives of their inhabitants, thereby countering overly technical and top-down forms of intervention that fail to engage with local complexity and richness. In opposition to unequal and segregated space structured according to capitalist principles, interaction fosters a new horizontal spatiality as a foundation for the promotion of the Right to the City (Biagi 2019).

Designed without architects and shaped by conditions of socio-spatial injustice, but also by the energies and creative powers of its inhabitants—both as individuals and as a collective confronting multiple daily challenges—BDT emerged from diverse and rich experiences of socio-spatial appropriation and production. Numerous tactics of everyday life marked the struggle for personal and collective subsistence, resistance, self-determination and agency, all of which were observed throughout the fieldwork. Close and sustained interaction with the field, within the context of the author's research and GESTUAL's collective action-research projectin BDT, fostered an in-depth understanding of the neighbourhood's socio-spatial selfproduction. The encounters and exchanges within the research—between the author, the academic group, neighbourhood leaders and inhabitants, who were considered as experts and reflective subjects, together with other actors—created an opportunity for dialogue established as an (inter)action of knowledges. This is a combination of local, scientific, technical and experiential forms of knowledge (Ferlicca and Pedro 2024), forming a process of mutual learning oriented towards the co-producation and iterative transformation of both knowledge and action, in line with the pedagogue Paulo Freire, who defines dialogue as 'the encounter of those addressed to the common task of learning and acting' (Freire 2005:90). The interactive nature of the field research methodology enabled the co-production of a 'knowledge in action', continually calling for renewal through encounters and exchanges, a praxis in which reflection and action in solidarity mutually and continuosly illuminate one another (Freire 1985).

The knowledge of 'experience' (Bondía 2002) gained in BDT also provided an opportunity to rethink the author's position as a researcher (and a person), which should be shaped by—and contribute to—other and multiple perspectives of observation and action. The interactivity of the methodology developed was not merely a research tool: rather, the research in BDT was established as an open and ongoing process oriented towards transformation. (Inter)action was fostered through an almost 'deambulatory' attitude of listening to the field—places and people alike—encouraging exchanges of roles in order to establish a new, transformative dialogue. The fieldwork was carried out through the author's diving into the everyday life of the self-produced neighbourhood of BDT—what Ascensão (2013) refers to as 'ethnographic presence'—and was marked by processes of 'affectation' (Favret-Saada 1990) and a strong commitment to the field. According to Bondía (2002), this particular attitude of openness on the part of the researcher, his/her *experience*, is necessary to enable a privileged channel of communication and a dialogue' with the research participants. Dialogue with BDT transformed the author's perception, as well as her theoretical and technical knowledge of self-produced neighbourhoods, through the encounters and exchanges with inhabitants' everyday life and expertise. Author's *experience* also had an impact on her own vulnerability, building her capacity to be affected by singularities and differences, and allowing new encounters and exchanges to happen. A situated militant knowledge in action and co-production—rooted in a commitment to ways of thinking and living in everyday life and oriented towards contributing to the transformation of society and the city—was also built in the process. Guided by (inter)action and therefore transformative, this form of knowledge is driven by imaginative processes and projects that stem from recognising other perspectives beyond single and hegemonic narratives, and from acknowledging that, at the same time, we all share the same world. In doing so, it seeks to blur borders in the world we imagine and build. These borders are shaped by powers of separation—between knowledges, formal and informal, city and non-city, centre and periphery, 'us' and 'them', researcher and researched, ... —but they can nevertheless be blurred by all agents who cross and intersect them, and '(inter)act' their trajectories.

Acknowledgements This work is financed by national funds through FCT—Fundação para a Ciência e a Tecnologia, I.P., under the Strategic Project with the references UIDB/04008/2020 and UIDP/04008/2020, and author's grant with the reference SFRH/BD/129102/2017.

References

Allegra M, Di Giovanni C (2024) 50 Anos de políticas de habitação. In: Ferrão J, Delicado A (eds) Portugal Social em Mudança. Instituto de Ciências Sociais da Universidade de Lisboa, Lisbon, pp 61–71

Área Metropolitana de Lisboa (2022) Diagnóstico das Condições Habitacionais Indignas: Habitação PRR-AML 2021/2026. https://www.aml.pt/susProjects/susWebBackOffice/uploadFiles/wt1wwpgf_aml_sus_pt_site/componentText/SUS63752D8AC7909/DCHI_SUMARIO_EXECUTIVO_AML_.PDF. Accessed: 31 Jan 2025

Arma R (2022) Resistência urbana pelas imagens: duas experiências no Bairro da Torre. Urbe: Revista Brasileira de Gestão Urbana 14:1-15. https://periodicos.pucpr.br/Urbe/article/view/29208. Accessed: 19 Dec 2025

Arma R (2023) O direito à habitação como direito à cidade: Os processos de realojamento dos moradores do Bairro da Cruz Vermelha e do Bairro da Torre. Cidades 47:15–35

Arma R, Andrade dos Santos C (2025) Right to the City (photo)voices: participatory photography with children in Greater Lisbon. Sophia Journal 8(1):149–168

Ascensão E (2013) A barraca pós-colonial: Materialidade, memória e afeto na arquitetura informal. In: Domingos N, Peralta E (eds) Cidade e Império: dinâmicas coloniais e reconfigurações pós-coloniais. Edições 70, Lisboa, pp 415–462

Ascensão E, Leal M (2019) Spatial analysis, research. http://expertsproject.ics.ulisboa.pt/analise.html. Accessed: 31 Jan 2025

Augé M (1992) Non-Lieux: Introduction à une anthropologie de la surmodernité. Seuil, Paris

Bertaux D (1997) Les récits de vie: Perspective ethnosociologique. Nathan Université, Paris

Biagi F (2019) Henri Lefebvre: Una teoria critica dello spazio. Jaca Book, Milano

Bondía JL (2002) Notas sobre a experiência e o saber de experiência. Revista Brasileira De Educação 19:20–28

Carolino J, Biagi F (2022) Claiming the city through the Cape Verdean Festivities of Kola San Jon in Lisbon: a Lefebvrian case study. International Critical Thought 12(1):133–156

Comissão Nacional de Acompanhamento do PRR (2025) Relatório1: 2025. https://cnaprr.pt/wp-content/uploads/2025/06/Relatorio-1_2025-CNAPRR.pdf. Accessed: 19 Dec 2025

Costa AF (1999) Sociedade de bairro: Dinâmicas sociais de identidade cultural. Celta, Oeiras

Davis M (2006) Planet of slums. Verso, London/New York

De Certeau M (1984) The practice of everyday life. University of California Press, Berkeley

Favret-Saada J (1990) Être Affecté. Gradhiva: Revue d'Histoire et d'Archives de l'Anthropologie 8:3–9

Ferlicca F, Pedro BH (2024) Insurgent planning practices and university-community engagement in popular urbanization: the urban planning commission in the land reclamation of Guernica, Buenos Aires. In: Rocco R, Silvestre G (eds) Insurgent planning practice. Agenda Publishing, Newcastle upon Tyne, pp 115–132

Ferrarotti F (2007) Las historias de vida como método. Convergencia: Revista de Ciencias Sociales 14(44):15–40

Freire P (1985) Extensão ou comunicação? Paz e Terra, Rio de Janeiro

Freire P (2005) Pedagogy of the oppressed. Continuum, New York

Geertz C (1973) The interpretation of cultures: selected essays. Basic Books, New York

Guerra I (2010) Participar porquê e para quê?: Reflexões em torno dos efeitos da democracia local na equidade e na legitimidade dos eleitos. Revista Crítica de Ciências Sociais 91:121–134

Harvey D (1980) Social justice and the city. Blackwell Publishers, Cambridge, MA/Oxford

Harvey D (2004) Spaces of hope. Edinburgh University Press, Edinburgh

Harvey D (2008) The right to the city. New Left Review 53:23–40

Holston J (1998) Spaces of insurgent citizenship. In: Sandercock L (ed) Making the invisible visible: a multicultural planning history. University of California Press, Oakland, pp 37–56

Instituto da Habitação e da Reabilitação Urbana (2018) Levantamento Nacional das Necessidades de Realojamento Habitacional. https://www.portaldahabitacao.pt/documents/20126/58203/

Relatorio_Final_Necessidades_Realojamento.pdf/d67e9ed8-93aa-d65a-eb36-b586389b2295?t=1550355947650 Accessed: 31 Jan 2015

Latour B (2008) Como falar do corpo: A dimensão normativa dos estudos sobre a ciência. In: Nunes JA, Roque R (eds) Objectos Impuros: Experiencias de estudo sobre a Ciencia. Afrontamento, Porto, pp 39–61

Lefebvre H (1968) Le droit à la ville. Anthropos, Paris

Lefebvre H (1974) La production de l'espace. Anthropos, Paris

Nascimento DM (2011) A autoconstrução na produção do espaço urbano. In: de Mendonça JG, de Moura HCS (eds) Estado e capital imobiliário. Convegências atuais na produção do espaço urbano brasilerio. Ed. C/Arte, Belo Horizonte, pp 217–230

Olivier de Sardan JP (1995) Anthropologie et développement: Essai en socio-anthropologie du changement social. APAD, Marseille

Olivier de Sardan JP (2008) La rigueur du qualitatif. Les contraintes empiriques de l'interprétation socio-anthropologique. Academia-EME, Louvain-la-Neuve

Raposo I (1999) Urbaniser villages et maison. PhD Thesis. Institut d'Urbanisme de Paris, Université de Paris XII, Val de Marne, Paris

Raposo I (2012) Bairros de génese ilegal: Metamorfoses dos modelos de intervenção. In: Mendes M, Sá T, Crespo JL, Ferreira CH (eds) A cidade entre bairros. Caleidoscópio, Lisboa, pp 107–119

Raposo I (2017) Pelo Direito à Habitação e à Cidade: A construção de redes alargadas de discussão. In: Rodrigues M, Fontes M, Silva M, Fontes A (eds) A cidade da participação. Edições Afrontamento, Porto, pp 21–25

Raposo I, Jorge S, Viegas S, Melo V (2012) Luanda e Maputo: Inflexões suburbanísticas da cidade socialista à cidade-metrópole neoliberal. Urbe: Revista Brasileira De Gestão Urbana 4(2):189–205

Tsallis A, Oliveira Moraes M, Balbino dos Santos B, Sibelle Freires Aires J, Renata Ribeiro Bravo J, de Oliveira Vianna da Silva K, Ribeiro da Silva L, Lorena do Nascimento Santos L, Araújo de Medeiros Brito M, Teodoro Sampaio R, Cristina de Azevedo da Fonseca S, de Magalhães Bredariol T (2022) Sobre afecTAR: Del campo a la escritura como laboratório. SciComm Report 2(1):12–13

The manufacturer's authorised representative in the EU is Springer Nature Customer Service Centre GmbH, Europaplatz 3, 69115 Heidelberg, Germany. If you have any concerns regarding our products, please contact ProductSafety@springernature.com

Printed and bound by CPI Group (UK) Ltd, Croydon, CR0 4YY
17/09/2026
02218432-0009